Innovations in Information Systems Modeling:
Methods and Best Practices

Terry Halpin
Neumont University, USA

John Krogstie
SINTEF, Norway

Erik Proper
Radboud University Nijmegen, The Netherlands

INFORMATION SCIENCE REFERENCE
Hershey · New York

Director of Editorial Content: Kristin Klinger
Director of Production: Jennifer Neidig
Managing Editor: Jamie Snavely
Assistant Managing Editor: Carole Coulson
Typesetter: Larissa Vinci
Cover Design: Lisa Tosheff
Printed at: Yurchak Printing Inc.

Published in the United States of America by
Information Science Reference (an imprint of IGI Global)
701 E. Chocolate Avenue, Suite 200
Hershey PA 17033
Tel: 717-533-8845
Fax: 717-533-8661
E-mail: cust@igi-global.com
Web site: http://www.igi-global.com

and in the United Kingdom by
Information Science Reference (an imprint of IGI Global)
3 Henrietta Street
Covent Garden
London WC2E 8LU
Tel: 44 20 7240 0856
Fax: 44 20 7379 0609
Web site: http://www.eurospanbookstore.com

Library of Congress Cataloging-in-Publication Data

Innovations in information systems modeling : methods and best practices / Terry Halpin, John Krogstie, and Erik Proper, editors.
p. cm.

Includes bibliographical references and index.
Summary: "This book presents cutting-edge research and analysis of the most recent advancements in the fields of database systems and software development"--Provided by publisher.

ISBN 978-1-60566-278-7 (hardcover) -- ISBN 978-1-60566-279-4 (ebook)
1. System design. 2. Database management. 3. Computer software--Development. I. Halpin, T. A. II. Krogstie, John. III. Proper, Erik.

QA76.9.S88I554 2009
005.1--dc22

2008040204

British Cataloguing in Publication Data
A Cataloguing in Publication record for this book is available from the British Library.

Innovations in Information Systems Modeling: Methods and Best Practices is part of the IGI Global series named *Advances in Database Research (ADR)* Series, ISBN: 1537-9299

Editorial Advisory Board

Table of Contents

Detailed Table of Contents

When modeling information systems, one often encounters subtyping aspects of the business domain that can prove challenging to implement in either relational databases or object-oriented code. In practice, some of these aspects are often handled incorrectly. This chapter examines a number of subtyping issues that require special attention (e.g. derivation options, subtype rigidity, subtype migration), and discusses how to model them conceptually. Because of its richer semantics, the main graphic notation used is that of second generation Object-Role Modeling (ORM 2). However, the main ideas could be adapted for UML and ER, so these are also included in the discussion. A basic implementation of the proposed approach has been prototyped in Neumont ORM Architect (NORMA), an open-source tool supporting ORM 2.

This chapter focuses on formally representing life cycle semantics of part-whole relations in conceptual data models by utilizing the temporal modality. This chapter approaches this by resorting to the temporal conceptual data modeling language ERVT and extend it with the novel notion of status relations. This enables a precise axiomatization of the constraints for essential parts and wholes compared to mandatory parts and wholes, as well as introduction of temporally suspended part-whole relations. To facilitate usage in the conceptual stage, a set of closed questions and decision diagram are proposed. The long-term objectives are to ascertain which type of shareability and which lifetime aspects are possible for part-whole relations, investigate the formal semantics for sharability, and how to model these kind of differences in conceptual data models.

This chapter will extend the ORM conceptual modeling language with constructs for capturing the relevant parts of an application ontology in a list of concept definitions. It will give the adapted ORM meta model and provide an extension of the accompanying Conceptual Schema Design Procedure (CSDP) to cater for the explicit modeling of the relevant parts of an application- or domain ontology in a list of concept definitions. The application of these modeling constructs will significantly increase the perceived quality and ease-of-use of (web-based) applications.

Section II
Modeling Approaches

This chapter presents experiences and reflections from using the EKD Enterprise Modeling method in a number of European organizations. The EKD modeling method is presented. The chapter then focuses on the EKD application in practice taking six cases as an example. Our observations and lessons learned are reported concerning general aspects of Enterprise Modeling projects, the EKD modeling language, the participative modeling process, tool support, and issues of Enterprise Model quality. It also discussed a number of current and emerging trends for development of Enterprise Modeling approaches in general and for EKD in particular.

This chapter discussed how an Enterprise Modeling approach, namely C3S3P , has been applied in an automotive supplier company. The chapter concentrates on the phases of the C3S3P development process such as Concept Study, Scaffolding, Scoping, and Requirements Modeling. It also presents the concept of task pattern which has been used for capturing, documenting and sharing best practices concerning business processes in an organization. Within this application context the authors have analyzed their experiences concerning stakeholder participation and task pattern development. They have also described how they have derived four different categories of requirements from scenario descriptions for the task patterns and from modeling of the task patterns.

Chapter VI
Methodologies for Active Knowledge Modeling 106
John Krogstie, NTNU, Norway
Frank Lillehagen, Commitment, Norway

Innovative design is the most important competitive factor for global engineering and manufacturing. Critical challenges include cutting lead times for new products, increasing stakeholder involvement, facilitating life-cycle knowledge sharing, service provisioning, and support. Current IT solutions for product lifecycle management fail to meet these challenges because they are built to perform routine information processing, rather than support agile, innovative work. Active Knowledge Modeling (AKM) provides an approach, methodologies, and a platform to remedy this situation. This chapter describes the AKM-approach applied by manufacturing industries and consultants to implement pragmatic and powerful design platforms. A collaborative product design methodology describes how teams should work together in innovative design spaces. How to configure the AKM platform to support such teams with model-configured workplaces for the different roles is described in the visual solutions development methodology. The use of this approach is illustrated through a case study and is compared with related work in the enterprise modeling arena to illustrate the novelty of the approach

Chapter VII
Data Modeling and Functional Modeling: Examining the Preferred Order of Using UML
Class Diagrams and Use Cases 122
Peretz Shoval, Ben-Gurion University of the Negev, Israel
Mark Last, Ben-Gurion University of the Negev, Israel
Avihai Yampolsky, Ben-Gurion University of the Negev, Israel

In the analysis phase of the information system development, the user requirements are studied, and analysis models are created. In most UML-based methodologies, the analysis activities include mainly modeling the problem domain using a class diagram, and modeling the user/functional requirements using use cases. Different development methodologies prescribe different orders of carrying out these activities, but there is no commonly agreed order for performing them.
In order to find out whether the order of analysis activities makes any difference, and which order leads to better results, a comparative controlled experiment was carried out in a laboratory environment. The subjects were asked to create two analysis models of a given system while working in two opposite orders. The main results of the experiment are that the class diagrams are of better quality when created as the first modeling task, and that analysts prefer starting the analysis by creating class diagrams first.

Section III
Modeling Frameworks, Architectures, and Applications

Chapter VIII
OntoFrame: An Ontological Framework for Method Engineering.......... 144
Mauri Leppänen, University of Jyväskylä, Finland

A large number of strategies, approaches, meta models, techniques and procedures have been suggested to support method engineering (ME). Most of these artifacts, here called the ME artifacts, have been constructed, in an inductive manner, synthesizing ME practice and existing ISD methods without any theory-driven conceptual foundation. Also those ME artifacts which have some conceptual groundwork have been anchored on foundations that only partly cover ME. This chapter presents an ontological framework, called OntoFrame, which can be used as a coherent conceptual foundation for the construction, analysis and comparison of ME artifacts. Due to its largeness, the article here describes its modular structure composed of multiple ontologies. For each ontology, this article highlights its purpose, subdomains and theoretical foundations. It also outlines the approaches and process by which OntoFrame has been constructed and deploy OntoFrame to make a comparative analysis of existing conceptual artifacts.

A process-oriented framework (QoMo) is pre-sented that aims to further the study of analysis and support of processes for modelling. The framework is strongly goal-oriented, and expressed largely by means of formal rules. The concepts in the framework are partly derived from the SEQUAL framework for quality of modelling. A number of modelling goal categories is discussed in view of SE-QUAL/QoMo, as well as a formal approach to the descrip-tion of strategies to help achieve those goals. Finally, a pro-totype implementation of the framework is presented as an illustration and proof of concept.

This chapter presents the basic ideas underlying Service Oriented Architecture as well as a brief overview of current research into the phenomena also known as SOA. SOA is defined, and principal components of one proposed SOA framework are discussed. The more relevant historical background behind the move toward SOA is presented, including SOA antecedents such as Web Services, SOAP, and CORBA, and enabling technologies such as XML and EJB. A basis for understanding SOA is presented, based on Krafzig, Banke, and Slama's (2005) three-level hierarchical perspective. The common SOA components including UDDI, Application Programming Interface, Service Bus, Service Contract, Interface, Implementation, Data, and Business Logic are also presented. Finally, relevant research in four categories is presented, including implementation strategies, patterns and blueprints, tool development, standards proposals or modifications (including middleware), and ontological or meta-model development or modification.

In the Web Engineering area, many methods and frameworks to support Web Information Systems (WISs) development have already been proposed. Particularly, the use of frameworks and container-based architectures is state-of-the-practice. This chapter presents a method for designing framework-based WISs called FrameWeb, which defines a standard architecture for framework-based WISs and a modeling language that extends UML to build diagrams that specifically depict framework-related components. Considering that the Semantic Web has been gaining momentum in the last few years, this chapter also proposes an extension to FrameWeb, called S-FrameWeb, that aims to support the development of Semantic WISs.

Section IV
Selected Readings

This chapter discusses the idea that even though information systems development (ISD) approaches have long advocated the use of integrated organisational views, the modelling techniques used have not been adapted accordingly and remain focused on the automated information system (IS) solution. Existing research provides evidence that business process simulation (BPS) can be used at different points in the ISD process to provide better integrated organisational views that aid the design of appropriate IS solutions. Despite this fact, research in this area is not extensive; suggesting that the potential of using BPS for the ISD process is not yet well understood. The chapter uses the findings from three different case studies to illustrate the ways BPS has been used at different points in the ISD process. It compares the results against IS modelling techniques, highlighting the advantages and disadvantages that BPS has over the latter. The research necessary to develop appropriate BPS tools and give guidance on their use in the ISD process is discussed.

In this chapter a formal agent based approach for the modeling and verification of intelligent information systems using Coloured Petri Nets is presented. The use of a formal method allows analysis techniques such as automatic simulation and verification, increasing the confidence on the system behavior. The agent based modelling allows separating distribution, integration and intelligent features of the system, improving model reuse, flexibility and maintenance. As a case study an intelligent information control system for parking meters price is presented.

With the design of reference models, an increase in the efficiency of information systems engineering is intended. This is expected to be achieved by reusing information models. Current research focuses mainly on configuration as one principle for reusing artifacts. According to this principle, all variants of a model are incorporated in the reference model facilitating adaptations by choices. In practice, however, situations arise whereby various requirements to a model are unforeseen: Either results are inappropriate or costs of design are exploding. This chapter introduces additional design principles that aim toward giving more flexibility to both the design and application of reference models.

Information systems development methodologies and associated CASE tools have been considered as cornerstones for building quality in an information system. The construction and evaluation of methodologies are usually carried out by evaluation frameworks and metamodels - both considered as meta-methodologies. This chapter investigates and reviews representative metamodels and evaluation frameworks for assessing the capability of methodologies to contribute to high-quality outcomes. It presents a summary of their quality features, strengths and weaknesses. The chapter ultimately leads to a comparison and discussion of the functional and formal quality properties that traditional meta-methodologies and method evaluation paradigms offer. The discussion emphasizes the limitations of both methods and meta-methods to model and evaluate software quality properties such as computability and implementability, testing, dynamic semantics capture, and people's involvement. This analysis along with the comparison of the philosophy, assumptions, and quality perceptions of different process methods used in information systems development, provides the basis for recommendations about the need for future research in this area.

Preface

Exploration, evaluation, and enhancement of modeling methods remain challenging areas within the field of information systems engineering. The benefits of model-driven development and information architectures are gaining wide recognition, not only by software companies and IT service providers, but also by industry groups such as the Object Management Group (OMG), with its promotion of Model Driven Architecture (MDA). In the interests of both agility and productivity, software engineers are increasingly exploiting the potential of tools for generating larger portions of their application code from high level models. In tune with these challenges and trends, this book provides an up-to-date coverage of central topics in information systems modeling and architectures by leading researchers in the area, including attention to both methodologies and best practices.

The contributions extend the state of the art, clarify fundamental concepts, evaluate different approaches and methodologies, and report on industrial applications of the techniques and methods discussed. With eleven chapters presented by top researchers from eight countries around the globe, the book provides a truly international perspective on the latest developments in information systems modeling, methods, and best practices. The range of topics explored is broad, while at the same time key topics are investigated at appropriate depth. This work should appeal to both practitioners and academics who wish to keep abreast of the latest developments in information systems modeling and information architecture.

Many of the topics in this book are typically not covered by traditional textbooks in analysis and design modeling. Modeling is used across a number of tasks in connection to information systems, but it is rare to see and easily compare all these uses of diagrammatical models as knowledge representation in the same book, highlighting both commonalities and differences between different kinds of modeling. The appropriateness of a modeling technique or language is often highlighted by the proponents of the technique, with little substantial evidence. This book focuses on the latest knowledge of and insights into this area gained through empirical and analytical evaluations of techniques used in practice. By providing a snapshot on the state of the art in modeling across a number of domains, this book should have lasting value as a reference in this field.

The chapters are arranged in three sections: *Conceptual Information Modeling; Modeling Approaches;* and *Modeling Frameworks, Architectures, and Applications*.

OVERVIEW OF CHAPTERS

Section I focuses on modeling of information at a truly conceptual level, where data models are presented in a way that nontechnical users of the relevant business domain can readily understand and validate. All three chapters in this section make reference to Object-Role Modeling (ORM), a fact-oriented modeling approach that represents information in terms of facts that assert that one or more objects play roles in relationships. In contrast to Entity-Relationship Modeling (ER) and the class diagramming technique within the Unified Modeling Language (UML), ORM models are attribute-free: this enables ORM models to be easily validated by verbalization and population, while at the same time conferring greater semantic stability (e.g. if one wishes in ER or UML to add an attribute of, or relationship to, an existing attribute, the latter attribute has to be remodeled as an object or relationship; in ORM the existing attribute would have been modeled as a fact type, and one may now simply add another fact type, without disrupting the original model).

Other benefits of ORM include its rich graphical constraint language, and its detailed modeling procedures for constructing conceptual schemas, validating them, and transforming them to attribute-based schemas such as ER schemas, UML class schemas, relational database schemas, or XML schemas. Chapters I and II include discussion of ER and/or UML to facilitate the application of their main ideas to the other approaches as well.

Each of the three chapters in Section I propose extensions to existing modeling approaches. Chapter I, entitled "Enriched Conceptualization of Subtyping", by Terry Halpin, suggests ways to address complex subtyping aspects that are often encountered in industrial application domains. Three classes of subtypes are distinguished: asserted, derived, and semiderived. Asserted subtypes are simply asserted or declared, whereas derived subtypes are fully defined in terms of other types that already exist. Semiderived types allow instances that are simply declared as well as instances that are derived. For subtypes that are derived or semiderived, care is needed to keep the types in sync with their derivation rules. Rigid types are distinguished from role types. An instance of a rigid type (e.g. Person) must remain in that type for the duration of its lifetime, whereas instances of role types (e.g. Student, Employee) may move into and out of those types at different times. Ways to declare rigidity are proposed for both ORM and UML, and the importance of keeping such declarations in sync with dynamic constraints such as changeability settings on association roles is highlighted. Various data model patterns are then identified to deal with modeling temporal aspects such as maintaining history of instances as they migrate from one type to another.

Chapter II, entitled "Essential, Mandatory, and Shared Parts in Conceptual Data Models", by Alessandro Artale and C. Maria Keet, focuses on formally representing life cycle semantics of part-whole relations in conceptual data models by utilizing the temporal modality. Their approach exploits the temporal conceptual data modeling language ERVT, extending it with the novel notion of status relations. This enables a precise axiomatization of the constraints for essential parts and wholes compared to mandatory parts and wholes, as well as the introduction of temporally suspended part-whole relations. A taxonomy is presented to cover both mereological and meronymic part-whole relations, and various notions such as essential parthood and essential wholehood are investigated. To facilitate usage in the conceptual stage, the authors propose a set of closed questions and a decision diagram to be used in communicating with the domain expert in order to determine just what kinds of shareability and lifetime aspects are intended for specific parthood fact types. Their work also suggests one way to provide a formal semantics for shareability based on description logics, and how to model these kinds of differences in conceptual data models.

Chapter III, entitled "Extending the ORM Conceptual Schema Language and Design Procedure with Modeling Constructs for Capturing the Domain Ontology", by Peter Bollen, extends the fact-oriented ORM approach, primarily to cater for lists of concept definitions that explain the meaning of the terms (object types) and relevant predicates (relationships) referred to in the information model. Extensions are proposed not only for the underlying metamodel but also for the modeling procedure itself, with particular attention to subtype definitions and naming conventions. The conceptual schema design procedure (CSDP) for ORM is augmented to ensure that appropriate attention is given to specifying the relevant explanations and definitions of the relevant noun and verb phrases used in expressing the model, thus improving the likelihood that users of the information system will understand the intended semantics of the model.

Section II focuses on modeling approaches, and consists of four chapters looking at broader modeling approaches including enterprise and process modeling in combination with data modeling.

In chapter IV, entitled "EKD - An Enterprise Modeling Approach to Support Creativity and Quality in Information Systems and Business Development" Janis Stirna and Anne Persson present experiences and reflections from using the EKD Enterprise Modeling method in a number of European organizations. The EKD (Enterprise Knowledge Development) modeling method is presented. The chapter then focuses on the application of EKD in practice, taking six cases as examples. The observations and lessons learned are reported concerning general aspects of Enterprise Modeling projects, the EKD modeling language, the participative modeling process, tool support, and issues of Enterprise Model quality. They also discuss a number of current and emerging trends for development of Enterprise Modeling approaches in general and for EKD in particular.

In chapter V, entitled "Integrated Requirement And Solution Modeling: An Approach Based On Enterprise Models" by Anders Carstensen, Lennart Holmberg, Kurt Sandkuhl, and Janis Stirna, the authors discuss how a particular enterprise modeling approach, namely C3S3P, has been applied in an automotive supplier company. The chapter concentrates on the phases of the C3S3P development process such as Concept Study, Scaffolding, Scoping, and Requirements Modeling. They also present the concept of task patterns which have been used for capturing, documenting and sharing best practices concerning business processes in an organization. Within this application context they have analyzed the experiences concerning stakeholder participation and task pattern development. They have also described how to derive four different categories of requirements from scenario descriptions for the task patterns and from modeling of the task patterns.

In chapter VI, entitled "Methodologies for Active Knowledge Modeling" by John Krogstie and Frank Lillehagen, the authors look deeper at AKM (Active Knowledge Modeling), which is the basis for among others the C3S3P-approach. Innovative design is the most important competitive factor for global engineering and manufacturing. Critical challenges include cutting lead times for new products, increasing stakeholder involvement, facilitating life-cycle knowledge sharing, service provisioning, and support. Current IT solutions for product lifecycle management fail to meet these challenges because they are built to perform routine information processing, rather than support agile, innovative work. AKM provides an approach, methodologies, and a platform to remedy this situation.

This chapter describes the AKM-approach applied by manufacturing industries and consultants to implement pragmatic and powerful design platforms. A collaborative product design methodology describes how teams should work together in innovative design spaces. How to configure the AKM platform to support such teams with model-configured workplaces for the different roles is described in the visual solutions development methodology. The use of this approach is illustrated through a case study in the same areas as the previous chapter, but here the resulting workplaces are presented in much more detail. The AKM-approach is also compared with related work in the enterprise modeling arena to illustrate the novelty of the approach.

The final chapter in this section, chapter VII, "Data Modeling and Functional Modeling - Examining the Preferred Order of Using UML Class Diagrams and Use Cases" by Peretz Shoval, Mark Last, and Avi Yampolsky, returns to more traditional software modeling, investigating the preferred order of two modeling techniques that are popular in UML. In the analysis phase of the information system development, the user requirements are studied, and analysis models are created. In most UML-based methodologies, the analysis activities include mainly modeling the problem domain using a class diagram, and modeling the user/functional requirements using use cases. Different development methodologies prescribe different orders of carrying out these activities, but there is no commonly agreed order for performing them.

In order to find out whether the order of analysis activities makes any difference, and which order leads to better results, a comparative controlled experiment was carried out in a laboratory environment. The subjects were asked to create two analysis models of a given system while working in two opposite orders. The main results of the experiment are that the class diagrams are of better quality when created as the first modeling task, and that analysts prefer starting the analysis by creating class diagrams first.

Section III is concerned with modeling frameworks, architectures, and applications. It essentially takes a broader perspective on the art of modeling. Individual chapters either focus on fundamental aspects of modeling methods, such as the construction of methods (Chapter VIII) or the study of heuristics and strategies to influence the quality of models (Chapter IX), or they explore modeling in specific application areas, such as service oriented software architectures (Chapter X), or web information systems (Chapter XI).

Chapter VIII, entitled "OntoFrame - An Ontological Framework for Method Engineering" by Mauri Leppänen, presents an ontological framework, called OntoFrame, that can be used as a coherent conceptual foundation for the construction, analysis and comparison of method engineering artifacts. A modular structure is used leading to multiple ontologies. For each ontology, the purpose, sub-domains and theoretical foundations are highlighted, as well as the approaches and process by which OntoFrame has been constructed. Furthermore, OntoFrame is applied to make a comparative analysis of existing conceptual artifacts.

Chapter IX, entitled "Concepts and Strategies for Quality of Modeling", by Patrick van Bommel, Stijn Hoppenbrouwers, Erik Proper, and Jeroen Roelofs, presents a process-oriented framework (QoMo) which aims to further the study of analysis and support of the processes involved in modeling. The resulting framework is strongly goal-oriented, and is expressed largely by means of formal rules. The chapter includes the discussion of a prototype implementation of the framework, which serves as an illustration and proof of concept.

Chapter X, entitled "Service Oriented Architecture - A Research Review from the Software and Applications Perspective", by John Erickson and Keng Siau, presents the basic ideas underlying service oriented architecture as used in the context of software and application architectures. Relevant historical background behind the move of software architectures towards a service oriented architectural style is presented, including antecedents such as Web Services, SOAP, and CORBA, and enabling technologies such as XML and EJB. The common components of a service-oriented software architecture, including UDDI, Application Programming Interface, Service Bus, Service Contract, Interface, Implementation, Data, and Business Logic are also presented. Finally, relevant research in four categories is presented, including implementation strategies, patterns and blueprints, tool development, standards proposals or modifications (including middleware), and ontological or meta-model development or modification.

In Chapter XI, "Designing Web Information Systems for a Framework-based Construction", by Vítor Estêvão Silva Souza, Ricardo de Almeida Falbo, and Giancarlo Guizzardi discusses a framework-based strategy for the construction of web-based information systems (WIS). More specifically, this chapter presents a design method called FrameWeb which defines a standard architecture for framework-based Web information systems and a modeling language that extends UML to build diagrams that specifically depict framework-related components.

Section IV ends the book with four chapters related to methodological aspects of modeling. In Chapter XII, "Business Process Simulation: An Alternative Modelling Technique for the Information System Development Process" by Tony Elliman, Tally Hatzakis, and Alan Serrano, another usage area for formal process models is illustrated. The article discuss the idea that even though information systems development (ISD) approaches have long advocated the use of integrated organizational views, the modeling techniques used have not been adapted accordingly and remain focused on the automated information system (IS) solution. Existing research provides evidence that business process simulation (BPS) can be used at different points in the ISD process in order to provide better integrated organizational views that aid the design of appropriate IS solutions. Despite this fact, research in this area is not extensive, which suggests that the potential of using BPS for the ISD process is not yet well understood. The article uses the findings from three case studies in order to illustrate the ways that BPS has been used at different points in the ISD process. It compares the results against IS modeling techniques, highlighting the advantages and disadvantages that BPS has over the latter. The research necessary to develop appropriate BPS tools and give guidance on their use in the ISD process is discussed

In Chapter XIII "An Agent Based Formal Approach for Modeling and Verifying Integrated Intelligent Information Systems" by Leandro Dias da Silva, Elthon Allex da Silva Oliveira, Hyggo Almeida, and Angelo Perkusich, illustrates a further usage of more formal modeling techniques. A formal agent-based approach for the modeling and verification of intelligent information systems using colored Petri nets is presented. The use of a formal method allows analysis techniques such as automatic simulation and verification, increasing the confidence on the system behavior. The agent-based modeling allows separating distribution, integration, and intelligent features of the system, improving model reuse, flexibility, and maintenance. As a case study, an intelligent information control system for parking meters price is presented.

In Chapter XIV "Design Principles for Reference Modeling: Reusing Information Models by Means of Aggregation, Specialization, Instantiation, and Analogy" by Jan vom Brocke, the focus is on generic process models, rather than concrete process models which are dealt with in an earlier chapter. With the design of reference models, an increase in the efficiency of information systems engineering is intended. This is expected to be achieved by reusing information models. Current research focuses mainly on configuration as one principle for reusing artifacts. According to this principle, all variants of a model are incorporated in the reference model, facilitating adaptations by choices. In practice, however, situations arise whereby various requirements to a model are unforeseen: either results are inappropriate or costs of design are rising strongly. This chapter introduces additional design principles aiming at giving more flexibility to both the design and application of reference models

Finally, in Chapter XV "Examining the Quality of Evaluation Frameworks and Meta-Modeling Paradigms of IS Development Methodologies", Eleni Berki gives an historical account on information systems development methodologies and associated CASE tools, and as such provides a backdrop on the framework presented in Chapter VIII. The construction and evaluation of methodologies are usually carried out by evaluation frameworks and meta-models, both considered as meta-methodologies. This chapter investigates and reviews representative meta-models and evaluation frameworks for assessing the capability of methodologies to contribute to high-quality outcomes. It presents a summary of their quality features, strengths, and weaknesses.

Section I
Conceptual Information Modeling

Chapter I
Enriched Conceptualization of Subtyping

Terry Halpin
Neumont University, USA

ABSTRACT

When modeling information systems, one often encounters subtyping aspects of the business domain that can prove challenging to implement in either relational databases or object-oriented code. In practice, some of these aspects are often handled incorrectly. This chapter examines a number of subtyping issues that require special attention (e.g. derivation options, subtype rigidity, subtype migration), and discusses how to model them conceptually. Because of its richer semantics, the main graphic notation used is that of second generation Object-Role Modeling (ORM 2). However, the main ideas could be adapted for UML and ER, so these are also included in the discussion. A basic implementation of the proposed approach has been prototyped in Neumont ORM Architect (NORMA), an open-source tool supporting ORM 2.

INTRODUCTION

In the wider sense, an information system corresponds to a business domain or universe of discourse rather than an automated system. As the name suggests, the universe of discourse is the world, or context of interest, about which we wish to discourse or talk. Most business domains involve some subtyping, where all instances of one type (e.g. Manager) are also instances of a more encompassing type (e.g. Employee). In this example, Manager is said to be a subtype of Employee (a supertype).

Various information modeling approaches exist for modeling business domains at a high level, for example Entity-Relationship Modeling (ER) (Chen, 1976), the Unified Modeling Language (UML) (Object Management Group, 2003a, 2003b; Rumbaugh, Jacobson & Booch, 1999), and Object-Role Modeling (ORM) (Halpin, 2006, 2007; Halpin & Morgan, 2008). These modeling approaches provide at least basic subtyping support. In industrial practice however, certain aspects of subtyping are often modeled or implemented incorrectly. This is sometimes due to a lack of appropriate modeling constructs (e.g. derivations to/from subtypes, subtype rigidity declarations), or to a lack of an obvious way to implement a subtyping pattern (e.g. historical subtype migration). This paper proposes solutions to some of these issues. Because of its richer semantics, the main graphic notation used is that of ORM 2 (second generation ORM), as implemented in NORMA, an open source ORM 2 tool. However, the main ideas could be adapted for UML and ER.

The next section overviews basic subtyping and its graphical depiction in ORM, UML, and ER, and identifies the condition under which formal derivation rules are required. The section after that proposes three varieties of subtyping (asserted, derived, and semiderived). The subsequent section distinguishes rigid and role subtypes, relates them to changeability settings on fact type roles, and discusses a popular party pattern. The next section discusses various patterns for modeling history of subtype or role migration. The final section notes implementation issues, summarizes the main results, suggests future research topics, and lists references.

BASIC SUBTYPING AND THE NEED FOR DERIVATION RULES

Figure 1(a) shows a simple case of subtyping in ORM 2 notation. Patients are identified by their patient numbers and have their gender recorded. Patient is specialized into MalePatient and FemalePatient. Pregnancy counts are recorded for, and only for, female patients. Prostate status is recorded only for male patients. In ORM 2, object types (e.g. Patient) are depicted as named, soft rectangles. A logical predicate is depicted as a named sequence of role boxes, each connected by a line to the object type whose instances may play that role. The combination of a predicate and its object types is a fact type—the only data structure in ORM (relationships are used instead of attributes). If an object type is identified by a simple fact type (e.g. Gender has GenderCode) this may be abbreviated by placing the reference mode in parentheses.

A bar spanning one or more roles depicts a uniqueness constraint over those roles (e.g. **Each** Patient has **at most one** Gender). A large dot depicts a mandatory constraint (e.g. **Each** Patient has **some** Gender). The circled dot with a cross through it depicts an exclusive-or constraint (**Each** Patient **is a** MalePatient **or is a** FemalePatient **but not both**). Overviews of ORM may be found in Halpin (2005b, 2006, 2007), a detailed treatment in Halpin & Morgan (2008), and a metamodel comparison between ORM, ER, and UML in Halpin (2004). Various dialects of ORM exist, for example Natural language Information Analysis Method (NIAM) (Wintraecken, 1990) and the Predicator Set Model (PSM) (ter Hofstede et al., 1993).

Figure 1(b) shows the same subtyping arrangement in UML. In UML, the terms "class" and "subclass" are used instead of "object type" and "subtype"". The "{P}" notation is the author's nonstandard addition to UML to indicate that an attribute is (at least part of) the preferred identifier for instances of the class. ORM and UML show subtypes outside their supertype(s), and depict the "is-a" relationship from subtype to supertype by an arrow.

The Barker ER notation (Barker, 1990), arguably the best industrial ER notation, instead uses an Euler diagram, placing the subtype shapes within the supertype shape, as shown in Figure 1(c). In spite of its intuitive appeal, the Barker ER subtyping notation is less expressive than that

Figure 1. Partitioning patient into subtypes in (a) ORM, (b) UML, and (c) Barker ER

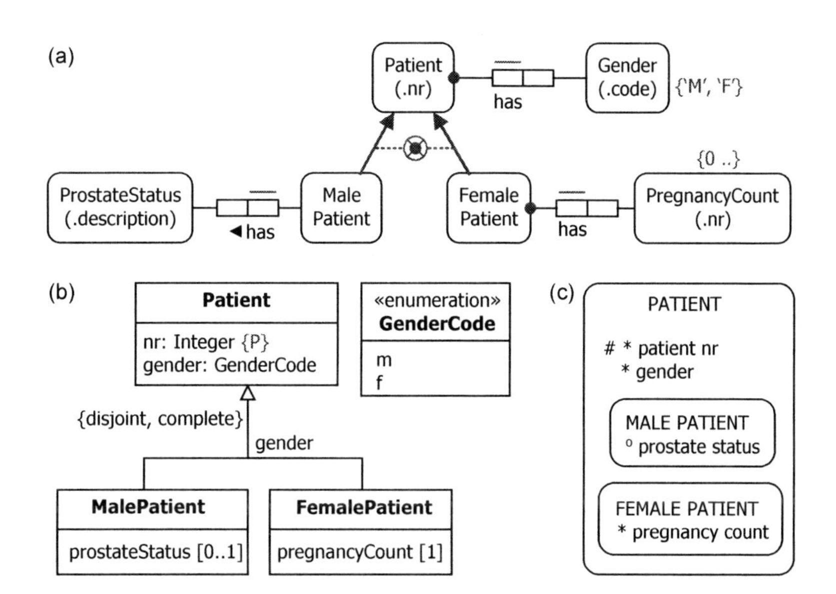

of ORM or UML (e.g. it cannot depict multiple inheritance).

The patient example illustrates the three main purposes of subtyping: (1) to indicate that some properties are specific to a given subtype (e.g. prostate status is recorded only for male patients); (2) to permit reuse of supertype properties (e.g. gender is specified once only (on Patient) but is inherited by each of its subtypes); (3) to display taxonomy (e.g. patients are classified into male and female patients).

In this example, the taxonomy is captured in two ways: (1) the subtyping; (2) the gender fact type or attribute (this may be needed anyway to record the gender of male patients with no prostate data). Both ORM and UML allow the possible values for gender (via gender code) to be declared. All of the diagrams in Figure 1 are conceptually incomplete, since they provide no formal connection between the two ways of displaying the classification scheme for patient. For example, there is nothing to stop us from assigning the gender code 'F" to patient 101 and then assigning the prostate status 'OK' for that patient. Even including "gender" as a discriminator to the subtyping, as allowed in UML (see Figure 1(b)) and some other versions of ER, will not suffice because there is still no formal connection between gender codes and the subtypes.

ORM traditionally solved this problem by requiring every subtype to be defined formally in terms of role paths connected to its supertype(s). For example, the ORM schema in Figure 2(a) adds the subtype definitions: **Each** MalePatient **is a** Patient **who** has Gender 'M'; **Each** FemalePatient **is a** Patient **who** has Gender 'F'. In ORM, an asterisk indicates "derived". In this example, the subtype definitions are derivation rules for deriving the subtypes. In previous versions of ORM, all subtypes had to be derived. ORM 2 removes this restriction, so an asterisk is added to indicate the subtype is derived.

The subtypes and fact types in ORM schema in Figure. 2(a) are populated with sample data. The population of the subtypes is derivable from the subtype definitions. The exclusive-or constraint is also derivable (as indicated by the asterisk) from these definitions given the mandatory and uniqueness constraints on the gender fact type.

While UML does not require subtype definitions, one could add them as notes in a language like the Object Constraint Language (OCL)

Figure 2. Adding definitions for derived subtypes in (a) ORM and (b) UML

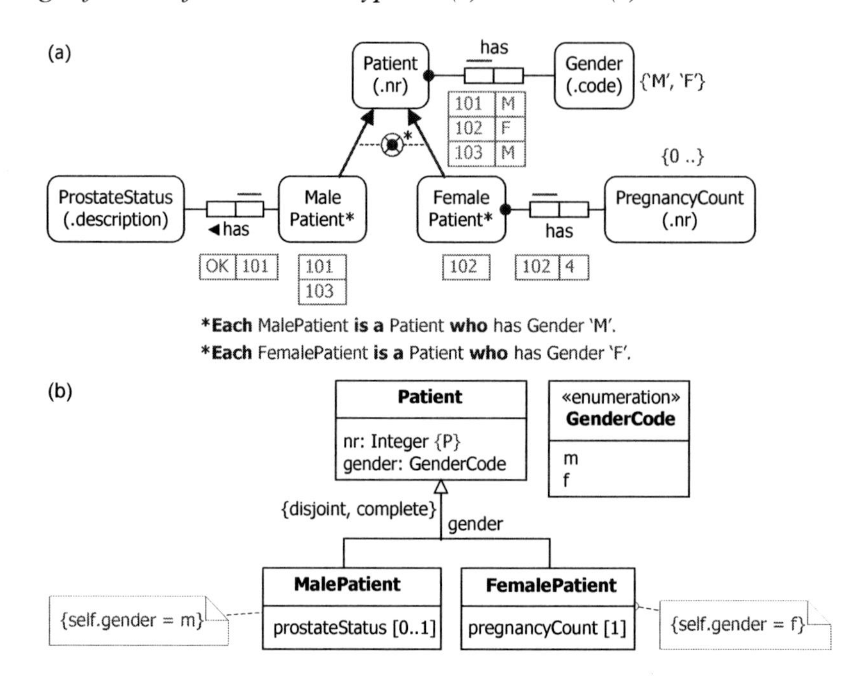

(Warmer & Kleppe, 2003), as shown in Figure. 2(b). Adding the constraints shown effectively redefines gender for the specific subclasses. A similar refinement technique is used in the MADS (Modeling of Application Data with Spatio-temporal features) approach (Parent et al., 2006, p. 47). Industrial versions of ER typically have no facility for defining subtypes, but could be extended to support this.

As discussed in the next section, subtype definitions/restrictions are not the only way to align multiple ways of depicting classification schemes. The main point at this stage is that *if a taxonomy is specified in two ways (via both subtypes and fact types/attributes) then derivation rules or constraints must be provided to formally align these two mechanisms.*

ASSERTED, DERIVED, AND SEMIDERIVED SUBTYPES

In previous versions of ORM, all subtypes had to be derived. We recently relaxed this restriction to permit three kinds of subtype: asserted, derived, and semiderived. An *asserted subtype* (or declared subtype) is simply declared without a definition. Asserted subtypes have always been permitted in UML and ER.

For example, if a gender fact type or attribute is excluded, then the patient subtypes may be simply asserted as shown in Figure 3. In this case, the exclusive-or constraint indicating that Patient is partitioned into these two subtypes must be explicitly declared, since it is not derivable. In ORM 2, this is shown by the lack of an asterisk beside the exclusive-or constraint. In this case, the classification scheme is depicted in only one way (via subtyping), so there is no need to provide any derivation rules.

Suppose however, that we still wish to query the system to determine the gender of patients. In this case, we may derive the gender from subtype membership. In Figure 4(a) the ORM fact type Patient is of Gender is derived (as noted by the asterisk) by means of the derivation rule shown. In Figure 4(b) the UML gender attribute is derived (as indicated by the slash) by means of

Figure 3. The subtypes are simply asserted rather than being derived

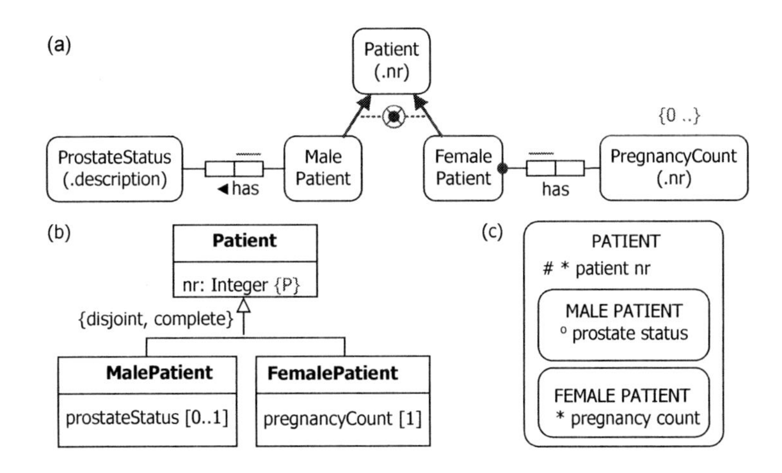

Figure 4. The subtypes are asserted, and gender is derived

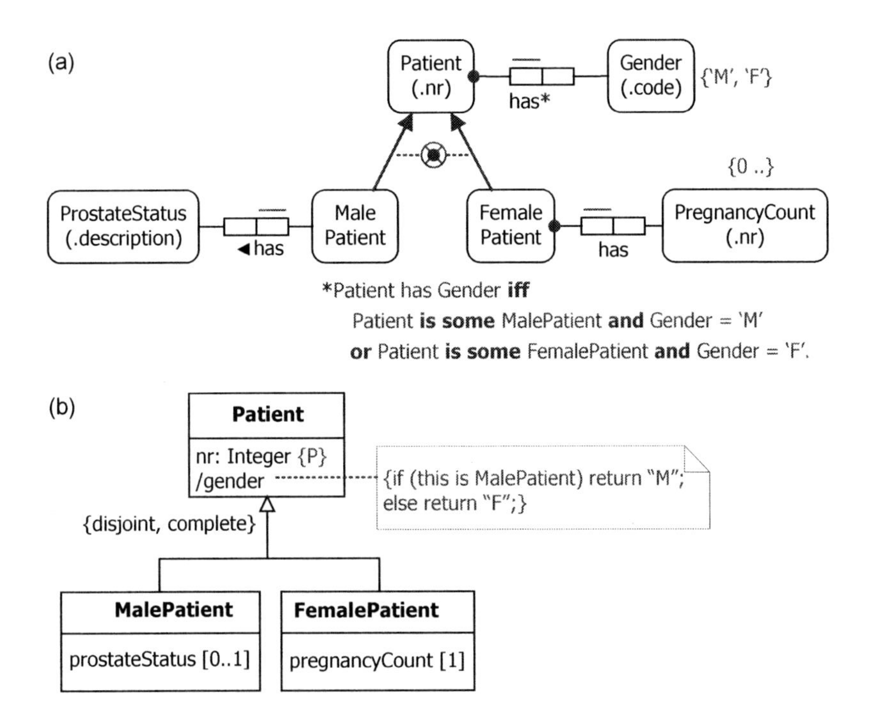

the derivation rule shown (here using C#). The UML derivation rule shows just one way to derive gender (e.g. we could instead provide overriding gender functions on the subclasses).

A *derived subtype* is fully determined by the derivation rule that defines it. For example, the subtypes in Figure. 2 are derived (from gender), not asserted. Notice that Figure 4 is the reverse of the situation in Figure. 2. Conceptually, a constraint applies between gender and the subtypes, and different modeling choices are available to satisfy this constraint (e.g. derive the subtypes from gender, or derive gender from the subtypes). Industrial ER typically does not support derivation rules in either direction.

Recently we introduced *semiderived subtypes* to ORM 2 to cater for rare cases such as that shown in Figure 5(a). Here we have incomplete knowledge of parenthood. If we know that person *A* is a parent of person *B* who is a parent of person *C*, then we may derive that *A* is a grandparent. If we know that someone is a grandparent without knowing the children or grandchildren, we can simply assert that he/she is a grandparent. The population of the subtype may now be partly derived and partly asserted.

In ORM 2, the semiderived nature is depicted by a "+" (intuitively, half an asterisk, so half-derived). We use the same notation for fact types, which may also be classified as asserted, derived, or semiderived. A semiderived status is much more common for fact types than for subtypes. We note in passing that the parenthood fact type has a spanning uniqueness constraint (hence is many:many), an alethic acyclic constraint, and a deontic intransitive constraint (by "parenthood" we mean biological parenthood).

Currently UML has no notation for semi-derived (e.g. see Figure 5(b)). The situation could be handled in UML by introducing an association or attribute for asserted grandparentood, adding a partial derivation rule for derived grandparent-hood, and adding a full derivation rule to union the two (cf. use of extensional and intensional predicates in Prolog). This is how we formerly handled such cases in ORM 1.

RIGID SUBTYPES AND ROLE SUBTYPES

Recent proposals from the ontology engineering community have employed type metaproperties to ensure that subtyping schemes are well formed from an ontological perspective. Guarino and Welty (2002) argue that every property in an ontology should be labeled as rigid, non-rigid, or anti-rigid. *Rigid* properties (e.g. being a person) necessarily apply to all their instances for their entire existence. *Non-rigid* properties (e.g. being hard) necessarily apply to some but not all their instances. *Anti-rigid* properties (e.g. being a patient) apply contingently to all their instances. One may then apply a meta-constraint (e.g. anti-rigid properties cannot subsume rigid properties) to impose restrictions on subtyping (e.g. Patient cannot be a supertype of Person).

Later Guizzardi, Wagner, Guarino, and van Sinderen (2004) proposed a UML profile that stereotyped classes into kinds, subkinds, phases, roles, categories, roleMixins and mixins, together with a set of meta-constraints, to help ensure that UML class models are ontologically well-formed. This modeling profile is used by Guizzardi (2005) in his doctoral thesis on ontological foundations for conceptual information models.

While we believe that the above research provides valuable contributions to ontology engineering, we have some reservations about its use

Figure 5. In the ORM schema (a) the subtype Grandparent is semiderived

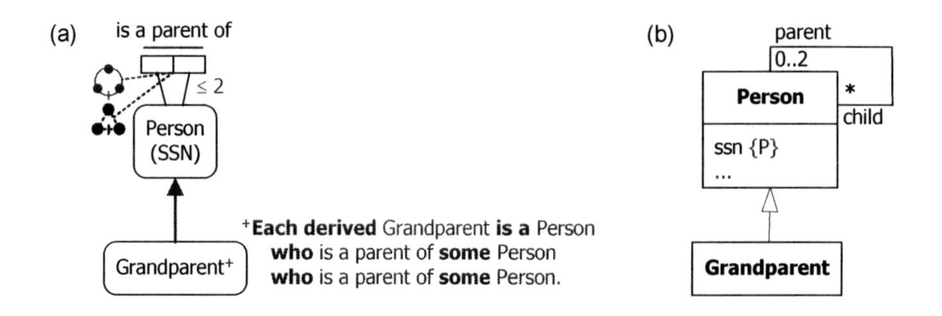

in industrial information systems modeling. Our experience with industrial data modelers suggests that the 7-stereotype scheme would seem overly burdensome to the majority of them. To be fair, we've also had pushback on the expressive detail of ORM, to which we've replied "Well, the world you are modeling is that complex—do you want to get it right or not?" Perhaps the same response could be made in defense of the 7-stereotypes.

At any rate, a simpler alternative that we are currently considering for ORM 2 *classifies each subtype as either a rigid subtype or role subtype*. A *type is rigid* if and only if each instance of that type must belong to that type for its whole lifetime (in the business domain being modeled). Examples include Person, Cat, Animal, Book. In contrast, any object that may at one time be an instance of a *role type* might not be an instance of that type at another time during its lifetime (in the business domain). Here we use "role" liberally to include a role played by an object (e.g. Manager, Student, Patient—assuming these are changeable in the business domain) as well as a phase or state of the object (e.g. Child, Adult, FaultyProduct—assuming changeability).

Though this rigid/role classification scheme applies to any type, we typically require this distinction to be made only for subtypes (our main purpose is to control subtype migration, as discussed shortly; also we wish to reduce the classification burden for modelers). As a simple example, Figure 6 shows how Dog and Cat might be depicted as rigid subtypes in ORM and UML. The rigidity notation tentatively being considered for ORM is square bracketing of the subtype name (violet for alethic as here; blue with "o" for deontic, e.g. changing from male to female might be possible but forbidden). For UML we have chosen a rigid stereotype. Our next example identifies a case where the rigidity of a root type (here Animal) should also be declared.

Notice that rigidity is a *dynamic constraint* rather than a static constraint since it restricts state changes (e.g. no dog may change into a cat). Currently, ORM is being extended to cater for a variety of dynamic constraints using a formal textual language to supplement the ORM graphical language (Balsters et al. 2006), and it is possible that rigidity might end up being captured textually in ORM rather than graphically as shown here.

In the above example, the subtypes are asserted. If instead they are derived, the relevant fact type/attribute used in their definition may be constrained by an appropriate changeability setting with impact on subtype rigidity. In Figure 7(a) the fact type Animal is of AnimalKind is made unchangeable (an animal can't change its kind), as indicated by the square brackets (this notation is tentative). In Figure 7(b) the defining animal kind attribute is constrained to be readOnly (prior to UML 2, this was called "frozen").

In either case, the unchangeability of animal kind combined with the rigidity of Animal implies that the subtypes are rigid. If we were instead to assert the subtypes and derive animal kind from subtype membership, the changeability/rigidity settings would still need to be kept in sync. Notice that even if we declare gender to be unchangeable in Figure 4, MalePatient and FemalePatient are not rigid unless Patient is rigid (and that depends on the business domain).

UML 2 (Object Management Group 2003a, 2003b) recognizes four changeability settings: unrestricted, readOnly, addOnly, and removeOnly. ORM 2 is currently being extended to enable declaration of fact type changeability (updateability

Figure 6. Rigid subtypes depicted in (a) ORM and (b) UML

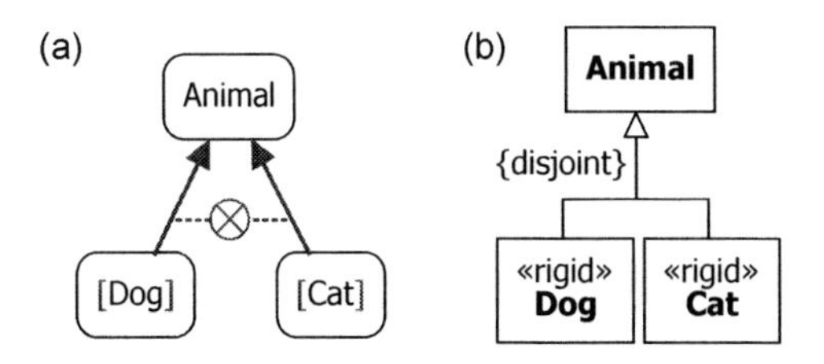

Figure 7. Rigidity of subtypes is now derived (given that Animal is rigid)

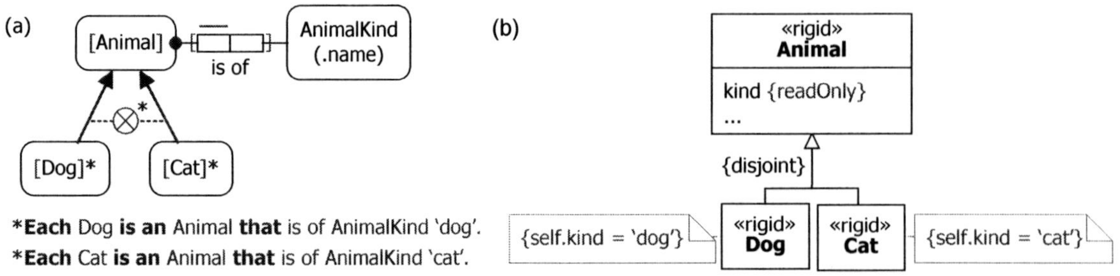

(a) [Animal] is of AnimalKind (.name) [Dog]* [Cat]*

***Each** Dog **is an** Animal **that** is of AnimalKind 'dog'.
***Each** Cat **is an** Animal **that** is of AnimalKind 'cat'.

(b) «rigid» **Animal** kind {readOnly} ... {disjoint} «rigid» **Dog** «rigid» **Cat** {self.kind = 'dog'} {self.kind = 'cat'}

and deleteability). Barker ER uses a diamond to indicate non-transferable relationships, but this may not be used for attributes.

To avoid explicitly declaring role subtypes as such, we propose that subtypes may be assumed to be role subtypes by default. This is similar to the default assumption in MADS that is-a clusters are dynamic (Parent et al. 2006, p. 44), but it is unclear whether MADS provides any graphical way to override the default. Unlike rigid subtypes, *migration between role subtypes is often permitted.* As a simple example, a person may play the role of child, teenager, and adult at different times in his/her lifetime (see Figure 8). In the next section we discuss various ways to maintain history of objects as they migrate between role subtypes.

An extension to ER to distinguish between "static subtypes", "dynamic subtypes" and "roles" has been proposed by Wieringa (2003, pp. 96–99), but this proposal is problematic, as it conflates roles with role occurrences, and reified roles (e.g. employee) with states of affairs (e.g. employment).

Now consider Figure 9, which shows the well known data model pattern for party in ORM, UML and Barker ER notations. Here Party is partitioned into Person and Organization. Ontologically, Person and Organization are substance sortals (they carry their own natural, intrinsic principle of identity). If "Party" simply means "Person or Organization" (a disjunction of sortals), then Party is a mixin type and there is no problem.

But what if "Party" has the sense of a role type (e.g. Customer)? If we replace "Party" by "Customer" in Figure 9, then Guizzardi (2005, p. 281) claims the schema is not well-formed because a rigid universal (e.g. Person) cannot be a subtype of an anti-rigid one (e.g. Customer). For information modeling purposes however, if each person or organization in the business domain must be a customer, then it's acceptable to specialize Customer into Person and Organization, even though ontologically this is incorrect (in the real world of which the business domain is just a part, not all persons are customers).

Our definition of rigid type is relative to the business domain. In the case just described, Cus-

Figure 8. Migration between role subtypes is allowed

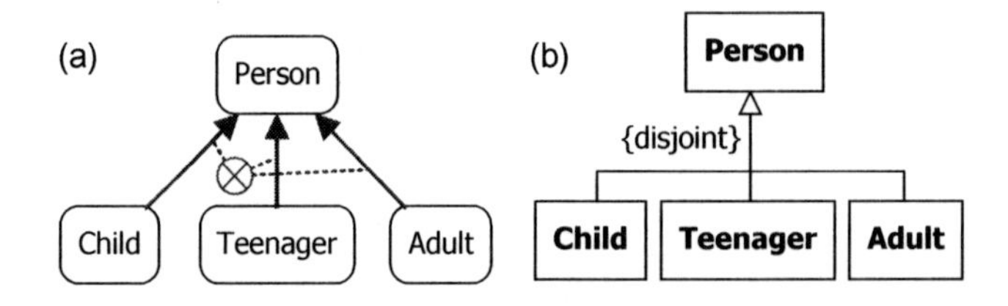

Figure 9. The Party pattern

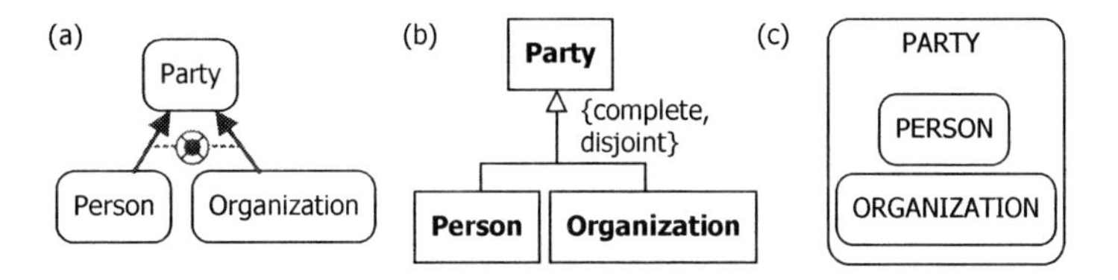

tomer is a rigid type in this sense, even though it is not rigid in the ontological sense. *Information models of business domains can be well formed even though they are not proper ontologies.*

If however our business domain includes (now or possibly later) some people or organizations that are not customers, then we do need to remodel, since Customer is no longer rigid even in our sense. One of many possible solutions using Party as a mixin type is shown in Figure 10. This solution differs from that of Guizzardi (2005, p. 282), where Person and Customer have no common supertype. Our original formalization of ORM, which made top level entity types mutually exclusive by default, requires the introduction of a supertype such as Party. This can be pragmatically useful (e.g. by allowing a simple global identification scheme for all parties).

However to avoid unnatural introduction of supertypes, we and a colleague long ago allowed the mutual exclusion assumption to be overridden by explicitly declaring an overlap possibility (depicted by overlapping "O"s) between top level types (Halpin & Proper, 1995). The same symbol is now used for this purpose in the MADS approach. To reduce notational clutter, and as a relaxation for ORM 2, we now allow overlap possibility between top level types to be implicitly deduced from the presence of a common subtype. With this understanding, the Party supertype could be removed from Figure 10.

Such a relaxation however should be used with care. For example, a UML class diagram produced by a UML expert depicted the classes Cashier and Customer. When we asked the expert whether it was possible for a customer to be a cashier, he said "Maybe". However nothing on the class diagram indicated this possibility, just as it did not reveal whether a customer could be a cashier transaction (another class on the diagram). The class diagram was little more than a cartoon with informal semantics.

It is sometimes useful in the modeling process to delay decisions about whether some service will be performed in an automated, semi-automated, or manual manner. For example, we can decide later whether cashiers will be ATMs and/or humans. Until we make that decision however, it is safer to allow for all possibilities (e.g. by explicitly declaring an overlap possibility between Cashier and Customer). Otherwise, one should explicitly indicate if one's current model is to be interpreted informally.

A radically different "two-layered approach" by Parsons and Wand (2000) allows instances to be asserted without requiring them to belong to a type. While interesting, this approach seems unattractive on both conceptual grounds (e.g. even with surrogate identifiers, instances must also be assigned natural identifiers, which are typically definite descriptions that invoke types) and implementation grounds.

HISTORY OF SUBTYPE MIGRATION OR ROLE MIGRATION

Some decades ago, we met our first application where we had to retain history of objects as they passed through various roles (e.g. applicant,

Figure 10. Remodeling is needed when Customer is a role type

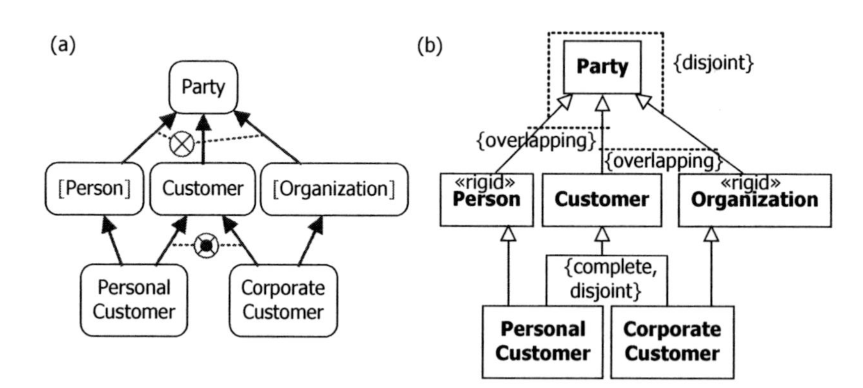

employee, past-employee etc.). Although such cases often arise in industry, we are unaware of their discussion in published papers on conceptual modeling. We now discuss some patterns we developed for dealing with such *historical subtype migration*.

The simplest pattern deals with linear state transitions. For example, in Figure 11(a) each role has specific details, and we wish to maintain these details of a person (e.g. favorite toy, favorite pop group) as he/she passes from one role to another.

One appropriate pattern for this case is to start with a supertype that disjoins all the roles, and then successively subtype to smaller disjunctions, as shown in Figure 11(c). We call this the *decreasing disjunctions pattern*. Depending on the business domain, a simple name may be available for the top supertype (e.g. Person).

An alternative solution is to model the child, teenager, and adult phases as life roles that are *once-only* (never repeated), and treat the playing of a life role by a person as an object itself. The life role playing type may then be subtyped into the three phases, and the relevant details attached to each, as shown in Figure 12. For completeness, reference schemes have been added as well as subtype definitions. This is an example of what we call this the *once-only role playing pattern*.

As a variation of this pattern, the objectified association LifeRolePlaying may instead be modeled directly as a coreferenced type, participating in the fact types LifeRolePlaying is by Person and LifeRolePlaying is of LifeRole. This makes the pattern usable in approaches such as industrial ER that do not support objectified associations.

Like the decreasing disjunctions pattern, the once-only role playing pattern assumes that each subtype is a once-only role, but unlike the previous pattern it does not specify the linear order of the roles. In ORM 2's Fact Oriented Modeling Language (FORML), this transition order may be specified as a dynamic textual constraint on the fact type Person plays LifeRole as follows, using "previous" in the derivable sense of "most recent". As this is a case of adding rather than replacing facts, "added" is used instead of "new".

For each Person,
in case previous lifeRole =
'Child': **added** lifeRole = 'Teenager'
'Teenager': **added** lifeRole = 'Adult'
end cases.

Sometimes role playing may be *repeated*, allowing loops in the role state graph. For example, the state chart in Figure 13 specifies the allowed marital state transitions. A person may play the roles of Married, Widowed, and Divorced more

Figure 11. Using the decreasing disjunctions pattern to retain subtype-specific details as a person changes roles

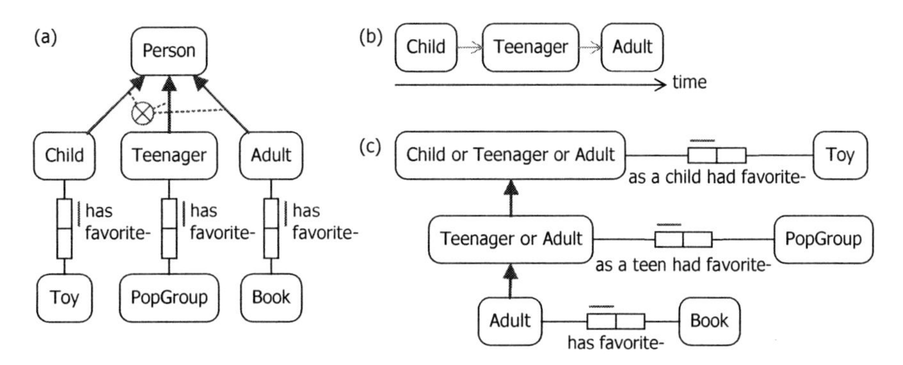

than once. Many other examples of this kind occur in business domains.

Suppose that historical details for each state are to be maintained. Such cases can be handled by what we call the *repeatable role playing pattern*. If at most one of these roles can be played by the same individual at any given time, then a start time of appropriate temporal granularity may be included to help identify any given role playing. For example, assuming that any given person cannot begin or end more than one marital state on the same day, the basic pattern shown in Figure 14 may be used.

A circled bar on the ORM schema depicts an external uniqueness constraint (e.g., the combination of person, role, and start date refers to at most one role playing). A double bar in ORM indicates the preferred identifier. The two external uniqueness constraints are captured as notes in UML. The preferred external uniqueness constraint is captured graphically in Barker ER using "#" and "|" symbols, but the other external uniqueness

Figure 12. Using the once-only role playing pattern to model historical role details

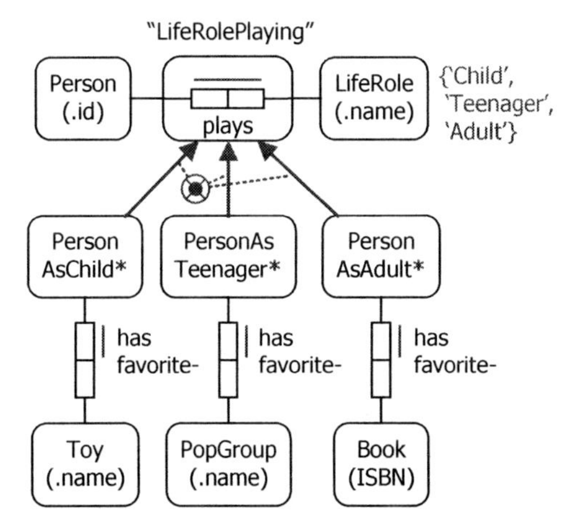

*Each PersonAsChild is a LifeRolePlaying involving LifeRole 'Child'.
*Each PersonAsTeenager is a LifeRolePlaying involving LifeRole 'Teenager'.
*Each PersonAsAdult is a LifeRolePlaying involving LifeRole 'Adult'.

Figure 13. An example of repeating roles

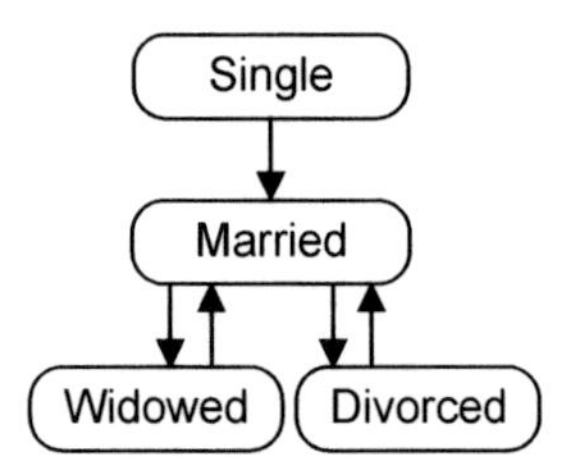

constraint involving end date is lost.

Here the Role and RolePlaying types may be suitably renamed (e.g. MaritalRole, MaritalRole-Playing) and constrained, and relevant subtypes (e.g. PersonAsSingle) may be introduced to record role playing specific details as needed. Figure 15 shows an example in the ORM notation, where subtype specific details may be attached directly to the subtypes.

The constraint on marital role transitions may be added as a textual rule in FORML as shown below, using "previous" in the derivable sense of "most recent". For discussion of a similar rule applied to a simpler schema, see Balsters et al. (2006).

For each Person,
in case previous maritalRolePlaying.maritalRole =
'S': **added** maritalRolePlaying.maritalRole = 'M'
'M': **added** maritalRolePlaying.maritalRole **in** ('W', 'D')
'W': **added** maritalRolePlaying.maritalRole = 'M'
'D': **added** maritalRolePlaying.maritalRole = 'M'
end cases.

Instead of using the start date as part of the preferred identifier for the role playing type, a simple numeric identifier may be used, as shown in Figure 16. So long as this identifier remains visibly part of human communication, it may be

Figure 14. A repeatable role playing pattern shown in (a) ORM, (b) UML and (c) Barker ER

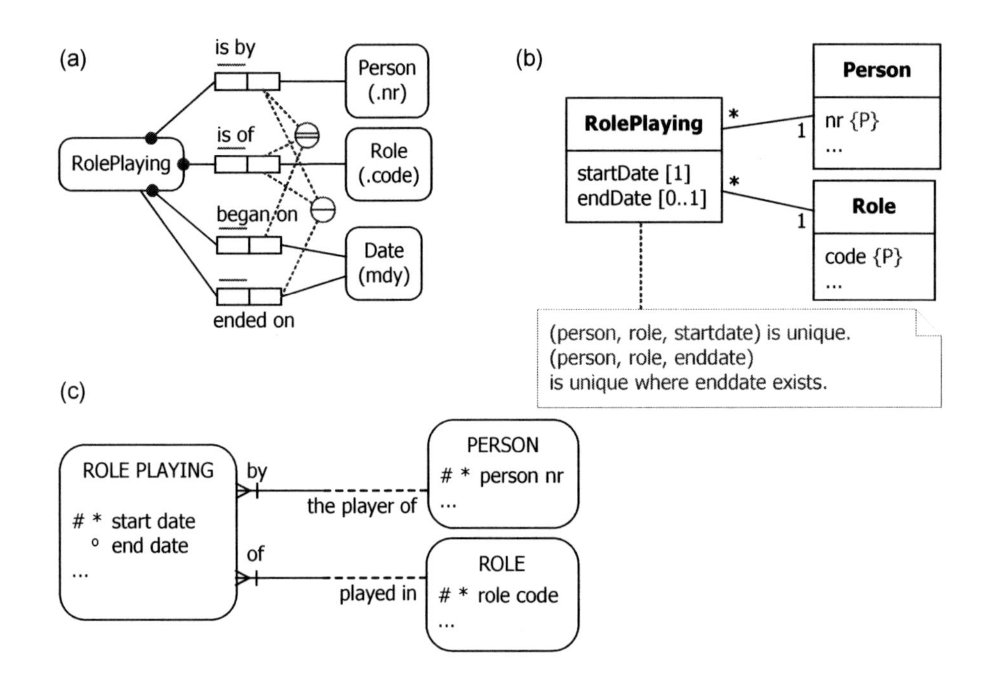

Figure 15. Using a repeatable role playing pattern to model historical role details

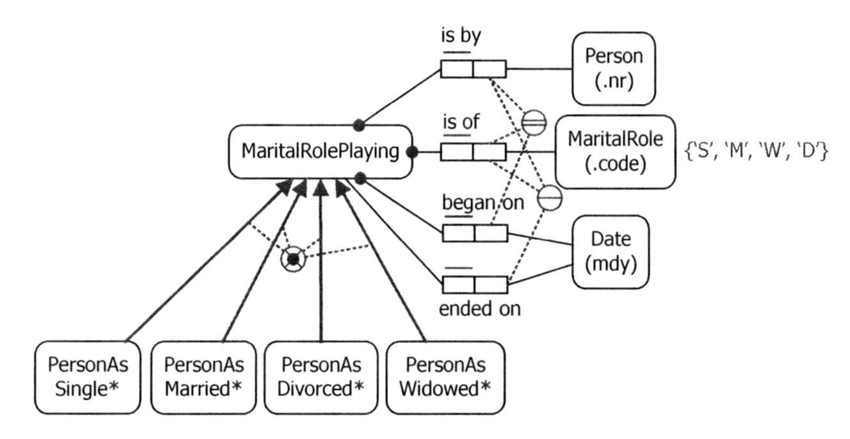

used even if the external uniqueness constraints do not apply.

Basic data model patterns for role playing have been described elsewhere using barker ER notation, for example Hay (2006) and Silverston (2001), but these ignore the fundamental impact of the state graph topology (e.g., whether looping is allowed) on the model requirements.

Yet another variation is to include an ordinal number instead of the start date as part of the identifier, as shown in Figure 17. For example, we might identify one particular role playing as the first (ordinal number = 1) marriage role of person 23. Again, this identification scheme works even if the secondary external uniqueness constraints involving start and end dates are removed (e.g. allowing someone to marry twice on the same date).

These latter patterns may be adapted for UML or ER in similar ways to those shown in Figure 14. One further option in UML 2 (Object Management Group, 2003a, 2003b) is to use its {nonunique} property to flag the association ends of an objectified relationship for role playing as

Figure 16. An alternative repeatable role playing pattern

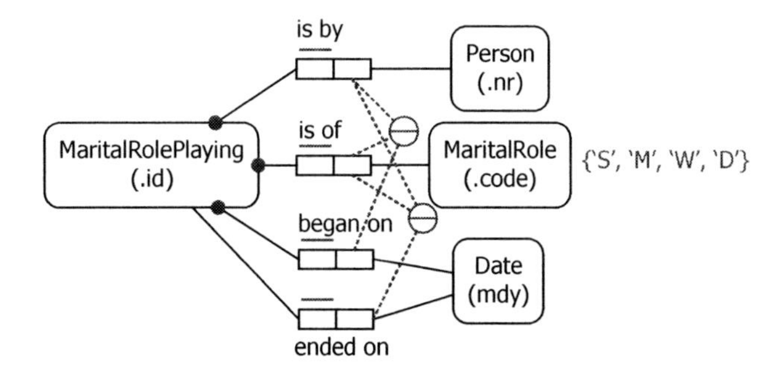

returning bags rather than sets, relying on hidden internal identifiers to provide the identification

CONCLUSION

This chapter discussed various ways to enrich the modeling of subtyping, especially within ORM 2. With options for asserted, derived, and semiderived subtypes, classification schemes may be specified in two ways, and derivation rules/constraints are then needed to keep these consistent, regardless of the direction of derivation. A lean ontological extension was proposed based on rigid and role subtypes, mainly to control subtype migration, with appropriate mechanisms for synchronizing subtype rigidity with fact type/attribute changeability. The party pattern was analyzed to highlight a fundamental difference between information models and ontologies, and a refinement was suggested for determining whether top level types are mutually exclusive. The decreasing disjunctions data model pattern and two versions (once-only and repeatable) of the role playing pattern were provided to model historical details about migration from one subtype or role to another.

The proposals for ORM 2 discussed in this chapter are being implemented in NORMA (Curland & Halpin, 2007), an open-source plug-in to Visual Studio that can transform ORM models into relational, object, and XML structures. Relational mapping of subtypes has been largely discussed elsewhere (e.g. Halpin & Morgan, 2008), and the dynamic aspects (e.g. non-updatability for rigid properties) can be handled by appropriate triggers. Object models treat all classes as rigid. To cater for migration between role subtypes as well as multiple inheritance when mapping to single-inheritance object structures (e.g. C#), is-a relationships are transformed into injective associations, an approach that bears some similarity to the twin pattern for implementing multiple inheritance (Mössenböck, 1999).

Further research is needed to refine both the graphical and textual languages of ORM 2 for advanced subtyping aspects, and to improve the code generation capabilities of NORMA to ensure an optimal implementation of these features.

REFERENCES

Balsters, H., Carver, A., Halpin, T., & Morgan, T. (2006). Modeling Dynamic Rules in ORM. In R.

Figure 17. Another alternative repeatable role playing pattern

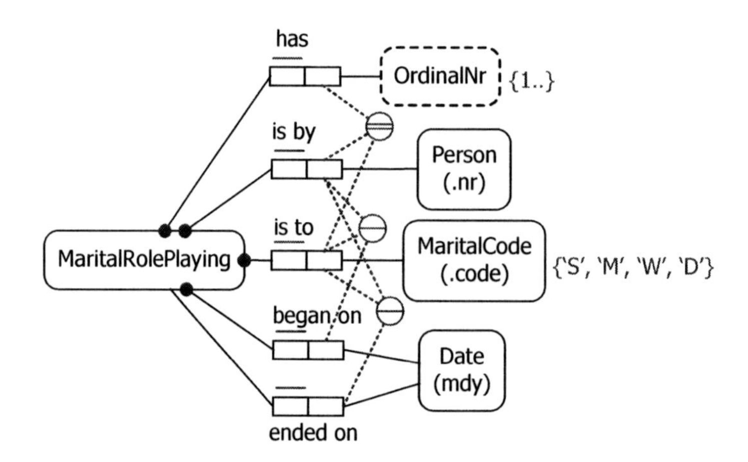

Meersman, Z. Tari, P. Herrero et al. (Eds.), *On the Move to Meaningful Internet Systems 2006: OTM 2006 Workshops*, (pp. 1201-1210). Montpellier: Springer LNCS 4278.

Barker, R. (1990). *CASE*Method: Entity relationship modelling.* Wokingham: Addison Wesley.

Chen, P. P. (1976). The entity-relationship model—towards a unified view of data. *ACM Transactions on Database Systems, 1*(1), 9–36.

Curland, M., & Halpin, T. (2007). Model Driven Development with NORMA. In *Proc. 40th Int. Conf. on System Sciences (HICSS-40)*, CD-ROM, IEEE Computer Society.

Guarino, N., & Welty, C. (2002). Evaluating Ontological Decisions with OntoClean. *Communications of the ACM, 45*(2), 61-65.

Guizzardi, G., Wagner, G., Guarino, N., & van Sinderen, N. (2004). An Ontologically Well-Founded Profile for UML Conceptual Models. In A. Persson & J. Stirna (Eds.), *Proc. 16th Int. Conf. on Advanced Inf. Sys. Engineering, CAiSE2004* (pp. 112-126). Springer LNCS 3084.

Guizzardi, G. (2005). *Ontological Foundations for Structural Conceptual Models*. CTIT PhD Thesis Series, No. 05-74, Enschede, The Netherlands.

Halpin, T., & Proper, H. (1995). Subtyping and polymorphism in object-role modeling. *Data & Knowledge Engineering, 15*(3), 251–281.

Halpin, T. (2004). Comparing Metamodels for ER, ORM and UML Data Models. In K. Siau (Ed.), *Advanced Topics in Database Research, 3*, 23-44. Hershey PA: Idea Publishing Group.

Halpin, T. (2005a). Higher-Order Types and Information Modeling. In K. Siau (Ed.), *Advanced Topics in Database Research, 4*, 218-237. Hershey PA: Idea Publishing Group.

Halpin, T. (2005b). ORM 2. In R. Meersman, Z. Tari, P. Herrero et al. (Eds.) *On the Move to Meaningful Internet Systems 2005: OTM 2005 Workshops* (pp. 676-687). Cyprus: Springer LNCS 3762.

Halpin, T. (2006). Object-Role Modeling (ORM/NIAM). In P. Bernus, K. Mertins, & G. Schmidt (Eds.), *Handbook on Architectures of Information Systems, 2nd edition* (pp. 81-103). Heidelberg: Springer.

Halpin, T. (2007). Fact-Oriented Modeling: Past, Present and Future. In J. Krogstie, A. Opdahl & S. Brinkkemper (Eds.), *Conceptual Modelling in Information Systems Engineering* (pp. 19-38). Berlin: Springer.

Halpin, T., & Morgan T. (2008). *Information Modeling and Relational Databases, 2nd edn*. San Francisco: Morgan Kaufmann.

Hay, D. (2006). *Data Model Patterns: A Metadata* Map. San Francisco: Morgan Kaufmann.

ter Hofstede, A., Proper, H. & Weide, th. van der (1993). Formal definition of a conceptual language for the description and manipulation of information models. *Information Systems, 18*(7), 489-523.

Mössenböck, H. (1999). Twin—A Design Pattern for Modeling Multiple Inheritance. Online: www.ssw.uni-linz.ac.at/Research/Papers/Moe99/Paper.pdf.

Object Management Group (2003a). *UML 2.0 Infrastructure Specification*. Online: www.ong.org/uml

Object Management Group (2003b). *UML 2.0 Superstructure Specification*. Online: www.omg.org/uml.

Parent, C., Spaccapietra, S., & Zimányi, E. (2006). *Conceptual Modeling for Traditional and Spatio-Temporal Applications*. Berlin: Springer-Verlag.

Parsons, J. & Wand, Y. (2000). Emancipating Instances from the Tyranny of Classes in Information Modeling. *ACM Transactions on Database Systems, 5*(2), 228-268.

Rumbaugh, J., Jacobson, I., & Booch, G. (1999). *The Unified Language Reference Manual*. Reading, MA: Addison-Wesley.

Silverston, L. (2001). *The Data Model Resource Book: Revised Edition*, New York: Wiley.

Warmer, J., & Kleppe, A. (2003). *The Object Constraint Language, 2nd Edition*. Addison-Wesley.

Wieringa, R. J. (2003). *Design Methods for Reactive Systems*. San Francisco: Morgan Kaufmann.

Wintraecken J. (1990). *The NIAM Information Analysis Method: Theory and Practice*. Deventer: Kluwer.

Chapter II
Essential, Mandatory, and Shared Parts in Conceptual Data Models

Alessandro Artale
Free University of Bozen-Bolzano, Italy

C. Maria Keet
Free University of Bozen-Bolzano, Italy

ABSTRACT

This chapter focuses on formally representing life cycle semantics of part-whole relations in conceptual data models by utilizing the temporal modality. The authors approach this by resorting to the temporal conceptual data modeling language $\mathcal{ER}_{VT}$ and extend it with the novel notion of status relations. This enables a precise axiomatization of the constraints for essential parts and wholes compared to mandatory parts and wholes, as well as introduction of temporally suspended part-whole relations. To facilitate usage in the conceptual stage, a set of closed questions and decision diagram are proposed. The long-term objectives are to ascertain which type of shareability and which lifetime aspects are possible for part-whole relations, investigate the formal semantics for sharability, and how to model these kind of differences in conceptual data models.

INTRODUCTION

Modeling part-whole relations and aggregations has been investigated and experimented with from various perspectives and this has resulted in advances and better problem identification to a greater or lesser extent, depending on the conceptual modeling language (Artale et al., 1996a; Barbier et al., 2003; Bittner & Donnelly, 2005; Borgo & Masolo, 2007; Gerstl & Pribbenow, 1995; Guizzardi, 2005; Keet, 2006b; Keet & Artale, 2008; Motschnig-Pitrik & Kaasbøll, 1999; Odell,

1998; Sattler, 1995). Nowadays, part-whole relations receive great attention both in conceptual modeling community (e.g., the Unified Modeling Language, UML, the Extended Entity Relationship, EER, and the Object-Role Modeling, ORM) as well as in the semantic web community (e.g. the Description Logic based language OWL).

Several issues, such as transitivity and types of part-whole relations, are being addressed successfully with converging approaches from an ontological, logical, and/or linguistic perspectives (Borgo & Masolo, 2007; Keet & Artale, 2008; Varzi, 2004; Vieu & Aurnague, 2005). On the other hand, other topics, such as horizontal relations among parts and *life cycle semantics* of parts and wholes, still remain an open research area with alternative and complimentary approaches (Bittner & Donnelly, 2007; Guizzardi, 2005; Motschnig-Pitrik & Kaasbøll, 1999). For instance, how to model differences between an Information System for, say, a computer spare parts inventory compared to one for transplant organs? Indeed, organs are at the time before transplantation not on the shelf as are independently existing computer spare parts, but these organs are part of another whole and can only be part of another whole *sequentially*. For a university events database, one may wish to model that a seminar can be part of both a seminar series and a course, *concurrently*. Another long-standing issue is how to represent essential versus mandatory parts and wholes (Artale *et al.*, 1996a). The solution proposed in Guizzardi (2005) as an extension to UML class diagrams is not easily transferable to other modelling/representation languages.

In this chapter we study representation problems related to the notion of *sharability* between parts and wholes. In particular, we are interested in representing that parts (i) cannot be shared by more than one whole; (ii) cannot exist without being part of the whole; (iii) can swap wholes in different ways. Clearly, these rich variations in shareability of parts cannot be represented in any of the common, industry-grade, UML class diagram, EER, or ORM CASE tools. In order to reach such a goal, we take a fist step by aiming to answer these main questions:

- Which type of sharability and which lifetime aspects are possible?
- What is the formal semantics for sharability?
- How to model these kind of differences in a conceptual data model?

To address these questions, we merge and extend advances in representing part-whole relations as in UML class diagrams with formal conceptual data modeling for temporal databases (temporal EER) and ORM's usability features. The various shareability constraints are reworded into a set of modeling guidelines in the form of closed questions and a decision diagram to enable easy navigation to the appropriate sharability case so as to facilitate its eventual integration in generic modeling methodologies.

Concerning the formalization of the sharability notion and the relationships between the lifespans of the involved entities, we use the temporally extended Description Logic $\mathcal{DLR}_{\mathcal{US}}$ (Artale *et al.*, 2002). Indeed, while DLs have been proved useful in capturing the semantics of various conceptual data models and to provide a way to apply automatic reasoning services over them (Artale *et al.*, 2007a; Berardi *et al.*, 2005; Calvanese *et al.*, 1998b, 1999; Franconi & Ng, 2000; Keet, 2007), temporal DLs have been applied to the same extent for temporal conceptual models (Artale & Franconi, 1999; Artale *et al.*, 2003, 2002, 2007b). The formalization we present here is based on the original notion of *status relations* that captures the evolution of a relation during its life cycle. Furthermore, a set of $\mathcal{DLR}_{\mathcal{US}}$ axioms are provided and proved to be correct with respect to the semantic we provide for each particular sharability relation.

The remainder of the chapter is organised as follows. We start with some background in section 2, where we review the state of the art of representing part-whole relations in the four main conceptual modeling languages, being UML class diagrams, EER, ORM, and Description Logic languages. The problems regarding shareability of parts are summarised in section 2.4 and the basic preliminaries of the temporal $\mathcal{DLR}_{US}$ and $\mathcal{ER}_{VT}$ languages are given in section 3. Our main contribution is temporalising part-whole relations to give a clear and unambiguous foundation to notions such as essential part and concurrent/sequentially shared part (section 4). Modeling guidelines will be presented in section 5 so as to transform the theory into usable material. Last, we look ahead by highlighting some future trends (section 6) and close with conclusions (section 7).

PART-WHOLE RELATIONS IN CONCEPTUAL MODELING LANGUAGES

Part-whole relations have been investigated from different starting points and with different aims. At one end of the spectrum we have philosophy with a sub-discipline called *mereology* and its sub-variants mereotopology and mereogeometry (Borgo & Masolo, 2007; Simons, 1987; Varzi, 2004, 2006a) that focus on the nature of the *part_of* relation and its properties, such as transitivity of *part_of*, reflexivity, antisymmetry, and parts and places, or take as point of departure natural language (mereonomy). From a mathematical perspective, there is interest in the relation between set theory and mereology (*e.g.*, Pontow & Schubert (2003) and references therein). At the other end of the spectrum we have application-oriented engineering solutions, such as aggregation functions in databases and data warehouses. Investigation into and use of representation of part-whole relations in conceptual data modeling languages lies somewhere in the middle: on the one hand, there is the need to model the application domain as accurately as possible to achieve a good quality application, yet, on the other hand, there is also the need to achieve usability and usefulness to indeed have a working information system. In this section, we take a closer look at four such conceptual modeling languages and how they fare regarding the part-whole relation: UML class diagrams (in section 2.1), EER and ORM (section 2.2), and Description Logics (section 2.3). It must be noted, however, that we primarily focus on notions such as exclusive and shareable parts among wholes and this overview is as such not a comprehensive introduction to all aspects of part-whole relations[1].

Aggregation in UML Class Diagrams

Part-whole relations in UML class diagrams are represented with *aggregatio*n associations. We first consider the UML specification of 2005 (OMG, 2005) and subsequently look at several significant extensions and formalizations that seek to address its shortcomings in particular regarding exclusiveness, sharability of parts among wholes, and thereby thus also the life cycle semantics of parts and wholes.

UML specification. UML (OMG, 2005) offers aggregation in two modes for UML class diagrams: composite and shared aggregation. Composite aggregation, denoted with a filled diamond on the whole-side of the association (see Figure 1), is defied as:

a strong form of aggregation that requires a part instance be included in ***at most one composite at a time****. If a composite is deleted, all of its parts are normally deleted with it. Note that a part can (where allowed) be removed from a composite before the composite is deleted, and thus not be deleted as part of the composite. Compositions define transitive* ***asymmetric*** *relationships – their links form a directed, acyclic graph. (OMG, 2005) (emphases added)*

The composite object is responsible for the existence and storage of the parts (OMG, 2005), which means an implicit *ontological commitment at the conceptual level*: the parts are existentially dependent on the whole (which, in turn, implies mandatoryness), and not that when a whole is destroyed its parts can exist independently. However, the "at a time" suggests that at another instant of time the same part could be part of another whole; may the part switch wholes instantaneously? In addition, the description for composite aggregation neither says that the whole is, or should be, existentially dependent on its part(s) nor that it has the part mandatorily. The difference between existential dependence and mandatory parts and wholes is refined well by Guizzardi (2005), as we will see in the next paragraph. There are three issues to take into account. First, to represent the difference between, say, a heart that must be part of a vertebrate animal but can be changed as long as the animal has a heart, whereas a brain cannot be transplanted and thus is deemed essential to the animal. Second, and a weaker constraint than essential/mandatory, is that we can have that a part indeed must be part of a whole, but either the part *p* (or whole *w*) can continue to exist as long as it is part of (has as part) some *w* (*p*). Third, it is not clear if *w* has as part *p only*. More general, let *A* be the whole with parts *B*, *C*, and *D* in a UML class diagram as in Figure 1-A, then each part is associated to the whole through a separate binary composite aggregation, as if *A* is a whole where its instance *a* is made up of a collection of instances of type *B*, and/or made up of a collection of instances of type *C* and/or *D*, making *A* a different type of entity for each way of aggregating its parts, which cannot be the intention of the representation because that does not have a correspondence with the portion of the real world it is supposed to represent. In addition, the description of composite aggregation says it is an "asymmetric" relationship, which is in mereological theories (Varzi, 2004) always attributed to *proper parthood*[2]. Thus, what needs to be represented (at least), is that instances of *B*, *C*, and *D* *together* make up the instance of the whole entity *A*, as in Figure 1-B, and prevent a modeler to create something like Figure 1-A. This difference is not mentioned in the UML specification, and one is left to assume it is a "semantic variation point" (OMG, 2005) which of the readings should be used. Of course, the Object Constraint Language (OCL) aids disambiguation, but is optional and one can only 'hope' for appropriate use as opposed to forcing the modeler to make such modeling decisions in the modeling stage.

Shared aggregation is denoted with an open diamond on the whole-side of the aggregation association, which has it that "precise semantics ... varies by application area and modeler" (OMG,

Figure 1. A: Ontologically ambiguous UML composite aggregation as separate binary associations; B: the composite A is composed of parts B, C, and D

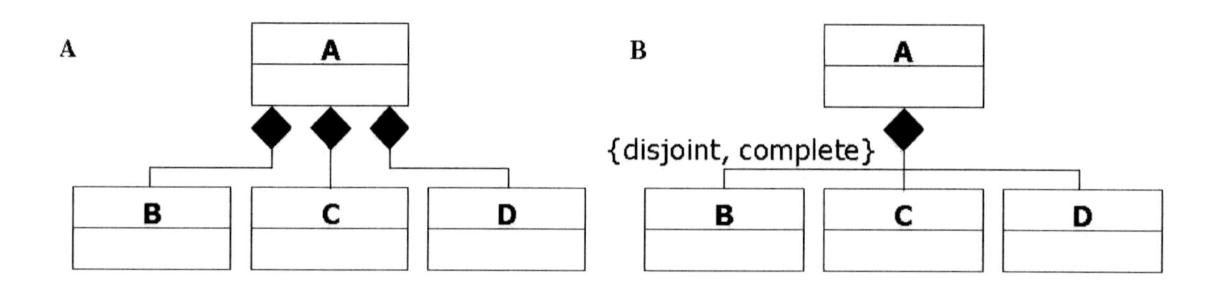

2005), and presumably can be used for any of the types of part-whole relations and shareability described in (Barbier *et al.*, 2003; Guizzardi, 2005; Johansson, 2004; Keet, 2006b; Keet & Artale, 2008; Motschnig-Pitrik & Kaasbøll, 1999; Odell, 1998; Winston *et al.*, 1987). The main difference with composite aggregation is that shared aggregation poses no constraint on multiplicity with respect to the whole it is part of. Thus, the part may be *directly shared* by more than one whole—be it *at the same time* or *sequentially*—where those wholes are instances of either the same class or different classes.

Thus, this raises at least four possibilities, but neither one can be represented in standard UML class diagrams in a way so as to be able to distinguish between them. Let p be a part of type P ($p \in P$) and w_i stand for wholes such that $w_1, w_2 \in W$, $w_a \in W'$, $\neg(W = W')$, and t_1, t_2 are points in time such that $t_1 < t_2$, then one can have that

1. p is part of w_1 at time t_1 and of w_2 at time t_2; e.g., a heart is transplanted from—was structural part of—one human w_1 into another w_2, or a car engine that used to be part of a FerrariCar w_1 used in Formula 1 racing and is put in another FerrariCar w_2, but the heart (car engine) cannot be part of both humans (cars) at the same time.
2. p is part of w_1 and w_2 at time t_1; e.g., an ethics course is part of multiple BSc curricula.
3. p is part of w_1 at time t_1 and of w_a at time t_2; e.g., phosphorylation (a phosphor atom part p is exchanged between two molecules w_1 and w_a of different type) or South Tyrol used to be part of (located in) the Austro-Hungarian Empire w_1 and is now part of the Republic of Italy w_a.
4. p is part of w_1 and w_a at time t_1; e.g., a seminar is part of both the Language Colloquia and of the Knowledge Representation course, or a cello player is member of a chamber ensemble w_1 and member of the Royal Philharmonic Orchestra w_a.

The examples for the four cases are fairly straightforward for transplant databases, car mechanics information systems, university databases, geographic information systems, and employment information systems, yet cannot be represented unambiguously with UML—nor with commonly used other conceptual data modeling languages, as we will see in the next sections—other than using shared aggregation for all of them.

A range of scenarios of life cycles of participating objects are possible, each with different behavior in the system, which thus ought to have their implementation-independent counterpart in the conceptual model. However, overall, the ambiguous specification and modeling freedom in UML does not enable making implicit semantics explicit in the conceptual model, and rather fosters creation of unintended models. This has been observed by several researchers, who have proposed a range of extensions to UML class diagrams.

Formalizations of aggregation in UML class diagrams. UML does not have a formal semantics, which demands from the researchers who propose extensions to also give the formal semantics. A near-complete formalization is proposed by Berardi *et al.* (2005), who developed a First Order Logic (FOL) as well as Description Logic (DL) encoding of UML with $\mathcal{DLR}_{ifd}$; that is, for each UML model, there is an equi-satisfiable $\mathcal{DLR}_{ifd}$ knowledge base. However, UML's aggregation has not been addressed other than 'shared aggregation' (no formalisation is provided to account for additional constraints and composite aggregation). In Berardi et al's (2005) formalisation of shared aggregation in UML class diagrams, we have $\forall x, y(G(x, y) \rightarrow C_1(x) \wedge C_2(y))$, where G is a binary predicate for the aggregation and C_1 and C_2 are concepts. That is, a straightforward binary relation as if it were a mere UML association relation. The avoidance to map UML's intuitive part-whole relation is partially due to the ambiguous semantics of aggregation in UML and partially because adequately representing parthood in DL has its own issues (see below). In contradistinction, others have been more precise on the aggregation, but they omitted a formalization of UML class diagram language. For instance, Barbier *et al.* (2003) represent several constraints on the part-whole relation using OCL, which is, however, not immediately transferrable to other conceptual modeling languages. They formulate the various add-ons as a `context` in OCL and add a meta-model fragment for whole-part relations where the attribute `aggregation` is removed from the `AssociationEnd` meta-class, `Whole` added as a subclass of `Relationship` and has two disjoint subclasses `Aggregation` and `Composite`. Motschnig-Pitrik & Kaasbøll (1999) and Guizzardi (2005) present a First Order Logic formalization and corresponding adornments for the graphical notation in UML class diagrams using new icons and labels, which are effectively *in addition* to the axiomatizations of mereological theories and perceived to be necessary for conceptual modeling of mereological and meronymic relations. Guizzardi adds, among others, the notion of essential part *EP*, which he defines as (Guizzardi (2005): p165):

Definition 5.11 (essential part): An individual x is an essential part of another individual y iff, y is existentially dependent on x and x is, necessarily, a part of y: $EP(x, y) =_{def} ed(y, x) \wedge \Box(x \leq y)$. *This is equivalent to stating that* $EP(x, y) =_{def} \Box(\varepsilon(y) \rightarrow \varepsilon(x)) \wedge \Box(x \leq y)$, *which is, in turn, equivalent to* $EP(x, y) =_{def} \Box(\in(y) \rightarrow \varepsilon(x) \wedge (x \leq y))$. *We adopt here the mereological continuism defended by (Simons, 1987), which states that the part-whole relation should only be considered to hold among existents, i.e.,* $\forall x,y(x \leq y) \rightarrow \varepsilon(x) \wedge \varepsilon(y)$. *As a consequence, we can have this definition in its final simplification (47).* $EP(x,y) =_{def} \Box(\varepsilon(y) \rightarrow (x \leq y))$

where ε denotes existence, $\leq$ a partial order, and $\Box$ necessity. The weaker version is mandatory parthood *MP*, which is defined as (Guizzardi, 2005, p167):

Definition 5.13 (mandatory part): An individual x is a mandatory part of another individual y iff, y is generically dependent of an universal U that x instantiates, and y has, necessarily, as a part an instance of U: (49). $MP(U, y) =_{def} \Box(\varepsilon(y) \rightarrow \exists U, x)(x < y))$.

Observe that in this setting, essential parts are also *immutable*—"stability in identity and number" (Barbier *et al.*, 2003) of the part—and *inseparable* ("$IP(x, y) =_{def} \Box(\varepsilon(x) \rightarrow (x \leq y))$" in (Guizzardi, 2005)). There are finer-grained details between essential parts (or wholes) and immutable parts (wholes) that are caused by the kind of classes that participate in the part-whole relation (Guizzardi, 2005, 2007; Artale *et al.*, 2008); that is, the former has participating classes that are rigid, whereas for immutable parts (wholes) the class is not rigid (indeed, anti-rigid). The notion of a class' metaproperty concerning rigidity—i.e., being rigid, non-rigid, semi-rigid, or anti-rigid—is important for designing good subtype hierarchies (Guarino & Welty, 2000) and its use with UML and ORM2 is actively being investigated regarding how to incorporate it and to what extent (Guizzardi, 2005; Halpin, 2007).

An example of adorning UML class diagrams is depicted in Figure 2 that demonstrates the proposed representation for the *sub_quantity_of* relation with an additional symbol, OCL constraint, and stereotypes. Motschnig-Pitrik and Kaasbøll, on the other hand, focus on *gradations* of exclusiveness between part and whole. This corresponds partially to Guizzardi's mandatoryness and (in)separability of the part from the whole, as can be observed from one of the definitions, such as total exclusiveness (Motschnig-Pitrik & Kaasbøll (1999): p785):

Total exclusiveness. A part-of reference is totally exclusive if there exists exactly one immediate

Figure 2. Part-whole relations among quantities. Essential parts are indicated with `essential = true`*, which implies a composite aggregation (filled diamond), which is of the type "Q" for quantities. The stereotypes ("≪≫") add further constraints to the permitted types of classes. (Source: Guizzardi (2005)).*

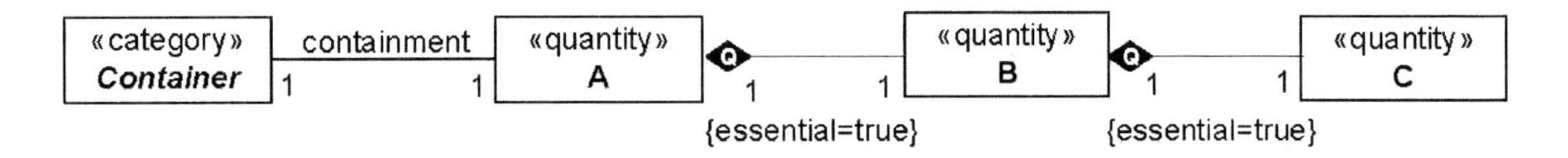

part-of link from a part-type P to a whole-type W and, for each instance p of P, there exists at most one instance w of W such that p part-of w. More formally, let:
p_k *instance-of P*, $k \in [1..n]$,
w_i *instance-of W*, $i \in [1..n']$, $w_{i'}$ *instance-of W*, $i' \in [1..n]$,
wx_j *instance-of WX*, $j \in [1..n'']$
then
P totally-exclusive part-of W $\Leftrightarrow \forall WX$
((P part-of W $\wedge$ *P part-of WX)*$\Rightarrow$*(W = WX* $\vee$ *W part-of WX))* $\wedge$
$((p_k\ part\text{-}of\ w_i \wedge p_k\ part\text{-}of\ w_{i'})) \Rightarrow (i = i')) \wedge$
$((p_k\ part\text{-}of\ w_i \wedge p_k\ part\text{-}of\ wx_j) \Rightarrow (w_i = wx_j \vee w_i\ part\text{-}of\ wx_j))$

One can add further gradations in *sharability*. Motschnig-Pitrik & Kaasbøll (1999) distinguish between "*degree of sharing* of parts among whole objects" and "*degree of dependence* between some part object and some whole object(s)" where the former acts out as static constraints and the latter concerns the life-cycle of objects. To summarize and comment on Motschnig-Pitrik & Kaasbøll (1999), there are six cases that each get their own modeling construct in a UML class diagram.

- Total exclusiveness: there exists exactly one immediate part-of relation from *P* to *W* (thus *P cannot* have another part-of relation to a *W'*), there is *at most one* instance *w* s.t. *p* part-of *w*. Thus, with this constraint, the *w* can also exist without having as part *p*, hence neither essential nor mandatory participation from that side.
- Arbitrary sharing: "*A part-of link from P to W is shared if there may exist further shared or intraclass exclusive (see below) links from P to whole-types WX, WY, etc., and if, for each instance p of P, there may exist more than one instance of W :* $w_1, w_2, \ldots$ *such that p part-of* w_1*, p part-of* w_2*, etc.*". Thus, with this combination where *p* can be part of $w_i \in W_j$ where $0 \leq i \leq n$ and $1 \leq j \leq n$; i.e., *p* can be part of zero or more *w*s that are instances of one or more *W*s, and thereby subsumes the next four options.
- Interclass exclusiveness: there exists exactly one type-level part-of relation from *P* to *W* and for each $p_1, \ldots, p_m \in P$ there may exist $w_1, \ldots, w_n \in W$ such that we have p_1 part-of $w_k, \ldots, p_1$ part-of $w_{n'}, \ldots, p_n$ part-of $w_1, \ldots, p_h$ part-of w_l (with $h \leq m$, $k \leq n$, and $l \leq n$). Or, simply a 0:*n* relation between p_i and w_i, where $w_i \in W$ and $0 \leq i \leq n$.
- Intraclass exclusiveness: as for interclass, but then ≥1 part-of relations to ≥1 different types of wholes.
- Selective exclusiveness: ≥1 part-of relations to different types of wholes, but only one of them may be instantiated. It is unclear if this means *at a time* or possibly *ever* during the life time of the part (an XOR constraint at the type level).
- Selectively intraclass exclusive: as for selective exclusiveness, but then also that all part-of relations have a max cardinality of 1.

In addition, the dependence/independence axis concerns "*A dependent part-of relationship between a part-type P and a whole-type W is one in which the existence of each part-object pi of type P depends on the existence of one and the same whole-object wi of type W throughout the lifetime of the part-object*", which is also called "lifetime-dependence", or in Guizzardi's terminology *essential* whole to the part, but which implies only a minimum cardinality of one on the W -side by Motschnig-Pitrik & Kaasbøll (1999).

Despite the problems with the UML class diagram specification as well as the limited extensions, the issues have been investigated to a greater extent than within other conceptual modeling languages. The next two sections focus on EER, ORM and DL languages.

Part-Whole Relations in (E)ER and ORM

It may be clear from the previous section that part-whole relations in UML class diagrams can have poorly defined semantics, but what about other conceptual modeling languages? Entity-Relationship (ER) does not have a separate constructor for the part-whole relations, despite the occasional (Shanks *et al.*, 2004) request. Neither does Object-Role Modeling (ORM) have a separate constructor for parthood relation. Are they better off than UML? What, if any, can already be represented from part-whole relations with ER or ORM? Here, we summarize ORM's difference with UML based on (Keet, 2006b).

Recollect that the UML specification inserts design and implementation considerations for composite aggregation, so that a part is *existentially dependent* on the whole, and not that when the whole is destroyed, the parts, explicitly, can have their own life or, explicitly, become part of another whole. Here there is a difference between UML and ORM intended semantics: with composite aggregation in the UML specification, part *p cannot* exist without that whole *w*, but ORM semantics of the suggested mapping (Halpin, 2001) says that 'if there is a relation between part *p* and whole *w*, then *p* must participate exactly once'. Put differently, *p* may indeed become part of some other whole w' after w ceases to exist as a whole, as long as there is *some* whole it is part of, but not necessarily *the same* whole. Hence, in contrast with UML, in ORM there is no implicit existential dependency of the part on the whole (see also Figure 3-B).

Compared to more and less comprehensive formalizations and extensions for aggregation in UML (Barbier *et al.*, 2003; Guizzardi, 2005; Motschnig-Pitrik & Kaasbøll, 1999; Berardi *et al.*, 2005), for ORM, richer representations of the semantics are possible already even without

Figure 3. Graphical representation of "aggregation" in UML and ORM. (Source: adapted from Halpin (1999) with part-whole relations as proposed in Keet (2006b))

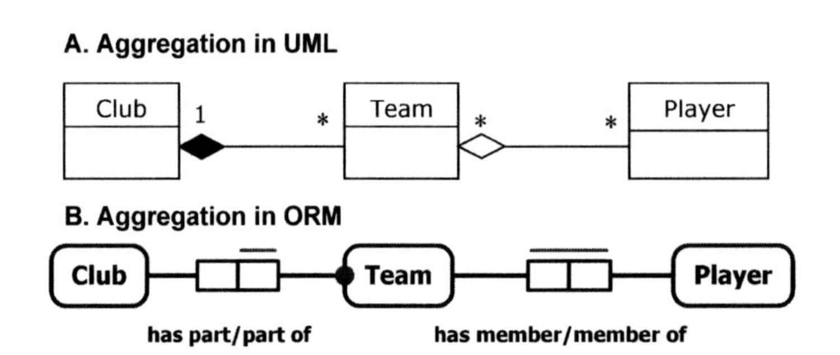

dressing up the ORM diagram with icons and labels. For instance, Motschnig-Pitrik & Kaasbøll (1999)'s new "selectively intraclass exclusive" constraint is an XOR constraint (dotted circle with cross) over the Part-roles of two or more fact types to different types of Whole in an ORM diagram. Suggestions to model several aspects of the part-whole without extending ORM were presented in (Keet, 2006b), which also includes several guidelines to ease selecting the appropriate part-whole relation and its mandatory and uniqueness constraints.

Description Logics

Description Logic (DL) languages are more often used as knowledge representation languages than as conceptual data modeling languages and if they are used for conceptual modeling, they are used in the background hidden from the modelers and domain experts because the formalisms are deemed not easily accessible. Thus far, the combination of DL languages and conceptual data modeling languages is primarily limited to the well-studied $\mathcal{DLR}$ family of DL languages (Artale *et al.*, 2006, 2007a; Berardi *et al.*, 2005; Calvanese *et al.*, 1998b, 1999; Calvanese & De Giacomo, 2003; Fillottrani *et al.*, 2006; Franconi & Ng, 2000; Keet, 2007), which provide not only a formal foundation for the mostly graphically-oriented conceptual data modeling languages, but also offer prospects of automated reasoning over conceptual data models to derive implicit relations, constraints, and inconsistencies and thereby contribute to better quality conceptual data models. Therefore, efforts in representing parthood in DL languages will be briefly summarized.

Research on part-whole relations for DLs date back to the early '90s, but thus far none of the DL languages that are being investigated and implemented (DL-Lite, $\mathcal{DLR}$, OWL, and $\mathcal{EL}$ families) have the part-whole relation as a first-class citizen. However, as neither UML nor ER nor ORM, implement part-whole relations properly, it might not matter that most DL languages do not have a comprehensive treatment of part-whole relations, at present. It is, however, being investigated.

We first address two early attempts. Artale *et al.* (1996a) experimented with adding a *has_part* relation as $\succcurlyeq$ with the transitive closure of a parthood relation (1). One can define, e.g., `Car` as having wheels that in turn have tires (2), such that it follows that cars have as part tires ($\texttt{Car} \sqsubseteq \exists{\succcurlyeq}.\texttt{Tire}$).

(1) $\succcurlyeq\ \doteq (\texttt{primitive-part})*$

(2) $\texttt{Car} \doteq \exists{\succcurlyeq}.(\texttt{Wheel} \sqcap \exists{\succcurlyeq}.\texttt{Tire})$

However, adding transitive closure makes languages of even low expressivity, such as ALC, already ExpTime-complete. Alternatively, one can define direct parthood $\prec_d$ (Sattler, 1995), but this should verify the immediate inferior, which makes the language undecidable (Artale *et al.*, 1996b), which is even less desirable for operational information systems. Schulz *et al.* (2000) have developed an elaborate workaround (with $\mathcal{ALC}$) so as to be able to simulate transitivity of parthood relations by remodeling the part-of relation as *is_a* hierarchies using so-called SEP triplets. The three core items are the **S**tructure-concept node that subsumes one (anatomical) entity, called **E**-node, and the parts of that entity (the **P**-node). An *is_a* hierarchy is then built up by relating the P-node of a whole concept *D* to the S-node of the part *C*, where in turn the P-node of *C* is linked to the S-node of *C*'s part. More formally, the definition of the whole *D* is (3), by which one can derive its anatomical proper part (a-pp) C as (4). Obviously, if this were to be used, this would require an intuitive user interface.

(3) $\texttt{D}_P \doteq \texttt{D}_S \sqcap \neg\texttt{D}_E \sqcap \exists\texttt{a-pp}.\texttt{D}_E$

(4) $\texttt{C}_E \sqsubseteq \exists\texttt{a-pp}.\texttt{D}_E$

Around the same time, Sattler (2000) showed that with some extensions to $\mathcal{ALC}$, it is possible to include more aspects of the parthood relation.

These are: transitive roles (that is, permit $R_+ \subseteq R$), inverse roles to have both part-of and has-part, role hierarchies to include subtypes of the parthood relation, and number restrictions to model the amount of parts that go in the whole. This brings us to the language called $\mathcal{SHIQ}$, which is a predecessor of the OWL-DL ontology language. In fact, the base language for the even more expressive OWL 2, $\mathcal{SROIQ}$, has constructors for all but one of the relational properties: antisymmetry, required for mereological part-of (Varzi, 2004), is not possible yet (Horrocks *et al.*, 2006).

The latest—and most comprehensive—attempt to represent parthood relations in a DL language is put forward by Bittner & Donnelly (2005), who approach the problem starting from a FOL characterization and subsequently limit its comprehensiveness and complexity to fit it into a DL language, although it is unclear if their $\mathcal{L}^{\sim Id \sqcup}$ is decidable. In their theory, called DL-PCC, several constraints and definitions cannot be represented. These are: impossibility to state that *component_of* (CP), *proper_part_of* (PP) and *contained_in* (CT) are irreflexive and asymmetric, and it is missing a discreteness axiom for CP or CT or a density axiom for PP (see Bittner & Donnelly (2005) for details and discussion). They include transitivity of the characterized parthood relations, but thereby do not have the option to state also that, e.g., a *directly_contained_in* relation is *in*transitive (the same problem as mentioned above for $\mathcal{ALC}$).

Artale *et al.* (1996a,b) have placed the requirement for adequately representing the part-whole relation in a wider context, where some outstanding issues of 12 years ago are still in need of a solution. For instance, (non)distributivity of part-whole relations[3], 'horizontal' relations between the parts, and disjoint covering over the parts. The latter is an issue with DL but not for database models, because DL languages adhere to the open world assumption whereas databases do not. For instance, if we have in a DL language a type-level (TBox-) statement where C has two parts D and E:

$$C \sqsubseteq \exists has\_part.D \sqcap \exists has\_part.E$$

then it may be that instances of C have more parts than only instances of D and E because the composite C is not fully defined. No DL language deals with an additional axiom that states that C is composed of—the mereological sum of—D and E only; what we can state is that C is defined by having D and E as parts ($C \doteq \exists has\_part.D \sqcap \exists has\_part.E$). In contradistinction, conceptual models and databases do adhere to a closed world assumption, thereby making instances of C uniquely composed of at least one instance of D and at least one instance of E. The status of differentiating between e.g. essential and mandatory part (see section 2.1) is unclear, unless we use a temporal DL such as $\mathcal{DLR}_{US}$ (see below).

Problems and Requirements for Modeling Shared and Composite Parts and Wholes

Summarizing the problems for adequately modeling the different ways that parts can be part of a whole in the main commonly used conceptual modeling languages—with or without extensions—, we have (1) the absence of adequately distinguishing between mandatory and essential parts and wholes and, vice versa, existence of p and/or w independently for some time (e.g., the relation is temporarily "suspended" or p is "scheduled" to become part of w), (2) lack of clarity how to represent that part p that can be part of more than one whole s.t. either $w_i \in W$ or ($w_j \in W$ and $w_k \in W'$) and (3) if these shared parts can be shared concurrently and/or sequentially among the wholes p being part of w_i and w_j. These issues can be reformulated in a set of requirements for conceptual data modeling languages if they want to be expressive enough to enable full shareability semantics of parts and wholes.

Requirements for modeling shareability of parts. Based on the literature review, we can formulate the following requirements for modeling shareability and, implicitly, life cycle semantics in conceptual modeling languages[4].

1. Arbitrarily shareable with no particular constraints;
2. Existentially dependent/essential part or whole (mutually, or not), and (im)possibility of independent existence of the part from the whole;
3. Being able to differentiate between mandatory and essential parts and wholes;
4. Change in whole; that is, a part p was part of w_1 (where $w_1 \in W$) and can become part of another whole w_2, be it that there is a (negligible) time that p exists independently before (after) being part of and be it that $w_2 \in W$ or $w_2 \in W'$;
5. Change in part; that is, the inverse of requirement nr.4 where the whole loses and gains a part during its lifetime;
6. Disjoint covering of parts being part of a whole.

Notions such as essential parts, change, and before and after indicate *temporality* of either the part-whole relation, or the participating parts and wholes, or both; hence, the more general requirement for *temporal conceptual data modeling* in order to address the above-mentioned requirements. For the current restricted scope, we may not need a full-fledged temporal knowledge representation language to adequately address the shareability of part and wholes. Before going into those details in section 3, we summarize the literature dealing with temporal parts.

Temporalizing part-whole relations. The most straightforward, yet also limited, way to temporalize part-whole relations is to turn a part-of predicate from a binary into a ternary relation, such that we have p part of w at time t: *part_of(p, w, t)*. To the best of our knowledge, almost all extant temporalizations of parthood take this approach (Bittner & Donnelly, 2007; Masolo *et al.*, 2003; Smith *et al.*, 2005) but do not go further to take advantage of a temporal knowledge representation language[5]. An exception is Barbier *et al.* (2003), who created an `oclUndefined` observer function to "assert that all parts of result do not exist before (`@pre`) the execution" of the creation of the whole instance *w*, which is intended for representing life time dependencies in UML class diagrams. They also tried representation of immutability, but this remained an open problem due to the lack of a full-fledged implementation of a temporal UML. Furthermore, Barbier *et al.* (2003); Opdahl *et al.* (2006) listed nine principle life cycle cases. We extend this here to 18 cases (see Figure 4) mainly since they represent two distinct perspectives: (i) fixed the lifespan of a whole, we are interested in characterizing the lifespan of its part (Figure 4-A) and, vice versa, (ii) fixed the lifespan of the part, we are interested in the temporal relations with the lifespan of its whole (Figure 4-B). We shall see in section 4.3 that these two views require distinct constraints, too. A curious feature of UML is the Boolean `readOnly` metaproperty that was initially proposed for attributes as a Changeability sort with constant symbols "frozen" and "changeable" (Álvarez & Alemán, 2000). OMG (2005) now constrains it such that "[i]f a navigable property is marked as `readOnly`, then it cannot be updated once it has been assigned an initial value" (OMG, 2005), where a property is "structural feature", such as attribute and association end but it is also suggested for representing rigid classes (Halpin, 2007). Unfortunately, its "semantics is undefined" (pp 241, 249, 251, 254, 280) for various cases, such as what should happen with a `readOnly` association end when a link is destroyed. Intuitively, it captures a property that holds globally during its entire existence, hence, is a candidate to use for representing at least some aspects of lifecycle semantics. Albert *et al.* (2003) provide an additional interpretation where objects instantiating `readOnly` classes can only participate in links created during creation of the object, but no links to that object can be added afterward, and subsequently use this for constraining the part-class in a composite aggregation. This seems too restrictive, however, because if we assume Brain to be such a composite part of Patient, then it would surely be possible in a hospital information system that brain o_1 of patient o_2 may have to be linked to, say, some instance $o_3 \in$ BrainScan and at a later point in time participate in a new aggregation association to $o_4 \in$ BrainTumor. The prospects for usage of `readOnly` may be more interesting if it is (1) applied either to the composite aggregation association or to the association ends because for the part-whole relation one has to consider also the temporal behavior of the relation and (2) its semantics would be defined precisely and have an effect in the modeling as opposed to in the software code only.

Regarding the ternary temporal part-whole relation, we have, for instance, Bittner and Donnelly's "temporal mereology" (2007), which was developed to deal with "portions of stuff", i.e., how to deal with subquantities (portions) of amounts of matter, such as gold, and mixtures, such as lemonade, in time (see also Figure 9 for types of part-whole relations). Limiting oneself to only ternary part-whole relations runs into rather complicated formalizations, whereas well-defined temporal logics—and those applied to temporal conceptual data modeling in particular—can hide at least some of the details, which enhances understandability and (re)usability of a conceptual model by modeler and domain expert alike. In order to arrive at this point, we will introduce such a formal temporal conceptual modeling language, which enables one to model essential and shareable parts in a precise and clear way. Moreover, in addition to the modeling enhancements for the conceptual models themselves (in section 4), we will add modeling guidelines to facilitate easy navigation and choosing the appropriate part-whole relation and object types (section 5).

TEMPORAL DATA MODELS

In order to capture the range of possibilities of shareability, composite, and essential parts and represent them in a convenient way in a conceptual modeling language, we introduce here representation languages able to capture time varying information. As outlined in the previous section, we need temporal constructs in a conceptual

Figure 4. Possible lifespans of the part with respect to the whole it is part of (A) and similarly for the whole's lifespan w.r.t its part (B)

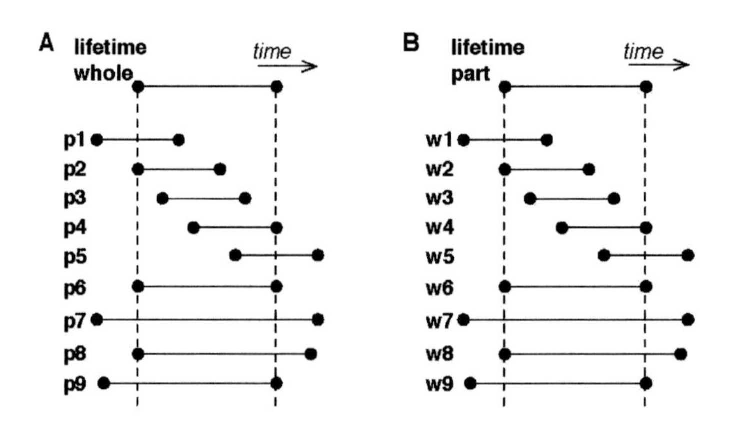

modeling language. Temporal EER have been developed (e.g., Artale *et al.* (2003, 2006, 2007b) and references therein) and a subset has been implemented in MADS (Parent *et al.*, 2006), but one may contend that, ideally, one should have a way to have the approach general enough so as to be transferrable to ORM and UML class diagrams, too. Artale *et al.*'s temporal EER, called $\mathcal{ER}_{VT}$, has a correspondence with the temporal Description Logic $\mathcal{DLR}_{US}$ (Artale *et al.*, 2002), which, in turn, gives a model-theoretic semantics to $\mathcal{ER}_{VT}$—any corresponding icon in the graphical diagrams can then be considered 'syntactic sugar' with a precise meaning. In addition, with an UML/EER/ORM to DL transformation we can then provide the sought-after genericity of the approach[6].

This section presents the formal background in modeling temporal varying information. Such a formalization will be used in the following sections when the basic conceptual data model is extended to capture particular properties of part-whole relations.

The Temporal Description Logic $\mathcal{DLR}_{US}$

The temporal description logic $\mathcal{DLR}_{US}$ (Artale *et al.*, 2002) combines the propositional temporal

Figure 5. Syntax and semantics of $\mathcal{DLR}_{US}$

$$
\begin{aligned}
C \rightarrow\ & \top \mid \bot \mid CN \mid \neg C \mid C_1 \sqcap C_2 \mid \exists^{\lessgtr k}[U_j]R \mid \\
& \Diamond^+ C \mid \Diamond^- C \mid \Box^+ C \mid \Box^- C \mid \oplus C \mid \ominus C \mid C_1 \,\mathcal{U}\, C_2 \mid C_1 \,\mathcal{S}\, C_2 \\
R \rightarrow\ & \top_n \mid RN \mid \neg R \mid R_1 \sqcap R_2 \mid U_i/n : C \mid \\
& \Diamond^+ R \mid \Diamond^- R \mid \Box^+ R \mid \Box^- R \mid \oplus R \mid \ominus R \mid R_1 \,\mathcal{U}\, R_2 \mid R_1 \,\mathcal{S}\, R_2
\end{aligned}
$$

$$
\begin{aligned}
\top^{\mathcal{I}(t)} &= \Delta^{\mathcal{I}} \\
\bot^{\mathcal{I}(t)} &= \emptyset \\
CN^{\mathcal{I}(t)} &\subseteq \top^{\mathcal{I}(t)} \\
(\neg C)^{\mathcal{I}(t)} &= \top^{\mathcal{I}(t)} \setminus C^{\mathcal{I}(t)} \\
(C_1 \sqcap C_2)^{\mathcal{I}(t)} &= C_1^{\mathcal{I}(t)} \cap C_2^{\mathcal{I}(t)} \\
(\exists^{\lessgtr k}[U_j]R)^{\mathcal{I}(t)} &= \{\, d \in \top^{\mathcal{I}(t)} \mid \sharp\{\langle d_1, \ldots, d_n \rangle \in R^{\mathcal{I}(t)} \mid d_j = d\} \lessgtr k \} \\
(C_1 \,\mathcal{U}\, C_2)^{\mathcal{I}(t)} &= \{\, d \in \top^{\mathcal{I}(t)} \mid \exists v > t.(d \in C_2^{\mathcal{I}(v)} \wedge \forall w \in (t, v).d \in C_1^{\mathcal{I}(w)})\} \\
(C_1 \,\mathcal{S}\, C_2)^{\mathcal{I}(t)} &= \{\, d \in \top^{\mathcal{I}(t)} \mid \exists v < t.(d \in C_2^{\mathcal{I}(v)} \wedge \forall w \in (v, t).d \in C_1^{\mathcal{I}(w)})\} \\
(\top_n)^{\mathcal{I}(t)} &\subseteq (\Delta^{\mathcal{I}})^n \\
RN^{\mathcal{I}(t)} &\subseteq (\top_n)^{\mathcal{I}(t)} \\
(\neg R)^{\mathcal{I}(t)} &= (\top_n)^{\mathcal{I}(t)} \setminus R^{\mathcal{I}(t)} \\
(R_1 \sqcap R_2)^{\mathcal{I}(t)} &= R_1^{\mathcal{I}(t)} \cap R_2^{\mathcal{I}(t)} \\
(U_i/n : C)^{\mathcal{I}(t)} &= \{\, \langle d_1, \ldots, d_n \rangle \in (\top_n)^{\mathcal{I}(t)} \mid d_i \in C^{\mathcal{I}(t)} \} \\
(R_1 \,\mathcal{U}\, R_2)^{\mathcal{I}(t)} &= \{\, \langle d_1, \ldots, d_n \rangle \in (\top_n)^{\mathcal{I}(t)} \mid \\
&\qquad \exists v > t.(\langle d_1, \ldots, d_n \rangle \in R_2^{\mathcal{I}(v)} \wedge \forall w \in (t, v).\, \langle d_1, \ldots, d_n \rangle \in R_1^{\mathcal{I}(w)})\} \\
(R_1 \,\mathcal{S}\, R_2)^{\mathcal{I}(t)} &= \{\, \langle d_1, \ldots, d_n \rangle \in (\top_n)^{\mathcal{I}(t)} \mid \\
&\qquad \exists v < t.(\langle d_1, \ldots, d_n \rangle \in R_2^{\mathcal{I}(v)} \wedge \forall w \in (v, t).\, \langle d_1, \ldots, d_n \rangle \in R_1^{\mathcal{I}(w)})\} \\
(\Diamond^+ R)^{\mathcal{I}(t)} &= \{\langle d_1, \ldots, d_n \rangle \in (\top_n)^{\mathcal{I}(t)} \mid \exists v > t.\, \langle d_1, \ldots, d_n \rangle \in R^{\mathcal{I}(v)}\} \\
(\oplus R)^{\mathcal{I}(t)} &= \{\langle d_1, \ldots, d_n \rangle \in (\top_n)^{\mathcal{I}(t)} \mid \langle d_1, \ldots, d_n \rangle \in R^{\mathcal{I}(t+1)}\} \\
(\Diamond^- R)^{\mathcal{I}(t)} &= \{\langle d_1, \ldots, d_n \rangle \in (\top_n)^{\mathcal{I}(t)} \mid \exists v < t.\, \langle d_1, \ldots, d_n \rangle \in R^{\mathcal{I}(v)}\} \\
(\ominus R)^{\mathcal{I}(t)} &= \{\langle d_1, \ldots, d_n \rangle \in (\top_n)^{\mathcal{I}(t)} \mid \langle d_1, \ldots, d_n \rangle \in R^{\mathcal{I}(t-1)}\}
\end{aligned}
$$

logic with the *Since* and *Until* operators and the (non-temporal) description logic $\mathcal{DLR}$ (Calvanese *et al.*, 1998a; Baader *et al.*, 2003) that serves as common foundational language for various conceptual data modeling languages (Calvanese *et al.*, 1998b, 1999). $\mathcal{DLR}_{US}$ can be regarded as an expressive fragment of the first-order temporal logic $L^{\{since,until\}}$ (Chomicki & Toman, 1998; Hodgkinson *et al.*, 2000).

The basic syntactical types of $\mathcal{DLR}_{US}$ are *classes* (also known as *entity types* or *object types*) and *n*-ary *relations* (*associations*) of arity ≥ 2. Starting from a set of *atomic classes* (denoted by *CN*), a set of *atomic relations* (denoted by *RN*), and a set of *role symbols* (denoted by *U*, comparable to an ORM-role or component of a UML association) we can define inductively (complex) class and relation expressions (see upper part of Figure 5), where the binary constructors ($\sqcap$, $\sqcup$, $\mathcal{U}$, $\mathcal{S}$) are applied to relations of the same arity, *i*, *j*, *k*, *n* are natural numbers, $i \leq n$, and *j* does not exceed the arity of *R*. Observe that for both class and relation expressions all the Boolean constructors are available. The selection expression $U_i/n : C$ denotes an *n*-ary relation whose *i*-th argument ($i \leq n$), named U_i, is of type *C*. (In ORM terminology, $U_i/n : C$ refers to the role U_i played by *C* in the fact type.) If it is clear from the context, we omit n and simply write ($U_i : C$). The projection expression $\exists^{\lessgtr k}[U_j]R$ is a generalization with cardinalities of the projection operator over argument U_j of relation *R*; the plain classical projection is $\exists^{\geq 1}[U_j]R$. It is also possible to use the pure argument position version of the language by replacing role symbols U_i with their corresponding position numbers *i*.

The model-theoretic semantics of $\mathcal{DLR}_{US}$ assumes a flow of time $\mathcal{T} = \langle \mathcal{T}_p, < \rangle$, where $\mathcal{T}_p$ is a set of time points (also called chronons) and < a binary precedence relation on $\mathcal{T}_p$, which is assumed to be isomorphic to $\langle \mathbb{Z}, < \rangle$. The language of $\mathcal{DLR}_{US}$ is interpreted in *temporal models* over $\mathcal{T}$, which are triples of the form $I = \langle \mathcal{T}, \Delta^I, \cdot^{I(t)} \rangle$, where Δ^I is a non-empty set of objects (the *domain* of *I*) and $\cdot^{I(t)}$ an *interpretation function* such that, for every $t \in \mathcal{T}$ ($t \in \mathcal{T}$ will be used as a shortcut for $t \in \mathcal{T}_p$), every class *C*, and every *n*-ary relation *R*, we have $C^{I(t)} \subseteq \Delta^I$ and $R^{I(t)} \subseteq (\Delta^I)^n$. The semantics of class and relation expressions is defined in the lower part of Figure 5, where $(u, v) = \{w \in \mathcal{T} \mid u < w < v\}$.

We will use the following equivalent abbreviations: $C_1 \sqcup C_2 \equiv \neg(\neg C_1 \sqcap \neg C_2)$; $C_1 \rightarrow C_2 \equiv \neg C_1 \sqcup C_2$; $\exists[U]R \equiv \exists^{\geq 1}[U]R$; $\forall[U]R \equiv \neg\exists[U]\neg R$; $R_1 \sqcup R_2 \equiv \neg(\neg R_1 \sqcap \neg R_2)$. Furthermore, the operators $\diamond^*$ (at some moment) and its dual $\square^*$ (at all moments) can be defined for both classes and relations as $\diamond^* C \equiv C \sqcup \diamond^+ C \sqcup \diamond^- C$ and $\square^* C \equiv C \sqcap \square^+ C \sqcap \square^- C$, respectively.

A *knowledge base* is a finite set Σ of $\mathcal{DLR}_{US}$ axioms of the form $C_1 \sqsubseteq C_2$ and $R_1 \sqsubseteq R_2$, and with R_1 and R_2 being relations of the same arity. An interpretation *I* satisfies $C_1 \sqsubseteq C_2$ ($R_1 \sqsubseteq R_2$) if and only if the interpretation of C_1 (R_1) is included in the interpretation of C_2 (R_2) at all time, i.e. $C_1^{I(t)} \subseteq C_2^{I(t)}$ ($R_1^{I(t)} \subseteq R_2^{I(t)}$), for all $t \in \mathcal{T}$. Various *reasoning services* can be defined in $\mathcal{DLR}_{US}$, such as satisfiability, logical implication and class (relation) subsumption (see Baader *et al.* (2003) for details). While $\mathcal{DLR}$ knowledge bases are fully able to capture atemporal EER schemas (Berardi *et al.*, 2005; Calvanese *et al.*, 1998a,b)—i.e. given an EER schema there is an equi-satisfiable $\mathcal{DLR}$ knowledge base—in the following Sections we show how $\mathcal{DLR}_{US}$ knowledge bases can capture temporal EER schemas with both timestamping and evolution constraints.

The Temporal Conceptual Model $\mathcal{ER}_{VT}$

In this section, the temporal EER model $\mathcal{ER}_{VT}$—which will be the basis to present our proposal—is briefly introduced (see (Artale & Franconi, 1999; Artale *et al.*, 2003) for full details). $\mathcal{ER}_{VT}$ supports timestamping for classes, attributes, and relationships. $\mathcal{ER}_{VT}$ is equipped with both a textual and

Definition 1.

Definition 1 ($\mathcal{ER}_{VT}$ Conceptual Data Model). *An $\mathcal{ER}_{VT}$ conceptual data model is a tuple: $\Sigma = (\mathcal{L}, \text{REL}, \text{ATT}, \text{CARD}, \text{ISA}, \text{DISJ}, \text{COVER}, \text{S}, \text{T}, \text{KEY})$, such that: $\mathcal{L}$ is a finite alphabet partitioned into the sets: $\mathcal{C}$ (class symbols), $\mathcal{A}$ (attribute symbols), $\mathcal{R}$ (relationship symbols), $\mathcal{U}$ (role symbols), and $\mathcal{D}$ (domain symbols) and*

1. *The set $\mathcal{C}$ of class symbols is partitioned into a set $\mathcal{C}^S$ of Snapshot classes (marked with an S), a set $\mathcal{C}^M$ of Mixed classes (unmarked classes), and a set $\mathcal{C}^T$ of Temporary classes (marked with a T). A similar partition applies to the set $\mathcal{R}$.*
2. *ATT is a function that maps a class symbol in $\mathcal{C}$ to an $\mathcal{A}$-labeled tuple over $\mathcal{D}$, $\text{ATT}(C) = \langle A_1 : D_1, \ldots, A_h : D_h \rangle$.*
3. *REL is a function that maps a relationship symbol in $\mathcal{R}$ to an $\mathcal{U}$-labeled tuple over $\mathcal{C}$, $\text{REL}(R) = \langle U_1 : C_1, \ldots, U_k : C_k \rangle$, and k is the arity of R.*
4. *CARD is a function $\mathcal{C} \times \mathcal{R} \times \mathcal{U} \mapsto \mathbb{N} \times (\mathbb{N} \cup \{\infty\})$ denoting cardinality constraints. We denote with $\text{CMIN}(C, R, U)$ and $\text{CMAX}(C, R, U)$ the first and second component of CARD.*
5. *ISA is a binary relationship $\text{ISA} \subseteq (\mathcal{C} \times \mathcal{C}) \cup (\mathcal{R} \times \mathcal{R})$. ISA between relationships is restricted to relationships with the same arity. ISA is visualized with a directed arrow.*
6. *DISJ, COVER are binary relations over $(2^{\mathcal{C}} \times \mathcal{C}) \times (2^{\mathcal{R}} \times \mathcal{R})$, describing disjointness and covering partitions, respectively, over a group of ISA that share the same superclass/superrelation. DISJ is visualized with a circled "d" and COVER with a double directed arrow.*
7. *S, T are binary relations over $\mathcal{C} \times \mathcal{A}$ containing, respectively, the snapshot and temporary attributes of a class;*
8. *KEY is a function, $\text{KEY} : \mathcal{C} \rightarrow \mathcal{A}$, that maps a class symbol in $\mathcal{C}$ to its key attribute. Keys are visualized as underlined attributes.*

a graphical syntax along with a model-theoretic semantics as a temporal extension of the EER semantics (Calvanese *et al.*, 1999). The formal foundations of $\mathcal{ER}_{VT}$ allowed also to prove a correct encoding of $\mathcal{ER}_{VT}$ schemas as knowledge base in $\mathcal{DLR}_{US}$ (Artale *et al.*, 2002, 2003).

The model-theoretic semantics associated with the $\mathcal{ER}_{VT}$ modeling language adopts the snapshot representation of temporal conceptual data models (Chomicki & Toman, 1998)[7].

Given such a set-theoretic semantics for the temporal EER (or, for that matter, UML class diagrams or ORM), some relevant modeling notions such as satisfiability, subsumption, and derivation of new constraints by means of logical implication have been defined rigorously (Artale *et al.*, 2007b).

Mapping $\mathcal{ER}_{VT}$ into $\mathcal{DLR}_{US}$

We briefly summarize how $\mathcal{DLR}_{US}$ is able to capture temporal schemas expressed in $\mathcal{ER}_{VT}$ —see (Berardi *et al.*, 2005; Artale *et al.*, 2006) for more details.

In the next sections we extend the formalism presented here to capture essential and sharable parts.

MODELING ESSENTIAL PARTS AND WHOLES

This section presents a formalization of the notion of essential part-whole relations. To formalize such properties of part-whole relations we will resort to the formalism introduced in the previ-

Definition 2.

Definition 2 ($\mathcal{ER}_{VT}$ Semantics). *Let Σ be an $\mathcal{ER}_{VT}$ schema. A* temporal database state *for the schema Σ is a tuple $\mathcal{B} = (\mathcal{T}, \Delta^{\mathcal{B}} \cup \Delta^{\mathcal{B}}_{D}, \cdot^{\mathcal{B}(t)})$, such that: $\Delta^{\mathcal{B}}$ is a nonempty set of abstract objects disjoint from $\Delta^{\mathcal{B}}_{D}$; $\Delta^{\mathcal{B}}_{D} = \bigcup_{D_i \in \mathcal{D}} \Delta^{\mathcal{B}}_{D_i}$ is the set of basic domain values used in the schema Σ; and $\cdot^{\mathcal{B}(t)}$ is a function that for each $t \in \mathcal{T}$ maps:*

- *Every basic domain symbol D_i into a set $D_i^{\mathcal{B}(t)} = \Delta^{\mathcal{B}}_{D_i}$.*
- *Every class C to a set $C^{\mathcal{B}(t)} \subseteq \Delta^{\mathcal{B}}$—thus* objects *are instances of classes.*
- *Every relationship R to a set $R^{\mathcal{B}(t)}$ of $\mathcal{U}$-labeled tuples over $\Delta^{\mathcal{B}}$—i.e. let R be an n-ary relationship connecting the classes $C_1, \ldots, C_n$, $\text{REL}(R) = \langle U_1 : C_1, \ldots, U_n : C_n \rangle$, then, $r \in R^{\mathcal{B}(t)} \rightarrow (r = \langle U_1 : o_1, \ldots, U_n : o_n \rangle \wedge \forall i \in \{1, \ldots, n\}. o_i \in C_i^{\mathcal{B}(t)})$. We adopt the convention: $\langle U_1 : o_1, \ldots, U_n : o_n \rangle \equiv \langle o_1, \ldots, o_n \rangle$, when $\mathcal{U}$-labels are clear from the context.*
- *Every attribute A to a set $A^{\mathcal{B}(t)} \subseteq \Delta^{\mathcal{B}} \times \Delta^{\mathcal{B}}_{D}$, such that, for each $C \in \mathcal{C}$, if $\text{ATT}(C) = \langle A_1 : D_1, \ldots, A_h : D_h \rangle$, then, $o \in C^{\mathcal{B}(t)} \rightarrow (\forall i \in \{1, \ldots, h\}, \exists a_i.\ \langle o, a_i \rangle \in A_i^{\mathcal{B}(t)} \wedge \forall a_i. \langle o, a_i \rangle \in A_i^{\mathcal{B}(t)} \rightarrow a_i \in \Delta^{\mathcal{B}}_{D_i})$.*

$\mathcal{B}$ is said a legal temporal database state *if it satisfies all of the constraints expressed in the schema, i.e. for each $t \in \mathcal{T}$:*

- *For each $C_1, C_2 \in \mathcal{C}$, if C_1 ISA C_2, then, $C_1^{\mathcal{B}(t)} \subseteq C_2^{\mathcal{B}(t)}$.*
- *For each $R_1, R_2 \in \mathcal{R}$, if R_1 ISA R_2, then, $R_1^{\mathcal{B}(t)} \subseteq R_2^{\mathcal{B}(t)}$.*
- *For each cardinality constraint $\text{CARD}(C, R, U)$, then:*
  $o \in C^{\mathcal{B}(t)} \rightarrow \text{CMIN}(C, R, U) \leq \#\{r \in R^{\mathcal{B}(t)} \mid r[U] = o\} \leq \text{CMAX}(C, R, U)$.
- *For $C, C_1, \ldots, C_n \in \mathcal{C}$, if $\{C_1, \ldots, C_n\}$ DISJ C, then,*
  $\forall i \in \{1, \ldots, n\}. C_i \text{ ISA } C \wedge \forall j \in \{1, \ldots, n\}, j \neq i. C_i^{\mathcal{B}(t)} \cap C_j^{\mathcal{B}(t)} = \emptyset$.
  (Similar for $\{R_1, \ldots, R_n\}$ DISJ R)
- *For $C, C_1, \ldots, C_n \in \mathcal{C}$, if $\{C_1, \ldots, C_n\}$ COVER C, then,*
  $\forall i \in \{1, \ldots, n\}. C_i \text{ ISA } C \wedge C^{\mathcal{B}(t)} = \bigcup_{i=1}^{n} C_i^{\mathcal{B}(t)}$.
  (Similar for $\{R_1, \ldots, R_n\}$ COVER R)
- *For each snapshot class $C \in \mathcal{C}^S$, then, $o \in C^{\mathcal{B}(t)} \rightarrow \forall t' \in \mathcal{T}. o \in C^{\mathcal{B}(t')}$.*
- *For each temporary class $C \in \mathcal{C}^T$, then, $o \in C^{\mathcal{B}(t)} \rightarrow \exists t' \neq t. o \notin C^{\mathcal{B}(t')}$.*
- *For each snapshot relationship $R \in \mathcal{R}^S$, then, $r \in R^{\mathcal{B}(t)} \rightarrow \forall t' \in \mathcal{T}. r \in R^{\mathcal{B}(t')}$.*
- *For each temporary relationship $R \in \mathcal{R}^T$, then, $r \in R^{\mathcal{B}(t)} \rightarrow \exists t' \neq t. r \notin R^{\mathcal{B}(t')}$.*
- *For each class $C \in \mathcal{C}$, if $\text{ATT}(C) = \langle A_1 : D_1, \ldots, A_h : D_h \rangle$, and $\langle C, A_i \rangle \in \text{S}$, then, $(o \in C^{\mathcal{B}(t)} \wedge \langle o, a_i \rangle \in A_i^{\mathcal{B}(t)}) \rightarrow \forall t' \in \mathcal{T}. \langle o, a_i \rangle \in A_i^{\mathcal{B}(t')}$.*
- *For each class $C \in \mathcal{C}$, if $\text{ATT}(C) = \langle A_1 : D_1, \ldots, A_h : D_h \rangle$, and $\langle C, A_i \rangle \in \text{T}$, then, $(o \in C^{\mathcal{B}(t)} \wedge \langle o, a_i \rangle \in A_i^{\mathcal{B}(t)}) \rightarrow \exists t' \neq t. \langle o, a_i \rangle \notin A_i^{\mathcal{B}(t')}$.*
- *For each $C \in \mathcal{C}, A \in \mathcal{A}$ such that $\text{KEY}(C) = A$, then, A is a snapshot attribute–i.e. $\langle C, A_i \rangle \in \text{S}$— and $\forall a \in \Delta^{\mathcal{B}}_{D}. \#\{o \in C^{\mathcal{B}(t)} \mid \langle o, a \rangle \in A^{\mathcal{B}(t)}\} \leq 1$.*

Definition 3.

Definition 3 (Reasoning Services). *Let Σ be a schema, $C \in \mathcal{C}$ a class, and $R \in \mathcal{R}$ a relationship. The following modelling notions can be defined:*

1. *C (R) is* satisfiable *if there exists a legal temporal database state $\mathcal{B}$ for Σ such that $C^{\mathcal{B}(t)} \neq \emptyset$ ($R^{\mathcal{B}(t)} \neq \emptyset$), for some $t \in \mathcal{T}$;*
2. *Σ is* satisfiable *if there exists a legal temporal database state $\mathcal{B}$ for Σ ($\mathcal{B}$ is also said a* model *for Σ);*
3. *C_1 (R_1) is* subsumed *by C_2 (R_2) in Σ if every legal temporal database state for Σ is also a legal temporal database state for C_1 ISA C_2 (R_1 ISA R_2);*
4. *A schema Σ' is* logically implied *by a schema Σ over the same signature if every legal temporal database state for Σ is also a legal temporal database state for Σ'.*

ous section. As a result, the $\mathcal{ER}_{VT}$ data model will be extended with the possibility to capture such part-whole properties while the description logic $\mathcal{DLR}_{US}$ will present a corresponding axiomatization for them. A basic building block to achieve the desired formalization is the notion of *status relations*. The formalization of status relations is an original contribution of this chapter. They are in analogy with status classes addressed by (Artale *et al.*, 2007b) and will be useful for modeling essential part-whole relations. We therefore start by introducing status relations in the following subsection and then we proceed by formalizing essential parts and wholes.

Status Relations

Status relations extend the notion of status classes (Spaccapietra *et al.*, 1998; Etzion *et al.*, 1998; Artale *et al.*, 2007b) to statuses for relations. Status classes—formalized in (Artale *et al.*, 2007b)—constrain the evolution of an instance's membership in a class along its lifespan. According to (Spaccapietra et al., 1998; Artale et al., 2007b), status modeling includes up to four different statuses *scheduled, active, suspended, disabled,* each one entailing different constraints.

Concerning status relations there are two options: (1) to derive a relation's status from the status of the classes participating in the relation, or (2) to explicitly define it on the relation itself, where the latter, in turn, puts constraints on the statuses of the classes. Since we are interested in modeling relations as first-class citizens, we choose to have a means to explicitly model the status of a relation. Therefore, as for classes, we have four different statuses for relations, too—*scheduled, active, suspended, disabled*—each illustrated with an example before we proceed to the formal characterization.

- Scheduled: a relation is scheduled if its instantiation is known but its membership will only become effective some time later. Objects in its participating classes must be either scheduled, too, be active, or suspended. For instance, a pillar for finishing the interior of the Sagrada Familia in Barcelona is scheduled to become part of that church, i.e., this *part_of* relation between the pillar and the church is scheduled.
- Active: the status of a relation is active if the particular relation fully instantiates the type-level relation: the part is part of the whole. For instance, the Mont Blanc mountain is part of the Alps mountain range, and the country Republic of Ireland is part of the European Union. Only active classes can participate in an active relation.

Definition 4.

Definition 4 **(Mapping $\mathcal{ER}_{VT}$ into $\mathcal{DLR}_{\mathcal{US}}$).** *Let* $\Sigma = (\mathcal{L}, \text{REL}, \text{ATT}, \text{CARD}, \text{ISA}, \text{DISJ}, \text{COVER}, \text{S}, \text{T}, \text{KEY})$ *be an* $\mathcal{ER}_{VT}$ *schema. The* $\mathcal{DLR}_{\mathcal{US}}$ *knowledge base,* $\mathcal{K}$*, mapping* Σ *is as follows.*

- *For each* $A \in \mathcal{A}$*, then,* $A \sqsubseteq \texttt{From}{:}\top \sqcap \texttt{To}{:}\top \in \mathcal{K}$;
- *If* C_1 ISA $C_2 \in \Sigma$*, then,* $C_1 \sqsubseteq C_2 \in \mathcal{K}$;
- *If* R_1 ISA $R_2 \in \Sigma$*, then,* $R_1 \sqsubseteq R_2 \in \mathcal{K}$;
- *If* $\text{REL}(R) = \langle U_1{:}C_1, \ldots, U_k{:}C_k \rangle \in \Sigma$*, then* $R \sqsubseteq U_1{:}C_1 \sqcap \ldots \sqcap U_k{:}C_k \in \mathcal{K}$;
- *If* $\text{ATT}(C) = \langle A_1 : D_1, \ldots, A_h : D_h \rangle \in \Sigma$*, then,* $C \sqsubseteq \exists[\texttt{From}]A_1 \sqcap \ldots \sqcap \exists[\texttt{From}]A_h \sqcap \forall[\texttt{From}](A_1 \to \texttt{To} : D_1) \sqcap \ldots \sqcap \forall[\texttt{From}](A_h \to \texttt{To} : D_h) \in \mathcal{K}$;
- *If* $\text{CARD}(C, R, U) = (m, n) \in \Sigma$*, then,* $C \sqsubseteq \exists^{\geq m}[U]R \sqcap \exists^{\leq n}[U]R \in \mathcal{K}$;
- *If* $\{C_1, \ldots, C_n\}$ DISJ $C \in \Sigma$*, then* $\mathcal{K}$ *contains:*
  $C_1 \sqsubseteq C \sqcap \neg C_2 \sqcap \ldots \sqcap \neg C_n$;
  $C_2 \sqsubseteq C \sqcap \neg C_3 \sqcap \ldots \sqcap \neg C_n$;
  ...
  $C_n \sqsubseteq C$;
- *If* $\{R_1, \ldots, R_n\}$ DISJ $R \in \Sigma$*, then* $\mathcal{K}$ *contains:*
  $R_1 \sqsubseteq R \sqcap \neg R_2 \sqcap \ldots \sqcap \neg R_n$;
  $R_2 \sqsubseteq R \sqcap \neg R_3 \sqcap \ldots \sqcap \neg R_n$;
  ...
  $R_n \sqsubseteq R$;
- *If* $\{C_1, \ldots, C_n\}$ COVER $C \in \Sigma$*, then* $\mathcal{K}$ *contains:*
  $C_1 \sqsubseteq C$;
  ...
  $C_n \sqsubseteq C$;
  $C \sqsubseteq C_1 \sqcup \ldots \sqcup C_n$;
- *If* $\{R_1, \ldots, R_n\}$ COVER $R \in \Sigma$*, then* $\mathcal{K}$ *contains:*
  $R_1 \sqsubseteq R$;
  ...
  $R_n \sqsubseteq R$;
  $R \sqsubseteq R_1 \sqcup \ldots \sqcup R_n$;
- *If* $\text{KEY}(C) = A$*, then,* $\mathcal{K}$ *contains:*
  $C \sqsubseteq \exists^{=1}[\texttt{From}]\Box^* A$;
  $\top \sqsubseteq \exists^{\leq 1}[\texttt{To}](A \sqcap [\texttt{From}] : C)$;
- *If* $C \in \mathcal{C}^S$*, then,* $C \sqsubseteq (\Box^* C) \in \mathcal{K}$ *(similar for* $R \in \mathcal{R}^S$*)*;
- *If* $C \in \mathcal{C}^T$*, then,* $C \sqsubseteq (\Diamond^* \neg C) \in \mathcal{K}$ *(similar for* $R \in \mathcal{R}^T$*)*;
- *If* $\langle C, A \rangle \in \text{S}$*, then,* $C \sqsubseteq \forall[\texttt{From}](A \to \Box^* A) \in \mathcal{K}$;
- *If* $\langle C, A \rangle \in \text{T}$*, then,* $C \sqsubseteq \forall[\texttt{From}](A \to \Diamond^* \neg A) \in \mathcal{K}$.

- Suspended: to capture a temporarily inactive relation. For example, an instance of a CarEngine is removed from the instance of a Car it is part of, for purpose of maintenance at the car mechanic. Note that at the moment of suspension, part p and w must be active, but can upon suspension of the relation be either active or become suspended too, but neither scheduled (see below constraints on scheduled) nor disabled.
- Disabled: to model expired relations that never again can be used. For instance, to represent the donor of an organ who has donated that organ and one wants to keep track of who donated what to whom: say, the heart p_1 of donor w_1 used to be a structural part of w_1 but it will never be again a part of it. The heart, p_1, then may have become participant in a new part-of relation with a new whole, w_2 where $w_1 \neq w_2$, but the original part-of between p_1 and w_1 remains disabled. Observe that participating objects can be member of the active, suspended or disabled class.

Status relations apply only to temporal relations (i.e. either temporary or mixed relations according to Definition 1). We assume that active relations involve only active classes and, by default, the name of a relation denotes already its active status—i.e. `Active-R` ≡ `R`. Disjointness and ISA constraints among the four status relations are analogous to the one for status classes and can be represented in $\mathcal{ER}_{VT}$ as illustrated in Figure 6. In addition to hierarchical constraints, the constraints shown in Box 1 hold (we present both the model-theoretic semantics and the correspondent $\mathcal{DLR}_{US}$ axioms considering, wlog, binary relations):

In the following we denote with Σ_{st} the above set of $\mathcal{DLR}_{US}$ axioms that formalize status relations. In analogy with the logical implications holding for status classes (Artale *et al.*, 2007b), we can derive the ones shown in Proposition 1 for status relations.

The proofs of these logical implications have been presented in (Artale & Keet, 2008).

Lifespan and related notions. The lifespan of an object with respect to a class describes the temporal instants (and thus intervals) where the object can be considered a member of that class. We can distinguish between the following notions: ExistenceSpan$_C$, LifeSpan$_C$, ActiveSpan$_C$, Begin$_C$, Birth$_C$, and Death$_C$ depending on the status of the class the object is member of. We briefly

Figure 6. Status relations (from status classes in Artale et al. (2007a))

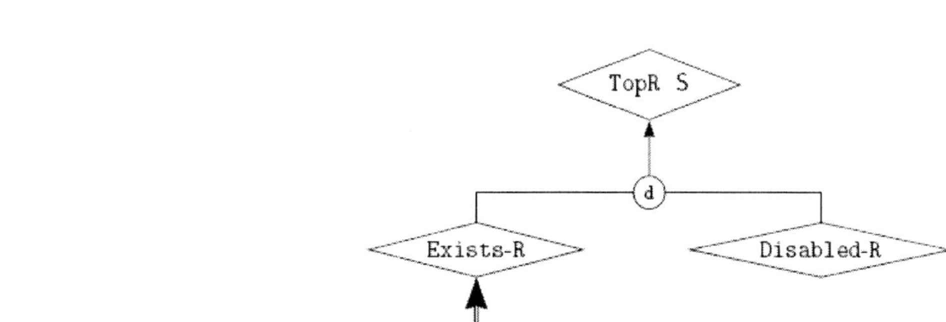

Box 1.

(ACT) *Active relations involve only active classes.*
$\langle o_1, o_2 \rangle \in \mathtt{R}^{\mathcal{B}(t)} \rightarrow o_i \in C_i^{\mathcal{B}(t)}, \ i = 1, 2$
$\mathtt{R} \sqsubseteq \mathtt{U_i}:\mathtt{C_i}, \ i = 1, 2$
(REXISTS) *Existence persists until Disabled.*
$\langle o_1, o_2 \rangle \in \mathtt{Exists\text{-}R}^{\mathcal{B}(t)} \rightarrow \forall t' > t.(\langle o_1, o_2 \rangle \in \mathtt{Exists\text{-}R}^{\mathcal{B}(t')} \vee \langle o_1, o_2 \rangle \in \mathtt{Disabled\text{-}R}^{\mathcal{B}(t')})$
$\mathtt{Exists\text{-}R} \sqsubseteq \Box^{+}(\mathtt{Exists\text{-}R} \sqcup \mathtt{Disabled\text{-}R})$
(RDISAB1) *Disabled persists.*
$\langle o_1, o_2 \rangle \in \mathtt{Disabled\text{-}R}^{\mathcal{B}(t)} \rightarrow \forall t' > t.\langle o_1, o_2 \rangle \in \mathtt{Disabled\text{-}R}^{\mathcal{B}(t')}$
$\mathtt{Disabled\text{-}R} \sqsubseteq \Box^{+}\mathtt{Disabled\text{-}R}$
(RDISAB2) *Disabled was Active in the past.*
$\langle o_1, o_2 \rangle \in \mathtt{Disabled\text{-}R}^{\mathcal{B}(t)} \rightarrow \exists t' < t.\langle o_1, o_2 \rangle \in \mathtt{R}^{\mathcal{B}(t')}$
$\mathtt{Disabled\text{-}R} \sqsubseteq \Diamond^{-}\mathtt{R}$
(RSUSP1) *Suspended was Active in the past.*
$\langle o_1, o_2 \rangle \in \mathtt{Suspended\text{-}R}^{\mathcal{B}(t)} \rightarrow \exists t' < t.\langle o_1, o_2 \rangle \in \mathtt{R}^{\mathcal{B}(t')}$
$\mathtt{Suspended\text{-}R} \sqsubseteq \Diamond^{-}\mathtt{R}$
(RSUSP2) *Suspended involve Active or Suspended Classes.*
$\langle o_1, o_2 \rangle \in \mathtt{Suspended\text{-}R}^{\mathcal{B}(t)} \rightarrow o_i \in C_i^{\mathcal{B}(t)} \vee o_i \in \mathtt{Suspended\text{-}C_i}^{\mathcal{B}(t)}, \ i = 1, 2$
$\mathtt{Suspended\text{-}R} \sqsubseteq \mathtt{U_i}:(\mathtt{C_i} \sqcup \mathtt{Suspended\text{-}C_i}), \ i = 1, 2$
(RSCH1) *Scheduled will eventually become Active.*
$\langle o_1, o_2 \rangle \in \mathtt{Scheduled\text{-}R}^{\mathcal{B}(t)} \rightarrow \exists t' > t.\langle o_1, o_2 \rangle \in \mathtt{R}^{\mathcal{B}(t')}$
$\mathtt{Scheduled\text{-}R} \sqsubseteq \Diamond^{+}\mathtt{R}$
(RSCH2) *Scheduled can never follow Active.*
$\langle o_1, o_2 \rangle \in \mathtt{R}^{\mathcal{B}(t)} \rightarrow \forall t' > t.\langle o_1, o_2 \rangle \notin \mathtt{Scheduled\text{-}R}^{\mathcal{B}(t')}$
$\mathtt{R} \sqsubseteq \Box^{+}\neg\mathtt{Scheduled\text{-}R}$

Proposition 1.

Proposition 1 (Status Relations: Logical Implications). *Given the set of axioms Σ_{st} (ACT-RSCH2), an n-ary relation (where $n \geq 2$) $R \sqsubseteq U_1 : C_1 \sqcap \ldots \sqcap U_n : C_n$, the following logical implications hold:*

(RACT) *Active will possible evolve into Suspended or Disabled.*
$\Sigma_{st} \models \mathtt{R} \sqsubseteq \Box^{+}(\mathtt{R} \sqcup \mathtt{Suspended\text{-}R} \sqcup \mathtt{Disabled\text{-}R})$
(RDISAB3) *Disabled will never become active anymore.*
$\Sigma_{st} \models \mathtt{Disabled\text{-}R} \sqsubseteq \Box^{+}\neg\mathtt{R}$
(RDISAB4) *Disabled classes can participate only in disabled relations.*
$\Sigma_{st} \models \mathtt{Disabled\text{-}C_i} \sqcap \Diamond^{-}\exists[\mathtt{U_i}]\mathtt{R} \sqsubseteq \exists[\mathtt{U_i}]\mathtt{Disabled\text{-}R}$
(RDISAB5) *Disabled relations involve active, suspended, or disabled classes.*
$\mathtt{Disabled\text{-}R} \sqsubseteq \mathtt{U_i}:(\mathtt{C_i} \sqcup \mathtt{Suspended\text{-}C_i} \sqcup \mathtt{Disabled\text{-}C_i})$, *for all* $i = 1, \ldots, n$.
(RSCH3) *Scheduled persists until active.*
$\Sigma_{st} \models \mathtt{Scheduled\text{-}R} \sqsubseteq \mathtt{Scheduled\text{-}R} \, \mathcal{U} \, \mathtt{R}$
(RSCH4) *Scheduled cannot evolve directly to Disabled.*
$\Sigma_{st} \models \mathtt{Scheduled\text{-}R} \sqsubseteq \oplus \neg\mathtt{Disabled\text{-}R}$
(RSCH5) *Scheduled relations do not involve disabled classes.*
$\mathtt{Scheduled\text{-}R} \sqsubseteq \mathtt{U_i}:\neg\mathtt{Dibabled\text{-}C_i}$, *for all* $i = 1, \ldots, n$.

report their definition as presented in (Artale et al., 2007b):

$\textsc{ExistenceSpan}_C(o) = \{t \in \mathcal{T} \mid o \in \text{Exists-C}^{\mathcal{B}(t)}\}$
$\textsc{LifeSpan}_C(o) = \{t \in \mathcal{T} \mid o \in \text{C}^{\mathcal{B}(t)} \cup \text{Suspended-C}^{\mathcal{B}(t)}\}$
$\textsc{ActiveSpan}_C(o) = \{t \in \mathcal{T} \mid o \in \text{C}^{\mathcal{B}(t)}\}$
$\textsc{Begin}_C(o) = \min(\textsc{ExistenceSpan}_C(o))$
$\textsc{Birth}_C(o) = \min(\textsc{ActiveSpan}_C(o)) \equiv \min(\textsc{LifeSpan}_C(o))$
$\textsc{Death}_C(o) = \max(\textsc{LifeSpan}_C(o))$

For atemporal classes, $\textsc{ExistenceSpan}_C(o) \equiv \textsc{LifeSpan}_C(o) \equiv \textsc{ActiveSpan}_C(o) \equiv \mathcal{T}$. This concludes the preliminaries. In the next section we will use the notions introduced so far for representing essential parts-whole relations.

Essential Parts and Wholes

Recollecting Guizzardi's (2005) contribution on the formalization of the difference between mandatory and essential parts and wholes we can say that: a part is mandatory if the whole cannot exist without it, which can also be verbalized as "the whole has a mandatory part"—i.e. a standard mandatory constraint on the role played by the whole in a part-whole relation. In a symmetric way we can define *mandatory wholes*. A part is *essential* if it is mandatory and cannot change without destroying the whole, i.e. "the whole has an essential part" (in an analogous way we can define *essential wholes*). Furthermore, we say that a part is *exclusive* if it can be part of at most one whole (similarly for *exclusive wholes*). In this section we provide a formalization using $\mathcal{DLR}_{US}$ axioms of such mandatory, essential and exclusive parts and wholes. Starting from Figure 4, Figure 7-A, shows the various temporal relations that can hold between a whole and its essential part, i.e. the lifespan of the whole is fixed and we consider the different lifespans for its essential parts (Figure 7-B considers a fixed part and the cases for its essential whole)[8].

Let $\texttt{partOf} \sqsubseteq \texttt{part:P} \sqcap \texttt{whole:W}$ be a generic part-whole relation, the following $\mathcal{DLR}_{US}$ axioms give a formalization of mandatory and exclusive parts and wholes:

(ManP)	$\texttt{W} \sqsubseteq \exists[\texttt{whole}]\texttt{partOf}$	*Has Mandatory Part*
(ManW)	$\texttt{P} \sqsubseteq \exists[\texttt{part}]\texttt{partOf}$	*Has Mandatory Whole*
(ExlP)	$\texttt{P} \sqsubseteq \exists^{\leq 1}[\texttt{part}]\texttt{partOf}$	*Is Exclusive Part*
(ExlW)	$\texttt{W} \sqsubseteq \exists^{\leq 1}[\texttt{whole}]\texttt{partOf}$	*Is Exclusive Whole*

To capture essential parts and wholes, in addition to the above axioms, we will use appropriate subsets of the following axioms:

(ConPO)	$\texttt{Suspended-partOf} \sqsubseteq \bot$	*Continuous Parts*
(DisP)	$\texttt{Disabled-partOf} \sqsubseteq \texttt{part}:\texttt{Disabled-P}$	*Disabled Part*
(DisW)	$\texttt{Disabled-partOf} \sqsubseteq \texttt{whole}:\texttt{Disabled-W}$	*Disabled Whole*
(SchPO)	$\texttt{partOf} \sqsubseteq \Diamond^{-}\texttt{Scheduled-partOf}$	*Scheduled Part-Whole*
(SchP)	$\texttt{Scheduled-partOf} \sqsubseteq \texttt{part}:\texttt{Scheduled-P}$	*Scheduled Part*
(SchW)	$\texttt{Scheduled-partOf} \sqsubseteq \texttt{whole}:\texttt{Scheduled-W}$	*Scheduled Whole*

We can now show that the above axiomatization is sufficient to represent the various forms of mandatory and essential parts as shown in Figure 7.

The proof has been presented in (Artale & Keet, 2008). A similar result can be proved considering the various forms of essential wholes.

Thus, from the axiomatization presented above, the essential parts and wholes in a part-

Figure 7. Lifespans of essential parts w.r.t. the whole (A) and vv (B)

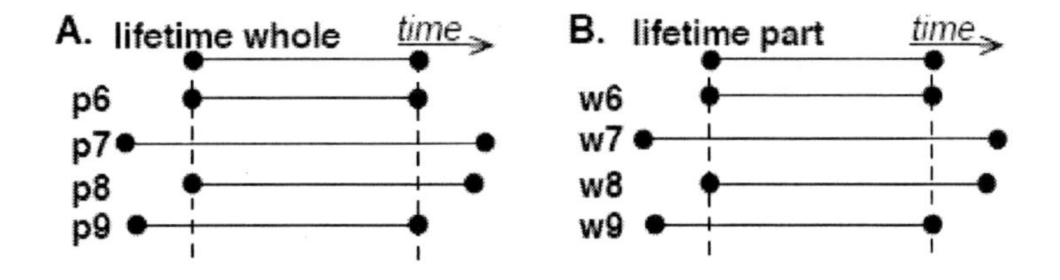

whole relation are always active, cannot be suspended, and when the strict case is allowed (i.e. either p6 or w6 holds) then they are either *both* member of their respective Scheduled class, or both Active, or both member of their respective Disabled classes. Hence, a change of membership from one of the two objects implies *instantaneous* change of the other in the same type of status class. Note that, in the literature, essential parts are often considered also exclusive. Our modeling of essential parts and wholes can be easily extended by adding to the axiomatization of Theorems 1-2 either the axiom (ExlP) or (ExlW) depending whether we want to capture exclusive essential parts or wholes.

This concludes the formal characterization of the principal combinations in life cycles between essential parts and wholes. The next section provides a solution sketch on how to deal both with suspended part-whole relations and with parts shared among (possibly different) wholes.

Shared Parts and Wholes

Sharing of parts and wholes permits many possibilities, some of which may be more useful in practice than others. Rather than enumerating, formalizing, and, where necessary, proving all theoretically possible options, we discuss the main typical cases and demonstrate they indeed can be characterized *by using the same formal apparatus* and principles that have been introduced in the previous sections. The first step is to add the possibility to suspend parts, wholes, and part-whole relations, and the second step to make explicit that some parts can be part of more than one whole of the same or different type.

Taking into account suspension. The first variation that can be added to the basic combinations of Figure 7, is that during some time either the participating part or the whole, or both, is suspended or the part-whole relation is suspended. This offers a wide range of possibilities. For instance, we have a word processing document management system where a particular paragraph (part *p*) is blocked—suspended—for use and the particular file (whole *w*) can only be published when the paragraph is member of the active class again. Differently, one could have a defunct pillar (member of `Suspended-Pillar`) in a historical building that is temporary removed for restoration (hence, a relational instance of part-of in `Suspended-partOf`) and the building collapses before the pillar gets restored. An example where both part and whole remain active but only the part-whole relation is suspended can occur for, e.g., a car mechanic's database that records the cars and the parts that are under service for cleaning, i.e., temporarily removed, and the part needs to be re-inserted in the car it was removed from, such as a car (o_w) and the car engine (o_p) that is structurally part of the car. Thus, the suspension of a relation does not necessarily impose constraints on the permissible memberships of

Theorem 1.

Theorem 1 (Essential Parts). *Let* partOf $\sqsubseteq$ part:P $\sqcap$ whole:W *be a generic part-whole relation satisfying* Σ_{st}, *then,*

1. *p7 holds if* (ManP), (ConPO), (DisW) *hold;*
2. *p9 holds if* (ManP), (ConPO), (DisW), (DisP) *hold;*
3. *p8 holds if* (ManP), (ConPO), (DisW), (SchPO), (SchP) *hold;*
4. *p6 holds if* (ManP), (ConPO), (DisW), (DisP), (SchPO), (SchP) *hold.*

Theorem 2.

Theorem 2 (Essential Wholes). *Let* partOf ⊑ part:P ⊓ whole:W *be a generic part-whole relation satisfying* Σ_{st}, *then,*

1. *w7 holds if* (ManW), (ConPO), (DisP) *hold;*
2. *w9 holds if* (ManW), (ConPO), (DisP), (DisW) *hold;*
3. *w8 holds if* (ManW), (ConPO), (DisP), (SchPO), (SchW) *hold;*
4. *w6 holds if* (ManW), (ConPO), (DisP), (DisW), (SchPO), (SchW) *hold.*

the part or whole in their respective status classes other than (RSusP2).

To structure and fully address all cases, one can—as a start—systematically apply suspension to the standard cases p6-p9 and w6-w9 as depicted in Figure 8-A and B, respectively. In addition, sp6′-sp9′ and sw6′-sw9′ in Figure 8-A and B denote the cases where, even though the whole (part) is suspended, the part (whole) still must remain active. To capture these cases (at least the non primed ones), we need to replace the (ConPo) axiom in both Theorem 1 and 2 with two additional axioms:

(SusP) Suspended-partOf ⊑ part: Suspended-P *Suspended Part*

(SusW) Suspended-partOf ⊑ whole: Suspended-W *Suspended Whole*

This change of axioms from (ConPO) to (SusP) and (SusW), however, does not immediately address our pillar and collapsed building example. That is, we know from the current axiomatization that when we have that at some time the pillar (part o_p) and its relation with the historical building (o_w ∈ W) become suspended (o_p ∈ Suspended-P and r ∈ Suspended-partOf), which is a legal

Figure 8. (A) Eight permutations for suspension from the viewpoint of the whole; (B) analogously from the part viewpoint; (C) Solid line as being part of the whole, dotted line as being not part of that whole but either the part-whole relation is suspended, or p is part of another whole, or both. For ps1-ps5 w_a, $w_b \in W$; the analogous situation where $w_a \in W$ and $w_b \in W'$ s.t. $W \neq W'$ is not drawn.

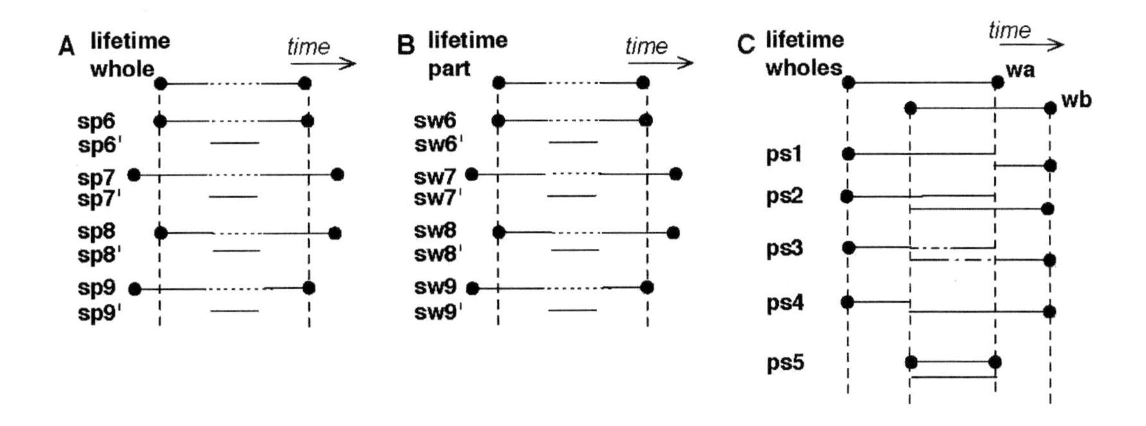

situation thanks to RSusp2 and Act, then when the whole becomes disabled due to the collapse ($o_w \in$ `Disabled-W`), then so must $r \in$ `Disabled-partOf` instantaneously, because of RDisab4. To combine the appropriate set of constraints and formally prove it to be correct with respect to the intended semantics is a topic of current work.

Explicit sharing of parts among wholes. We now turn to explicit sharing of parts, where the main variations are depicted in Figure 8-C. Cases psl-ps5 actually capture two possibilities: either $w_a, w_b \in W$ or $w_a \in W$, $w_b \in W'$ such that $W \neq W'$; henceforth, the latter will be identified with psl′-ps5′. Observe that there is a principal distinction between parts that are shared "sequentially" and parts that are shared in "parallel". Examples of the former (psl) could be the heart transplantation between humans w_a and w_b, and an example for psl′ a "multipurpose part" that is reused for another whole, e.g, a screw in a table that is used later for a bookshelf. The opposite case is where p becomes part of w_b as soon as it becomes active (ps4 and ps4′). An example of the latter, concurrent part-of (ps5′), is a seminar being part of both a seminar series and of a graduate course in a teaching database. An example of ps3 with the notion of one of the parthood relations, *contained_in*, is where the contained p switches containers w_a and w_b back and forth or a administrative region in a war zone area (or simply a longer historical time interval) so that it alternately belongs to one country or another. Colloquially, one can reformulate the constraints of psl, ps3, ps4, psl′, ps3′, and ps4′ as the part being sequentially part of more than one whole in some way, and for ps2, ps5, ps2′, and ps5′ as where the part can be part of more than one whole *concurrently.*

To formally represent psl-ps5 and psl′-ps5′, we may not need additional axioms, but recombine in various ways the 12 listed above. For instance, ManW says only that the part must participate in the part-whole relation but does not have a range restriction at all, meaning it could a whole of any type (W, W', ...); so for psl, this means just mandatory participation by the part and if w_a, $w_b \in W$ then there is no mandatory constraint on the whole (because w_b does not have to have the part), whereas if $w_a \in W$, $w_b \in W'$ and $W \neq W'$, then it may be the case that the whole must have a part, but one cannot know this *a priori*[9]. From a conceptual modeling perspective, however, this is undesirable, because one would want to be able to distinguish between the psl-ps5 and psl′-ps5′ series. To this end to be utterly explicit, we can add the following axiom:

$$\text{(DisjW)}\ \mathtt{W_i} \sqsubseteq \sqcap_{\mathtt{j=i+1}}^{\mathtt{n}} \neg \mathtt{W_j} \qquad \textit{Disjoint Wholes}$$

Given the full set of axioms, then psl-ps5 and psl′-ps5′ may be formally characterized and proven by taking different subsets of constraints. We have omitted them here due to their length and detail and for several cases it is not easy to find realistic examples. In addition, they obfuscate that, at least in some cases (psl, ps4) we actually deal with a contracted version of the pl-p5 and wl-w5 cases in Figure 4. It may now be clear that although p2-p4 have the same ratio of the lines as w7-w9, the base axioms for the p-series is distinct from that of the w-series in Theorems 1 and 2 and that they do not necessarily involve essential or mandatoryness but can be optional parts/wholes. This is even more flexible with pl, p5, wl, and w5, where the part (whole) is contingently part of the whole (part), or: they are "independent parts" and "independent wholes" for which one can fix either only a minimal set of constraints where, temporally, almost anything is allowed or choose to be utterly specific so as to capture that and only that life cycle option. We are currently working on defining and proving the meaningful and realistic cases of the suggested options that are depicted informally in Figure 8-C.

Finally, in $\mathcal{ER}_{VT}$ and in $\mathcal{DLR}_{US}$ we can specify cardinality constraints on the participation of classes into relations. This allows for expressing multiple sharing of parts/wholes like, for example, in specifying that cars must have exactly four wheels as parts:

$$\texttt{partOf} \sqsubseteq \texttt{part:Wheel} \sqcap \texttt{whole:Car} \text{ (typing of the part-of relation)}$$
$$\texttt{Car} \sqsubseteq = 4\ \texttt{[whole]partOf.Car} \text{ (each car has exactly four wheels)}$$
$$\texttt{Wheel} \sqsubseteq \leq 1\ \texttt{[part]partOf.Car} \text{ (each wheel is part of max 1 car)}$$

The constraints introduced in this section can represent all shareability constraints proposed earlier in the related literature (recollect section 2), meets the requirements as laid out in section 2.4, and refines shareability further with notions such as concurrently versus sequentially being part of a whole and temporary suspension of a part-whole relation.

Interaction with Types of Part-Whole Relations

Examples in the previous section for various cases of part-whole life cycles did mention different part-whole relations, such as (spatially) contained in, structural parthood, and location. How these types of part-whole relations interact precisely with the life cycle semantics is an open question and in this section we only provide a flavor of the issues.

Summarizing the taxonomy of types of part-whole relations, we have a diagrammatic rendering in together with the formal characterization of the leaf types, where *part_of* is the parthood relation from Ground Mereology whereas *mpart_of* is neither transitive nor intransitive; refer to (Keet & Artale, 2008) for details on its rationale and the formal characterization, and Keet (2006b) for additional modeling guidelines.

Considering the possible interactions between the part-whole relations and shareability, one directly can note that if something is physically a *proper part* of a whole, such as that a car engine is a proper part of the car, then obviously, this proper part cannot physically be *directly* part of another whole at the same time, and likewise for its subtypes proper containment and proper location. Put differently, in those cases we must enforce, at least, the (ExLP) axiom. In contrast, a proper sub-process can be simultaneously *involved_in* (part of) several grander processes; e.g., a key chemical reaction intersecting in two metabolic pathways. Likewise, we can have, say, a musician *m* who is concurrently *member_of* a string quartet and of the Royal Philharmonic Orchestra. The situation becomes more complicated with *subquantity_of* and so-called "portions of stuff". Provided one uses measurements for the quantities—say, a syringe full of dissolved morphine taken from the dissolved morphine stock in the bottle—then we can assert at the type level that the part-quantity must have its individual part-whole relation to the stock quantity as either member of `Suspended-partOf` (in case the liquid in the syringe can be put back in the bottle) or of `Disabled-partOf`. To address such issues fully requires additional temporal constraints, which has been addressed only in part by Bittner & Donnelly (2007) (cf. section 2.4). Last, with the *constitutes* relation we have *no* sharing. These kind of interactions, however, merit further research and a precise characterization of constraints.

MODELING GUIDELINES

Clearly, while the formal characterization in the previous section provides precise semantics to shareability, one cannot burden the conceptual modeler, let alone the domain expert, with such details. The first step toward modeling guidelines is for the visually-oriented user: either present Figures 7-8 and one can directly point to the appropriate option or use their respective informal descriptions as initial modeling heuristic, such as:

Figure 9. Taxonomy of basic mereological and meronymic part-whole relations; the part-whole relations in the left-hand branch are all transitive, but those in the right-hand branch not necessarily. s-part-of = structural part-of; f-part-of = functional part-of. Dashed lines indicate that the subtype has additional constraints on the participation of the entity types; ellipses indicate several possible finer-grained extensions to the basic part-whole relations. (Source: Keet & Artale (2008))

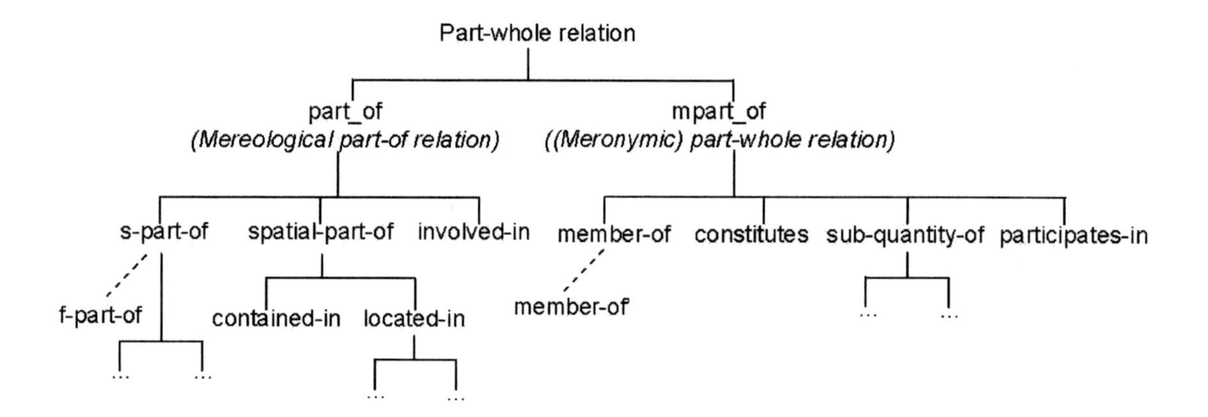

(EXAMPLE) Can an instance p of P exist before some w of type W it will become part of?

Yes ⇒ p1, p7, or p9

And can that instance p exist after that w is disabled/deleted?

Yes ⇒ p7

No ⇒ p1 or p9

That w can outlive that p?

Yes ⇒ p1

No ⇒ p9

And so forth for the other six cases. The alternatives are a set of questions alike implemented in VisioModeler 3.1, but then tailored to shareability of part-whole relations, or a decision diagram alike proposed for choosing the appropriate type of part-whole relation (Keet, 2006b). We combine these two approaches. The first step is to differentiate between the basic options of essential part/whole, mandatory part/whole, and shareability, through posing a set of closed questions. The questions are formulated in such a way so as to be both uniform in sentence structure and to simplify further processing of the answers. Also, in a real conceptual model, P and W are replaced by their respective object types in the conceptual model:

(A1) Can an instance p of P exist without some w of type W it is part of?

(A2) Can an instance p of P exist without the same w of type W it is part of?

(A3) Can an instance p of P be part of more than one whole w at some time?

(B1) Can an instance w of W exist when it does not has part some p of type P?

(B2) Can an instance w of W exist when it does not has part the same p of type P?

(B3) Can an instance w of W has part more than one part p of type P at some time?

Figure 10 shows the resultant fact types for the six questions when answered with "no". We choose to depict them in ORM2 for three reasons. First, ORM2 is more expressive than either UML class diagrams or EER and ORM has an established transformation to these other conceptual data modeling languages (Halpin, 2001), the usability approach with questions-to-ORM diagram interaction is already an established practice, and it simplifies adding additional icons for the shareability semantics; the proposed icons are listed in Figure 11 and added to sample $\mathcal{ER}_{VT}$ diagrams in Figure 12[10]. *In casu*, for A2/B2, the current graphical ORM2 language is extended with a filled box in the role for essential part or whole (hence, note the different effects of the "some" and "same" in the questions and representation). On can, of course, combine the "no" answers to A1-B3, two of which are depicted in Figure 10, too. Relating this back to the sets of constraints, then A1 corresponds to a simple ManW constraint and A2+B2 combines the constraints of p6+w6 (see Theorems 1 and 2).

Given that there is a whole list of questions, one can build in intermediate feedback loops, such as asking the modeler after all "yes" on A1-B3:

(FEEDBACK) With a "yes" on A1-B3, either the part or the whole, or both, can be shared. Is that true?

If the answer is, "no", then A1-B3 should be revisited; if the answer is in the affirmative, the modeler can proceed to the second set of questions. The second step is assessment of the sharing of parts, where we can reuse the answers on the previous five questions. For instance, when we have a mandatory but not essential participation on the part-side—a "no" for A1—it is obvious that the part cannot exist independently; i.e., then we must have any of the options p2, p3, or p4 or ps1-ps5′ so that asking questions to cover the remaining options has become irrelevant.

With the C1-C5 question series, one can extract from a domain expert if the sharing can/must be sequential or in parallel and if the wholes may be of a different type or not; this selection procedure is depicted in Figure 13 with additional explanatory notes that a CASE tool developer might want to include as an extra service:

(c1) Can an instance p of P be part of more than one whole w at the same time?
(c2) Regarding the wholes $w_1, \ldots, w_n$ that p can be part of, must $w_1, \ldots, w_n$ be instances of the same type W?
(c3) Can an instance p of P also be part of only one whole w of type W?
(c4) Can an instance p of P only become part of another whole w_2 after whole w_1 cease to be active as whole?
(c5) Will an instance p of P become part of another whole w_2 and cease to be part of w_1 as soon as w_2 becomes active as whole?

A sample diagrammatic representation for ps1′-ps5′ is included in Figure 14; UML and EER are currently less fine-grained and less expressive, but with a $\mathcal{DLR}$ in the background in the CASE tool, this can be added trivially, see e.g. the Icom tool (Fillottrani *et al.*, 2006). Formulating the same series for the perspective of the whole—starting with the "yes" on B3—is left as an exercise to the reader. Third, we add a further dimension:

Figure 10. Representations resulting from the answers to questions A1-B3 when answered with "no". Regarding semantics, A2 and B2 have an additional icon (rectangle in the role) to denote essential part/whole; see also Figure 11

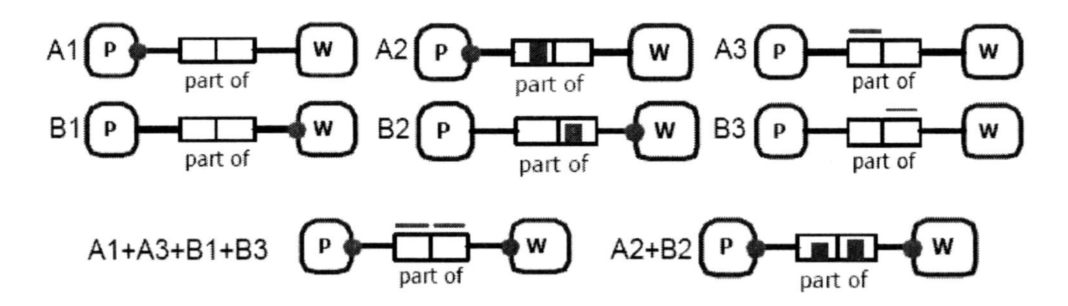

Figure 11. Suggested icons to denote the various aspects of shareability of parts and of the part-whole relation. Example use is demonstrated for ORM2 notation, but also can be added to EER or UML's association relation and classes. Note that ■ and ~ are new additions for all modeling languages, the arrows make explicit certain temporal behavior, whereas < and — are mainly useful in the light of further model development.

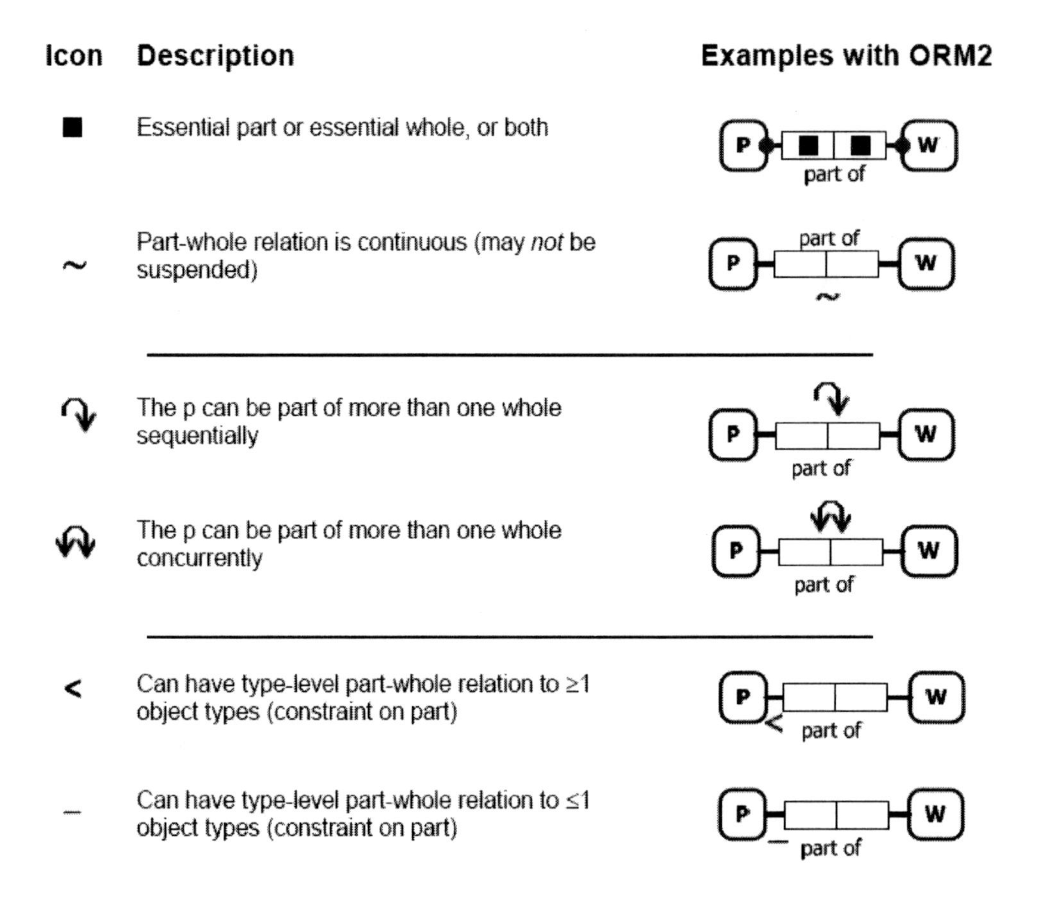

suspension of the part-whole relation or of the parts or wholes. By default, the part-whole relation can be suspended, except where it is explicitly disallowed, i.e. for those cases that include ConPO. To ascertain this, one has to ask at least the following questions:

(D1) Can the **part of** relation be suspended?
If "no" then we have any of the cases that have ConPO; if "yes", the constraints for the cases cannot include ConPO.

(D2) Can an instance *p* of **P** that is **part of** *w* of **W** become suspended?
If "no" then **P** must have a (strong) essential whole. If "yes" then **P** can have a weak essential whole or other shareability options (depending on the answers of prior questions).

(D3) If this *p* of **P** cannot be suspended, can the *w* of **W** it is **part of** become suspended?
If "no" then *p* has an essential whole (w6-w9), if "yes" then *p* has a strong essential whole (sw6'-sw9').

One could add cross-checks to prevent violation of the constraints, as has been proposed by Motschnig-Pitrik & Kaasbøll (1999) in a model-

Figure 12. Several examples of the suggested icons in conjunction with $\mathcal{ER}_{VT}$ for conceptual models that are ontology-inspired, intended for organ transplant databases, government administration software, and a hardware manufacturing database, respectively. (That is, it could be modeled differently for different application software, such as permitting coalition governments to create ministerial posts in name only, i.e., without a portfolio and ministry, to please a coalition partner so that the double arrow should be removed). The "T" in the part-of relation between Heart and Person makes explicit it is a temporal relation, as defined in $\mathcal{ER}_{VT}$ (Definition 1).

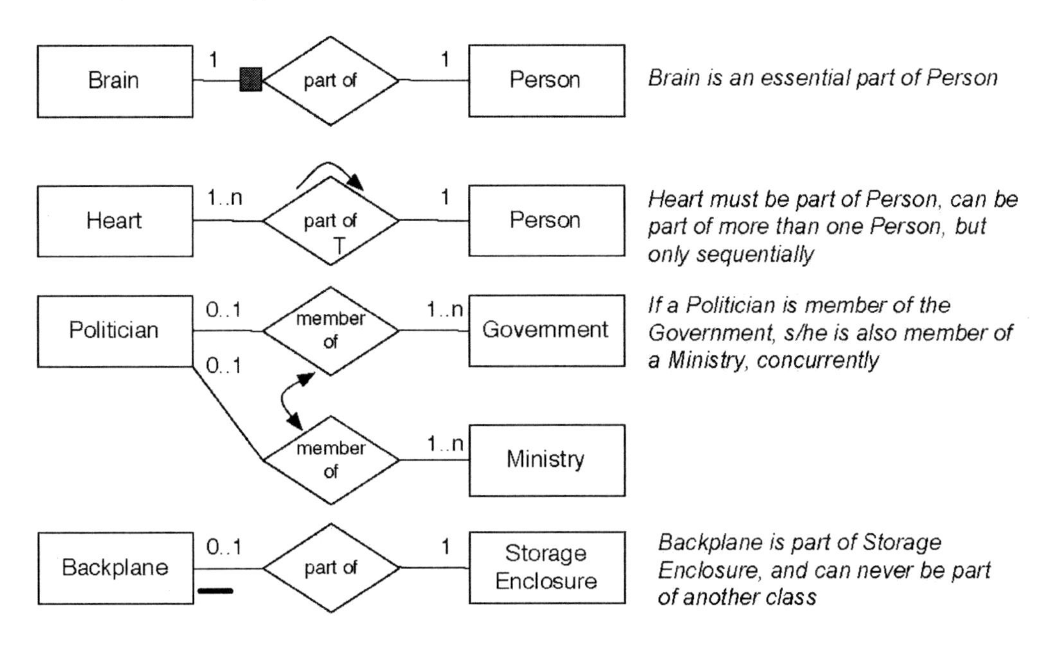

Figure 13. Decision diagram to assess on the appropriate type of shareability; see text for the complete questions in the decision diamonds

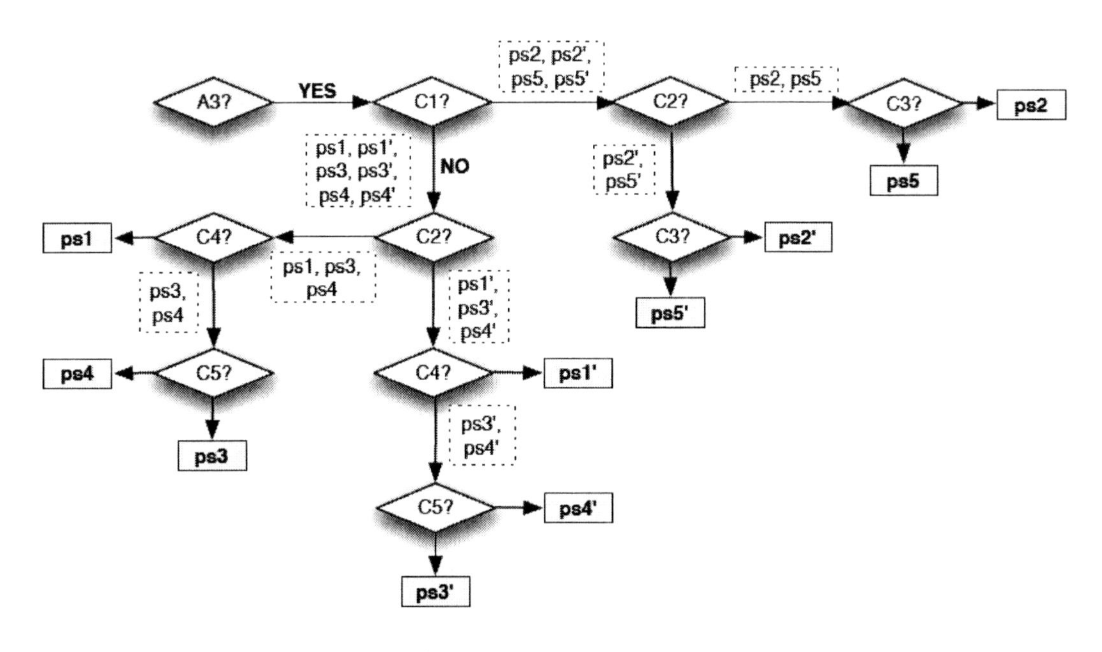

Figure 14. Graphical rendering of constraints ps1'-ps' in ORM2. Note that the double arrows in ps2' and ps5' are redundant in ORM because they can be captured with the regular constraints already, but are required for UML class diagrams and ER. Optionally, one could add yet another new icon so as to have the distinction between ps1' and ps4'. The < on P is obviously satisfied in the figure, but in the light of conceptual model evolution, useful to make explicit.

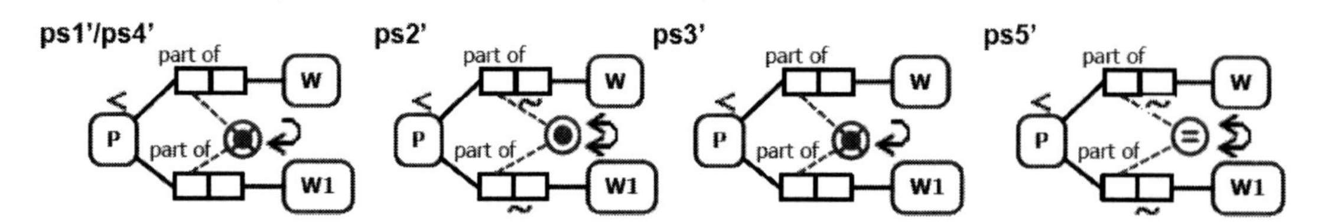

ing guideline mode. These are, however, already covered by the current decision diagram and specific set of constraints for each case and a simple check (set comparison) can be implemented where the manually modeled constraints are compared with the combinations of constraints from the previous section.

At the time of writing, however, question series C1-C5 and D1-D3 are for indicative purpose to gain insight into what actually are the salient cases that recur during the modeling process and which scenarios are more prevalent in a subject domain. Thus, they aid in stimulating domain experts and modelers to assess in more detail which life cycle semantics are more important and realistic, as opposed to providing a forest of constraints where possibly only a third or even less is practically used. After scoping relevance, one then could choose to implement only the useful subset of part-whole shareability.

FUTURE TRENDS

As the reader may have noticed from section 2, there is much recent research activity on part-whole relations. From one viewpoint, this is good because significant improvements in understanding of part-whole relations are being achieved. From a practitioner's viewpoint, however, this also means that most of these advances have not yet made it into the readily available CASE tools and are thereby primarily of direct use as 'paper exercise' in the analysis and modeling steps of the software development process. Considering some general requirements to implement support for various aspects of part-whole relations, an overall structure and implementation would have to meet several requirements such as to:

i. Ensure the representation is such that one can distinguish between parthood relations of a class (or its instances) and other generic properties (/relations/roles/associations), i.e. to make part-whole and whole-part relations first-class citizens;
ii. Identify unambiguously and model a 'minimal amount' of part-whole relations in the conceptual data models; sub-requirements (Keet & Artale, 2008) comprise representing at least the simplest parthood theory Ground Mereology (Varzi, 2004), expressing ontological categories and their taxonomic relations, having the option to represent transitive and intransitive relations, and to be able to specify the domain and range restrictions (/relata/entity types) for the classes participating in a relation;
iii. Provide a set of combinations of essential, uniqueness, and mandatoryness constraints applicable to the relation;

iv. Clarify and accommodate for other, sometimes called "secondary", properties of part-whole relations, such as functional dependence and completeness;
v. Ensure the inverse, has-part, relation is properly modeled as well;
vi. Transitivity of parthood relations is enabled where applicable and prohibited for non-transitive part-whole relations;
vii. Address the possibilities and consequences of horizontal interrelations between the parts of a whole;
viii. Develop an underlying unifying paradigm that relates the conceptual modeling language specific constructors, if possible.

At the time of writing, we are far away from meeting all these requirements. Point i is met only informally with UML's aggregation relation. Point ii can be met partially with the taxonomic structure as presented in (Keet, 2006b; Keet & Artale, 2008), and UML's stereotypes and/or conceptual model meta-modeling. So-called secondary properties, inverse relations, transitivity, antisymmetry, and horizontal relations are only to a limited extent possible in some languages (extensions of UML, ORM's ring constraints, but no DL has a constructor for antisymmetry), which could be extended and harmonized further. Full computational support not only for computer-aided conceptual modeling with the guidelines but also including automated satisfiability and consistency checking of a conceptual model, is likely to be difficult due to undecidability—e.g., $\mathcal{DLR}_{US}$ with time stamping for relations is undecidable whereas without the option to represent temporal persistence of n-ary relations, reasoning in $\mathcal{DLR}_{US}$ is an ExpTime-complete problem (Artale *et al.*, 2002). Dropping the evolution constraints from $\mathcal{DLR}_{US}$ allows timestamping on relations and is decidable (Artale *et al.*, 2007b). On the other hand, conceptual modeling emphasizes expressiveness, not computability; constraints can be dropped in a design-level specification and then one at least knows what has been removed and why.

Open problems. To give an indication of open problems on part-whole relations, i.e., transforming several requirements from the previous subsection into research and engineering questions, we outline a non-exhaustive list of avenues.

- DL: Which subtheory of mereology fits best with any of the extant DLs? Can one of the DL languages be extended, and if so, how and what about its complexity? What about property inheritance across the parthood relation? What are the difference between the intensional and extensional reasoning (behavior of the parthood relation at the TBox and ABox, respectively)?
- ER and ORM: what and how to add part-whole relations? What about developing more expressive versions of ER and ORM that include the parthood relation, alike the DL languages with difference in expressiveness and complexity? How to make it usable for the modeler?
- Applied parthood relations: what lessons can be learned from practical use of part-whole relations in specific subject domains such as bio-ontologies and geographical information systems? If any, can this be fed back into mereology to extend mereological theories? Does usage of the part-whole relation across subject domains reveal domain specific intricacies that cannot be generalised to domain-independent characteristics?

Notwithstanding these gaps, many aspects of the part-whole relations can be modeled with extant conceptual modeling languages or require only minor extensions. This situation changes if one were to require efficient reasoning over conceptual data models, but as long as the focus is on expressiveness to enable the subject domain as good as possible, then this poses no problem.

CONCLUSION

The main questions addressed in this chapter were: which type of shareability and which lifetime aspects are possible, what is the formal semantics for shareability, and how to model these kind of differences? In order to solve these issues, we merged and extended advances in representing part-whole relations in UML class diagrams, EER, ORM, and DL languages with formal conceptual data modeling for temporal databases (temporal EER) and ORM's usability features. First, the different semantics of part-whole shareability notions were formally characterized by availing of the temporal Description Logic $\mathcal{DLR}_{US}$, which was hitherto only used for temporal EER ($\mathcal{ER}_{VT}$), and having extended it with the original notion of *status relations* so as to capture unambiguously and at the conceptual layer the different life cycle semantics of parts and wholes and what happens to the part-whole relation. These formally defined constraints can represent all shareability constraints proposed earlier in the related literature, meets the requirements as laid out in section 2.4, and refines shareability further with notions such as concurrently versus sequentially being part of a whole and temporary suspension of a part-whole relation. By having used $\mathcal{DLR}_{US}$ as foundational mechanism to represent the different shareability semantics, the results obtained are easily transferrable to UML class diagrams and ORM/ORM2. Second, the shareability options were transformed into three complimentary conceptual modeling guidelines: visuals, a simple list of questions, and a decision diagram so as to easily navigate to the appropriate constraints and their shareability cases of the relations and its participating parts and wholes.

Several issues, however, have not been addressed, both from the theoretical and practical side. These include, but are not limited to, across-time part-whole relations, their precise interaction with various types of part-whole relations, and software support to simplify modeling of part-whole relations.

REFERENCES

Albert, M., Pelechano, V., Fons, J., Ruiz, M., & Pastor, O. (2003). Implementing UML Association, Aggregation, and Composition. A Particular Interpretation Based on a Multidimensional Framework. In J. Eder and M. Missikoff (Ed.), *Proceedings of CAiSE'03 LNCS 2681* (pp. 143-158). Berlin: Springer Verlag.

Álvarez, A. T., & Alemán, J. L. F. (2000). Formally modeling UML and its evolution: A holistic approach. In *Fourth International Conference on Formal methods for open object-based distributed systems IV* (pp. 183-206). Amsterdam: Kluwer Academic Publishers.

Artale, A., Calvanese, D., Kontchakov, R., Ryzhikov, V., & Zakharyaschev, M. (2007a). Complexity of Reasoning over Entity Relationship Models. *Proceedings of DL-07, CEUR WS, 250.*

Artale, A., Franconi, E., Guarino, N., & Pazzi, L. (1996a). Part-Whole Relations in Object-Centered Systems: an Overview. *Data and Knowledge Engineering, 20*(3), 347-383.

Artale, A., Franconi, E., & Guarino, N. (1996b). Open Problems for Part-Whole Relations. In: *Proceedings of 1996 International Workshop on Description Logics (DL-96)* (pp 70-73). Cambridge, MA: AAAI Press.

Artale, A., & Franconi, E. (1999). Temporal ER modeling with description logics. In *Proc. of the Int. Conf. on Conceptual Modeling (ER'99).* Berlin: Springer-Verlag.

Artale, A., Franconi, E., & Mandreoli, F. (2003). Description Logics for Modelling Dynamic Information. In J. Chomicki, R. van der Meyden, & G. Saake (Eds.), *Logics for Emerging Applications of Databases*. Berlin: Springer-Verlag.

Artale, A., Franconi, E., Wolter, F., & Zakharyaschev, M. (2002). A temporal description logic for reasoning about conceptual schemas and queries. In S. Flesca, S. Greco, N. Leone, G. Ianni (Eds.), *Proceedings of the 8th Joint European Conference on Logics in Artificial Intelligence (JELIA-02), LNAI, 2424* (pp. 98-110). Berlin: Springer Verlag.

Artale, A., Guarino, N., & Keet, C.M. (2008). Formalising temporal constraints on part-whole relations. In G. Brewka & J. Lang, J. (Eds.), *11th International Conference on Principles of Knowledge Representation and Reasoning (KR'08).* Cambridge, MA: AAAI Press.

Artale, A., & Keet, C. M. (2008). Essential and Mandatory Part-Whole Relations in Conceptual Data Models. *Proceedings of the 21st International Workshop on Description Logics (DL'08), CEUR WS Vol 353.* Dresden, Germany, 13-16 May 2008.

Artale, A., Parent, C., & Spaccapietra, S. (2006). Modeling the evolution of objects in temporal information systems. In: *4th International Symposium on Foundations of Information and Knowledge Systems (FoIKS-06), LNCS, 3861*,22-42. Berlin: Springer-Verlag.

Artale, A., Parent, C., & Spaccapietra, S. (2007b). Evolving objects in temporal information systems. *Annals of Mathematics and Artificial Intelligence (AMAI), 50*(1-2), 5-38.

Baader, F., Calvanese, D., McGuinness, D. L., Nardi, D., & Patel-Schneider, P. F. (Eds). (2003). *Description Logics Handbook.* Cambridge: Cambridge University Press.

Barbier, F., Henderson-Sellers, B., Le Parc-Lacayrelle, A., & Bruel, J.-M. (2003). Formalization of the whole-part relationship in the Unified Modelling Language. *IEEE Transactions on Software Engineering, 29*(5), 459-470.

Berardi, D., Calvanese, D., & De Giacomo, G. (2005). Reasoning on UML class diagrams. *Artificial Intelligence, 168*(1-2), 70-118.

Bittner, T., & Donnelly, M. (2005). Computational ontologies of parthood, component-hood, and containment, In L. Kaelbling (Ed.), *Proceedings of the Nineteenth International Joint Conference on Artificial Intelligence 2005 (IJCAI05)* (pp. 382-387). Cambridge, MA: AAAI Press.

Bittner, T., & Donnelly, M. (2007). A temporal mereology for distinguishing between integral objects and portions of stuff. In *Proceedings of the Twenty-second AAAI Conference on Artificial intelligence (AAAI'07)* (pp. 287-292). Cambridge, MA: AAAI Press.

Borgo, S., & Masolo, C. (in press). Full mereogeometries. *Journal of Philosophical Logic.*

Calvanese, D., & De Giacomo, G. (2003). Expressive description logics. In F. Baader, D. Calvanese, D. McGuinness, D. Nardi, & P. Patel-Schneider (Eds), *The Description Logic Handbook: Theory, Implementation and Applications* (pp. 178-218). Cambridge University Press.

Calvanese, C., De Giacomo, G., & Lenzerini, M. (1998a). On the decidability of query containment under constraints. In *Proceedings of the 17th ACM SIGACT SIGMOD SIGART Symposium on Principles of Database Systems (PODS'98)* (pp. 149-158).

Calvanese, D., Lenzerini, M., & Nardi, D. (1998b). Description logics for conceptual data modeling. In J. Chomicki & G. Saake, (Eds), *Logics for Databases and Information Systems.* Amsterdam: Kluwer.

Calvanese, D., Lenzerini, M., & Nardi, D. (1999). Unifying class-based representation formalisms. *Journal of Artificial Intelligence Research, 11*, 199-240.

Chomicki, J., & Toman, D. (1998). Temporal logic in information systems. In J. Chomiki & G. Saake

(Eds.), *Logics for databases and information systems*. Amsterdam: Kluwer.

Etzion, O., Gal, A., & Segev, A. (1998). Extended update functionality in temporal databases. In O. Etzion, S. Jajodia, & S. Sripada, (Eds.), *Temporal Databases -- Research and Practice, LNCS* (pp 56-95). Berlin: Springer-Verlag.

Fillottrani, P., Franconi, E., & Tessaris, S. (2006). *The new ICOM ontology editor. In 19th International Workshop on Description Logics (DL 2006),* Lake District, UK. May 2006.

Franconi, E., & Ng, G. (2000). The iCom tool for intelligent conceptual modeling. *7th International Workshop on Knowledge Representation meets Databases (KRDB'00)*, Berlin, Germany. 2000.

Gerstl, P., & Pribbenow, S. (1995). Midwinters, end games, and body parts: a classification of part-whole relations. *International Journal of Human-Computer Studies, 43*, 865-889.

Guarino, N., & Welty, C. (2000). A formal ontology of properties. In Dieng, R. (Ed.), *Proceedings of EKAW '00*. Berlin: Springer Verlag.

Guizzardi, G. (2005). *Ontological foundations for structural conceptual models*. PhD Thesis, Telematica Institute, Twente University, Enschede, the Netherlands.

Guizzardi, G. (2007). Modal Aspects of Object Types and Part-Whole Relations and the de re/de dicto distinction. *19th International Conference on Advances in Information Systems Engineering (CAiSE) LNCS 4495*. Berlin: Springer-Verlag.

Halpin, T. (1999). UML Data Models from an ORM Perspective (Part 8). *Journal of Conceptual Modeling*, 8, April 1999. Stable URL http://www.inceoncept.com/jcm.

Halpin, T. (2001). *Information Modeling and Relational Databases*. San Francisco: Morgan Kaufmann Publishers.

Halpin. T. (2007). Subtyping revisited. In B. Pernici,& J. Gulla (Eds.), *Proceedings of CAiSE'07 Workshops* (pp. 131-141). Academic Press.

Hawley, K. (2004). Temporal Parts. In E. N. Zalta, (Ed.), *The Stanford Encyclopedia of Philosophy (Winter 2004 Ed.)*. Stable URL http://plato.stanford.edu/archives/win2004/entries/temporal-parts/.

Hodgkinson, I. M., Wolter, F., & Zakharyaschev, M. (2000). Decidable fragments of first-order temporal logics. *Annals of pure and applied logic, 106*, 85-134.

Horrocks, I., Kutz, O., & Sattler, U. (2006). The Even More Irresistible SROIQ. In *Proceedings of the 10th International Conference of Knowledge Representation and Reasoning (KR2006)*, Lake District, UK, 2006.

Johansson, I. (2004). On the transitivity of the parthood relation. In Hochberg, H. and Mulligan, K. (eds.) *Relations and predicates* (pp. 161-181). Frankfurt: Ontos Verlag.

Keet, C. M. (2006a). *Introduction to part-whole relations: mereology, conceptual modeling and mathematical aspects* (Tech. Rep. No. KRDB06-3). KRDB Research Centre, Faculty of Computer Science, Free University of Bozen-Bolzano, Italy.

Keet, C. M. (2006b). Part-whole relations in Object-Role Models. 2nd International Workshop on Object-Role Modelling (ORM 2006), Montpellier, France, Nov 2-3, 2006. In Meersman, R., Tari, Z., Herrero, P. et al. (Eds.) *OTM Workshops 2006 LNCS, 4278*, 1116-1127. Berlin: Springer-Verlag.

Keet, C. M. (2007). Prospects for and issues with mapping the Object-Role Modeling language into DLRifd. *20th International Workshop on Description Logics (DL'07) CEUR-WS, 250*, 331-338. 8-10 June 2007, Bressanone, Italy.

Keet, C. M. (2008). A formal comparison of conceptual data modeling languages. *13th International Workshop on Exploring Modeling Methods in Systems Analysis and Design (EMMSAD'08) CEUR-WS, 337*, 25-39. Montpellier, France, 16-17 June 2008.

Keet, C. M., & Artale, A. (in press). Representing and Reasoning over a Taxonomy of Part-Whole Relations. *Applied Ontology – Special Issue on Ontological Foundations for Conceptual Models, 3*(1).

Masolo, C., Borgo, S., Gangemi, A., Guarino, N., & Oltramari, A. (2003). *Ontology Library.* WonderWeb Deliverable D18 (ver. 1.0, 31-12-2003). http://wonderweb.semanticweb.org.

Motschnig-Pitrik, R., & Kaasbøll, J. (1999). Part-Whole Relationship Categories and Their Application in Object-Oriented Analysis. *IEEE Transactions on Knowledge and Data Engineering, 11*(5), 779-797.

Object Management Group. (2005). *Unified Modeling Language: Superstructure. v2.0. formal/0507-04.* http://www.omg.org/cgi-bin/doc?formal/05-07-04.

Odell, J. J. (1998). *Advanced Object-Oriented Analysis & Design using UML.* Cambridge: Cambridge University Press.

Opdahl, A. L., Henderson-Sellers, B., & Barbier, F. (2001). Ontological analysis of whole-part relationships in OO-models. *Information and Software Technology, 43*(6), 387-399.

Parent, C., Spaccapietra, S., & Zimányi, E. (2006). *Conceptual modeling for traditional and spatio-temporal applications—the MADS approach.* Berlin: Springer Verlag.

Pontow, C., & Schubert, R. (2006). A mathematical analysis of theories of parthood. *Data & Knowledge Engineering, 59*, 107-138.

Sattler, U. (1995). A concept language for an engineering application with part-whole relations. In A. Borgida, M. Lenzerini, D. Nardi, & B. Nebel (Eds.), *Proceedings of the international workshop on description logics* (pp. 119-123).

Sattler, U. (2000). Description Logics for the Representation of Aggregated Objects. In W. Horn (Ed.) *Proceedings of the 14th European Conference on Artificial Intelligence (ECAI2000).* Amsterdam: IOS Press.

Schulz, S., Hahn, U., & Romacker, M. (2000). Modeling Anatomical Spatial Relations with Description Logics. In J. M. Overhage (Ed.), *Proceedings of the AMIA Symposium 2000* (pp. 779-83).

Shanks, G., Tansley, E., & Weber, R. (2004). Representing composites in conceptual modeling. *Communications of the ACM, 47*(7), 77-80.

Simons, P. (1987). *Parts: A study in Ontology.* Oxford: Clarendon Press.

Smith, B., Ceusters, W., Klagges, B., Köhler, J., Kumar, A., Lomax, J., Mungall, C., Neuhaus, F., Rector, A. L., & Rosse, C. (2005). Relations in biomedical ontologies. *Genome Biology,* 6, R46.

Spaccapietra, S., Parent, C., & Zimanyi, E. (1998). Modeling time from a conceptual perspective. In *Int. Conf. on Information and Knowledge Management (CIKM98).*

Varzi, A. C. (2004). Mereology. In E.N. Zalta (Ed.), *The Stanford Encyclopedia of Philosophy (Fall 2004 Ed.).* Stable URL http://plato.stanford.edu/archives/fall2004/entries/mereology/

Varzi, A. C. (2006a). Spatial reasoning and ontology: parts, wholes, and locations. In M. Aiello, I. Pratt-Hartmann, & J. van Benthem (Eds.), *The Logic of Space.* Dordrecht: Kluwer Academic Publishers.

Varzi, A. C. (2006b). A Note on the Transitivity of Parthood. *Applied Ontology, 1*, 141-146.

Vieu, L., & Aurnague, M. (2005). Part-of Relations, Functionality and Dependence. In M.

Aurnague, M. Hickmann, & L. Vieu (Eds.), *Categorization of Spatial Entities in Language and Cognition*. Amsterdam: John Benjamins.

Winston, M. E., Chaffin, R., & Herrmann, D. (1987). A taxonomy of part-whole relations. *Cognitive Science, 11*(4), 417-444.

ENDNOTES

1. Other recurring topics that will receive comparatively little attention are transitivity of part-whole relations (Johansson, 2004; Varzi, 2004, 2006b), analysis of types of part-whole relations (Gerstl & Pribbenow, 1995; Keet, 2006b; Keet & Artale, 2008; Odell, 1998; Winston *et al.*, 1987), horizontal relations between parts, and automated reasoning with part-whole relations. For a comprehensive introduction to such subtopics from different perspectives, see, e.g., (Artale *et al.*, 1996a; Guizzardi, 2005; Keet, 2006a; Simons, 1987).
2. Proper parthood is usually defined in terms of the parthood relation ($\forall x, y(proper\_part\_of(x, y) \equiv part\_of(x, y) \wedge \neg part\_of(y, x))$), but also can be taken as primitive and then to have parthood defined in terms of proper parthood ($\forall x, y(part\_of(x, y) \equiv proper\_part\_of(x, y) \vee x = y)$).
3. Downward distributive: there are properties of the whole that the parts inherit; upward distributive: the whole inherits properties from its parts. Alternatively, they are called property inheritance through parts and property refinement through parts.
4. In addition to these requirements, the usual ones for including basic properties of the parthood relation remain as well. That is, transitivity, reflexivity and antisymmetry for parthood, and transitivity, irreflexivity and asymmetry for proper parthood relations (see for a discussion and feasibility Keet & Artale (2008)).
5. Temporal parts in the sense of 4-dimensionalism (Hawley, 2004) is out of scope for common information systems modeling.
6. More precisely, UML without the part-of, EER, and a subset of ORM and ORM2 have a correspondence with DLR_{ifd} (Berardi *et al.*, 2005; Calvanese *et al.*, 1998a, 1999; Keet, 2007, 2008), which is DLR with additional identification and non-unary functional dependency constraints
7. Following the snapshot paradigm, T_p is a set of time points (or chronons) and $<$ is a binary precedence relation on T_p, *the flow of time* $\mathcal{T}_p = \langle T_p, < \rangle$ is assumed to be isomorphic to either $\langle \mathbb{Z}, < \rangle$ or $\langle \mathbb{N}, < \rangle$. Thus, standard relational databases can be regarded as the result of mapping a temporal database from time points in T to atemporal constructors, with the same interpretation of constants and the same domain.
8. Cases pl-p5 shown earlier in Figure 4 are not essential parts/wholes; we return to this point at the end of this section.
9. Optional participation (≥ 0) of either the part or the whole in the part-whole relation means that either (ManP) or (ManW) is not included in the list of constraints.
10. We are open to better suggestions (the icons have not been examined by modelers and domain experts). Basically, for each constraint ManP-DisjW there could be some explicit on in the diagrammatic modeling language, even though some are already covered in the language itself, such as XOR in ORM2. It is beyond the current scope to provide a fixed graphical and pseudo-NL syntax for all major conceptual modeling languages.

Chapter III
Extending the ORM Conceptual Schema Language and Design Procedure with Modeling Constructs for Capturing the Domain Ontology

Peter Bollen
Maastricht University, The Netherlands

ABSTRACT

In this chapter the authors extend the ORM conceptual modeling language with constructs for capturing the relevant parts of an application ontology in a list of concept definitions. The authors give the adapted ORM meta model and provide an extension of the accompanying Conceptual Schema Design Procedure (CSDP) to cater for the explicit modeling of the relevant parts of an application- or domain ontology in a list of concept definitions. The application of these modeling constructs will significantly increase the perceived quality and ease-of-use of (Web-based) applications.

INTRODUCTION

The objective of this book is to disseminate best practices and research outcomes of the information systems modeling (ISM) community to researchers, practitioners and students in the ISM field of knowledge. This chapter presents some extensions to an information modeling methodology called Object-Role Modeling (ORM) (Halpin, 2001).

ORM (including other fact-oriented languages (e.g. (Bakema, Zwart, & van der Lek, 1994; Lemmens, Nijssen, & Nijssen, 2007))) is a conceptual modeling approach that models the world in terms of objects and roles that they play (Halpin, 2001). ORM has a single fact encoding construct: the fact type, in contrast to other popular conceptual modeling methodologies, e.g. ER (Chen, 1976) and (E)ER (Teorey, Yang, & Fry, 1986) that contain at least two fact encoding constructs: the *attribute* and the *relationship* (Chen, 1976; Teorey et al., 1986).

The roots of ORM can be traced back to the early seventies, when the research focus in the ISM field was on database modeling languages (see for an overview of ORM's history the review article of Halpin (Halpin, 2007)). The objective of the researchers at that time was to define a truly conceptual modeling language for expressing database requirements, independently from the database implementation languages, that were existing at that time, e.g. CODASYL, hierarchical, or relational and object-oriented databases. ORM (or one of its ancestors at any point in time) and other contemporary fact-oriented modeling languages, have evolved over the past 30 years. The modeling constructs have evolved in order to enable the language to model an ever increasing range of domain requirements in a declarative way. Until the nineties, ORM and other fact-oriented modeling languages were mainly focused on modeling the requirements in the information perspective. In the past 20 years, the language has been extended with modeling constructs and methodology, that also cover the process- and event- perspectives in conceptual modeling (Balsters, Carver, Halpin, & Morgan, 2006; Bollen, 2007a, 2007b; Morgan, 2006, 2007; Prabhakaran & Falkenberg, 1988).

In the literature a number of definitions for ontology can be found: "the definition of the basic terms and relations comprising the vocabulary of a topic area" (Neches et al., 1991), " an ontology is a description of the concepts and relationships for an agent or a community of agents." (Gruber, 1993), " shared understanding of a domain that can be communicated between people and application systems." (Fensel, 2001), "an ontology is a formal conceptualization of a real world, sharing a common understanding of this real world." (Lammari & Metais, 2004, p.155). Burton-Jones et al. (2005) distinguish four types of material ontologies: *application-, domain-, generic-* and *representation* ontologies. Application ontologies specify definitions needed for a particular application, domain ontologies specify conceptualizations specific to a domain, generic ontologies specify conceptualizations generic to several domains and representation ontologies specify conceptualizations that underlie knowledge representation formalisms.

In the last 10 years, the penetration of the world-wide web into the heart of the business information systems, has lead to a renewed interest in conceptual modeling, albeit now from the perspective of 'connected' agents that communicate with each over via the internet. The research field that has attracted a lot of scholars and practitioners is the field of ontology, leading to standards for communication via the world wide web. Examples of these standards are the Web Ontology Language OWL (Bechhofer et al., 2004) and the Web Service Modeling Language WSML (Bruijn et al., 2005).

In this chapter we will extend the ORM conceptual modeling methodology and its *representation* ontology with additional modeling constructs to help us capture the relevant part of a *domain-* or *application ontology* in an implementation-independent way.

RELATED WORK

Weber and Zhang (Weber & Zhang, 1996) analyze the extent in which a pre-decessor to ORM: NIAM complies to the Bunge-Wand-Weber (BWW) model for ontological expressiveness.

Spyns, Meersman and Jarrar (Spyns, Meersman, & Jarrar, 2002) introduce the DOGMA ontology engineering approach in which they separate an ontology base from the ontological commitment that contains the specific rules of the application domain. DOGMA uses a binary version of ORM. DOGMA uses a browser: T-lex to create *application* ontologies out of (a) *domain(s)* ontology(ies) (Bach, Meersman, Spyns, & Trog, 2007; Trog, Vereecken, Christiaens, De Leenheer, & Meersman, 2006).

Dumas et al. (2002) give an approach for the ontology markup in the context of the generation of web forms. Their approach, however, is limited to tasks that do not involve complex reasoning, and they claim that the rules that can be captured in a conceptual modeling language such as ORM or UML suffice for capturing the ontology. De Troyer et al. (2003) state that semantic information is needed to solve possible conflicts between different ORM model chunks. They define 'reusable' concepts for designers, e.g. object concepts, relationships and tuples, the definition of the concepts however, is left to the designers: ' ..explaining in an informal way the meaning of the concept.'

Piprani (Piprani, 2007) applies ORM in a project in which different application ontologies (i.e. airports, carriers, transport agencies) are integrated into a uniform domain ontology for air traffic.

The rest of this article is organized as follows: In section 2 we will introduce the list of concept definitions and we will show how such a list of concept definitions can be used to augment a conceptual schema with a *domain- or application* ontology (Burton-Jones et al., 2005). In section 3 we will show how the ORM modeling construct of the subtype can be partially contained in *the list of concept definitions* and how the list of concept definitions can be used to capture the precise semantics of naming conventions. In section 4 we will give a significant part of the augmented ORM meta model including the accompanying *representation* ontology. In section 5 an extension of the ORM conceptual schema design procedure (CSDP) is given. Finally, in section 6, conclusions and a brief future outlook will be given.

THE LIST OF CONCEPT DEFINITIONS

Brasethvik and Gulla (2001) give an approach for semantic document classification and retrieval. During the process of constructing domain models from document collections terms are selected from the proposed list and defined in cooperation with domain users and other stakeholders from the domain (Brasethvik & Gulla, 2001, p. 50-51). In the next section we will illustrate how such a list of terms or concept definitions can help us capture an *application-* or *domain* ontology in the conceptual modeling process and how it can be integrated with an existing ORM conceptual schema. We will use as running example the University Enrollment UoD (see section 2.1).

Example 1: University Enrollment Part 1

The University Enrollment example will be used to illustrate the modeling concepts throughout this article. The University of Vandover offers a number of majors in education. Students can choose between majors in *Science, History* and *Economics* among others.

In figure 1 an information example is given of a university enrollment document (example 1). In this example the Vandover University wants to record information about the major for each of its students. It is assumed that the *student ID* can be used to identify a specific student among the union of students that are (and have been) enrolled in the Vandover University and that a *major name* can be used as identifier for a specific major among the set of majors that are offered by the Vandover University.

In figure 2 we have given the standard ORM-(I) conceptual schema for the Vandover University enrollment UoD.

Figure 1. Information example Vandover University Enrollment (example 1)

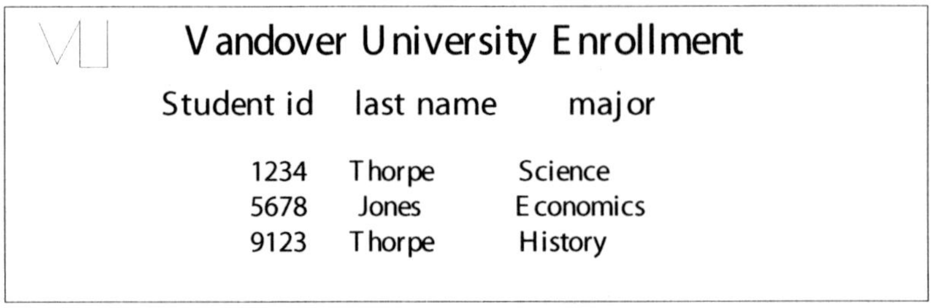

VU Vandover University Enrollment

Student id	last name	major
1234	Thorpe	Science
5678	Jones	Economics
9123	Thorpe	History

Figure 2. Standard ORM conceptual schema for University Enrollment (example 1)

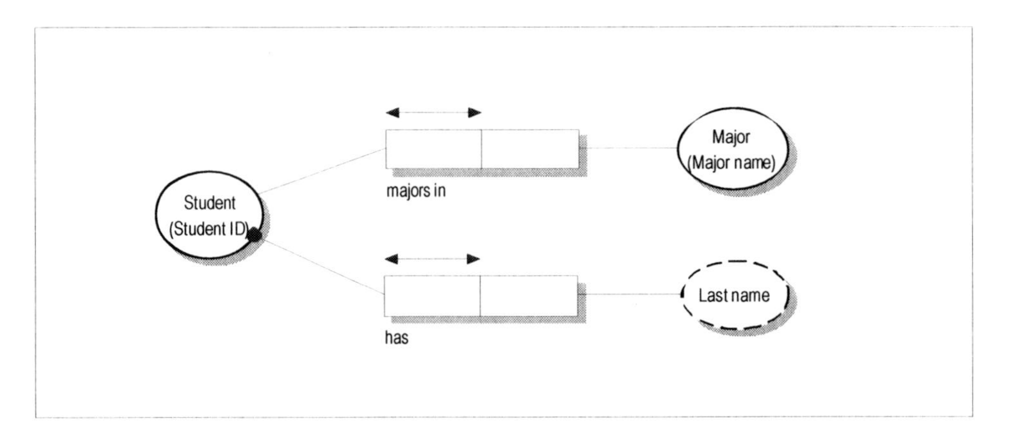

Adding the List of Concept and Definitions to the ORM Conceptual Schema

In the standard ORM conceptual schema for the University enrollment application subject area in figure 2, there is no definition given for the vocabulary of the University Enrollment topic area. To make the ontology of the application explicit, we need to incorporate a definition of the concepts in this ORM conceptual schema. Such a list of concept definitions should at least include the ORM object types, i.e. value types, (nested and unnested) entity types and reference types. We recommend to incorporate all of them into the list of concept definitions of the application UoD (see section 3). For *each* Entity type in the UoD, we must incorporate the name class(es) to be used as references for entity types (*Student ID*, *Major name*). The definition of a name class that can be used to identify an entity type must be given precisely, thereby explicitly focusing on the context in which the instances of a name class can be considered as identifiers for the defined entity type. For example the definition of the name class *student ID* in figure 2 should contain a description of the context in which the Student ID can be used as an identifier for a student: only those students that have ever been, are or will be studying at Vandover University. Furthermore, we must incorporate name classes that are merely value types (for example *last name*). In the enrollment UoD it must at least contain the definition of the concepts *Student*, *Major*, *last name*, *major name* and *student ID:*

Student: a student is a person that studies at Vandover University.

Major: a major is a course program offered to students by Vandover University

Last name: a name class

Major name: a name class, instances of which can be used to identify a major among the union of majors offered by Vandover University

Student ID: a name class, instances of which can be used to identify a Student among the union of students that have ever been, are or will be enrolled at Vandover University

If we inspect an ORM conceptual schema, we will also see that the part that distinguishes one fact type from another is the predicate or the verb part. For those fact types in which the predicate contains more semantics than for example: ..has.., ..is part of.. it is recommended to give a definition. In our university enrolment example we could decide to add a concept definition for the verb: ..majors in... :

Majors in: the choice of a student for a specific major subject

Furthermore, we can add 'intermediate' concepts into the list of concepts definitions if this is necessary. In our university enrollment example we could add the 'intermediate' concept *major subject*.

The definition of the concept types in the list of concept definitions must specify how the knowledge forming the concept (*definiendum*) is to be constructed from the knowledge given in the definition itself and in the defining concepts (*definiens*). A defining concept should either be a different concept that must be defined in another place in the list of concepts or it should be defined in a common business ontology. Brasethvik and Gulla use such a list of concept definitions in the context of a 'shared' or 'common information space' in which the semantics of information is locally constructed and '.. reflects the 'shared agreement' on the meaning of the information' (Brasethvik & Gulla, 2001, p.47).

We can now conclude that in order to capture the ontology of an application subject we need to add a list of concept definitions to the 'standard' conceptual schema. We will denote a 'definiendum' concept that is listed in the list of concept definitions between brackets ('<, >'). We, furthermore, recommend to order the list of definitions in such a way, that every concept that is a definiendum for a concept that is ranked N, must be defined itself in a rank < N in the same list. We will call such an ordered list of concept definitions: a list of concept definitions in *order of comprehension*. In figure 3 we have given an example of such a ordered list of concept definitions for the university enrollment UoD.

With respect to the fact type predicates in ORM it is recommended to add (parts of) them to the list of concept definitions in case the meaning differs from what is generally understood in business domains.

In the Vandover University enrollment example the concept (in this example an entity type) *course program* is considered to be defined in a common business ontology for university education which implies that all 'agents' that are involved in this UoD have attached the same meaning to this concept.

RELATIONSHIP BETWEEN CONCEPT DEFINITION AND SUBTYPING IN ORM

Entity types that are generally related to each other by means of super-sub type relationships in ORM can alternatively be captured in a list of definitions. In this section we will illustrate what this means for the existing modeling conventions

Figure 3. List of concept definitions in order of comprehension for University Enrollment application (example 1)

Concept	Definition
Student	*a person that studies at Vandover University*
Student ID	*a name class, instances of which can be used to identify a <Student> among the union of <students> that have ever been, are or will be enrolled at Vandover University*
Major	*a course program offered to <students> by Vandover University*
Major name	*a name class, instances of which can be used to identify a <major> among the union of <majors> offered by Vandover University*
major subject	*a specific specialization in field of study*
Majors in	*the choice of a< student> for a specific <major subject>*
Last name	*a name class*

in ORM regarding super/sub types or specialization and/or generalization. To illustrate these modeling issues we will extend our University enrollment example.

Example 2: University Enrollment Part 2

We will now extend our UoD with a second 'real-life' example of communication. This example contains additional background information on some students. The extended UoD is given in figure 4.

In the extended Universe of Discourse (examples 1 and 2 in combination) we see that for some students we will record only their University of origin. These students are generally known as exchange students. For some of the other students we will record the name of the high school from which they have graduated. In figure 5 the ORM-(I) conceptual schema for this extended Universe of Discourse is given.

For another group of *non-exchange* students no specific details will be recorded. The non-exchange students are are referred to as *regular* students. In figure 5 we have given the ORM conceptual schema for this UoD including the subtypes and a subtype defining fact type (Halpin, 2001, p.253). If we carefully inspect the ORM conceptual schema in figure 5 together with the user example from figure 4 we see that the examples on which this conceptual schema is based do *not* contain an explicit recording of the *Student Category.* It is during the CSDP step 1 that the verbalization of an example leads to natural language sentences that subsequently can be abstracted into (elementary) fact types. The addition of the *student category* fact type

Figure 4. Additional example Vandover University Enrollment

Additional student information

Student ID	University of origin	High school
1234	Harvard	--
5678	--	St. Paul new York

Figure 5. ORM conceptual schema for extended University Enrollment example

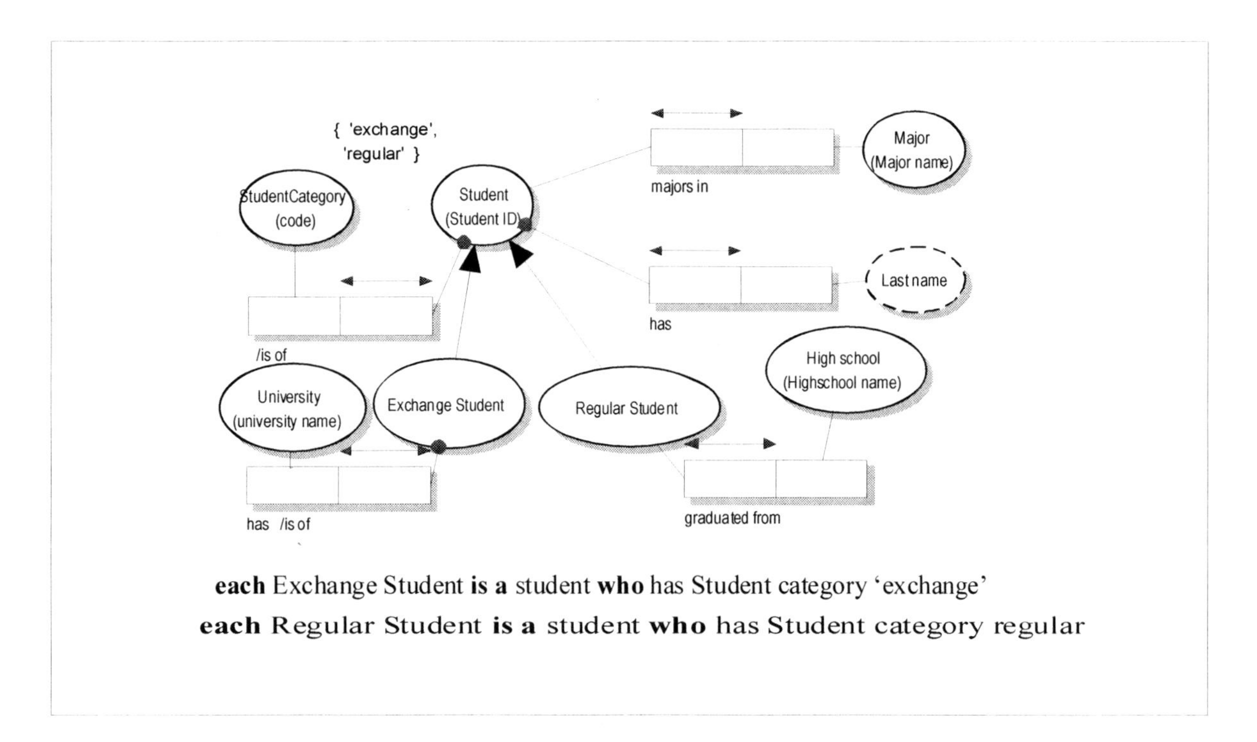

in the ORM conceptual schema in figure 5 is therefore not justified from the point of view of the application UoD because no explicit recording of the student category can be traced to the user examples.

We will now illustrate how the incorporation of a list of concept definitions will allow us to capture intentional subtype definitions in the list of definitions. This augmented ORM conceptual schema is given in figure 6.

Definition of Naming Conventions in ORM

Another reason for incorporating a list of definitions into an ORM conceptual schema is the necessity of an explicit definition of the context in which a naming convention holds. In current ORM conceptual schemas it can only be encoded which name class can be used to identify an instance of an entity type in general. Even in the explicit recording of a ORM referencing scheme it can only be expressed that an entity type is identified by (exactly) one (set of) instance(s) of one (or more) value type(s). What is not captured in such a referencing scheme (Halpin, 2001:186-189) is a description of the context in which a name class (e.g *Student ID, University name, High school name*) can be used to identify instances of a concept type (e.g. *Student, University, High school*). In figure 6 this is illustrated in the list of concept definitions for the name classes *student ID, major name*, *university name* and the name class *high school name.* If we study the definitions for the latter concepts we see that we have incorporated the explicit context under which the names from the name class can be used as identifiers for entities of the entity types *Student* and *Major.* A further reason to incorporate a precise definition of the semantics of a naming convention will be illustrated by an example in which one or more entity types need a compound reference

Figure 6. Augmented ORM conceptual schema for extended University Enrollment example

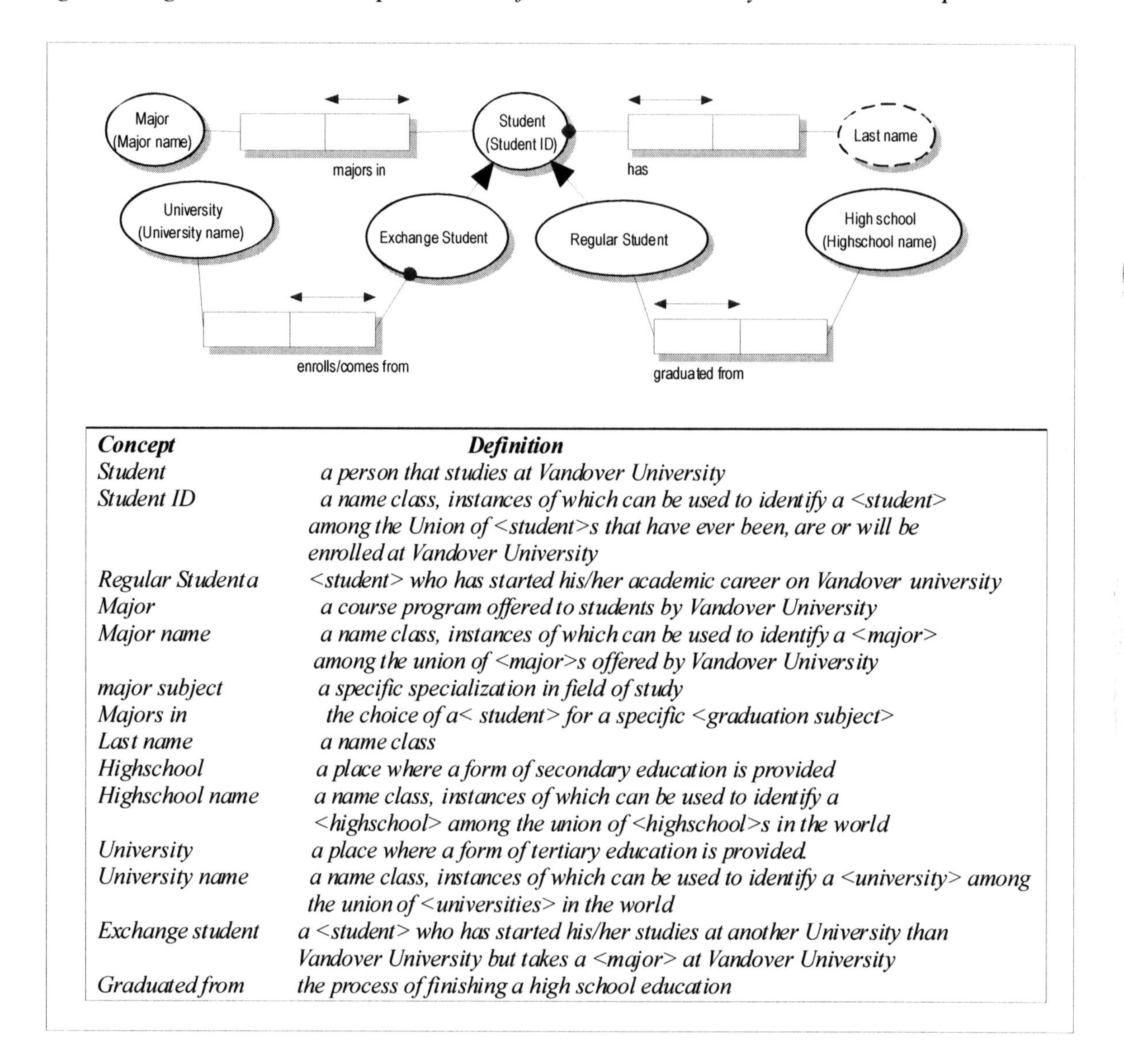

Concept	*Definition*
Student	*a person that studies at Vandover University*
Student ID	*a name class, instances of which can be used to identify a <student> among the Union of <student>s that have ever been, are or will be enrolled at Vandover University*
Regular Studenta	*<student> who has started his/her academic career on Vandover university*
Major	*a course program offered to students by Vandover University*
Major name	*a name class, instances of which can be used to identify a <major> among the union of <major>s offered by Vandover University*
major subject	*a specific specialization in field of study*
Majors in	*the choice of a< student> for a specific <graduation subject>*
Last name	*a name class*
Highschool	*a place where a form of secondary education is provided*
Highschool name	*a name class, instances of which can be used to identify a <highschool> among the union of <highschool>s in the world*
University	*a place where a form of tertiary education is provided.*
University name	*a name class, instances of which can be used to identify a <university> among the union of <universities> in the world*
Exchange student	*a <student> who has started his/her studies at another University than Vandover University but takes a <major> at Vandover University*
Graduated from	*the process of finishing a high school education*

type. We will illustrate the necessity of recording these explicit naming convention semantics by an extension of our running example.

Example 3: University Enrollment Part 3

We assume that Vandover University has merged with Ohao University. In order to streamline the enrollment operations it is decided to centralize them. This means that a student after the merger can no longer be identified by the existing student ID because a given *student ID* can refer to a student in the former Ohoa university *and* to a different student in the (former) Vandover University. To capitalize on the existing naming conventions it is decided to add the qualification *O* (for Ohao) or *V* (for Vandover) to the existing student ID. This extension is the *university code.* In figure 7 we have given the adapted ORM conceptual schema

Figure 7. ORM conceptual schema including list of concept definitions for Enrollment example

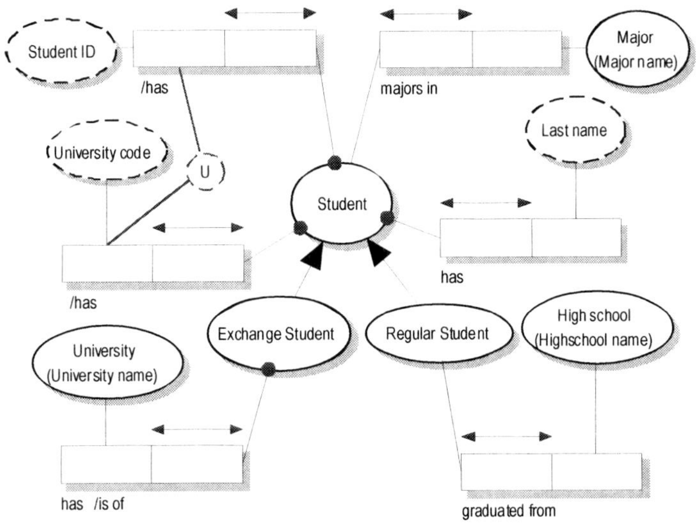

Concept	*Definition*
Student	*a person that studies at Vandover or Ohao University*
Student ID	*a value type*
University code	*a value type*
	a student at the integrated Vandover and Ohao university is identified by a combination of <student ID> and <University code>
Regular Student	*a <student> who has started his/her academic career on Vandover or Ohao university*
Major	*a course program offered to students by the Vandover/Ohao University*
Major name	*a name class, instances of which can be used to identify a <major> among the union of <major>s offered by Vandover/Ohao University*
major subject	*a specific specialization in a field of study*
Majors in	*the choice of a< student> for a specific <major subject>*
Last name	*a name class*
Highschool	*a place where a form of secondary education is provided*
Highschool name	*a name class, instances of which can be used to identify a <highschool> among the union of <highschool>s in the world*
University	*a place where a form of tertiary education is provided.*
University name	*a name class, instances of which can be used to identify a <university> among the union of <universities> in the world*
Exchange student	*a <student> who has started his/her studies at another University than Vandover University but takes a <major> at Vandover University*
Graduated from	*the process of finishing a high school education*

including a compound reference scheme (Halpin, 2001, p.189) and an adapted list of definitions in which we have now incorporated the precise definition of the naming conventions and the conditions under which a name class is valid.

THE META MODEL FOR THE AUGMENTED ORM

In this section we will give a (segment of an) ORM meta model in which we have incorporated the list of concept definitions that allows us to capture the ORM representation ontology (Burton-Jones et al., 2005).

We will use one of the two the ORM meta models (model A) that were presented in (Cuyler & Halpin, 2003). One of the distinctive features of this meta-model is that it allows for inactive subtypes. Furthermore it enforces a subtype definition for each distinguished subtype. In our augmented meta-model we will relax the requirement in which it is stated that each subtype must have a subtype-definition in terms of a definition that is defined on a subtype defining fact type. We will furthermore remove inactive subtypes from the conceptual schema but we will add their definitions in the list of concepts and definitions.

In figure 8 we have copied the relevant segment of the ORM meta model from (Cuyler & Halpin, 2003, p.2). We will illustrate how we can adapt this meta model for the list of concept definitions. We can derive the status of the concept types by inspecting the definition in the list of concepts. If the definition is a refinement of another concept type then such a concept type is a *subtype*. If the definition is the description of a condition under which instances of a concept can be used to identify entity types then such a concept type is a *name class*. A concept type can be involved in a definition fact type, but this is however not mandatory for every concept type. Furthermore, the subtype definition is no longer mandatory for

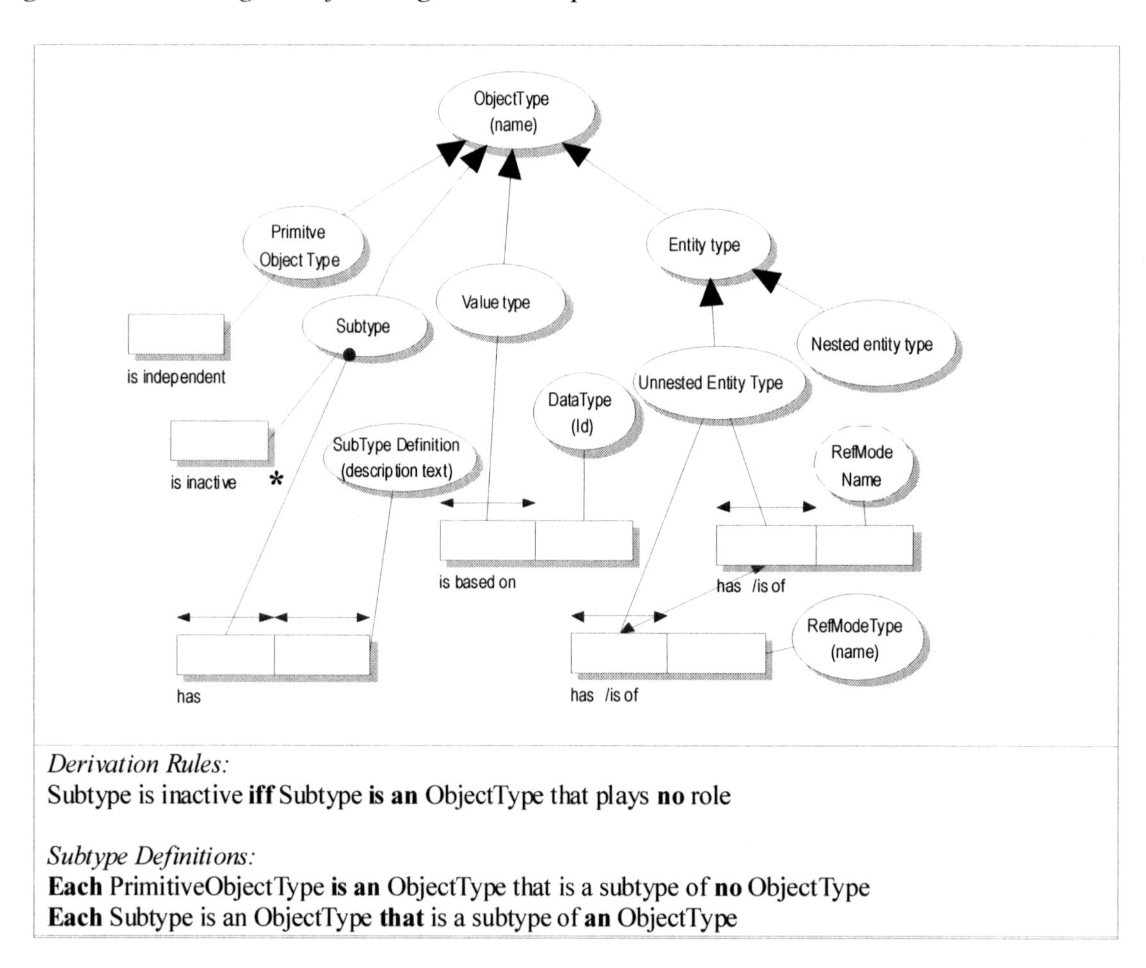

Figure 8. Relevant segment of existing ORM conceptual meta schema

Figure 9. Augmented (segment of) ORM conceptual meta schema

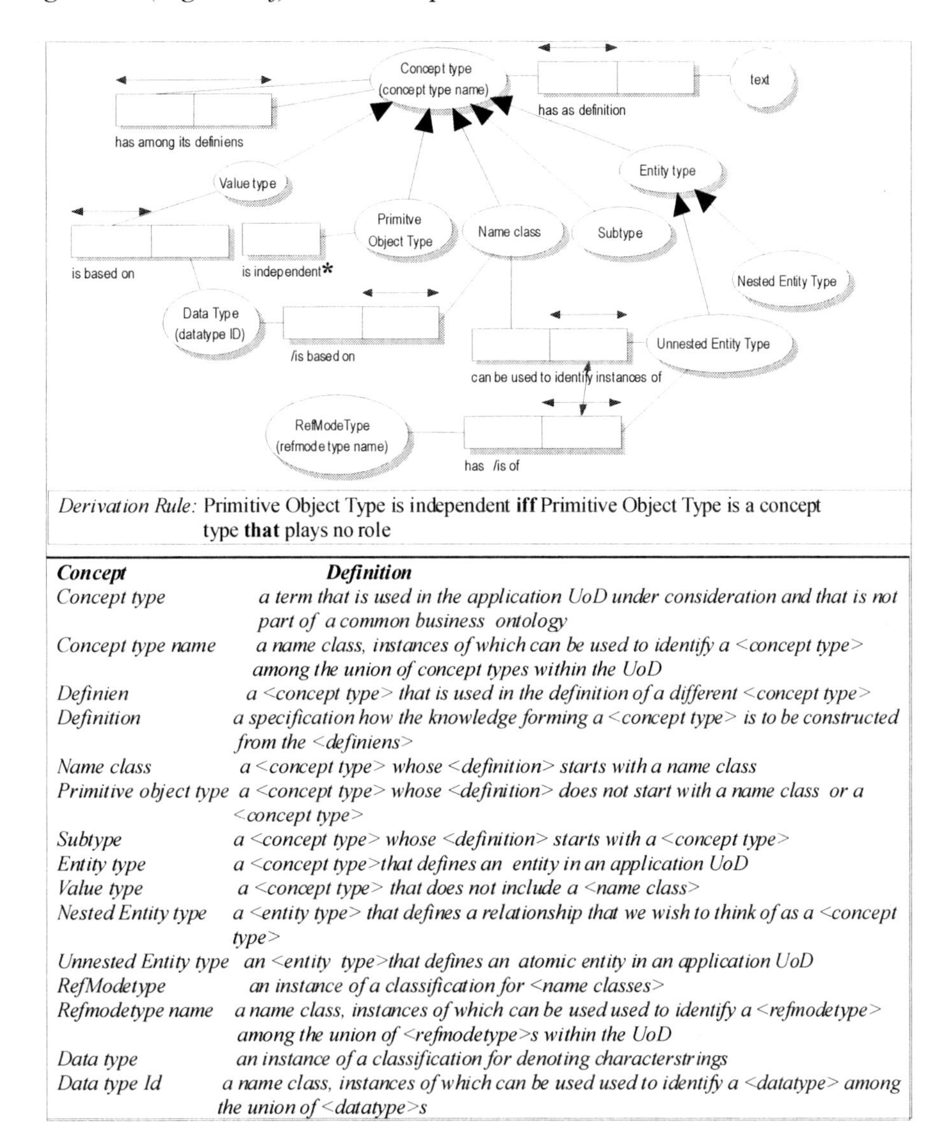

Derivation Rule: Primitive Object Type is independent **iff** Primitive Object Type is a concept type **that** plays no role

Concept	***Definition***
Concept type	*a term that is used in the application UoD under consideration and that is not part of a common business ontology*
Concept type name	*a name class, instances of which can be used to identify a <concept type> among the union of concept types within the UoD*
Definien	*a <concept type> that is used in the definition of a different <concept type>*
Definition	*a specification how the knowledge forming a <concept type> is to be constructed from the <definiens>*
Name class	*a <concept type> whose <definition> starts with a name class*
Primitive object type	*a <concept type> whose <definition> does not start with a name class or a <concept type>*
Subtype	*a <concept type> whose <definition> starts with a <concept type>*
Entity type	*a <concept type>that defines an entity in an application UoD*
Value type	*a <concept type> that does not include a <name class>*
Nested Entity type	*a <entity type> that defines a relationship that we wish to think of as a <concept type>*
Unnested Entity type	*an <entity type>that defines an atomic entity in an application UoD*
RefModetype	*an instance of a classification for <name classes>*
Refmodetype name	*a name class, instances of which can be used used to identify a <refmodetype> among the union of <refmodetype>s within the UoD*
Data type	*an instance of a classification for denoting characterstrings*
Data type Id	*a name class, instances of which can be used used to identify a <datatype> among the union of <datatype>s*

a concept type that is a subtype but is contained in the general definition for a concept type.

ADAPTING THE CSDP FOR THE LIST OF CONCEPT DEFINITIONS

ORM and other fact-oriented conceptual modeling languages differ from other conceptual modelling approaches like (E)ER (Chen, 1976; Teorey et al., 1986) and UML (Booch, Rumbauch, & Jacobson, 2005) not only with respect to the number of fact-encoding constructs, but also in terms of the availability of a modeling procedure that guides an analyst in the creation of a conceptual application schema in dialogue with (a) domain user(s). To synchronize the extended fact-oriented language that we have discussed in this chapter with ORM's conceptual schema design procedure (Halpin, 2007, p.26) , we will present an adapted modeling procedure[a] in table 1 that ensures that the list of concept definitions is created in for every application.

CONCLUSION AND FUTURE TRENDS

In this chapter we have given additional modeling concepts for ORM together with the relevant increments for ORM's CSDP to cater for the explicit modeling of a application domain's ontology. The new modeling constructs for the list of concept definitions that we have added allow us to capture the relevant part of a domain ontology of an application UoD. We have chosen the modeling constructs in such a way that the overall complexity of the ORM meta-model has not increased, but is reduced by using the list of concept definitions to capture subtype definitions. The augmentation of ORM with a list of concept definitions also preserves the methodological pureness of ORM in which every fact type in the ORM conceptual schema can be traced to (parts of) a user example. A further improvement to current ORM is the incorporation of the complete naming convention semantics in the list of concept definitions. The practical relevance of this extension of a well-established conceptual modeling language with the list of concept definitions is in the 'networked' society and business-world in which a traditional conceptual schema has to be 'upgraded' to cater for communication with potential external agents, e.g. customers, suppliers, web-service brokers, whose identity is not yet known to us at design time.

In line with semantic web developments, the conceptual schema needs a communication part that contains 'definition' instances to be shared with the potential agents in order for them to be able to communicate effectively and efficiently with a ('web-based') business application in which the 'traditional' allowed communication patterns and their state (transition) constraints will not be violated. This will significantly increase the perceived quality and ease-of-use of such a (web-based) application, since it has established a

Table 1. Augmented ORM CSDP

Step	Description
1.	Transform familiar information examples into elementary facts
2.	**Add entity types, reference schemes and relevant (parts) of predicates to the list of concept definitions.** Apply quality checks.
2.	Draw the fact types, and apply a population check
3.	Check for entity types that should be combined. **Add the definitions for the newly defined atomic entity types and nested entity types and their reference modes.** Note any arithmetic derivations **Add the definitions of the 'underlying' derivation business process**
4.	Add uniqueness constraints and check arities of fact types
5.	Add mandatory role constraints and check arities for logical derivations
6.	Add value, set comparison and subtyping constraints. **Add the subtype definition onto the list of concept definitions**
7.	Add other constraints and perform final checks

semantic bridge with the potential external users, allowing them to communicate in a direct way with the business application, by preventing semantic ambiguities from occurring in the first place.

Another advantage of using (extended) fact-oriented modeling languages is that a business organization is not forced to remodel the application or domain ontology every time a new 'implementation' standard has been defined. Business organizations can capitalize on the 'ORM' conceptual modeling investment, for the foreseeable future by applying the appropriate mappings between a fact-oriented application ontology and the implementation standard of the time.

REFERENCES

Bach, D., Meersman, R., Spyns, P., & Trog, D. (2007). *Mapping OWL-DL into ORM/RIDL*. Paper presented at the OTM 2007/ORM 2007.

Bakema, G. P., Zwart, J. P., & van der Lek, H. (1994). Fully communication oriented NIAM. In G. Nijssen & J. Sharp (Eds.), *NIAM-ISDM 1994 Conference* (pp. L1-35). Albuquerque NM.

Balsters, H., Carver, A., Halpin, T., & Morgan, T. (2006). *Modeling dynamic rules in ORM*. Paper presented at the OTM 2006/ORM 2006.

Bechhofer, S., Harmelen, F. v., Hendler, J., Horrocks, I., McGuinness, D., Patel-Schneider, P., et al. (2004). *OWL web ontology language reference*. Retrieved. from http://www.w3.org/TR/owl-ref/.

Bollen, P. (2007a). *Fact-oriented Business Rule Modeling in the Event Perspective*. Paper presented at the CAISE 2007.

Bollen, P. (2007b). *Fact-oriented modeling in the data-, process- and event perspectives*. Paper presented at the OTM 2007, ORM 2007.

Booch, G., Rumbauch, J., & Jacobson, I. (2005). *Unified Modeling Language User Guide* (2nd ed.): Addison-Wesley Professional.

Brasethvik, T., & Gulla, J. (2001). Natural language analysis for semantic document modeling. *Data & Knowlege Engineering, 38*, 45-62.

Bruijn, J. d., Lausen, H., Krummenacher, R., Polleres, A., Predoiu, L., Kifer, M., et al. (2005). *WSML working draft 14 march 2005*: DERI.

Burton-Jones, A., Storey, V., Sugumaran, V., & Ahluwalia, P. (2005). A semiotic metrics suite for assessing the quality of ontologies. *Data & Knowlege Engineering, 55*, 84-102.

Chen, P. (1976). The Entity-Relationship Model : Toward a Unified View. *ACM Transactions on Database Systems, 1*(1), 9 - 36.

Cuyler, D., & Halpin, T. (2003). *Meta-models for Object-Role Modeling*. Paper presented at the International Workshop on Evaluation of Modeling Methods in Systems Analysis and Design (EMMSAD '03).

De Troyer, O., Plessers, P., & Casteleyn, S. (2003). *Solving semantic conflicts in audience driven web-design*. Paper presented at the WWW/Internet 2003 Conference (ICWI 2003).

Dumas, M., Aldred, L., Heravizadeh, M., & Ter Hofstede, A. (2002). *Ontology Markup for Web Forms Generation*. Paper presented at the WWW '02 Workshop on Real-World applications of RDF and the semantic web.

Fensel, D. (2001). *Ontologies: Silver Bullet for Knowledge Management and Electronic Commerce*: Springer Verlag.

Gruber, T. (1993). A translation approach to portable ontologies. *Knowledge Acquisition, 5*(2), 199- 220.

Halpin, T. (2001). *Information Modeling and Relational Databases; from conceptual analysis to logical design*. San Francisco, California: Morgan Kaufmann.

Halpin, T. (2007). Fact-oriented modeling: past, present and future. In J. Krogstie, A. Opdahl &

S. Brinkkemper (Eds.), *Conceptual modeling in information systems engineering* (pp. 19 - 38). Berlin: Springer Verlag.

Lammari, N., & Metais, E. (2004). Building and maintaining ontologies: a set of algorithms. *Data & Knowlege Engineering, 48*, 155- 176.

Lemmens, I., Nijssen, M., & Nijssen, G. (2007). *A NIAM2007 conceptual analysis of the ISO and OMG MOF four layer metadata architectures.* Paper presented at the OTM 2007/ ORM 2007.

Morgan, T. (2006). *Some features of state machines in ORM.* Paper presented at the OTM2006/ ORM 2006 workshop.

Morgan, T. (2007). *Business Process Modeling and ORM.* Paper presented at the OTM 2007/ ORM 2007.

Neches, R., Fikes, R., Finin, T., Gruber, T., Patil, R., Senator, T., et al. (1991). Enabling technology for knowledge sharing. *AI magazine, fall 1991*, 36-56.

Piprani, B. (2007). *Using ORM in an ontology based approach for a common mapping across heterogeneous applications.* Paper presented at the OTM2007/ORM 2007 workshop.

Prabhakaran, N., & Falkenberg, E. (1988). Representation of Dynamic Features in a Conceptual Schema. *Australian Computer Journal, 20*(3), 98-104.

Spyns, P., Meersman, R., & Jarrar, M. (2002). Data modelling versus Ontology engineering. *SIGMOD record: special issue on semantic web and data management, 31*(4), 12-17.

Teorey, T., Yang, D., & Fry, J. (1986). A logical design methodology for relational databases using the extended E-R model. *ACM Computing Surveys, 18*(2), 197-222.

Trog, D., Vereecken, J., Christiaens, S., De Leenheer, P., & Meersman, R. (2006). *T-Lex; A role-based ontology engineering tool.* Paper presented at the OTM 2006/ORM 2006 workshop.

Weber, R., & Zhang, Y. (1996). An analytical evaluation of NIAM's grammar for conceptual schema diagrams. *Information Systems Journal, 6*(2), 147-170.

ENDNOTE

[a] The newly added (sub)-steps in the CSDP are printed in bold font in table 1.

Section II
Modeling Approaches

Chapter IV
EKD:
An Enterprise Modeling Approach to Support Creativity and Quality in Information Systems and Business Development

Janis Stirna
Jönköping University, Sweden

Anne Persson
University of Skövde, Sweden

ABSTRACT

This chapter presents experiences and reflections from using the EKD Enterprise Modeling method in a number of European organizations. The EKD modeling method is presented. The chapter then focuses on the EKD application in practice taking six cases as an example. The authors' observations and lessons learned are reported concerning general aspects of Enterprise Modeling projects, the EKD modeling language, the participative modeling process, tool support, and issues of Enterprise Model quality. They also discuss a number of current and emerging trends for development of Enterprise Modeling approaches in general and for EKD in particular.

INTRODUCTION

Enterprise Modeling (EM), or Business Modeling, has for many years been a central theme in information systems engineering research and a number of different methods have been proposed. There are two main reasons for using EM (Persson & Stirna, 2001):

- Developing the business. This entails developing business vision, strategies, redesigning the way the business operates, developing the supporting information systems, etc.

- Ensuring the quality of the business. Here the focus is on two issues: (1) sharing the knowledge about the business, its vision and the way it operates, and (2) ensuring the acceptance of business decisions through committing the stakeholders to the decisions made.

Examples of EM methods can be found in (Bajec & Krisper, 2005; Dobson, Blyth & Strens 1994; Castro et al., 2001; Johannesson et al., 1997; Willars, 1993; Bubenko, 1993; Bubenko, Persson & Stirna, 2001, F3 Consortium, 1994; Fox, Chionglo, & Fadel, 1993; Krogstie et al., 2000; Loucopoulos et al., 1997; Yu & Mylopoulos, 1994). Examples of application domains for EM can be found in (Wangler, Persson & Söderström, 2001, Wangler & Persson, 2003; Wangler et al., 2003; Niehaves & Stirna, 2006; Stirna, Persson & Aggestam, 2006; Gustas, Bubenko & Wangler, 1995; Kardasis et al. 1998).

Since the beginning of the 1990-ies, the authors of this paper have been involved in the development, refinement and application of the Enterprise Knowledge Development (EKD) method for EM. During this time we have applied the method in a fair number of cases in a variety of organizations, during which observations have been collected and later analyzed. We have also performed Grounded Theory (Glaser & Strauss, 1967) studies focusing on the intentional and situational factors that influence participatory EM and EM tool usage (Persson & Stirna, 2001; Persson, 2001; Stirna, 2001). The synthesized results of cases and other studies are reported in this paper. The paper focuses on issues related to the EKD modeling language, the EKD modeling process, quality aspects of EM models, and EKD tool support.

The remainder of this paper is organized as follows. In section 2 we present the EKD Enterprise Modeling method. Section 3 describes a number of cases of applying of method. Observations from applying EKD are presented in Section 4, while Section 5 discusses the findings and provides some directions for future work.

ENTERPRISE KNOWLEDGE DEVELOPMENT (EKD)

In Scandinavia, methods for Business or Enterprise Modeling (EM) was initially developed in the 1980-ies by Plandata, Sweden (Willars, 1988), and later refined by the Swedish Institute for System Development (SISU). A significant innovation in this strand of EM was the notion of business goals as part of an Enterprise Model, enriching traditional model component types such as entities and business processes. The SISU framework was further developed in the ESPRIT projects F3 – "From Fuzzy to Formal" and ELEKTRA – "Electrical Enterprise Knowledge for Transforming Applications". The current framework is denoted EKD – "Enterprise Knowledge Development" (Bubenko, Persson & Stirna, 2001; Loucopoulos, et al., 1997).

EKD – Enterprise Knowledge Development method is a representative of the Scandinavian strand of EM methods. It defines the modeling process as a set of guidelines for a participatory way of working and the language for expressing the modeling product.

The EKD Modeling Language

The EKD modeling language consists of six sub-models: Goals Model (GM), Business Rules Model (BRM), Concepts Model (CM), Business Process Model (BPM), Actors and Resources Model (ARM), as well as Technical Components and Requirements Model (TCRM). Each sub-model focuses on a specific aspect of an organization (see table 1).

The GM focuses on describing the goals of the enterprise. Here we describe what the enterprise and its employees want to achieve, or to avoid, and when. The GM usually clarifies questions, such as: where should the organization be moving; what are the goals of the organization; what are the importance, criticality, and priorities of these goals; how are goals related to each other; which problems hinder the achievement of goals?

The BRM is used to define and maintain explicitly formulated business rules, consistent with the GM. Business Rules may be seen as operationalization or limits of goals. The BRM usually clarifies questions, such as: which rules affect the organization's goals; are there any policies stated; how is a business rule related to a goal; how can goals be supported by rules?

The CM is used to strictly define the "things" and "phenomena" one is talking about in the other models. We represent enterprise concepts, attributes, and relationships. The CM usually clarifies questions, such as: what concepts are recognized in the enterprise; which are their relationships to goals, activities, processes, and actors; how are they defined; what business rules and constraints monitor these objects and concepts?

The BPM is used to define enterprise processes, the way they interact and the way they handle information as well as material. A business process is assumed to consume input in terms of information and/or material and produce output of information and/or material. In general, the BPM is similar to what is used in traditional dataflow diagram models. The BPM usually clarifies questions, such as: which business activities and processes are recognized in the organization, or should be there, to manage the organization in agreement with its goals? How should the business processes, tasks, etc. be performed (workflows, state transitions, or process models)? Which are their information needs?

The ARM is used to describe how different actors and resources are related to each other and how they are related to components of the GM, and to components of the BPM. For instance, an actor may be responsible for a particular process in the BPM or the actor may pursue a particular goal in the GM. The ARM usually clarifies questions, such as: who is/should carry out which processes and tasks; how is the reporting and responsibility structure between actors defined?

The TCRM becomes relevant when the purpose of EKD is to aid in defining requirements for the development of an information system. Attention is focused on the technical system that is needed to support enterprise's goals, processes, and actors. Initially one needs to develop a set of high level requirements or goals, for the information system as a whole. Based on these, we attempt to structure the information system in a number of subsystems, or technical components. TCRM is an initial attempt to define the overall structure and properties of the information system to support the business activities, as de-fined in the BPM. The TCRM usually clarifies questions, such as: what are the requirements for the information system to be developed; which requirements are generated by the business processes; which potential has emerging information and communication technology for process improvement?

The modeling components of the sub-models are related within a sub-model (intra-model relationships), as well as with components of other sub-models (inter-model relationships). Figure 1 shows *inter-model relationships*. The ability to trace decisions, components and other aspects throughout the enterprise is dependent on the use and understanding of these relationships. When developing a full enterprise model, these relationships between components of the different sub-models play an essential role. For instance, statements in the Goals Model need to be defined more clearly as different concepts in the Concepts Model. A link is then specified between the corresponding Goals Model component and the concepts in the Concepts Model. In the same way, goals in the Goals Model motivate particular processes in the Business Processes Model. The processes are needed to achieve the goals stated. A link therefore is defined between a goal and the process. Links between models make the model traceable. They show, for instance, why certain

Table 1. Overview of the sub-models of the EKD method (Stirna, Persson & Sandkuhl, 2007)

	Goals Model (GM)	Business Rules Model (BRM)	Concepts Model (CM)	Business Process Model (BPM)	Actors and Resources Model (ARM)	Technical Component & Requirements Model(TCRM)
Focus	Vision and strategy	Policies and rules	Business ontology	Business operations	Organizational structure	Information system needs
Issues	What does the organization want to achieve or to avoid and why?	What are the business rules, how do they support organization's goals?	What are the things and "phenomena" addressed in other sub-models?	What are the business processes? How do they handle information and material?	Who are responsible for goals and process? How are the actors interrelated?	What are the business requirements to the IS? How are they related to other models?
Com-po-nents	Goal, problem, external constraint, opportunity	Business rule	Concept, attribute	Process, external process, information set, material set	Actor, role, organizational unit, individual	IS goal, IS problem, IS requirement, IS component

Figure 1. Overview of the sub-models of the EKD method

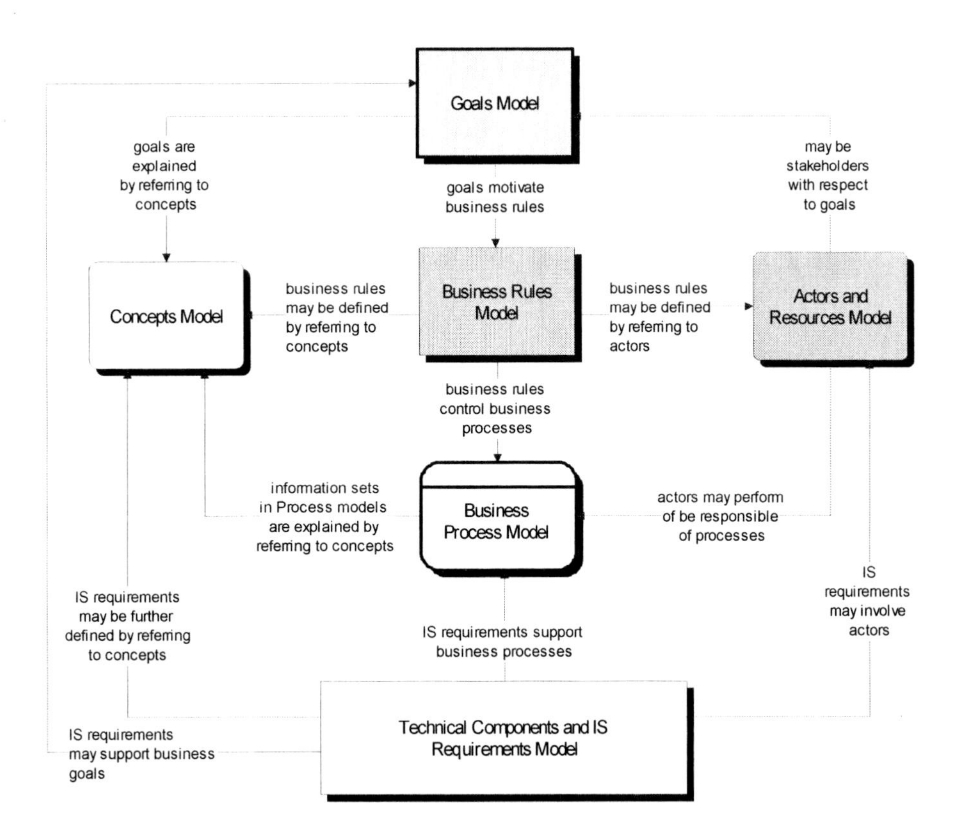

processes and information system requirements have been introduced.

The ability to trace decisions, components and other aspects throughout the enterprise is dependent on the use and understanding of the relationships between the different sub-models addressing the issues in table 1. While different sub-models address the problem domain from different perspectives, the inter-model links ensure that these perspectives are integrated and provide a complete view of the problem domain. They allow the modeling team to assess the business value and impact of the design decisions. An example of inter-model links is shown in Figure 2.

There are two alternative approaches to notation in EKD:

1. A fairly simple notation, suitable when the domain stakeholders are not used to modeling and the application does not require a high degree of formality.

Figure 2. Fragment of an enterprise model with inter-model links (Stirna, 2001)

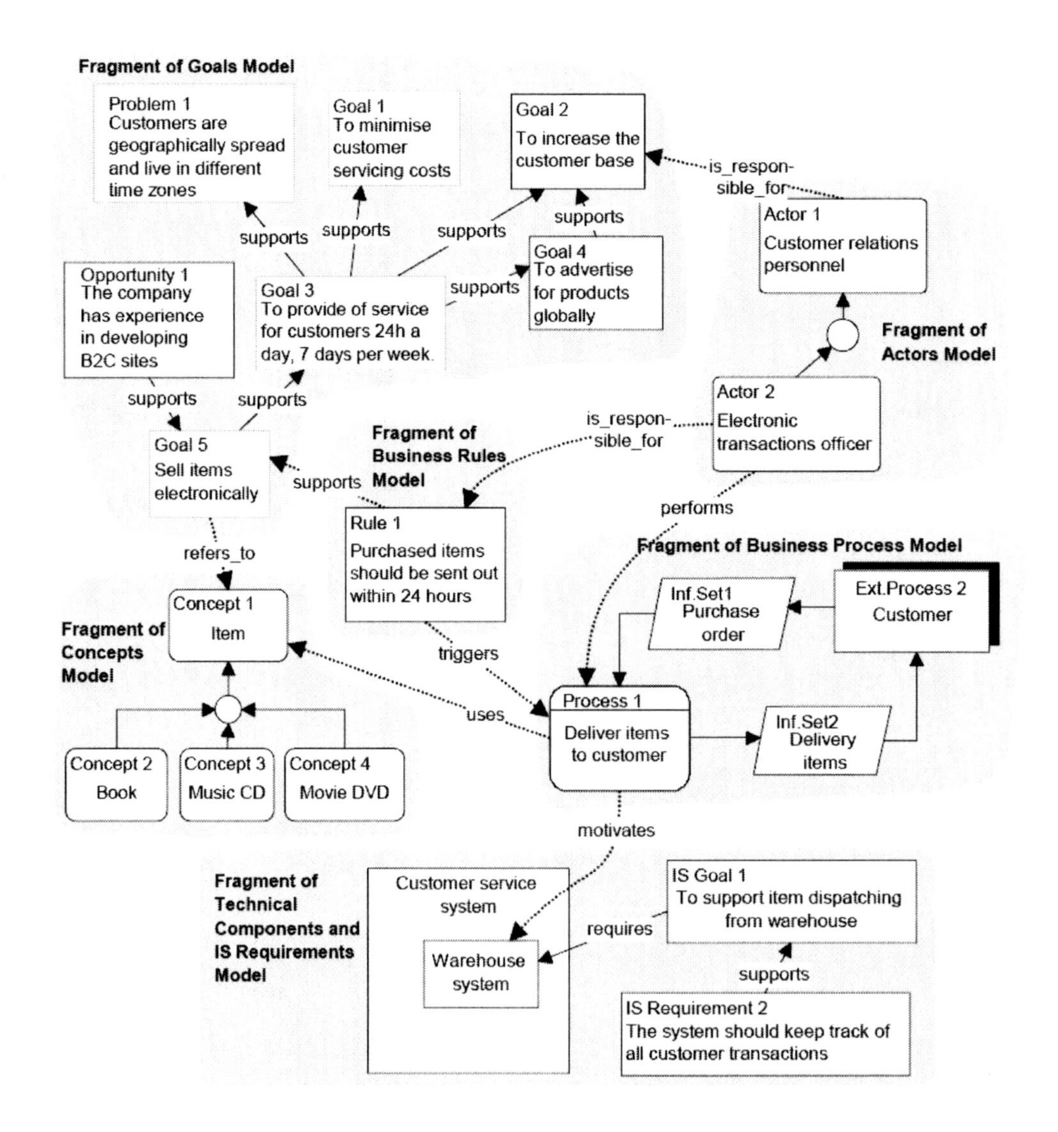

2. A semantically richer notation, suitable when the application requires a higher degree of formality and/or the stakeholders are more experienced with modeling.

E.g., in Figure 2 the Goals Model fragment uses the simple notation where relationships between model components are depicted by an arrow and a verb that is meant to facilitate the understanding of the relationship between the components.

The modeling situation at hand should govern the choice of notation, which will be shown in the subsequent discussion about the method. The full notation of EKD can be found in (Bubenko, Persson & Stirna, 2001).

The EKD Modeling Process

In order to achieve results of high quality, the modeling process is equally important as the modeling language used. There are two aspects of the process, namely the approach to participation and the process to develop the model.

When it comes to gathering domain knowledge to be included in Enterprise Models, there are different approaches. Some of the more common ones are interviews with domain experts, analysis of existing documentation, observation of existing work practices, and facilitated group modeling.

EM practitioners and EKD method developers have advocated a participatory way of working using facilitated group modeling (see e.g. (Bubenko, Persson & Stirna, 2001, F3 Consortium, 1994; Persson, 2001; Nilsson, Tolis & Nellborn, 1999). In facilitated group modeling, participation is *consensus-driven* in the sense that it is the domain stakeholders who "own" the model and govern its contents. In contrast, *consultative* participation means that analysts create models and domain stakeholders are then consulted in order to validate the models. In the participatory approach to modeling, stakeholders meet in modeling sessions, led by a facilitator, to create models collaboratively. In the modeling sessions, models are often documented on large plastic sheets using paper cards. The "plastic wall" (Figure 6) is viewed as the official "minutes" of the session, for which every domain stakeholder in the session is responsible. There are two main arguments for using the participatory approach, if possible (Persson, 2001):

1. The quality of a model is enhanced if the models are created in collaboration between stakeholders, rather than resulting from a consultant's interpretation of interviews with domain experts.
2. The adoption of a participatory approach involves stakeholders in the decision making process, which facilitates the achievement of acceptance and commitment. This is particularly important if the modeling activity is focused on changing some aspect of the domain, such as its visions/strategies, business processes and information system support.

In a modeling session, the EKD process populates and refines the sub-model types used in that particular session gradually and in parallel (Figure 3).

When working with a model type, driving questions are asked in order to keep this parallel modeling process going. E.g. when working with the Business Process Model:

- What does this information set contain? Go to the Concepts Model to define it and return to the Business Process Model.
- Who performs this activity? Go to the Actors and Resources model to define it and return to the Business Process Model.
- Why is this activity performed? Go the Goals Model to define it and return to the Business Process Model.

Figure 3. The EKD modeling process

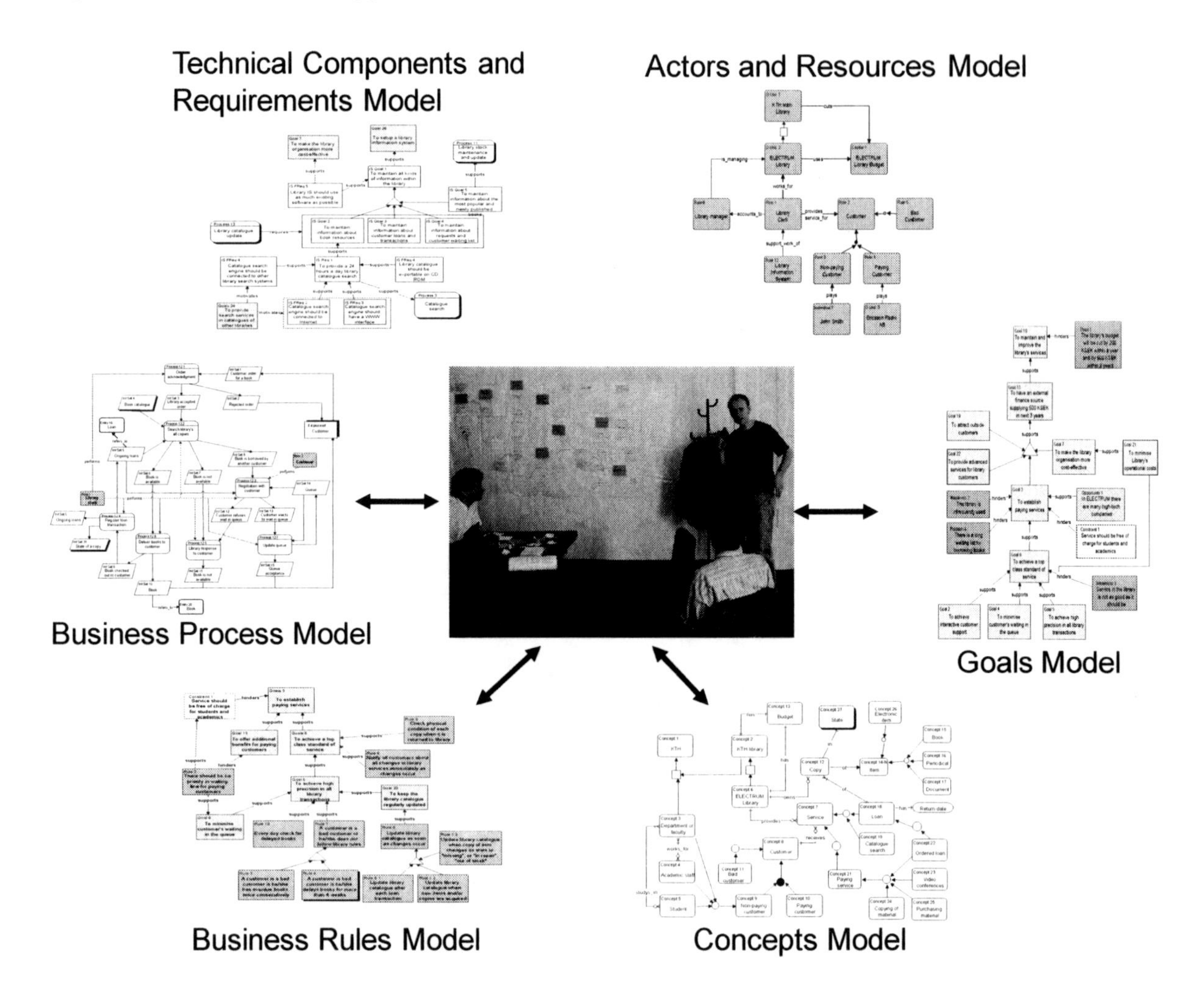

Figure 4 illustrates driving questions and their consequences on establishing inter-model links in the enterprise model. This process has three goals: (1) define the relevant inter-model links, (2) to drive the modeling process forward, and (3) to function as quality assurance of the model. It is argued that shifting between model types while focusing on the same domain problem enhances the participants' understanding of the problem domain and the specific problem at hand.

More about the modeling process used in EKD and about facilitating modeling group sessions can be found in (Nilsson, Tolis & Nellborn, 1999) and (Persson, 2001).

CASES OF EKD APPLICATION

The authors of this paper have applied versions of EKD in a variety of projects and application areas since the beginning of the 1990-ies. In this section we describe the most illustrative applications, which were, for the most part, carried out within international research projects financed by the European Commission. The applications took place in the years 1993-2007. During the applications, the modeling process and its outcome were observed and analyzed. The synthesis of these analyses is reported in this paper together with

Figure 4. Working with inter-model links through driving questions

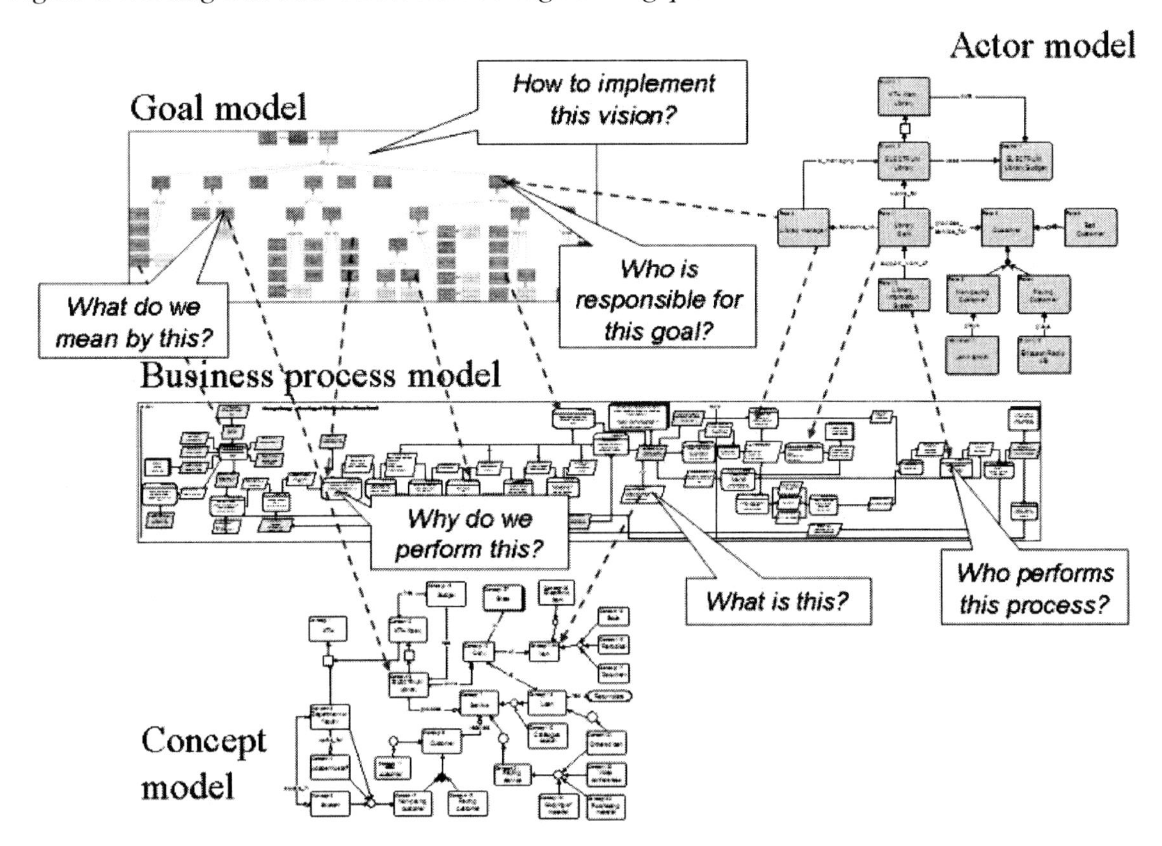

results from Grounded Theory studies focusing on the intentional and situational factors that influence participatory EM and EM tool usage (Persson & Stirna, 2001, Persson, 2001, Stirna, 2001). An overview of the most illustrative applications is given in Table 2.

Apart from these projects, EKD and its earlier versions have also been used in a number of smaller problem solving and organizational design cases at organizations such as e.g. Strömma AB (S), Ericsson (S), Systeam AB (S), Livani District (LV), Riga Technical University (LV), University of Skövde (S) and RRC College (LV).

British Aerospace, UK

During the ESPRIT project F3 – "From Fuzzy to Formal" in 1992-1994 an application of the F³ EM method (F3 Consortium, 1994), a previous version of EKD, was made at British Aerospace, UK, within the Computer Aided Engineering and Technical Computing Department. The department developed software applications adhering to different standards for a number of user departments. The software applications were concerned with aerodynamics, flight tests, stress, etc., and ran on different platforms.

The software packages supporting development (development environments, project management tools, word processors, case tools, configuration management, risk management) were available on certain platforms only. Modeling was focused on defining the requirements (functional and non-functional) for an environment that would enable the developers to use the available tools in an integrated way, to avoid repetition of effort and to increase development quality and productivity.

Table 2. Overview of application cases

Organization	Domain	Period in time	Problems addressed
British Aerospace, UK	Aircraft development and production	1992-1994	Requirements Engineering
Telia AB, Sweden	Telecommunications industry	1996	Requirements validation Project definition
Volvo Cars AB, Sweden	Car manufacturing	1994-1997	Requirements engineering
Vattenfall AB, Sweden	Electrical power industry	1996-1999	Change management Process development Competence management
Riga City Council, Latvia	Public administration	2001-2003	Development of vision and supporting processes for knowledge management
Verbundplan GmbH, Austria	Electrical power industry	2001-2003	Development of vision and supporting processes for knowledge management

The goal of the study in the From Fuzzy to Formal project was to make an assessment of the EM method and to provide input for implementing the above mentioned software development platform. A number of facilitated modeling sessions were carried out with stakeholders from British Aerospace.

The models were documented using Business Modeler, a tool for EM developed by the Swedish Institute for Systems Development 1991-1994.

Telia AB, Sweden

Two application cases were carried out in 1996 at Telia AB, Sweden's largest telecommunications company (Persson & Horn, 1996).

In the first case, the F[3] EM method (F3 Consortium, 1994), a previous version of EKD, was used for reviewing requirements specifications. Natural language specifications from two software development projects were analyzed and their contents modeled using EM. Walkthroughs of the models were carried out with the owners of the specifications, focusing on information that was missing from the specifications and on identified contradictions. This case study has been reported in (Persson, 1997).

In the second case the EM method was used for defining the problem area at the outset of a new development project. Facilitated modeling sessions were carried out with the stakeholders involved. The tool used for documenting the models was Micrografx FlowCharter.

Volvo Cars AB and Volvo Trucks AB, Sweden

An application took place at Volvo during 1994-1997. The focus of the project was the business process of purchasing, for which a support system was to be developed or acquired. The case study had two parts, the Volvo Car Motor sub-project and the Volvo Truck Motor sub-project. The motivation for the division into sub-projects was that they were conducted one after the other and not in parallel. Furthermore the preconditions for the two sub-projects were different.

Both sub-projects involved the full implementation of the F[3] EM methodology. The result of the two sub-projects is a set of models, which describes differences and similarities between the views of the two sub-projects on the purchasing system problem. Based on the models, a requirements specification was developed. The Micrografx FlowCharter tool was used to document the models.

Vattenfall AB, Sweden

An application took place 1996-2000 at Vattenfall AB, Sweden's largest electricity supply company, within two FP4 ESPRIT projects ELKD – Electrical Knowledge Development (No R20818) and ELEKTRA – Electrical Enterprise Knowledge for Transforming Applications (No 22927). The main objective of Vattenfall within these projects was to restructure its human resource management system as well as to close the gap between business planning and competence planning. Detailed information about these activities is available in (Bergenheim et al., 1997; Bergenheim et al., 1998).

The project was structured into five pilot projects each focusing on certain aspects of competence management at Vattenfall. There were different kinds of results of these pilot projects such as problem elicitation and analysis, concept clarification and analysis, as well as designs of business processes. These results were then consolidated in a number or EKD modeling sessions in order to finalize the vision and design of the future human resource management system at Vattenfall.

We used Micrografx FlowCharter tool and ESI Toolset (Singular, 1998) for documenting the modeling product together with the BSCW tool for communicating within the project.

Riga City Council, Latvia

An application case took place at Riga City Council (RCC), Latvia, in 2001 and 2002 within the FP5 IST programme project "Hypermedia and Pattern Based Knowledge Management of Smart Organizations" (no IST-2000-28401). For a complete report of the application see (Kaindl et al., 2001; Mikelsons et al., 2002).

The objective of the application was to develop and deploy a knowledge management (KM) system. Hence, the purpose of EKD modeling was to develop a specification and an adoption plan for a KM system at RCC. The case was structured into a number of sub-projects taking place at various departments of the RCC – Drug Abuse Prevention Center, Traffic Department, School Board, Municipal Police, Department of Environment, and Department of Real Estate. Each of these used EKD to elaborate and resolve specific issues related to KM. Across the cases ca 60 stakeholders, 4 modeling facilitators and 7 modeling technicians were involved in the modeling process. Results of the sub-projects were later integrated in order to develop KM strategy of the RCC.

The Micrografx FlowCharter tool was used for documenting the modeling product and the BSCW tool for communicating within the project.

Verbundplan GmbH, Austria

An application case was carried out in 2001 and 2002 at Verbundplan GmbH, Austria, which is the consulting branch of the largest energy producer in Austria. It took place in parallel with the RCC case within the FP5 IST programme project "Hypermedia and Pattern Based Knowledge Management of Smart Organizations". For a complete report of the application see (Kaindl et al., 2001; Dulle, 2002).

Similarly to the RCC case, the purpose of EKD modeling at Verbundplan was to establish the vision, KM process as well as capture business requirements for a KM system to be used at Verbundplan.

EKD modeling was performed in three sub-projects: repairing damages in hydro power plants, risk management, and project identification. Results from these sub-projects contributed to establishing the corporate KM process and the KM system.

Models were documented by Micrografx FlowCharter tool. The modeling team communicated by using BSCW tool. At the later stages of the project models were part of corporate knowledge repository supported by the Requirements Engineering Through Hypertext (RETH) tool and its web-export functionality (Kaindl, Kramer & Hailing, 2001).

Skaraborgs Sjukhus, Sweden

The application case at Skaraborgs Sjukhus (SKaS) took place during 2003, within the project Efficient Knowledge Management and Learning in Knowledge Intensive Organizations (EKLär), supported by Vinnova, Sweden. SKaS is a cluster of hospitals in Western Sweden working together with primary care centers and municipal home care.

The objective of the project was to develop a KM system and routines to support knowledge sharing among various actors in the healthcare process (e.g. nurses in primary care and municipal home care). More about this project is available in (Stirna, Persson & Aggestam, 2006).

The purpose of EM was to develop a knowledge map that describes the contents in and structure of the knowledge repository. The knowledge map is in the form of an EKD Concepts Model. The specifics of this project was that although the resulting Concepts model could be considered as relatively small, it was refined numerous times and constantly updated throughout the project in order to reflect the stakeholders' understanding of the knowledge domain. This model essentially serves as a "blueprint" for the knowledge repository at SKaS. iGrafx Flowcharter was used to document the model.

OBSERVATIONS FROM EKD IN PRACTICE

Throughout our practical application of the EKD Enterprise Modeling method, we have collected a large number of observations. Some are of a general nature and some are more specifically related to the EKD modeling language, the EKD modeling process, the quality of the modeling process and outcome, as well as tool support for the method.

General Observations

It is our experience that the EKD Enterprise Modeling method has the potential to provide good results in terms of high quality models, improved understanding of the problem among domain stakeholders, improved communication between stakeholders and personal commitment among stakeholders to the modeling result. However, it is fair to say that the achievement of these results is more resource-consuming than may appear at first. This is due to the perceived simplicity of the method. It may appear that anyone can use the method with a minimum of training and practical experience. Our personal experience as well as research (see e.g. (Persson, 2001) has shown that this is a false perception. We recommend that the following preconditions are fulfilled before attempting to use the method in a real life situation that has any real importance to the organization concerned:

- The modeling team must be given a clearly stated mission to pursue.
- Sufficient time and other resources must be allocated to the activity for the internal project group and for other people in the organization so that they can engage in the modeling work.
- The modeling team must be given authority to design or re-design organizational as well as technical processes, procedures, concepts, and rules.
- The team must be well-balanced in terms of relevant knowledge about to problem at hand.
- There is a skilled and experienced modeling facilitator available.

In addition to these conditions, each particular situation should be assessed in order to decide whether or not it is appropriate to use the EKD method. We have found that the characteristics in Table 2 distinguish appropriate from inappropriate situations.

These characteristics are mainly related to the fact that EKD uses a participatory approach to modeling. More on general recommendations for using participatory modeling can be found in (Stirna, Persson & Sandkuhl, 2007; Persson & Stirna, 2001; Nilsson, Tolis & Nellborn, 1999).

The EKD Modeling Language

There are six model types in EKD, each with its particular focus. Depending on the application context or situation, some become more important than others. However, it is fair to say that whichever the application the Goals Model, the Business Process Model, the Concepts Model, and The Actors and Resources Model dominate EKD usage. These sub-models answer to the Why, How, What and Who questions that needs to be asked about an enterprise regardless of situation, be it systems development, process development or strategic development. In fact, throughout our EKD work, we have used the Business Rules Model and the Technical Components Model very little and more research is needed in order to assess their usefulness in practice.

Even though EKD has its own modeling language, it allows replacing the modeling language of a sub-model with a similar modeling language addressing the same modeling problem. It also allows adding other sub-models. Both of these adaptations can be made as long as the inter-model relationships in EKD are kept intact. This feature is useful when the situation in general is appropriate for using EKD, but when there are specific needs with regard to modeling capacity that the method cannot cater for.

There are two alternative notations in EKD, one that is simple and one that is more semantically rich. In the main portion of our work we have used the simple notation in modeling sessions, due to the fact that the stakeholders involved have to a large extent not been experienced modelers. We have found this to be a successful approach. In fact, if the facilitator is experienced there is no need for training the domain stakeholders involved. The facilitator will instead introduce the ideas of modeling and the notation little by little. Some situations, however, require greater formality. We suggest that the introduction of more formality is made "off line" after the modeling sessions using interviews with the relevant stakeholders.

The feature of inter-model links is something that we have not seen in other methods. The method suggests that they are useful for ensuring the quality of models and for driving the modeling

Table 2. Characteristics that distinguish appropriate from inappropriate situations for EKD usage

Appropriate situations	Inappropriate situations
consensus-oriented organizational culture	authoritative organizational culture
management by objectives	management by directives
when agreement among stakeholders needs to be ensured	constant "fire-fighting"
when reliable information is otherwise difficult to obtain (e.g. multiple stakeholder perspectives need to be consolidated, wicked or ill-defined problems)	strong sense of hidden agendas
	trivial problem
	lack of skilful modellers

process forward. We strongly agree that this is the case. The links support reasoning about the model and is a means by which the facilitator can validate the models and the decisions of the modeling group. Also, shifting the focus of the group from one sub-model to another creates a "distraction" that many times can resolve deadlocks in the group.

Modeling Process

The EKD modeling process, if carried out in a professional manner, has the inputs and outputs described in Figure 5.

The features of the EKD process that have been most useful throughout the cases are: 1) the parallel development of sub-models using inter-model links and 2) the participatory approach to modeling.

In general our opinion is that modeling languages need to be combined with a suggested process for populating the models. This is, however, something that is seldom provided, but EKD *does* provide this. The danger of not providing a process is this could imply that modeling in practice is fairly simple. In our experience it takes a long time to become a skilled modeler, and also to become a skilled modeling facilitator. The training of modelers and modeling facilitators should not be taken lightly and an organization that is planning to develop this competency should have a long-term strategy for this purpose. More about competency requirements for modeling can be found in (Persson, 2001).

Another critical issue is the planning of an EM project/activity. In addition to ordinary project preparations we want to stress that the domain experts that are going to be involved in modeling sessions need to be properly prepared in order to achieve the desired results. It is highly desirable that method experts have a strong influence on the composition of the modeling team. Once the modeling participants to be included in the different modeling sessions have been chosen, they need to be prepared for what will happen during the sessions. This is particularly critical in organizations where the employees are not used to modeling in general and particularly to modeling in a group. Before the modeling session each individual modeling participant has to:

- understand the objective of the modeling session,
- agree upon the importance of this objective,

Figure 5. EKD inputs and outputs

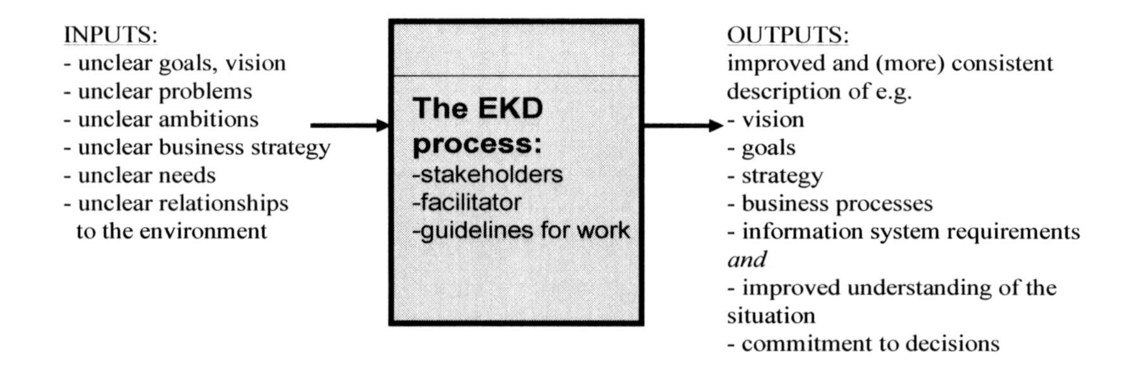

- feel personally capable to contribute to a positive result, and
- be comfortable with the rest of the team (including the facilitator).

More about preparing for EM can be found in (Stirna, Persson & Sandkuhl, 2007).

Tool Support

In order to be useful, the EM process needs to be supported by computerized tools – EM tools. Exactly which types of tools and which software packages are useful is determined by (1) the organization's intentions (e.g. will the models be kept "alive") and (2) situational properties (e.g. the presence of skillful tool operators, availability of resources). More about how to select and introduce EM tools in organizations is available in (Stirna, 2001). In the remainder of this section we will discuss several types of tools and their use to support EM.

Group meeting facilitation tools are used to support the modeling session. Tools such as GroupSystems[1] or DigiMod[b] are becoming more sophisticated and gaining popularity in the business community. However, they still lack specific support for participatory modeling, e.g. guidance the modeling process (c.f. for instance (Nurcan & Rolland, 1999), or "close to reality" graphic resolution. Our recommendation is to use a large plastic sheet and colored notes do document the model during a modeling session (see fig.6). The advantages of this approach is that it can be set-up in almost any room with a sufficiently large and flat wall, and that it allows the modeling participants to work on the model without disturbing each other. On the contrary, if a computerized tool and a large projection screen are used, then the participants have to "queue" in order to enter their contributions via the tool operator. This usually slows down the creative process. In addition, the "plastic wall" is also cheap and does not require technicians to set it up.

After the modeling session the contents of the plastic may be captured by a digital camera. If the model needs to be preserved (e.g. included in reports, posted on the intranet), it needs to be documented in a computerized modeling tool (see fig. 7).

This category of tools includes simple drawing tools as well as a more advanced model develop-

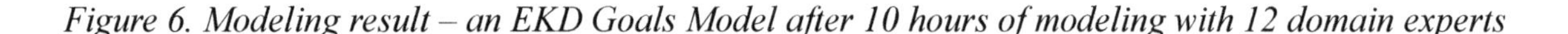

Figure 6. Modeling result – an EKD Goals Model after 10 hours of modeling with 12 domain experts

ment and management tools. In "stand-alone" projects only the drawing support might be needed and, if so, simple drawing tools such as Microsoft Visio and iGrafx FlowCharter[3] have proven to be useful and cost-effective (c.f. for instance Persson & Stirna, 2001; Stirna, 2001). In other cases, e.g. when enterprise models need to be communicated to large audiences over intranet or linked with existing information systems, more advanced EM tools are to be used. Common examples in this category of tools are Aris[4] (IDS Scheer) and Metis[5] (Troux Technologies).

Computerized tools may also be used during the modeling workshops, if their objective is to "polish" the existing models rather than build new ones. In this case the modeling participants make relatively few changes which can be communicated to a tool operator who then introduced them in the model.

Apart from the modeling tools an EM project usually needs group communication and collaboration tools. For this purpose we have been successfully using Basic Support for Collaborative Work (BSCW) tool[6].

Concerning the business requirements for EM tools, the most common requirements address integration of EM tools with MS Office software, model visualization and presentation requirements (often in web-format) as well as reporting and querying requirements. Apart from these we have also observed a growing need to connect models to information systems thus making the models executable. Extended presentation of requirements for EM tools is available in (Stirna, 2001).

Quality of Enterprise Models

Persson (2001) states that the main criteria for successful application of EKD are (1) the quality of the produced models is high, (2) the result is useful and actually used after the modeling activity is completed, and (3) the involved stakeholders are satisfied with the process and the result.

High quality in models means that they make sense as a whole and that it is possible to implement them as a solution to some problem. Successful EM is when the result of modeling, e.g., a new strategy or a new process, is effectively implemented in the organization.

The EM process produces two results that are potentially useful:

Figure 7. The model depicted in Figure 6, now documented in Microsoft Visio

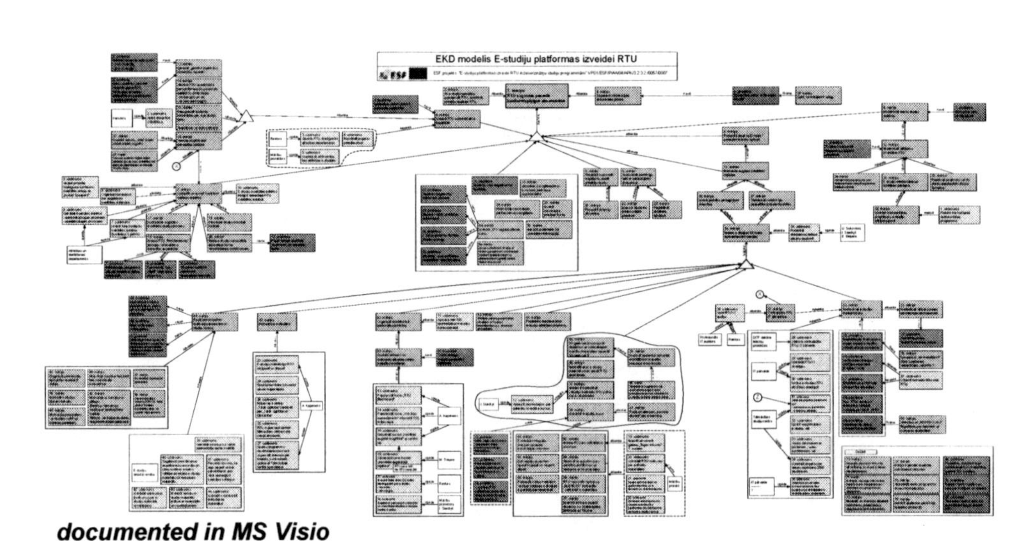

- the produced models, which are used as input to further development or implementation activities, and
- the changed thinking and the improved knowledge of the participants.

Despite substantial attention from both researchers and practitioners to research issues of business process and data model quality (c.f. for instance, Davies et al., 2004, Moody & Shanks, 2003, Maes & Poels, 2006, Mendling, Reijers & Cardoso, 2007, Krogstie, Sindre & Jørgensen, 2006, Rosemann, 2006) many EM projects produce models of really bad quality, thus making them essentially useless. This is best illustrated by this interview quote:

"I claim that only 30 % of what is produced today is at all acceptable as a platform to stand on [for further development work]... Most of what is done today, especially with regard to business processes, is garbage." (Persson, 2001, p. 197).

Perhaps one of the reasons for this situation is that many EM projects do not really know what kind of model quality they should strive towards. This then leads to either insufficient model quality, which undermines the usefulness of the model, or too high quality which is not necessary to solve the problem at hand. The required quality level of models is usually determined by the project goals and the specific goals that apply to each modeling session. E.g. some modeling sessions might be geared towards creatively gathering ideas in which case model quality is of lesser importance. In other cases when models are intended to be used as part of, for instance, a requirements specification, the models have to adhere to certain quality rules.

Larsson & Segerberg (2004) provide an initial investigation of whether the model quality criteria of Moody & Shanks (2003) are applicable to Enterprise Models and concludes that the factors completeness, correctness, flexibility, integration, simplicity, understandability, and usability are applicable. This work should be further extended towards specific modeling guidelines for conducting EM and for documenting the modeling result. An example of one such guideline addressing understandability and applicability is to strive towards SMART goals in the Goal Model, meaning that every goal should be specific (S), measurable (M), accepted (A), realistic(R), and time framed (T). This guideline would contribute to increasing the understandability and usability of the model. A guideline for improving simplicity would be to improve the graphical presentation of large and interconnected models, i.e. "spaghetti models", rearranging them into sub-models and by eliminating unnecessary relationships. In order for this guideline to be efficient, tool support that automates some of the activities, by, example, wizards for reviewing and querying relationships would be required.

Our vision is that in order for an EM method to be useful in practice it needs to incorporate model quality assurance guidelines in terms of (1) improving the quality of the modeling result and (2) facilitating the modeling process in such a way that high quality models are achieved. A problem in practice is the lack of appropriate tool support that guides the user to achieve high quality models.

CURRENT AND FUTURE DEVELOPMENT OF EKD

For decades the IS engineering community has used modeling for various purposes. We would like to think that the field is reaching saturation in terms of method quality and experience of applying them. In reality, it seems that we have only found a set of reasonably useful modeling languages and notations. Even here we do not have one solution to all situations – the usefulness of different modeling languages largely depends on the objective and purpose of modeling. Most of our knowledge is about modeling formalisms and we know considerably less about the modeling process. The modeling methods and tools that we use mainly support us with the presentation capabilities. This is because the modeling process to a large extent is dependent on the organization's intentions and the application context.

One way to address the challenge of modeling in different situational contexts is to develop modeling methods that attempt to prescribe all appropriate modeling situations. In this case the method developer tries to predict and write as many modeling instructions as possible. The other path is to define a set of general practices and principles of modeling and then empower a skilful development team to make the right decisions based on what the organization and the situation requires. The agile development approaches, such as, for instance, Agile Modeling (Ambler, 2002) are taking this path.

EKD has the potential to support agile software development because EKD's participative way of working contributes to the agile way of working by providing additional modeling guidelines. Initial attempts of combining EKD with Agile Modeling are available in Stirna & Kirikova, 2008), but more investigation and experimentation is needed.

Another emerging trend in EM is to use Enterprise Models as tools for work and corporate governance – they are linked to components of organization's information, and are essentially controlling those components. In this case we are able to talk about Active Knowledge Models (Krogstie & Jørgensen, 2004; Jørgensen, Karlsen & Lillehagen, 2007) and about model-configured enterprise architectures. As a contribution to supporting organizational governance we are investigating the possibilities of supporting the development and implementation of Balanced Scorecards with EKD (Niehaves & Stirna, 2006). In this context guidance for quality assurance and improvement of Enterprise Models as well as adequate modeling tool support becomes particularly important.

Despite the advancements in the areas of modeling methods and tools, their impact in practice is largely dependant on how an EM approach is adopted and institutionalized. In practice EM usage often follows the phases of initial interest, pilot project, and subsequent institutionalization. The most challenging is the final phase because at this stage the organization should presumable have enough competence to perform modeling without external support. In cases when this is not so, modeling struggles to make positive impact and is gradually forgotten. Hence, the method development community should pay more attention to supporting the process of adopting the modeling way of working in order to make modeling "normal" business practice for collaborative problem solving. In order for the institutionalization efforts to become efficient we need to develop a more efficient support for training the modeling personnel, such as, modelers, facilitators and tool experts. Training approaches should also acknowledge the challenge of transferring expert modeling knowledge between people. E.g., it is not enough to attend a course to become a skillful modeling facilitator. Extensive personal practical experience and mentoring by an expert in the field will most likely be needed as well.

CONCLUDING REMARKS

This paper has presented observations collected during more than 10 years of applying the EKD EM method in a variety of settings. These observations will influence further development of the method and its supporting tools.

As a conclusion to this paper we would like to point out that organizations trying to adopt EKD or similar approaches tend to perceive them as a "magic tool" that relieves its users from thinking and acting or an approach that always leads to a software system. Instead they need to understand the following about EKD:

- It is an integrated collection of methods, techniques, and tools that supports the process of analyzing, planning, designing, and changing the business.
- It is an approach that supports thinking, reasoning, and learning about the business.
- It is an approach that leads to more complete and consistent business or information system designs.

REFERENCES

Ambler, S. (2002). *Agile modeling: Effective practices for extreme programming and the unified process* (1st ed.). John Whiley & Sons Inc.

Bajec, M., & Krisper, M. (2005). A methodology and tool support for managing business rules in organisations. *Information Systems, 30*(6), pp423-443.

Bergenheim, A., Wedin, K., Waltré, M., Bubenko Jr., J. A., Brash, D., & Stirna, J. (1997). *BALDER – Initial Requirements Model for Vattenfall.* Stockholm, Vattenfall AB

Bergenheim, A., Persson, A., Brash, D., Bubenko, J. A. J., Burman, P., Nellborn, C., & Stirna, J. (1998). *CAROLUS - System Design Specification for Vattenfall*, Vattenfall AB, Råcksta, Sweden

Bubenko, J. A. j., Persson, A. & Stirna, J. (2001). *User Guide of the Knowledge Management Approach Using Enterprise Knowledge Patterns, deliverable D3, IST Programme project Hypermedia and Pattern Based Knowl-edge Management for Smart Organisations*, project no. IST-2000-28401, Royal Institute of Technology, Sweden.

Bubenko Jr., J. A. (1993). Extending the Scope of Information Modelling. *Fourth International Workshop on the Deductive Approach to Information Systems and Databases*, Lloret, Costa Brava (Catalonia), Sept. 20-22, 1993. Department de Llenguatges i Sistemes Informatics, Universitat Politecnica de Catalunya, Report de Recerca LSI/93-25, Barcelona.

Castro, J., Kolp, M., Mylopoulos, J., & Tropos, A. (2001). A Requirements-Driven Software Devel-opment Meth-odology. In *Proceedings of the 3rd Conference on Advanced Information Systems Engineering* (CAiSE 2001), 108-123, Springer LNCS 2068, Interlaken, Switzerland.

Davies, I., Green P., Rosemann, M., & Gallo S., (2004) Conceptual Modelling - What and Why in Current Practice. In P. Atzeni et al. (Eds), *Proceedings of ER 2004*, Springer, LNCS 3288.

Dobson, J., Blyth, J., & Strens, R. (1994). Organisational Requirements Definition for Information Technology. In *Proceedings of the International Conference on Requirements Engineering 1994*, Denver/CO.

Dulle, H. (2002). *Trial Application in Verbundplan, deliverable D5, IST Programme project HyperKnowledge – Hypermedia and Pattern Based Knowledge Management for Smart Organisations*, project no. IST-2000-28401, Verbundplan, Austria

F3-Consortium (1994). *F3 Reference Manual*, ESPRIT III Project 6612, SISU, Stockholm.

Fox, M. S., Chionglo, J. F., & Fadel, F. G. (1993). A common-sense model of the enterprise. In *Proceedings of the 2nd Industrial Engineering Research Conference*, Institute for Industrial Engineers, Norcross/GA.

Glaser, B. G., & Strauss, A. L. (1967). *The Discovery of Grounded Theory: Strategies for Qualitative Research*, Weidenfeld and Nicolson, London.

Gustas, R., Bubenko Jr., J. A., & Wangler, B. (1995). Goal Driven Enterprise Modelling: Bridging Pragmatic and Semantic Descriptions of Information Systems. *5th European - Japanese Seminar on Information Modelling and Knowledge Bases*, Sapphoro, May 30-June 3, 1995.

Johannesson P., Boman, M., Bubenko, J., & Wangler, B. (1997). *Conceptual Modelling*, 280 pages, Prentice Hall International Series in Computer Science, Series editor C.A.R. Hoare, Prentice Hall.

Jørgensen, H. D, Karlsen D., & Lillehagen F. (2007). Product Based Interoperability – Approaches and Rerequirements. In Pawlak et al. (Eds.), *Proceedings of CCE'07*, Gesellschaft für Informatik, Bonn, ISBN 978-3-88579-214-7

Kardasis, P., Loucopoulos, P., Scott, B., Filippidou, D., Clarke, R., Wangler, B., & Xini, G. (1998). *The use of Business Knowledge Modelling for Knowledge Discovery in the Banking Sector, IMACS-CSC'98*, Athens, Greece, October, 1998.

Kaindl, H., Kramer, S., & Hailing, M. (2001). An interactive guide through a defined modeling process. In *People and Computers XV, Joint Proc. of HCI 2001 and IHM 2001*, Lille, France, 107-124

Kaindl H., Hatzenbichler G., Kapenieks A., Persson A., Stirna J., & Strutz G. (2001). *User Needs for Knowledge Management, deliverable D1, IST Programme project HyperKnowledge - Hypermedia and Pattern Based Knowledge Management for Smart Organisations*, project no. IST-2000-28401, Siemens AG Österreich, Austria

Krogstie, J., Lillehagen, F., Karlsen, D., Ohren, O., Strømseng, K., & Thue Lie, F. (2000). *Extended Enterprise Methodology*. Deliverable 2 in the EXTERNAL project, available at http://research.dnv.com/external/deliverables.html.

Krogstie, J., & Jørgensen, H. D. (2004). Interactive Models for Supporting Networked Organizations. *Proceedings of CAiSE'2004, Springer LNCS 3084*, ISBN 3-540-22151-4

Krogstie, J., Sindre, G., & Jørgensen, H. (2006). Process Models Representing Knowledge for Action: a Revised Quality Framework. *European Journal of Information Systems, 15*(1), 91–102

Larsson, L., & Segerberg R., (2004). *An Approach for Quality Assurance in Enterprise Modelling.* MSc thesis. Department of Computer and Systems Sciences, Stockholm University, no 04-22

Loucopoulos, P., Kavakli, V., Prekas, N., Rolland, C., Grosz, G., & Nurcan, S. (1997). Using *the EKD Approach: The Modelling Component*, UMIST, Manchester, UK.

Maes, A., & Poels, G. (2006). Evaluating Quality of Conceptual Models Based on User Perceptions. In D. W. Embley, A. Olivé, & S. Ram (Eds.), *ER 2006, LNCS 4215*, 54 – 67, Springer

Mendling, J., Reijers H. A., & Cardoso J. (2007). What Makes Process Models Understandable? In G. Alonso, P. Dadam, & M. Rosemann, (Eds.), *International Conference on Business Process Management (BPM 2007), LNCS 4714*, 48–63. Springer.

Mikelsons, J., Stirna, J., Kalnins, J. R., Kapenieks, A., Kazakovs, M., Vanaga, I., Sinka, A., Persson, A., & Kaindl, H. (2002). *Trial Application in the Riga City Council, deliverable D6, IST Programme project Hypermedia and Pattern Based Knowledge Management for Smart Organisations*, project no. IST-2000-28401. Riga, Latvia.

Moody, D. L., & Shanks, G. (2003). Improving the quality of data models: Empirical validation of a quality management framework. *Information Systems (IS) 28*(6), 619-650, Elsevier

Nilsson, A. G., Tolis, C., & Nellborn, C. (Eds.) (1999). *Perspectives on Business Modelling: Understanding and Changing Organisations.* Springer-Verlag

Niehaves, B., & Stirna, J. (2006). Participative Enterprise Modelling for Balanced Scorecard Implementation. In *14th European Conference on Information Systems (ECIS 2006)*, Gothberg, Sweden.

Nurcan, S., & Rolland, C. (1999). Using EKD-CMM electronic guide book for managing change in organizations. In *Proceedings of the 9th European-Japanese Conference on Information Modelling and Knowledge Bases*, Iwate, Japan.

Persson, A., & Horn, L. (1996). *Utvärdering av F3 som verktyg för framtagande av tjänste- och produktkrav inom Telia.* Telia Engineering AB, Farsta, Sweden, Doc no 15/0363-FCPA 1091097.

Persson, A. (1997). Using the F^3 Enterprise Model for Specification of Requirements – an Initial Experience Report. *Proceedings of the CAiSE '97 International Workshop on Evaluation of Modeling Methods in Systems Analysis and Design (EMMSAD)*, June 16-17, Barcelona, Spain.

Persson, A., & Stirna, J. (2001). Why Enterprise Modelling? -- An Explorative Study Into Current Practice, CAiSE'01. *Conference on Advanced Information System Engineering*, Springer, ISBN 3-540-42215-3

Persson, A. (2001). *Enterprise Modelling in Practice: Situational Factors and their Influence on Adopting a Participative Approach*, PhD thesis, Dept. of Computer and Systems Sciences, Stockholm University, No 01-020, ISSN 1101-8526.

Rosemann, M. (2006). Potential Pitfalls of Process Modeling: Part A. *Business Process Management Journal, 12(2)*, 249–254

Rosemann, M. (2006). Potential Pitfalls of Process Modeling: Part B. *Business Process Management Journal, 12*(3):377–384

Singular Software. (1998). *"Ikarus": Design of the ESI Toolset, deliverable,* ESPRIT project No 22927 ELEKTRA, Singular Software, Greece

Stirna, J. (2001). *The Influence of Intentional and Situational Factors on EM Tool Acquisition in Organisations*, Ph.D. Thesis, Department of Computer and Systems Sciences, Royal Institute of Technology and Stockholm University, Stockholm, Sweden.

Stirna, J., Persson, A., & Aggestam, L. (2006). Building Knowledge Repositories with Enterprise Modelling and Patterns - from Theory to Practice. In *proceedings of the 14th European Conference on Information Systems (ECIS)*, Gothenburg, Sweden, June 2006.

Stirna, J., Persson, A., & Sandkuhl, K. (2007). Participative Enterprise Modeling: Experiences and Recommendations. *Proceedings of the 19th International Conference on Advanced Information Systems Engineering (CAiSE 2007)*, Trondheim, June 2007.

Stirna, J., & Kirikova M. (2008). How to Support Agile Development Projects with Enterprise Modeling. In P. Johannesson & E. Söderström (Eds.), *Information Systems Engineering - from Data Analysis to Process Networks.* IGI Publishing, ISBN: 978-1-59904-567-2

Wangler, B., Persson, A., & Söderström, E. (2001). Enterprise Modeling for B2B integration. In *International Conference on Advances in Infrastructure for Electronic Business, Science, and Education on the Internet*, August 6-12, L'Aquila, Italy (CD-ROM proceedings)

Wangler, B., & Persson, A. (2002). Capturing Collective Intentionality in Software Development. In H. Fujita & P. Johannesson, (Eds.), *New Trends in Software Methodologies, Tools and Techniques.* Amsterdam, Netherlands: IOS Press (pp. 262-270).

Wangler, B., Persson, A., Johannesson, P., & Ekenberg, L. (2003). Bridging High-level Enterprise Models to Implemenation-Oriented Models. in H. Fujita & P. Johannesson (Eds.), *New Trends in Software Methodologies, Tools and Techniques.* Amsterdam, Netherlands: IOS Press.

Willars, H. (1988). *Handbok i ABC-metoden.* Plandata Strategi.

Willars, H. et al (1993). TRIAD Modelleringshandboken N 10:1-6 (in Swedish), SISU, Electrum 212, 164 40 Kista, Sweden.

Yu, E. S. K., & Mylopoulos, J. (1994). From E-R to "A-R" - Modelling Strategic Actor Relationships for Business Process Reengineering. In *Proceedings of the 13th International Conference on the Entity-Relationship Approach*, Manchester, England

ENDNOTES

1. See http://www.groupsystems.com/
2. See http://www.ipsi.fraunhofer.de/concert/index_en.shtml?projects/digimod
3. See http://www.igrafx.com/products/flowCharter/
4. See http://www.ids-scheer.com
5. See http://www.troux.com
6. See http://bscw.fit.fraunhofer.de/

Chapter V
Integrated Requirement and Solution Modeling:
An Approach Based on Enterprise Models

Anders Carstensen
Jönköping University, Sweden

Lennart Holmberg
Kongsberg Automotive, Sweden

Kurt Sandkuhl
Jönköping University, Sweden

Janis Stirna
Jönköping University, Sweden

ABSTRACT

This chapter discusses how an Enterprise Modeling approach, namely C3S3P[1], has been applied in an automotive supplier company. The chapter concentrates on the phases of the C3S3P development process such as Concept Study, Scaffolding, Scoping, and Requirements Modeling. The authors have also presented the concept of task pattern which has been used for capturing, documenting and sharing best practices concerning business processes in an organization. Within this application context they have analyzed their experiences concerning stakeholder participation and task pattern development. The authors have also described how they have derived four different categories of requirements from scenario descriptions for the task patterns and from modeling of the task patterns.

INTRODUCTION

Every information system (IS) engineering project needs to have a clear vision and purpose, and to know what kind of properties the developed product should possess. This usually is the focus of requirements engineering (RE) activities that are being performed in early stages of an IS development project. The main objective of this process is not only to put forward a number of features that the product should have, but also to connect them to the business needs of the customer organization in such a way that each product requirement is traceable to some business objective of the organization. This explicit connection between IS requirements and business goals helps avoiding unnecessary rework and increases the business value of the product. Moreover, in the process of eliciting and linking the business needs and IS requirements, the development team and the stakeholders usually have to tackle a number of "wicked" or "ill-structured" problems (Rittel & Webber, 1984) typically occurring in organizations.

Enterprise Modeling (EM) seeks to solve organizational design problems in, for instance, business process reengineering, strategy planning, enterprise integration, and information systems development (Bubenko & Kirikova, 1999). The EM process typically leads to an integrated and negotiated model describing different aspects (e.g. business goals, concepts, processes) of an enterprise. A number of EM approaches (c.f., for instance (Bubenko, Persson, & Stirna, 2001; Castro, Kolp, & Mylopoulos, 2001; F3-Consortium, 1994; Loucopoulos et al., 1997; van Lamsweerde, 2001; Yu & Mylopoulos, 1994)) have been suggested. To document the models and to support the EM processes computerized tools are used. They differ in complexity from simple, yet cost-effective, drawing tools such as Microsoft Visio and iGrafx FlowCharter to more advanced tools such as Aris (IDS Scheer) and METIS (Troux Technologies).

The participative approach to EM contributes to the quality of the requirement specification as well as increases the acceptance of decisions in the organizations, and is thus recommended by several EM approaches (c.f. for instance (Bubenko & Kirikova, 1999; Persson & Stirna, 2001; Stirna, Persson, & Sandkuhl, 2007)). The participative approach suggests that the modeling group consists of stakeholders and domain experts who build enterprise models following guidance given by a modeling facilitator. An alternative expert driven approach suggests interviews and questionnaires for fact gathering and then creation of an enterprise model in an analytical way.

EM and especially the participative way of working is highly useful in situations when the development team needs to capture and consolidate the user needs and then to propose an innovative solution to them. In such situations one of the main challenges is to establish traceability between the user needs, such as, for instance, goals, processes, and requirements, and the designed solutions to these needs in terms of tasks, methods, and tools. Furthermore, in such situation the user needs can be met in many different ways, which requires early stakeholder validation of the envisioned solution. A common example of such an application context is an innovative research and development (R&D) project aiming to develop new methods and tools. In this chapter we present an EU supported R&D project MAPPER (Model-based Adaptive Product and Process Engineering) that has successfully overcome these challenges. More specifically, the objective of this chapter is *to report how a specific EM approach, namely C3S3P*[2]*, was applied in an automotive supplier company in order to elicit requirements for a reconfigurable IS to support collaborative engineering and flexible manufacturing processes*. More specifically, we will address the following questions:

- How were the stages of C3S3P followed to develop requirements and what where the experiences and lessons learned in each of them?

- What kinds of model elements were used in the project and how were they supported by the METIS[3] tool?

The remainder of the chapter is organized as follows. Section two presents the application case at Kongsberg Automotive. Section three discusses EM and requirements modeling with the C3S3P approach and introduces the concept of task pattern. Section four presents the result of EM activities at a company – Kongsberg Automotive AB. This section discusses the requirements model and the proposed solutions to the requirements. Section five summarizes our experiences concerning stakeholder participation and the C3S3P approach. Conclusions and issues for future work are discussed in section six.

BACKGROUND TO EM APPLICATION AT KONGSBERG AUTOMOTIVE

The application case is taken from the EU FP6 project MAPPER, which aims at enabling fast and flexible manufacturing by providing methodology, infrastructure and reusable services for participative engineering in networked manufacturing enterprises. A core element of the MAPPER approach is reconfigurable enterprise models including an executable environment for these models, which are considered active knowledge models. See, for instance, (Krogstie & Jørgensen, 2004; Lillehagen & Krogstie, 2002; Stirna et al., 2007) for more information about developing active knowledge models.

The application case focuses on distributed product development with multi project lifecycles in a networked organization from automotive supplier industry – Kongsberg Automotive (KA). Within this project the main partner at KA is Business Area Seat Comfort. Its main products are seat comfort components (seat heater, seat ventilation, lumber support and head restraint), gear shifts and commercial vehicle components.

The case is focused on the department Advanced Engineering within the Business Area Seat Comfort. Development of products includes identification of system requirements based on customer requirements, functional specification, development of logical and technical architecture, co-design of electrical and mechanical components, integration testing, and production planning including production logistics, floor planning and product line planning. This process is geographically distributed involving engineers and specialists at several KA locations and suppliers from the region. A high percentage of seat comfort components are product families, i.e. various versions of the components exist and are maintained and further developed for different product models and different customers. KA needs fast and flexible product development in concurrently performed forward-development processes. Hence, general requirements regarding infrastructure and methodology are:

- To support geographical distribution and flexible integration of changing partners;
- To enable flexible engineering processes reflecting the dynamics of changing customer requirements and ad-hoc process changes, and at the same time well-defined processes for coordinated product development;
- To coordinate a large number of parallel product development activities competing for the same resources;
- To allow richness of variants while supporting product reuse and generalization.

The purpose of the requirements modeling activity is to further specify and elicit these general requirements. Regarding the service and infrastructure, the requirements will address the collaboration infrastructure, which has to support networked manufacturing between different locations of KA and KA's suppliers. Furthermore, services for performance and coordination of work, management of projects, tasks, and workplaces are subject of requirements analysis. Regarding the methodology, KA has to express and qualify the desired methodology support for sustainable collaboration, multi-project portfolio management, organizational learning, and EM.

ENTERPRISE MODEL BASED REQUIREMENTS ENGINEERING

The approach for requirements elicitation and specification used in the application case (see section two "Background to EM Application at Kongsberg Automotive") was not based on the traditional way of producing documents with text and illustrations. The requirements were captured in a (partial) enterprise model, which at the same time, to a large extent, can be used as executable solution model during the later phases of the project. This way of heavily using networked manufacturing EM reflects the philosophy of *model-adapted* product and process engineering: the requirements model will be adapted to support real-world projects in the application case by configuring the solution for those projects.

The rest of this section presents the requirement model development process, the concept of task patterns for capturing organizational knowledge in a reusable way and the elements of the requirements model. The actual requirements model including illustrative examples of the model is subject of section four "Requirements Model".

Requirement Model Development Process

Development of the requirements model included a number of phases and iterations. The development process of the requirements model can to a large extent be considered as an enterprise knowledge modeling process, which was inspired by the POPS* (Lillehagen, 2003) and the C3S3P approaches (Krogstie & Jørgensen, 2004; Stirna et al., 2007). The POPS* approach recommends considering at least four interrelated perspectives when developing enterprise knowledge models: Processes in the enterprise, Organization structure, Products of the enterprise, and Systems supporting the processes. Depending on the purpose and application context of the enterprise knowledge modeling, additional perspectives might be necessary, which is indicated by the asterisk in "POPS*".

C3S3P distinguishes between seven stages called Concept-study, Scaffolding, Scoping, Solutions-modeling, Platform integration, Piloting in real-world projects and Performance monitoring and management. The requirements modeling presented in this chapter relates to the first stages of C3S3P:

- The concept study and scaffolding phase aiming at understanding of visual knowledge modeling and at creating shared knowledge, views and meanings for the application subject;
- The scoping phase focusing on creation of executable knowledge model pilots supporting the application case;
- The requirement modeling phase aiming to identify and consolidate requirements for platform, methodology, approach and solution.

The above phases will be discussed in the subsections that follow.

POPS* and C3S3P both are independent from the modeling language used. In our case we used the METIS enterprise architecture framework (MEAF), which is based on the GEM (Generic Enterprise Modeling) language.

Concept Study and Scaffolding Phase

At the beginning of this phase, a number of preparation steps had to be taken, which were contributing to a joint understanding of visual knowledge modeling – a prerequisite for the later phases. After having identified all team members, they were offered an introduction and basic training in METIS. We started with scaffolding workshops with all team members. The following roles were involved:

- Manager – the owner of the application case who is responsible for establishing it at KA, assigning the right personnel resources, arranging meetings, etc.;
- Planner – responsible for the way of working and for establishing a consensus between all partners, coordinating the different tasks, moderating the meetings, etc.;
- Modeling expert – provides expert knowledge about the method and the tool;
- Facilitator – experienced in using the selected modeling process and tool and facilitates model construction and capturing of knowledge in the models;
- Coach – supports the modeling process by coaching the modelers;
- Modeler – develops the enterprise models in the tool during the modeling process;
- Domain expert – provides knowledge about the problem domain.

After the preparations, an iterative process started, which at the end resulted in three major versions of the scaffolding model. The first modeling step in the first iteration was to clearly define the purpose of the model to be developed. Within the scaffolding phase, the purpose was to model the current situation in the problem area as seen by the different stakeholders from KA, such as R&D manager, engineers with different specializations, purchaser, customer responsible, etc. Starting from the POPS* approach, relevant perspectives were identified. During this step he initial POPS* perspectives Process, Organization, Product and Systems were supplemented with other perspectives like Objectives, Technical Approaches or Skills. Definition of additional perspectives was not performed just once, but had to be repeated in every iteration. Between the workshops, the facilitator and the modeling experts checked the jointly developed models and added details such as textual descriptions.

Scoping Phase

Focus in the scoping phase was on developing initial versions of solution models that specified the intended future way of working in the application case, i.e. the future Process of Innovation (POI) in Advanced Engineering of KA's business area seat comfort. The resulting solution models should be executable in the MAPPER infrastructure, which required all modeling perspectives to be defined on such a level of detail that they contained all model elements required for execution. In comparison to the scaffolding model, the solution model had to fulfil higher demands with respect to completeness and consistency, e.g. the complete process flow had to be modeled and not just the essential tasks illustrating the way of working.

Due to the needed level of completeness and detail, the way of modeling had to be revised. Instead of only using joint modeling workshops, we first created textual scenario descriptions for all relevant task patterns, which included information about the intended way of working, involved roles, and tools or documents used. They also contained statements explicitly identifying requirements. Based on each scenario description, the facilitator and modeling coach developed a model, which was then refined together with the stakeholders from KA.

The work during the scoping phase was structured by dividing POI into nine main task patterns and grouping these task patterns into three pilot installations.

Requirements Modeling Phase

The requirements modeling and scoping ran partly in parallel. The purpose of the requirements modeling phase was to make requirements to the MAPPER infrastructure and methodology explicit. This phase started with identification of the types of requirements to be captured and with agreeing on how the requirements should be represented in the models. We decided to distinguish between platform, methodology, approach, and solution requirements and to represent them in the model. More details about requirement types and representation are in the subsection "Task Patterns" below.

Requirements identification was performed in joint workshops with stakeholders from KA and in two ways: (1) analyze the textual scenario descriptions, extract the requirements described there, and add them to the model; (2) review the solution model for each task pattern and check all tasks on all refinement levels for presence of requirements. The resulting requirements were presented to all team members in joint modeling workshops.

Like in scoping, the sequence and structure of requirements modeling was governed by the division into task patterns and pilots.

Task Patterns

During the requirement model process, the intention was not only to divide the process under consideration into sub-tasks, but to produce adaptable models capturing best practices for reoccurring tasks in networked enterprises. The term *task pattern* was selected to indicate that these adaptable models are not only valid and applicable in a specific organization unit, but that they are also relevant for other departments and processes of the company. Some such task patterns might even be transferable to other enterprises in the same application domain. These best practices include the POPS* perspectives introduced in the above subsection "Requirement Model Development Process".

The concept of *task pattern* is defined as *a self-contained model template with well-defined connectors to application environments capturing knowledge about best practices for a clearly defined task* (Sandkuhl, Smirnov, & Shilov, 2007), where

- "Self-contained" means that a task pattern includes all perspectives, model elements and relationships between the modeled elements required for capturing the knowledge reflecting a best practice;
- "Model template" means that the task pattern has to be expressed in a defined modeling language and that no instances are contained in the task patterns;
- "Connectors" are model elements that facilitate the adaptation of the task pattern to target application environments;
- "Application environments" currently are limited to enterprise models.

Examples for task patterns are "establish material specification", "develop new test method", "prototype build" or "establish product specification" (Johnsen et al., 2007).

Figure 1 presents "develop new test method" as an example for a task pattern. The figure focuses on the process perspective, i.e. the tasks "check real need", "prepare draft", "evaluation of test method concepts" and "release new test method" are shown. On the lower part of the figure, the infrastructure resources are presented (right hand side) and the human resources involved in the process are included. All four tasks include a number of refinement levels, which are not shown in the figure. These refinements often define the sequence of the tasks, but in cases where the sequence is not important or not pre-definable just list the tasks to be performed. On the upper part of the figure, documents supporting the task pattern are included. Relationships between processes,

Figure 1. Example task pattern "develop new test method"

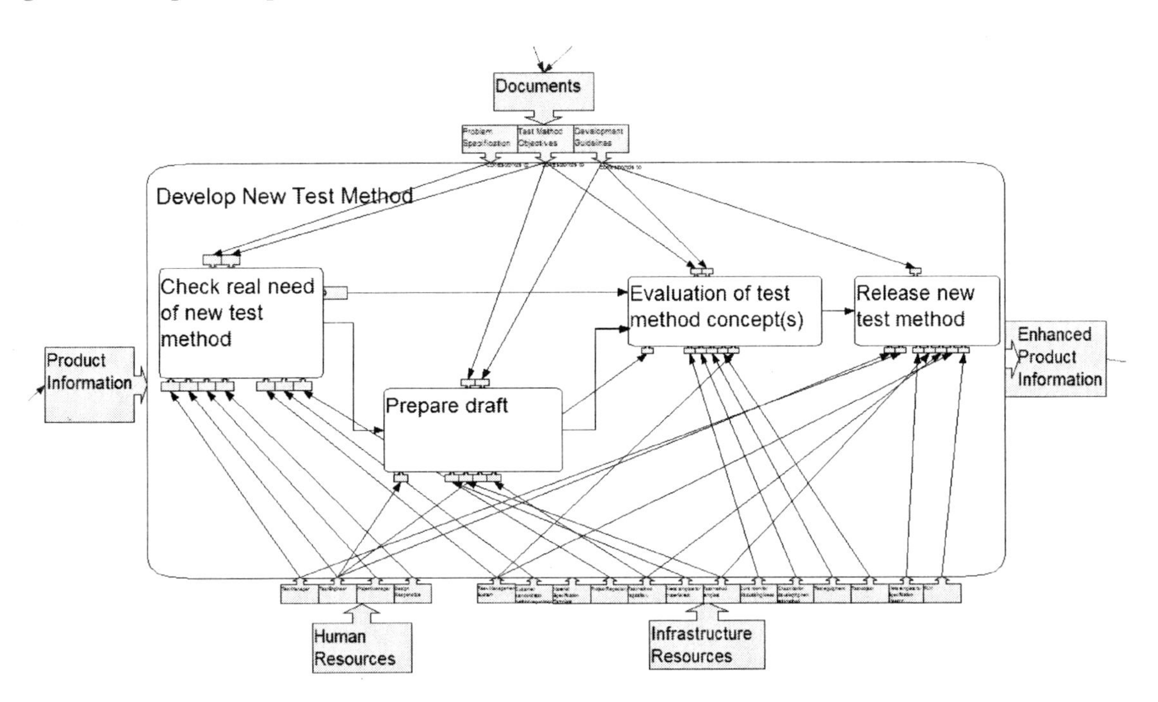

roles, systems and documents are visualized using arrows. The input to the task pattern is "product information", which during the task pattern is enriched and can be considered as one output. This input and output is shown on left and right hand side respectively of the process container.

The intention of this task pattern is to express the organizational knowledge of how to develop a new test method. The model includes model elements and their relations in all perspectives required to express this knowledge, i.e. the process (tasks and work flow for developing a new test method), organization (roles involved), products (expressed within the documents and as input to the task pattern) and systems (infrastructure resources). The model template used in this case is MEAF, a meta-model provided by the vendor of the METIS tool for modeling enterprise knowledge. The connectors of this task pattern are all model elements representing human resources, infrastructure resources and documents. These model elements would have to be mapped on corresponding model elements in the knowledge model of the target enterprise, where the task pattern shall be applied.

Requirement Model Elements

The requirements were expressed in two main elements: textual scenario descriptions and a requirement model. Both are available for the overall process under consideration, i.e. the POI, and for the task patterns contributing to the overall process. The task patterns and the POI model include four different types of requirements:

- Approach requirements are relevant in case different principal approaches for an activity in the application case exist. E.g. line vs. matrix organization of projects, and room-based vs. message-based support of asynchronous cooperation;
- Methodology requirements express any requirement related to methodology use and design, e.g. planned application of (parts of) the MAPPER baseline methodology or

required extensions or refinements of the baseline methodology;

- Platform requirements include requirements with respect to the MAPPER infrastructure. This part of the model will typically capture which task or activity needs support by what MAPPER service, IT tool or template;
- Solution requirements for implementation properties of the solution for a specific project. E.g. which service to use for a project if several equivalent services exist.

All requirements from the scenario descriptions are reflected in the requirements model, but

Figure 2. Requirements and Task Pattern model for the task pattern "Establish Material Specification". The numbered requirements 1, 2 and 3 correspond to the numbered requirements in Figure 3.

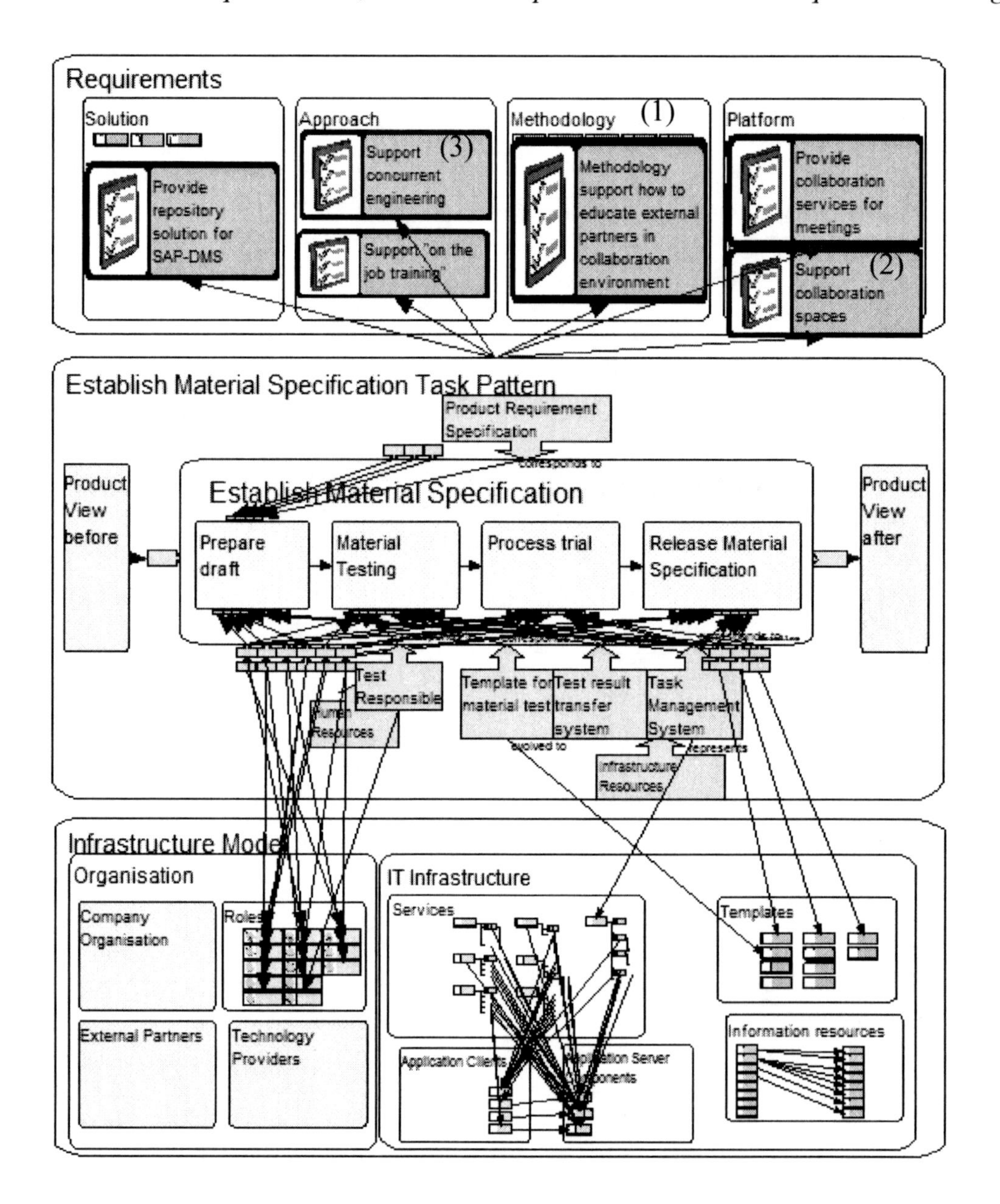

not necessarily vice versa, i.e. the requirements model contains additional requirements because the additional requirements emerged in the model development process, which were not clear or visible during the stage of scenario development.

REQUIREMENT MODEL

This section presents an overview of the requirements model and gives an additional example of a task pattern model. We also discuss how the requirements relate to a typical task pattern "Establish Material Specification". The requirements and the task pattern model are shown in figure 2.

- The top part shows a sub-model containing all the requirements relevant for the different task patterns. The requirements are divided into solution, approach, methodology and platform requirements (as introduced in the subsection "Task Patterns").
- The middle part consists of a sub-model with the task pattern "Establish Material Specification" that consists of the tasks: "Prepare draft", "Material testing", "Process trial" and "Release material specification". Each such task can be further refined into smaller subtasks.
- The bottom part shows the infrastructure model, which is a sub-model common for all task patterns. The organization part (to the left) includes the organizational structure of KA, the specific roles required for the task patterns, as well as the skills and competences required for these roles. The IT infrastructure part (to the right) includes all IT services and client applications of the projects infrastructure, and the information resources available at KA.

Figure 3. Part of the scenario description for task pattern Establish Material Specification. The numbered requirements 1, 2 and 3 correspond to the numbered requirements in Figure 2.

Sida 3 av 7

2) Material Testing
RDE receives samples of the material and initiates material testing: RDE creates a new test process based on a METIS model template. The person who will conduct the test is identified and assigned to the test process. The necessary documents are also linked to the test process. (In the future, KA would like to have not only internal material tests, but also the suppliers to conduct tests, i.e. external partners are integrated in the infrastructure with access to only the artifacts relevant for them.)
Requirement: Test process exists as a model template and can be instantiated by assigning persons to roles and documents to the process; different views and access rights for internal and external suppliers in the project repository.
(1) *Methodology requirement: Formation operation, sustainable collaboration. WP3 needed services from WP4; educating and helping supplier start using Metis modeling.*
Methodology requirement: Resource management includes identifying the person responsible for performing the material testing when done at KA; to manage all persons as well as test equipment resources involved; Scheduling the persons and test equipment in a kind of booking system.
Methodology requirement: Repository content management and view management includes to automatically set up and make available necessary documents (as e.g. checklists and test methods) for the person performing the test.

Sida 4 av 7

Platform requirement: Services for integration and communication; support for web communication with suppliers.
Platform requirement: Test result transfer system must exist; a system for transferring test results from the external tester to KA.
Platform requirement: Support collaboration spaces and Services for collaborated work; to initiate and perform external testing task pattern.
Platform requirement: Support model design and generated workplaces and Provide services for context management; Link documents to test process ; Automatically set up and make available necessary documents (as e.g. checklists and test methods) for the person performing the test. (2)
Approach requirement: Support concurrent engineering; Task pattern template. All task patterns must be possible to run in parallel. (3)

RDE receives the test results as soon as the person conducting the test has completed it. If the results are satisfactory, the process trial will be started. Otherwise, a new supplier will be selected and new material samples will be ordered in order to repeat "material testing".

3) Process Trial
RDE initiates a process trial process based on a METIS model template. The person who will be responsible for the trial is identified and assigned to the trial process. The necessary documents are also linked to the trial process. (Currently, there is

Figure 2 shows relationships between requirements and task patterns. If the requirement is used by the task pattern it has been related to the task pattern container (relations from middle part to top part). Relations between model objects and infrastructure objects (relations from middle part to bottom part) illustrate the use of the infrastructure objects, and are necessary for execution of the models. From the scenario description of the task pattern we can follow some of the requirements that are relevant for the task pattern, see figure 3. The figure shows the description of the task *Material testing* which is a part of the task pattern and is shown more closely in figure 4.

The model of the task *Material testing* follows the scenario description quite closely. We can see that most of the modeled sub-tasks are covered in the scenario text and there are several requirements that are tied to the task. Whereas the textual scenario description gives the first expression and an explanation of the task, the model of the task moves closer to the environment where the task is performed or executed. When modeling the scenario description it is necessary to take into account in which order the different sub-tasks have to be considered. It is also necessary to consider such things as what input and output do the subtasks have and what human and infrastructure resources are connected to the different subtasks. This way of working is also reflected in the way requirements have been elicited. At first when capturing the scenario descriptions the initial requirements have been captured. They have later been refined and additional requirements have been added. However the scenario descriptions have not been updated with the requirements added at a later stage. In the scenario description it is possible to trace to which task some of the requirements are related. This has not been preserved in the model since it has only been considered important to trace the requirements to the specific task patterns.

A requirement may relate to goals that KA has for the project work as well as relate to different task patterns that have been created. A model overview has been created that shows how the requirements relate to goals and task patterns, see figure 5. In the figure typical model artifacts have been enlarged specifically for this presentation in order to exemplify the different parts of the model. The model contains three different parts:

- A goal sub-model containing four different types of goals: Knowledge sharing goals,

*Figure 4. Model showing the task **Material testing** from task pattern Establish Material Specification*

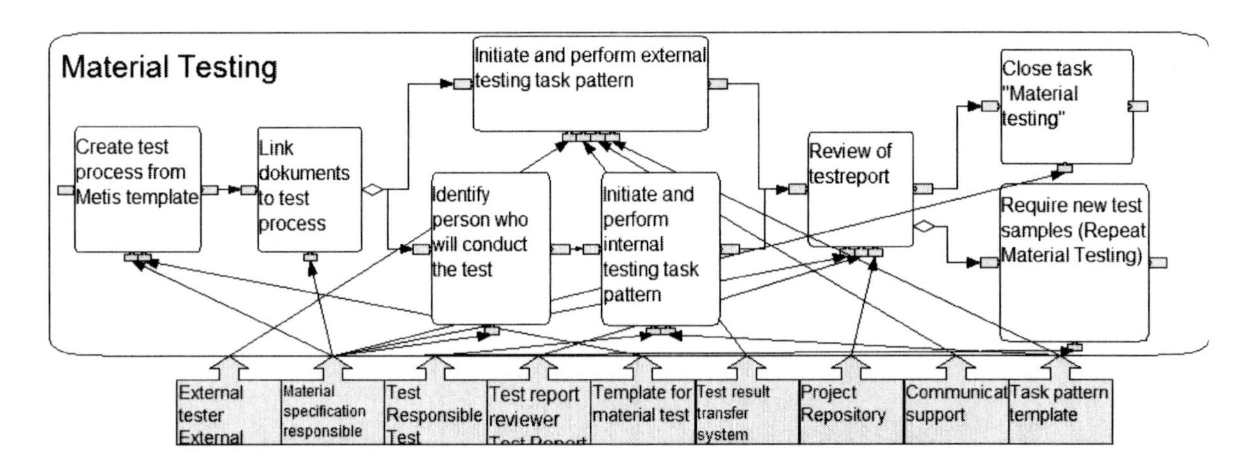

Process goals, Financial goals, and Product innovation goals. In the figure the contents of the container with the goals for Knowledge Sharing is exposed, while the others are only indicated. For the Knowledge sharing goals there is a main goal "Increase reuse and knowledge sharing" to which the other goals are sub goals;

- A requirements sub-model in which the requirements are divided in four different categories as mentioned previously: Solution, Approach, Methodology and Platform.

Figure 5. Overview of how requirements relate to goals and task patterns

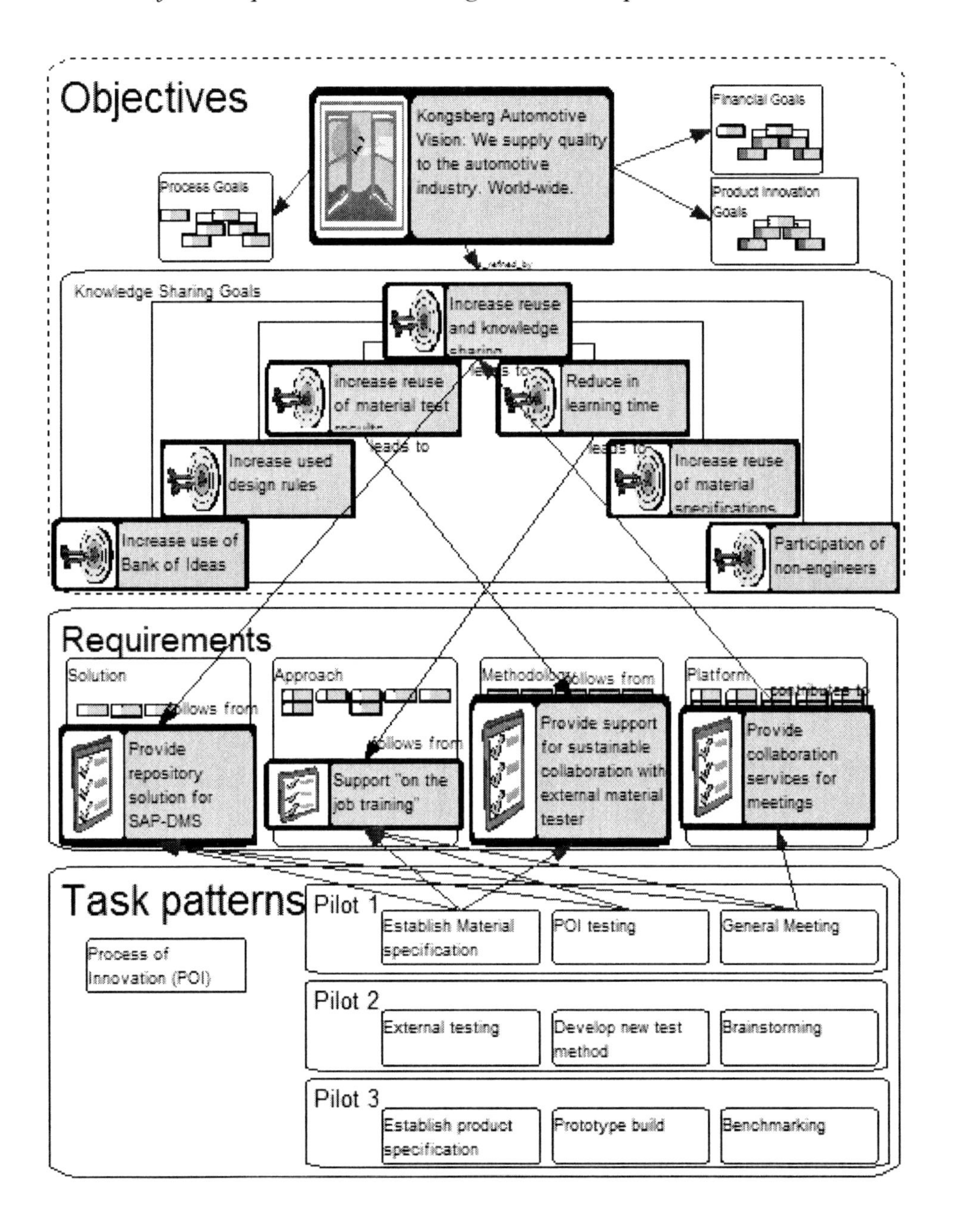

Some of the requirements (one for each category) have been enlarged;
- A model part summarizing the existing task patterns and how they are divided in different pilots.

The goal sub-model was constructed in the concept and scaffolding phase while the task patterns were developed in the subsequent scoping phase. The requirements were partly expressed during the scoping phase when writing scenario descriptions for the task patterns but were compiled and refined after all the nine task patterns had been completed. Only the relationships that connect to the emphasized requirements are shown in the figure and only for the task patterns of KA's pilot 1, in order not to overcrowd the figure with relationships. There exist similar relationships between the other requirements and the task patterns in all pilots and the goals in all containers in the goals model.

For the four highlighted requirements in figure 5 we will now discuss their background details to clarify the relationships with goals and task patterns. The requirement "Provide collaboration services for meetings" is associated to the task pattern "General meeting". During meetings there are lots of documents, drawings etc. to be handled and often there is a need to share the information with participants at distant locations. To be able to do this there is a need to be able to use services that distribute information in a convenient way. We also need easy to use solutions that can remind and connect people at different locations into meetings. Hence, the platform requirement for such collaboration services contributes to achieving the goal "Increase reuse and knowledge sharing".

Some activities such as testing of and even designing materials can be done by an external partner. Therefore there is a need for methodologies that support collaboration with external partners. One such requirement is the methodology requirement "Provide support for sustainable collaboration with external material tester". The requirement has an impact on task patterns such as "Establish Material Specifications" and "External testing" (relationship not shown in the figure) and follows from the goal "Increase reuse of material test results".

"On the job training" is an alternative, compared to sending the employees to courses at other locations, when it is more efficient to use the local facilities and expert knowledge at the company for the training. This may have a positive effect on the time that is needed for new team members to understand the way of working, the product structure and organization structure and thus an approach requirement such as "Support on the job training" follows from the goal to "Reduce in learning time". The task patterns "Establish Material specification" and "POI testing" concerns using specific methods and equipment for testing materials and prototypes and is thus relevant for the requirement "Support on the job training".

Several possibilities may exist to solve how to store and share different kinds of documents and data. Such a possibility can, for example, be facilitated with a system such as SAP or other electronic archive systems. The requirement "Provide repository solution for SAP-DMS" is relevant for KA since they are in the process of introducing SAP into there activities. Since most of the data is in some way stored in documents, drawings or directly stored into a database such a solution requirement will also contribute to the fulfillment of the goal "Increase reuse and knowledge sharing". The requirement is also relevant for all the task patterns since all the task patterns are managing information in various forms of documents.

EXPERIENCES

This section analyses our experiences concerning stakeholder participation in the C3S3P development process and task pattern development.

Experiences Regarding Stakeholder Participation

Participation of stakeholders from different departments of KA and from different MAPPER partners was considered a key success factor for creating adaptable knowledge models suitable for use in everyday practice. We noted that different participation levels were adequate for the different modeling phases (see the subsection "Requirement Model Development Process").

In the initial phase, scaffolding, nearly all stakeholders participated all the time during the model development process. This included presentations from and discussions with KA about the modeling domain, discussions and joint decisions about perspectives to be included in the model, meta-modeling, joint creation, editing and describing of the models with METIS. The produced models were not very large, but in terms of sharing knowledge and creating a joint understanding about the domain and the nature and the EM process, the modeling sessions with all stakeholders were highly valuable. They created a joint ownership of the result. Between the modeling sessions the facilitators refined the model and the meta-model. The subsequent modeling sessions always started by reviewing the model which contributed to a shared understanding and improving the quality of the model.

During the second phase, scoping, the way of working was changed and the level of participation reduced – before developing METIS models, we developed textual scenario descriptions for the task under consideration. The scenario descriptions were developed participatory, but the development of models based on the scenario description was done by the modeling facilitator alone. The reason for this change was the level of detail of the models required by the need to provide executable models. These models included a lot of detailed and technical information. Joint editing of such models was perceived too time consuming and did not really concern the stakeholders. The models produced by the facilitator were presented to the other stakeholders for validation and to strengthen the joint ownership of the results.

In the third phase, requirements modeling, the participation level increased again, as all requirements had to be qualified and described. Although most requirement model refinements were done in smaller working groups of 3-4 people, the participation from different stakeholders was of high importance.

We consider this participatory modeling not only positive for the quality of the modeling results itself, but also with regards to the modeling process and the internal dissemination at the use case partner. The way of modeling provided valuable input for the methodology development by confirming the importance of stakeholder participation, scoping, iterative refinement etc. Furthermore, the way of expressing requirements in four different perspectives on both detail levels (task patterns) and in an aggregated way on use case level (requirements matrix) emerged as new practice during the requirements modeling process. We consider this an important experience in requirement structuring and visualization. The involvement of many stakeholders from KA during the complete modeling process also helped to promote MAPPER internally. With a solid understanding of the potential and work practices of model adapted product and process development the KA stakeholders were able to disseminate MAPPER ideas in the organizations, thus contributing to an increased management attention and further internal developments.

Experiences with C3S3P

The development of requirement model and solution model was based on the C3S3P approach as introduced in the subsection "Requirement Model Development Process". The decision to use C3S3P was mainly based on previous positive experiences and strong recommendations from the industrial participants in the MAPPER project. From a scientific perspective, C3S3P is not extensively researched, i.e. there are not many experience reports, empirical groundings or theoretical foundations of C3S3P published. Although the general structure of C3S3P shows the "specify-design-implement-test-maintain" pattern, which can be found in numerous software and systems development processes, we need to reflect about the advantages and shortcomings in our case.

The general impression from the project is that C3S3P fits quite well to the industrial case under consideration. There were no substantial shortcomings regarding the proposed stages experienced. The main challenge was the need for large parts of the project team including project managers to learn the way of working with C3S3P while using it.

Concept-study and scaffolding were considered extremely valuable for an application case where the two main "parties", industrial experts and academic facilitators, did not have substantial knowledge about the other party's working area. During the concept-study it were mainly the industrial experts who explained their working area, e.g. how does seat heating work, what are the parts of the products produced, how does the innovation process look like, what "neighboring" processes and organization units exist in the company? The scaffolding phase changed the role distribution and required the academic facilitators to introduce general modeling principles, purpose and process of model development, modeling languages, modeling tools, etc. to the industrial experts. The importance of this phase for team formation and creating a shared understanding and ownership of the models has already been mentioned in the previous subsection "Experiences Regarding Stakeholder Participation". From a technical perspective, the results of concept-study and scaffolding may be considered as "throw-away prototype", as the scaffolding models themselves were not used in the later process.

The main results transferred to the next stage were (besides the shared understanding):

- A clear proposal of which organization units and process parts that should be the focus of future work. This was a direct contribution to scoping.
- Which perspectives to take into account during scoping and solution modeling. In addition to the POPS* perspectives mentioned in the subsection "Requirements Model Development Process", these perspectives were business objectives, requirements, and skills and competences of roles.

The step from solution modeling to platform integration included some unexpected activities. The main work planned was the integration of the established IT-infrastructure in the solution model and the configuration of the model for execution by assigning individuals to roles or providing the information needed in the tasks. However, quite some work had to be spent on optimizing and adjusting the solution model in order to make it executable or more performant in the execution environment selected. As modeling tool, model repository and execution engine were offered by the supplier as integrated platform, it was initially expected that the solution model should be executable "as is". This only confirms the importance and usefulness of the platform integration stage.

The important part in the piloting stage from our experience is adjustment of the solution model to needs of the individual users. It was during piloting, when a shift of perspective from process centric modeling to product-centric modeling was decided because the end users requested a tighter integration or process and product knowledge.

The main limitation of the experiences presented in this chapter is that the focus of our work was on the early phases of C3S3P. Concept-study, scaffolding, scoping and solution modeling were the main activities; platform integration and piloting in real-world projects were done only for small parts of the solution model; performance monitoring and management was not conducted at all.

Experiences Regarding Modeling Task Patterns

Even though the idea behind task patterns is to identify and capture reusable and generic practices, they are deeply rooted in the individual enterprise processes. Hence, for an external modeler to extract and model the relevant task patterns, an access to the experience and knowledge of KA's domain experts is an absolute requirement.

The existing scenario descriptions provided a good starting point in facilitating a top-down modeling approach. For each task pattern the outline was built initially. The modeling was not always performed strictly top-down. Sometimes defining the content of a task led to changes at a higher level, e.g. in some cases the headings from the outline were changed upon defining the scenario content. In other cases when revising the content a more suitable heading could be found. The order in the resulting outlines can be characterized by a generic sequence – *prepare*, *conduct* and *report*.

The textual descriptions originally focused mainly on task breakdown structure and were less expressive about task sequence, resources, inputs and outputs. While modeling task patterns, new ideas and questions emerged leading to refinements of the textual descriptions. This shifting of focus took several iterations.

The number of requirements for the execution environment where not multiplied by the number of task patterns because they shared certain requirements.

CONCLUDING REMARKS AND FUTURE WORK

We have analyzed how the C3S3P EM approach has been applied in an automotive supplier company and presented the phases of the C3S3P development process such as Concept Study, Scaffolding, Scoping, and Requirements Modeling. In the MAPPER project we have also introduced the concept of a task pattern for capturing, documenting and sharing best practices concerning business processes in organization. On the basis of this application case we have analyzed our experiences concerning stakeholder participation and task pattern development.

We will further develop the requirements model during the planned pilots in terms of requirement change management and tracking of the requirements document.

Future work in the application case will focus on the C3S3P modeling process and on task patterns. When starting the development of task patterns, we had in mind to capture knowledge about best practices in handy, flexible models that can be put in a library and used on-demand by selecting, instantiating and executing them. This basic idea has resulted in 9 task patterns. In order to transfer them to other organizations, adaptation mechanisms will have to be defined. Furthermore, we will have to investigate, from a methodological point of view how adaptation to substantial changes in the infrastructure should be best facilitated, how patterns can be combined and integrated, as well as how to support pattern evolution.

REFERENCES

Bubenko, J. A. jr., & Kirikova, M. (1999). Improving the Quality of Requirements Specifications by Enterprise Modeling. In A.G. Nillson, C., Tolis, & C. Nellborn (Eds.), *Perspectives on Business Modeling*. Springer, ISBN 3-540-65249-3.

Bubenko, J. A. jr., Persson, A., & Stirna, J. (2001). *User Guide of the Knowledge Management Approach Using Enterprise Knowledge Patterns, deliverable D3, IST Programme project Hypermedia and Pattern Based Knowledge Management for Smart Organisations*, project no. IST-2000-28401, Royal Institute of Technology, Sweden.

Castro, J., Kolp, M., & Mylopoulos, J. (2001). A Requirements-Driven Software Development Meth-odology. In *Proceedings of CAiSE'2001*, Springer LNCS 2068, ISBN 3-540-42215-3.

F3-Consortium (1994). *F3 Reference Manual*, ESPRIT III Project 6612, SISU, Stockholm.

Johnsen, S. G., Schümmer, T., Haake, J., Pawlak, A., Jørgensen, H., Sandkuhl, K., Stirna, J., Tellioglu, H., & Jaccuci, G. (2007). Model-based adaptive Product and Process Engineering. In M. Rabe & P. Mihok (Eds.), *Ambient Intelligence Technologies for the Product Lifecycle: Results and Perspectives from European Research.* Fraunhofer IRB Verlag, 2007.

Krogstie, J., & Jørgensen, H. D. (2004). Interactive Models for Supporting Networked Organizations. In *Proceedings of CAiSE'2004*, Springer LNCS, ISBN 3-540-22151.

Lillehagen, F. (2003). The Foundations of AKM Technology. In *Proceedings 10th International Conference on Concurrent Engineering (CE) Conference*, Madeira, Portugal.

Lillehagen, F., & Krogstie, J. (2002). Active Knowledge Models and Enterprise Knowledge Management. In *Proceedings of the IFIP TC5/WG5.12 International Conf. on Enterprise Integration and Modeling Technique: Enterprise Inter- and Intra-Organizational Integration: Building International Consensus, IFIP, 236*, Kluwer, ISBN: 1-4020-7277-5.

Loucopoulos, P., Kavakli, V., Prekas, N., Rolland, C., Grosz, G., & Nurcan, S. (1997). *Using the EKD Approach: The Modeling Component.* Manchester, UK: UMIST

Persson, A., & Stirna, J. (2001). An explorative study into the influence of business goals on the practical use of Enterprise Modeling methods and tools. In *Proceedings of ISD'2001*, Kluwer, ISBN 0-306-47251-1.

Rittel, H. W. J., & Webber, M. M. (1984). Planning Problems are Wicked Problems. In Cross (Ed.), *Developments in Design Methodology.* John Wiley & Sons.

Sandkuhl, K., Smirnov, A., & Shilov, N. (2007). Configuration of Automotive Collaborative Engineering and Flexible Supply Networks. In P. Cunningham & M. Cunningham (Eds.), *Expanding the Knowledge Economy – Issues, Applications, Case Studies.* Amsterdam, The Netherlands: IOS Press. ISBN 978-1-58603-801-4.

Stirna, J., Persson A., & Sandkuhl, K. (2007). Participative Enterprise Modeling: Experiences and Recommendations. In *Proceedings of CAiSE'2007.* Trondhiem, Norway: Springer LNCS.

van Lamsweerde, A. (2001). Goal-Oriented Requirements Engineering: A Guided Tour. In *Proceedings of the 5th International IEEE Symposium on Requirements Engineering.* IEEE.

Yu, E. S. K., & Mylopoulos, J. (1994). From E-R to "A-R" - Modeling Strategic Actor Relationships for Business Process Reengineering. In *Proceedings of the 13th International Conference on the Entity-Relationship Approach.* Manchester, England.

ENDNOTES

1. C3S3P is an acronym for *Concept study, Scaffolding, Scoping, Solution modeling, Platform integration, Piloting, Performance monitoring and management*

[2] C3S3P is used in ATHENA (http://www.athena-ip.org) and MAPPER projects (http://mapper.troux.com), see also Stirna, Persson and Sandkuhl (2007).

[3] See http://www.troux.com

Chapter VI
Methodologies for Active Knowledge Modeling

John Krogstie
NTNU, Norway

Frank Lillehagen
Commitment, Norway

ABSTRACT

Innovative design is the most important competitive factor for global engineering and manufacturing. Critical challenges include cutting lead times for new products, increasing stakeholder involvement, facilitating life-cycle knowledge sharing, service provisioning, and support. Current IT solutions for product lifecycle management fail to meet these challenges because they are built to perform routine information processing, rather than support agile, innovative work. Active Knowledge Modeling (AKM) provides an approach, methodologies, and a platform to remedy this situation. This chapter describes the AKM-approach applied by manufacturing industries and consultants to implement pragmatic and powerful design platforms. A collaborative product design methodology describes how teams should work together in innovative design spaces. How to configure the AKM platform to support such teams with model-configured workplaces for the different roles is described in the visual solutions development methodology. The use of this approach is illustrated through a case study and is compared with related work in the enterprise modeling arena to illustrate the novelty of the approach

INTRODUCTION

The Active Knowledge Modeling (AKM) (Lillehagen and Krogstie, 2008; Lillehagen, 2003) technology is about discovering, externalizing, expressing, representing, sharing, exploring, configuring, activating, growing and managing enterprise knowledge. Active and work-centric knowledge has some very important intrinsic properties found in mental models of the human mind, such as reflective views, repetitive flows, recursive tasks and replicable knowledge architecture elements. One approach to benefit from these intrinsic properties is by enabling users to perform enterprise modeling using the AKM platform services to model methods, and execute work using role-specific, model-generated and configured workplaces (MGWP). Visual knowledge modeling must become as easy for designers and engineers as scribbling in order for them to express their knowledge while performing work, learning and excelling in their roles. This will also enable users to capture contextual dependencies between roles, tasks, information elements and the views required for performing work.

To be active, a visual model must be available to the users of the underlying information system at execution time. Second, the model must influence the behavior of the computerized work support system. Third, the model must be dynamically extended and adapted, users must be supported in changing the model to fit their local needs, enabling tailoring of the system's behavior. Industrial users should therefore be able to manipulate and use active knowledge models as part of their day-to-day work (Jørgensen, 2001; Jørgensen, 2004; Krogstie, 2007).

Recent platform developments support integrated modeling and execution in one common platform, enabling what in cognitive psychology is denoted as "closing the learning cycle".

AKM APPROACH

The AKM approach has at its core a customer delivery process with seven steps. The first time an enterprise applies AKM technology, we recommend that these steps are closely followed in the sequence indicated. However, second and third time around work processes and tasks from the last five steps can be reiterated and executed in any order necessary to achieve the desired results.

The AKM approach is also about mutual learning, discovering, externalizing and sharing new knowledge with partners and colleagues. Tacit knowledge of the type that actually can be externalized, is most vividly externalized by letting people that contribute to the same end product actually work together, all the time exchanging, capturing and synthesizing their views, methods, properties, parameter trees and values, and validating their solutions. Common views of critical resources and performance parameters provide a sense of holism and are important instruments in achieving consensus in working towards common goals. The seven steps of the approach are shown in Figure 1. The steps are denoted C3S3P and have briefly been described earlier in (Stirna, Persson, & Sandkuhl, 2007). Concept testing, performing a proof-of-concept at the customer site, is not included in the figure. The solutions modeling stage is vital for creating holistic, multiple role-views supporting work across multi-dimensional knowledge spaces, which in turn yield high-quality solution models.

Description of Methodology Step

1. Concept testing is about creating customer interest and motivation for applying the AKM technology. This is done by running pilots and by assessing value propositions and benefits from applying the AKM approach.

Figure 1. The steps of the customer delivery process

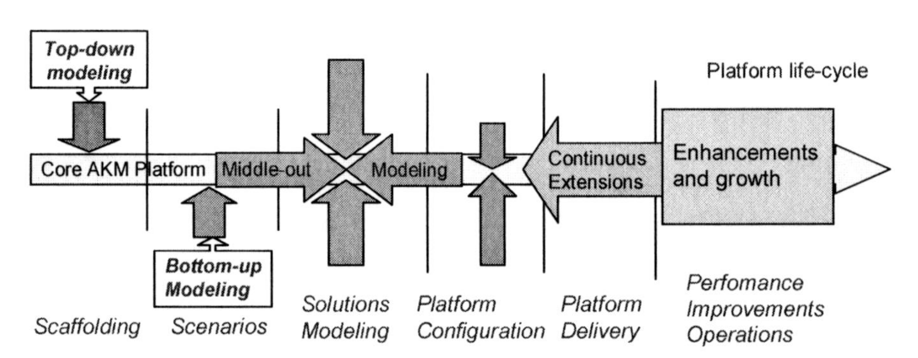

2. Scaffolding is about expressing stakeholder information structures and views, and relating them to roles, activities and systems to provide a model to raise the customer's understanding for modeling and inspire motivation and belief in the benefits and values of the AKM approach.
3. Scenario modeling is about modeling "best-practice" work processes. The focus is on capturing the steps and routines that are or should be adhered to when performing the work they describe. This is the core competence of the enterprise, and capturing these work-processes is vital to perform work, support execution and perform several kinds of analyses in the solutions modeling step.
4. Solutions' modeling is about cross-disciplinary and cross-functional teams working together to pro-actively learn and improve quality in most enterprise life-cycle aspects. The purpose is creating a coherent and consistent holistic model or rather structures of models and sub-models meeting a well-articulated purpose. Solutions' modeling involves top-down, bottom-up, and middle-out multi-dimensional modeling for reflective behavior and execution.
5. Platform configuration is about integrating other systems and tools by modeling other systems data models and other aspects often found as e.g. UML models. These are created as integral sub-models of the customized AKM platform, and their functionality will complement the CPPD methodology (see below) with PLM system functions, linking the required web-services with available software components
6. Platform delivery and practicing adapts services to continuous growth and change by providing services to keep consistency and compliance across platforms and networks as the user community and project networking expands, involving dynamic deployment of model-designed and configured workplace solutions and services
7. Performance improvement and operations is continuously performing adaptations, or providing services to semi-automatically re-iterate structures and solution models, adjusting platform models and re-generating model-configured and -generated workplaces and services, and tuning solutions to produce the desired effects and results.

Collaborative Product and Process Design (CPPD) as mentioned above is anchored in pragmatic product logic, open data definitions and practical work processes, capturing local innovations and packaging them for repetition and reuse. Actually most of the components, such as the Configurable Product Components – CPC and the Configurable Visual Workplaces – CVW, are based on documented industrial methodologies. CPPD mostly re-implements them, applying the principles, concepts and services of the AKM Platform. CVW in particular is the workplace for configuring workplaces. This includes functionality to

- Define the design methodology tasks and processes
- Define the roles participating in the design methodology
- Define the product information structures
- Define the views on product information structures needed for each task
- Perform the work in role-specific workplaces
- Extend, adapt and customize when needed

Industrial customers need freedom to develop and adapt their own methodologies, knowledge structures and architectures, and to manage their own workplaces, services and the meaning and use of data. The AKM approach and the CPPD methodology provide full support for these capabilities, enabling collaborative product and process design and concurrent engineering as will be partly illustrated in the next section.

CASE DESCRIPTION

The first industrial piloting of model-configured solutions applying the most recent AKM technology was started at Kongsberg Automotive (KA) in the Autumn of 2006 as one of three industrial scenarios of the EU project MAPPER, IST 015 627. The goals at KA are improved seat heating design, better product quality, fewer data errors, and improved ways of working to interpret and fulfill customer requirements, producing improved product specifications and supplier requirements.

The needs of Kongsberg are very similar to the needs expressed by companies in other industrial sectors, such as aerospace, construction and energy. Short term, Kongsberg's needs and goals are to:

- Capture and correctly interpret customer requirements,
- Create role-specific, simple to use and reconfigurable workplaces,
- Create effective workplace views and services for data handling,
- Improve the quality of specifications for customers and suppliers,
- Improve communications and coordination among stakeholders,
- Find a sound methodology for product parameterization, automating most of the tasks for product model customized engineering.

To fulfill these goals KA are investigating the AKM approach, adapting several methodologies and building knowledge model-based workplaces.

Material specifications are the core knowledge of collaboration between the customer, represented by Kongsberg Automotive (KA) and the supplier, represented by Elektrisola (E). As illustrated in Figure 2 the material specification is today managed as a document, typically created in Microsoft Word. The content in a specific version of the material specification is put together by one person in KA and approved by one person in E and both companies are filing one copy of the approved material specification. Of course over time additional customer requirements need to be communicated resulting in new parameter values in new versions of the document. The biggest disadvantages with the existing solution are:

- The content in the material specifications is not easily accessed and cannot contribute to

Figure 2. Illustrating the current work logic with the material specification document

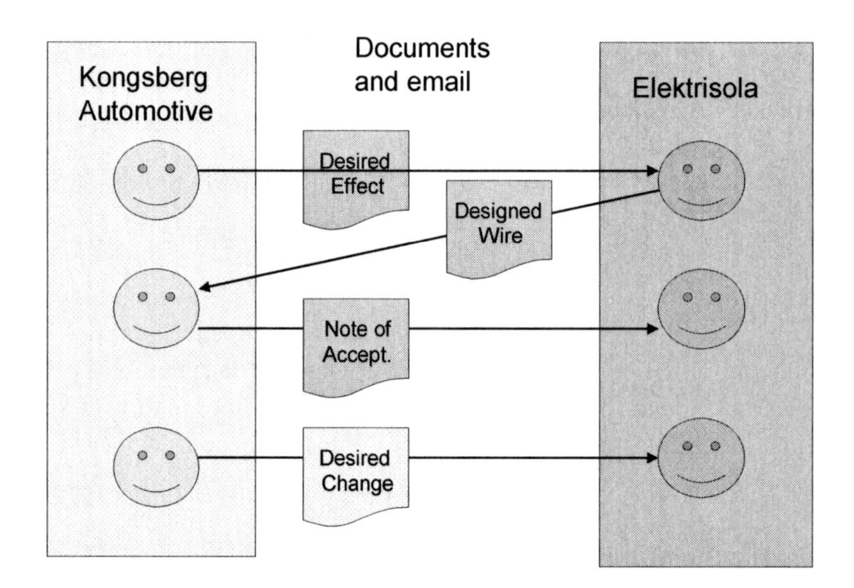

the two companies' operational knowledge architecture.

- The process and work logic to achieve a consistent specification is not captured, making integration with other processes impossible.
- The involvement and commitment from the supplier is not encouraged, there is no support for mutual adjustments in supply and demand.
- Keeping the material specifications updated in both companies is quite time consuming

The general approach as illustrated in Figure 3 has been to replace the document with an operational knowledge architecture built by using the Configurable Visual Workplace (CVW) module developed by AKM within MAPPER. A demo has been developed where two communicating workplaces, one at KA and one at E are modeled and configured. The biggest advantages with the model based knowledge architected solution are:

- The content in the material specifications will be easy to access by both companies and can be part of the each company's complete knowledge architecture, provided that the model based solution is replacing the document based solution for other applications within the companies.
- The involvement from the supplier will be encouraged and the supplier commitment will be more obvious.
- The time for updating the material specifications is expected to be reduced in both companies. There is no real need of filed paper copies anymore.

Current Workplaces

The solution currently supports these five Configurable Visual Workplaces (CVWs):

1. A KA customer- responsible workplace
2. An Elektrisola customer- responsible workplace
3. An Elektrisola smart-wire family designer workplace

Figure 3. Model-configured Workplaces driven by active knowledge architectures

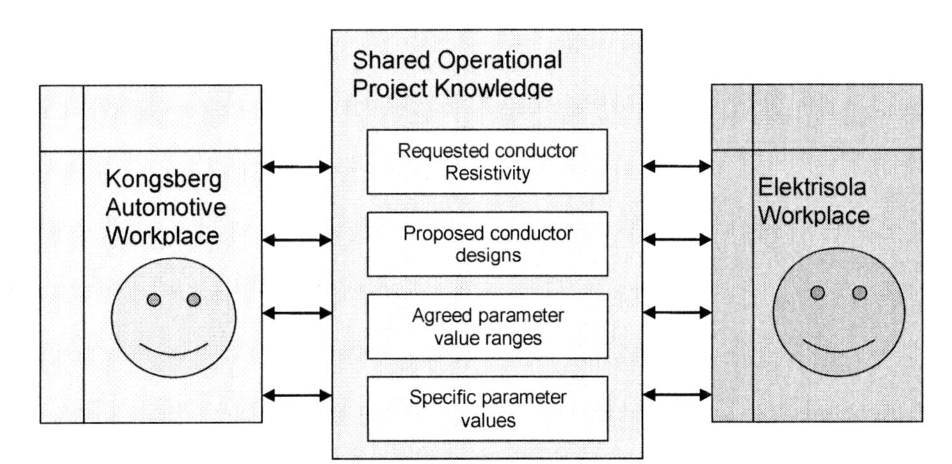

4. A KA material specification workplace
5. A KA heating conductor component designer workplace

All workplace model elements are stored in and share a common Active Knowledge Architecture (AKA). The AKA is used to communicate, configure, coordinate, context-preserve and perform work, acting as a "knowledge amplifying mirror". By extending the scope and methods of work more workplaces can be modeled and generated from the AKA. All workplaces that are ready for use have a similar appearance and behavior.

Scope

The content in the material specifications is continuously refined. The origin for the material specification is the conceptual design. The demonstrated workplaces shall be able to support the refinement of the material specification in the complete life cycle. The focus below is on the early conceptual design phase. In the specific use case the component responsible at the customer, Kongsberg Automotive (KA), will request heating conductor proposals from the supplier, Elektrisola (E), with a targeted property value for the conductor resistivity.

Workplaces and Role

The demonstrator is developed from both customer and supplier perspectives. The generic names of the roles involved are different from KA and E perspectives.

The roles involved from a KA perspective are:

- Component responsible (at KA)
- Component supplier (is E)

From E's perspective, the role of E is more specific than the role of KA:

- Customer (is KA)
- Product responsible (at E)

Models and Knowledge Architecture Contents

The logic of product variants, customer projects, families, specifications and parameters of interest for describing heat conductor design requests are modeled and pre-defined in the shared seat heat model. All demonstration scenarios are adding to or editing these contents as an example of automating the re-use of knowledge.

Figure 4. KA front-end

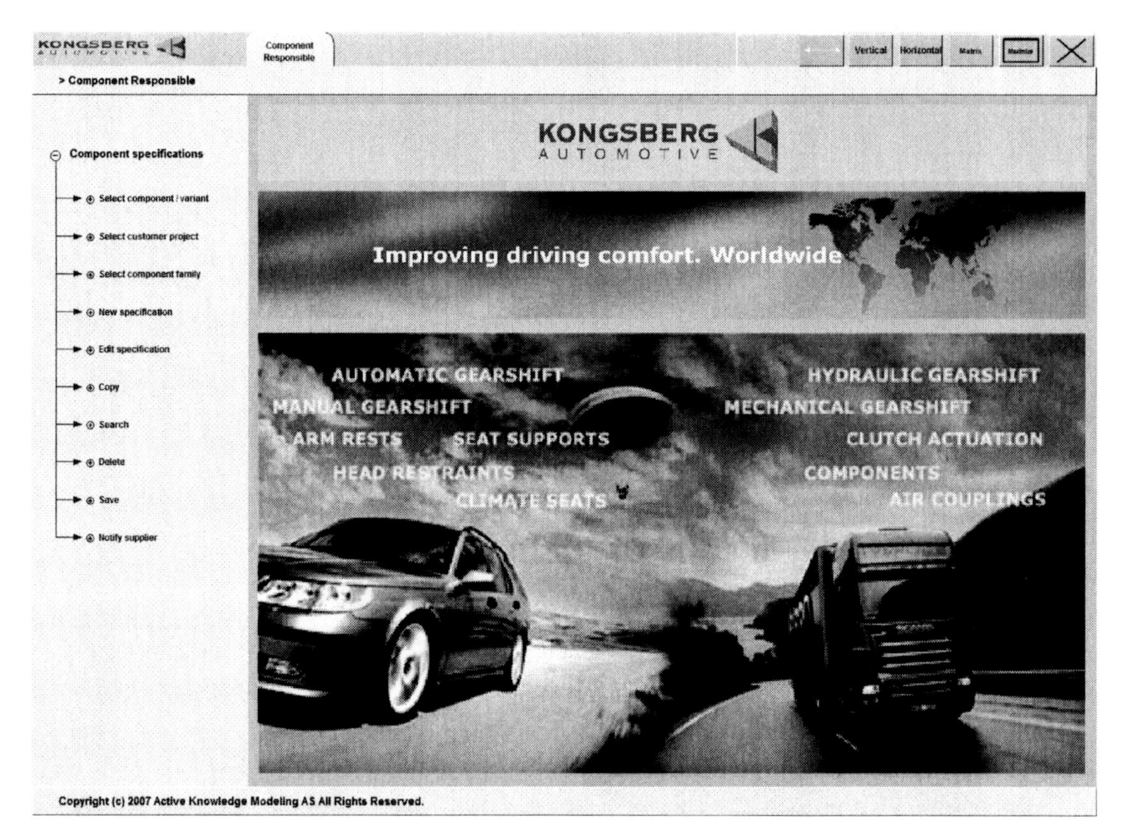

The *product variants* built are sinus-wires and smart wires. The variants are defined by their families, and the families by their specific occurrences.

Customer projects may be defined independent of product variants. In the pilot, there is a design collaboration of KA customer heat conductor and Elektrisola smart-wire product responsible. This is the customer project collaborative work process to coordinate the heat conductor design based on smart-wires.

The start-page of the KA-customer responsible workplace is shown in Figure4. All functionality shown in this and the following screen-shots are model-generated on top of more traditional enterprise models, as illustrated in Figure5. As we see here, one often combines pictures with the more classical box and arrow-modeling. When choosing 'select component variant' in top of the menu, the welcoming Kongsberg workplace front-end is replaced with a dialog box for component variant. In this example we select smart wires, and then choose 'Select Customer project' in the left menu, before defining a new project 'KA 1002'. The project name appears when you perform 'New Specification' or 'Select Component Family', reflecting requirements and life-cycle experiences. When choosing 'New Specification', one gives relevant name to Specification; e.g. 201015 and the screen in Figure 6 appears.

Choosing 'Edit Specification' one can give a nominal value to Resistivity; e.g. 0,176 Ohm/m. Wire resistivity appears as shown in Figure 7 below.

The next step in the co-operation, is to select 'Notify Supplier' - to make the supplier aware of the new specification. An email is set up, ready for completion and sending. Alternative notifications are easily modeled.

Figure 5. Underlying enterprise models for the generation of workPlaces

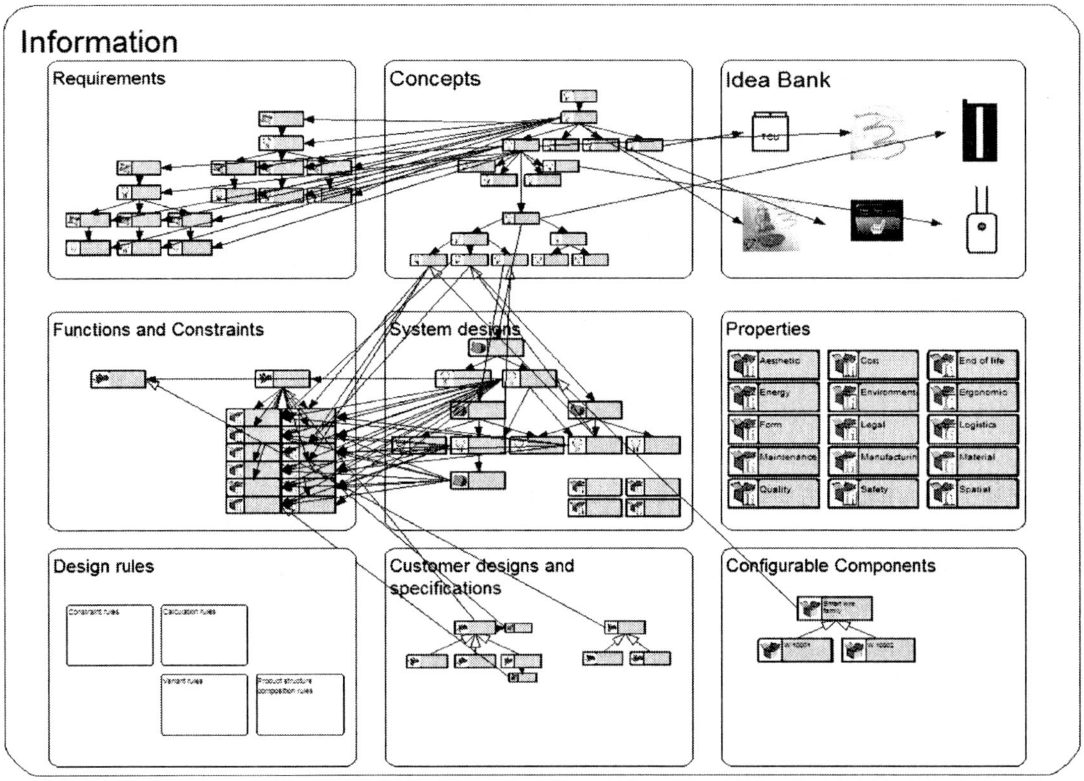

Figure 6. Customer project KA 1002, specification 201015 when first defined

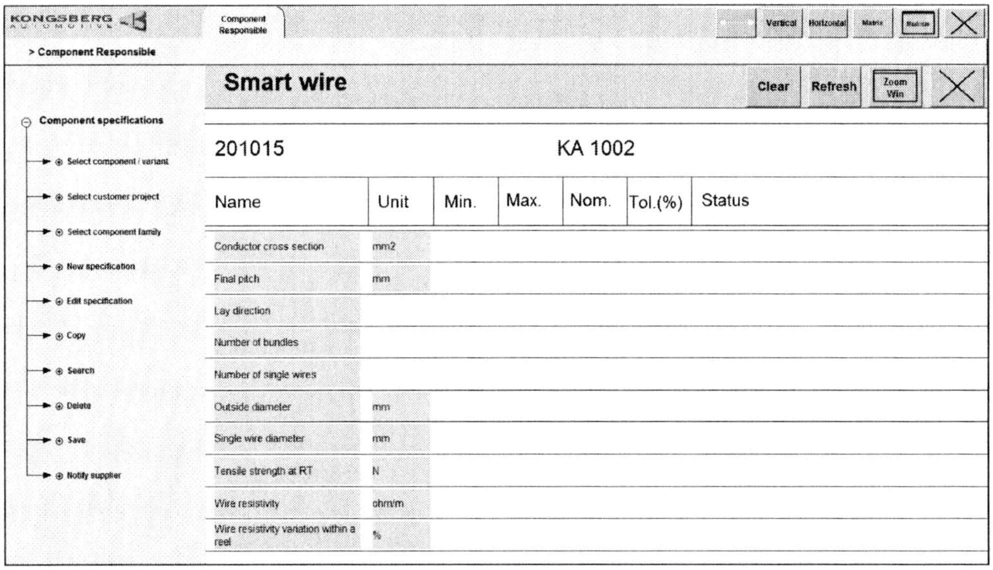

Smart wire

201015 KA 1002

Name	Unit	Min.	Max.	Nom.	Tol.(%)	Status
Conductor cross section	mm2					
Final pitch	mm					
Lay direction						
Number of bundles						
Number of single wires						
Outside diameter	mm					
Single wire diameter	mm					
Tensile strength at RT	N					
Wire resistivity	ohm/m					
Wire resistivity variation within a reel	%					

Figure 7. Editing specification customer project KA 1002

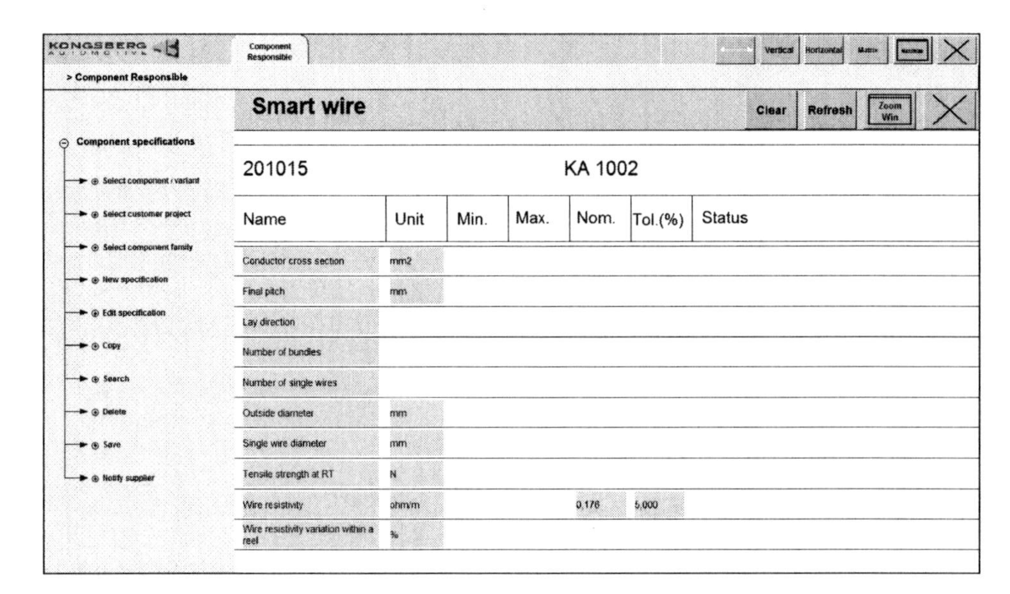

Figure 8. Elektrisola Customer Responsible Workplace

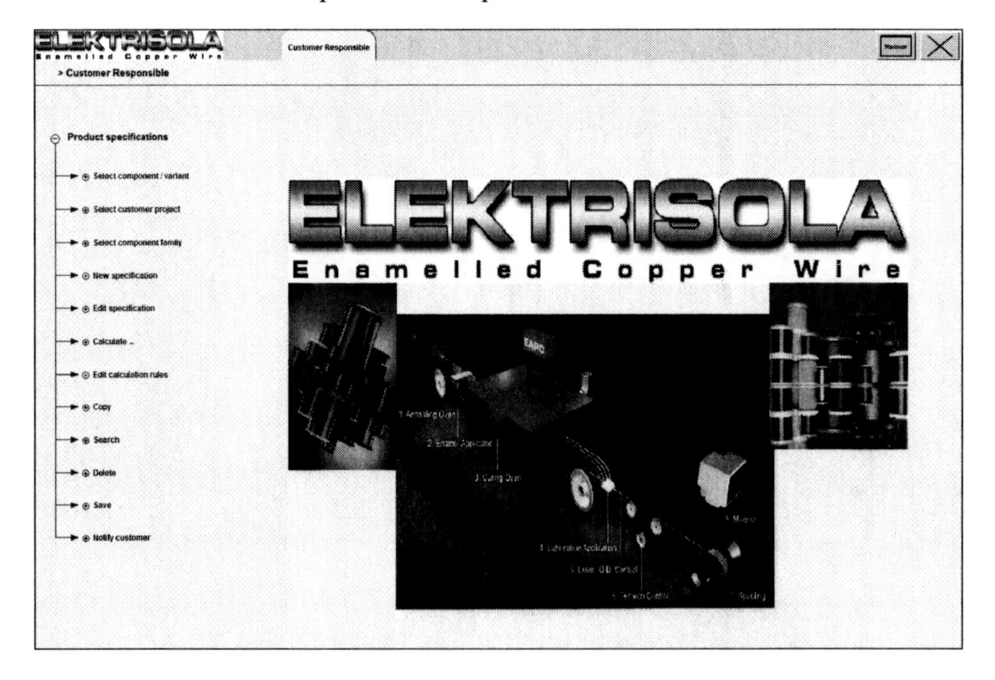

Figure 9. Addressing requirements for KA 1002

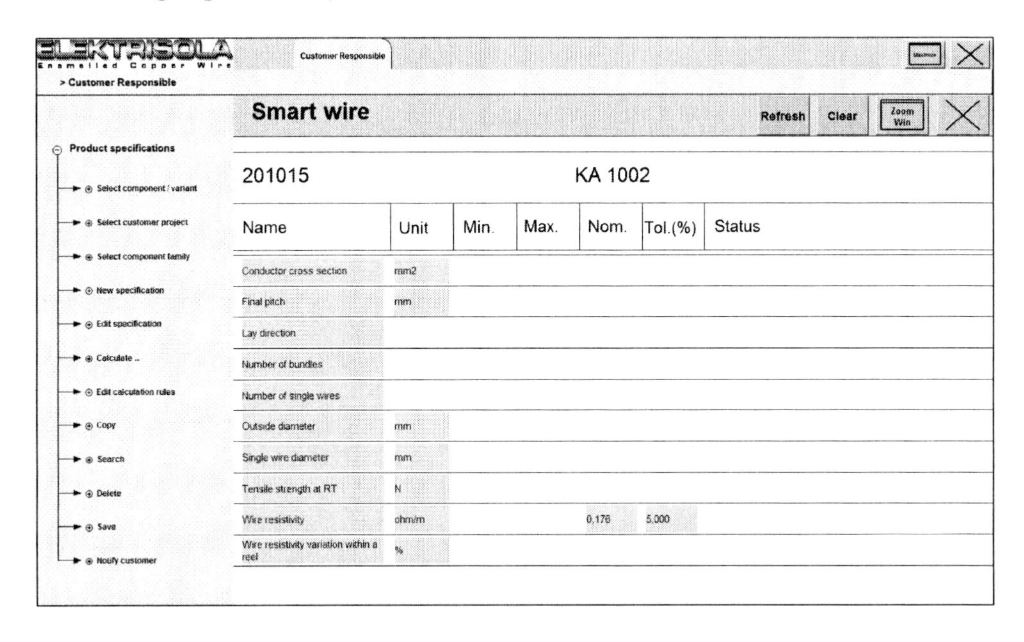

Figure 10. First attempt to address requirements

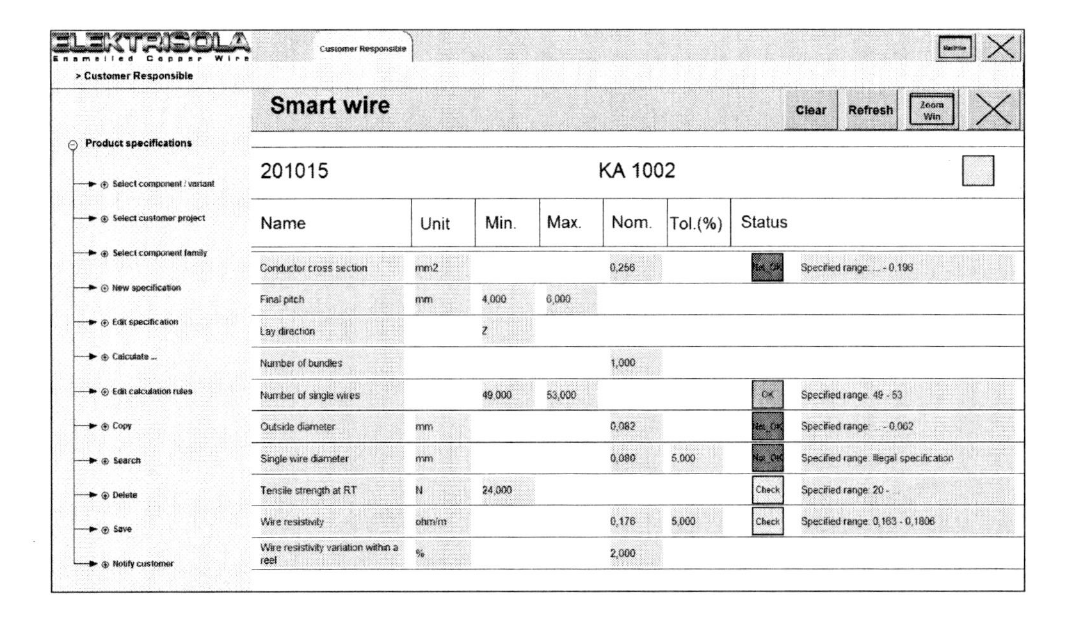

Figure 11. KA 1002 after adapting some properties

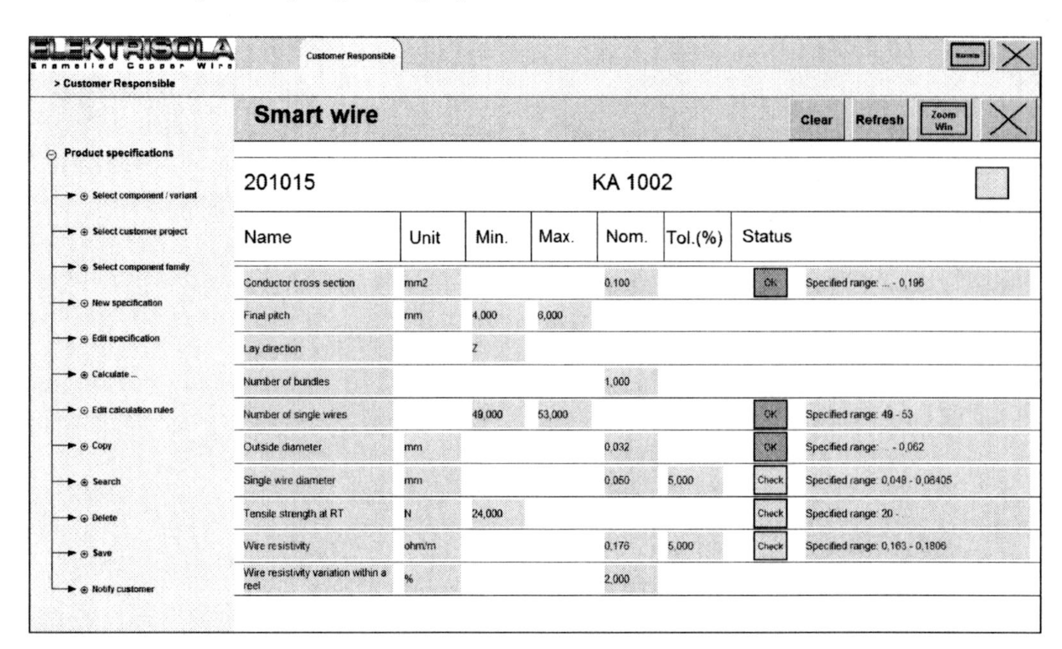

Figure 12. KA 1002 back at KA

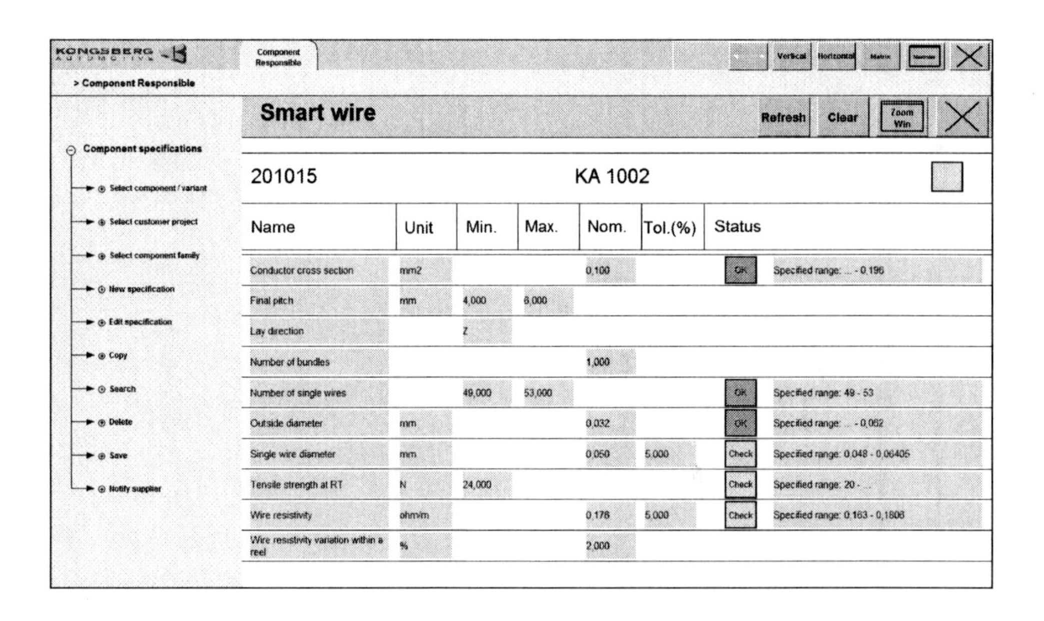

Figure 8 illustrates the front end of the Elektrisola customer–responsible workplace. The role name appears in the top left of the screen.

Choosing 'Select component/variant', in the top of the menu, one can choose Smart Wire, Select customer project, KA 1002 and select specification 201015 using the 'Search' functionality bringing up the screen given in Figure 9.

Now the request from the Kongsberg customer responsible must be responded to. In the example, we have chosen 'Edit Specification', and deliberately given some out of range values to show up in later comparisons, including setting single wire cross-section to 0,08. Choosing 'Calculate' one get the last two values of conductor cross section and outside diameter. Selecting 'Edit Calculation Rules' one can open e.g. an excel spreadsheet with formulae for calculating the outside diameter and the cross section. Figure 10 includes the results after the first requirements are given, indicating the status of different properties.

Updating problematic parameters, can give a picture as seen in Figure 11, where a complete design proposals with parameter values and their calculated ranges is found.

The customer responsible can then choose 'Save', which saves the specification of the proposed conductor design and 'Notify', to send an email to KA requestor, or sets a flag in the designer workplace. Figure 12 illustrates how the model looks like seen from the KA-customer–responsible workplace. The values stored for this concrete case go directly into the knowledge base to be able to compare future specifications with this and previous specifications.

The dynamic evolution and adaptation of work-generative content and context, the workplace composition and the user preferences are impossible to support by programming and compiling the logic. This is simply because any extension or adaptation of contents in one solution model and its views need to be reflected in other models and views that will be used to model-configure other workplaces. The tasks to be executed are totally dependent on the context created by interrelating work solution, workplace behavior and configuration models and views, and role-specific, model-configured workplaces.

Product and material requirements and supplier specifications of the Kongsberg Automotive seat comfort product line have been improved. The materials specification workplace is accepted by the users, but will get some additional services to manage and communicate design issues among customers and suppliers. The customer product specifications workplace will be further developed and related to three or more role-specific workplaces for product design;- the product family responsible, the customer product configuration responsible and the product portfolio responsible.

In the customer product specifications workplace colors are used to indicate the degree of requirements satisfaction, parameter consistency and the solution fit to meet the requirements. The more role-specific the workplaces, their tasks, views and data are built, the bigger the potential to further exploit the resulting knowledge architecture elements, and reuse the reflective views and task-structures.

RELATED WORK

Most enterprise modeling and enterprise architectural frameworks are descriptive frameworks designed to define views on the contents of specific enterprise domains. However, most of the frameworks and technologies supporting them lack capabilities and services for meta-data management, role-driven viewing, and integration with operational platforms and systems. A tighter integration of descriptive model views with operational and execution views, supported by model-configured platforms integrating systems, tools and new services, will achieve full support for model-driven and model-generated business process and work execution at all levels. An analysis of the frameworks concerning their appropriateness and usability for designing and developing architectural enterprise interoperability solutions leads to the following conclusions:

The architecture of the Zachman framework (Sowa & Zachman, 1992; Zachman, 1987) provides the set of concepts and principles to model, describe and categorize existing enterprise artifacts in a holistic way. It supports the integrated interoperability paradigm. Research has been reported to map Zachman's framework to the GERAM framework which is now an ISO standard (ISO 15704). The Zachman framework has with some success been used by the Ontario Government in Canada to design their EA approach, and has been used as a reference categorization structure for enterprise knowledge repositories, but is only focusing on analysis of static models.

GERAM (Bernus & Nemes, 1996) gives a very good overview of the different EM aspects and domains, but lacks some very important aspects such as meta-modeling capabilities, knowledge spaces and the importance of supporting modeling and execution by integrated platforms with repositories.

The GIM modeling framework (Chen & Doumeingts, 1996; Doumeingts, Vallespir & Chen, 1998) introduces the decision views which are not explicitly taken into account in other modeling frameworks. The decisional aspect is essential for establishing interoperability in the context of collaborative enterprises. To interoperate in such an environment, decision-making structures, procedures, rules and constraints need to be explicitly defined and modeled, so that decentralized and distributed decision-making can be performed and managed. The decisional model is now a European TS (technical specification) elaborated by CEN TC310 WG1. However, while providing a strong support for performance indicator management and decision making, the GIM framework has limited expressiveness and platform integration.

ARIS (Scheer, 1999) has strong top-down process modeling and integration capabilities, but lacks expressiveness for other aspects and the "big picture" created by a holistic approach. The different views of the ARIS-concept include different modeling languages, e.g. EPC for illustrating the collaborative business processes. But there are extensions needed concerning the requirements of modeling collaborative enterprises like new role-concepts or the problem of depicting internal and external views of the same business process.

The CIMOSA Framework (ESPRIT Consortium AMICE, 1993; Zelm, 1995) is a comprehensive reference framework at the conceptual layer, but lacks expressiveness for multiple dependencies of types and kinds of views, for evolving concepts, contents and capabilities, and for capturing context. It is the basis to establish the European Pre-standard ENV 40003 now published as a joint European and ISO standard known as EN/ISO 19439: Framework for enterprise modeling. Although the CIMOSA Framework does not explicitly consider interoperability issues, it can be a contribution to an integrated paradigm at the conceptual layer to establish interoperability.

The DoDAF Architecture Methodology (DoD, 2003a; DoD, 2003b) from the FEAC Institute is a comprehensive framework and methodology targeted specifically at systems engineering in the military forces, and covers all kinds of technical systems not just software systems. It has a strong categorization of enterprise knowledge contents. The DoDAF is along with TOGAF one of the most comprehensive enterprise architecture frameworks and provides a good understanding of the stakeholders and users and the minimum of information they must receive to align business and IT and get value out of their business and IT initiatives. However, as all the other enterprise architecture frameworks it contributes little to integrated platforms, model-driven design and to generation of interoperable solutions.

TOGAF (Open Group, 2005) from the Open Group has a good methodology for developing an approach (the Architecture Development Methodology – ADM), but again TOGAF is just a framework of descriptive and fairly abstract views. The Open Group is now cooperating with OMG (Object Management Group) to investigate synergies by integrating the OODA, SOAD (Service Oriented Analysis and Design) approaches and the ADM methodology.

The TEAF Methodology (TEAF, 2007) from the US Dept. of Commerce is specifically tuned to deliver a methodology to the US Government agencies and all administrative, legislative and financial systems, so this architecture is very rich for those application domains. A TEAF model developed in enterprise modeling tools such as e.g. Troux Architect (Troux, 2008) allows visual navigation of all core domains and their constructs and relationships to other constructs and domains, such as: strategies, proposed initiatives, present IT portfolio, present systems and their use, users and vendors, all systems, their capabilities and use, and support for searching, view-generation, reporting, and performing analysis, from simple "what-if" to network and impact analysis.

The ISO 15745 standard (ISO, 2003) allows dealing with interoperability of applications early in the stages of system requirements and design and through the lifecycle by using the Application Integration Framework. It provides a clear definition of manufacturing (enterprise) application as a set of processes, resources and information exchanges. The desired or as-built interface configurations can be described in a set of XML files using the XML schemas defined in the standard. This approach can work only if various standards used to specify interfaces are interoperable between them.

SUMMARY AND FUTURE WORK

The demand for the type of solution we see exemplified in the case being addressed with AKM technology will increase once western industry faces head on competition from India and China. Global networking will require full integration of enterprise teams across cultures, and collaborative design, global team-composition and learning must be supported. Creating active or living knowledge, and supporting customized working environments and approaches to industrial computing and knowledge and data sharing, exploiting the web as the new medium, also incorporating and exploiting the generative aspects of Web 2.0 solutions will be a necessity.

Involvement of stakeholders in sharing knowledge and data, is a key issue. Interrelating all stakeholder perspectives and life-cycle views from requirements, expectations and constraints on design to maintenance and decommissioning or re-engineering. Being able to interrelate and analyze, building the "big picture" and making parts of it active so that it drives execution, depends mainly on parallel team-working:

1. Designers and engineers must work with model builders on real customer product deliveries, and
2. Product and process methodologies are designed/modeled and applied (executing tasks) in concert with the pilot team.

This implies closing the gap between modeling and execution, supporting these capabilities:

- Interaction of users in developing common approaches, pragmatic best-practices and matching solutions,
- Integration of systems to create effective, well-balanced solutions.
- Interoperability of enterprises in performing networked collaborative business
- The discovery of multi-dimensional knowledge spaces and views
- The socialization, discovery and externalization of what would otherwise be tacit human knowledge

- The recognition that software is nothing else but knowledge.

A good metaphor is to think of the web as an intelligent mirror, a mirror "painted" on its back with software components, enabling us to interrelate the mental models of humans and the object-oriented models of computers. This intelligent mirror has memory, and behavior and can mimic more and more what humans think and do.

We have in the article illustrated how to move towards model-designed service-configurable enterprises, and shown how AKM technology can get us at least part of the way.

There is a potential for that the future will be model-designed and model-managed. IT and knowledge engineering will be able to move from object-orientation towards view-orientation, with industrial users developing their own model-configured, user-composed services and platforms.

Goals to strive towards:

1. Turning the web into a visual medium, enhancing our use of the left part of the human mind.
2. Providing generic software tools / components that can be a common platform for AKM driven higher level customer and networking platforms.
3. Providing a platform to integrate the increasingly more isolated communities of modern societies.
4. Providing increased citizen involvement in democratic activities to shape our future nations and global societies.
5. Giving the ownership of data back to the users, providing services for people to define data, calculate and share values, and control data availability.
6. Combining design, learning and problem-solving to make closed environments and industrial and other workspaces accessible by trainees.

REFERENCES

Bernus, P., & Nemes, L. (1996). A framework to define a generic enterprise reference architecture and methodology. *Computer Integrated Manufacturing Systems*, *9*(3), 179-191.

Chen, D., & Doumeingts, G. (1996). The GRAI-GIM reference model, architecture, and methodology. In P. Bernus et al. (Eds.), *Architectures for Enterprise Integration*. London: Chapman

DoD (Ed.). (2003a). *DoD architecture framework. Version 1.0. Volume I: Definitions and Guidelines.* Washington, DC: Office of the DoD Chief Information Officer, Department of Defense.

DoD (Ed.) (2003b). *DoD architecture framework. Version 1.0. Volume II: Product Descriptions.* Washington, DC: Office of the DoD Chief Information Officer, Department of Defense.

Doumeingts, G., Vallespir, B., & Chen, D. (1998). Decisional modelling using the GRAI grid. In P. Bernus, K. Mertins, & G. Schmidt (Eds.), *Handbook on Architectures of Information Systems*, (pp. 313-338). Berlin: Springer.

ESPRIT Consortium AMICE (Ed.). (1993). *CIMOSA: Open system architecture for CIM. Volume 1* (Research report ESPRIT, Project 688/5288) (2nd revised and extended edition). Berlin: Springer.

ISO (2003). *ISO 15745-1 Industrial automation systems and integration - Open systems application integration framework - Part 1: Generic reference description*. Geneva, Switzerland: International Organization for Standardization.

Jørgensen, H. D. (2001). Interaction as a Framework for Flexible Workflow Modelling, *Proceedings of GROUP'01*, Boulder, USA, 2001.

Jørgensen, H. D. (2004). *Interactive Process Models*, PhD-thesis, NTNU, Trondheim, Norway, 2004 ISBN 82-471-6203-2.

Krogstie, J. (2007) Modelling of the People, by the People, for the People. In *Conceptual Modelling in Information Systems Engineering.* Berlin: Springer Verlag. ISBN 978-3-540-72676-0. s. 305-318

Lillehagen, F. (2003). The foundation of the AKM Technology, In R. Jardim-Goncalves, H. Cha, A. Steiger-Garcao (Eds.), *Proceedings of the 10th International Conference on Concurrent Engineering* (CE 2003), July, Madeira, Portugal. A.A. Balkema Publishers

Lillehagen, F., & Krogstie, J. (2008). *Active Knowledge Modeling of Enterprises.* Heidelberg, Berlin, New York: Springer.

Open Group. (2005). The Open Group architectural framework (TOGAF) version 8. Retrieved December 2007 from http://www.opengroup.org/togaf/

Scheer, A.W. (1999). *ARIS, business process framework* (3rd ed.) Berlin: Springer.

Sowa, J. F., & Zachman, J. A. (1992). Extending and formalizing the framework for information systems architecture. *IBM Systems Journal, 31*(3).

Stirna, J., Persson, A., & Sandkuhl, K. (2007). Participative Enterprise Modeling: Experiences and Recommendations. In John Krogstie, Andreas L. Opdahl, Guttorm Sindre (Eds.), *Advanced Information Systems Engineering, 19th International Conference, CAiSE 2007,* Trondheim, Norway, June 11-15, 2007, Proceedings. Lecture Notes in Computer Science 4495 Springer, ISBN 978-3-540-72987-7

TEAF (2007) Retrieved December 2007 from http://www.eaframeworks.com/TEAF/index.html

Troux (2008) http://www.troux.com Last visited April 2008

Zachman, J. A. (1987). A framework for information systems architecture. *IBM Systems Journal, 26*(3), 276-291.

Zelm, M. (Ed.) (1995). *CIMOSA: A primer on key concepts, purpose, and business value* (Technical Report). Stuttgart, Germany: CIMOSA Association.

Chapter VII
Data Modeling and Functional Modeling:
Examining the Preferred Order of Using UML Class Diagrams and Use Cases

Peretz Shoval
Ben-Gurion University of the Negev, Israel

Mark Last
Ben-Gurion University of the Negev, Israel

Avihai Yampolsky
Ben-Gurion University of the Negev, Israel

ABSTRACT

In the analysis phase of the information system development, the user requirements are studied, and analysis models are created. In most UML-based methodologies, the analysis activities include mainly modeling the problem domain using a class diagram, and modeling the user/functional requirements using use cases. Different development methodologies prescribe different orders of carrying out these activities, but there is no commonly agreed order for performing them. In order to find out whether the order of analysis activities makes any difference, and which order leads to better results, a comparative controlled experiment was carried out in a laboratory environment. The subjects were asked to create two analysis models of a given system while working in two opposite orders. The main results of the experiment are that the class diagrams are of better quality when created as the first modeling task, and that analysts prefer starting the analysis by creating class diagrams first.

INTRODUCTION

The main goal of this research is to examine the better order of performing the two main activities in the analysis phase of UML-based software development processes: functional modeling with use cases, and domain (conceptual data) modeling with class diagrams. Though system development is usually an iterative process of refinement, the analysis stage of each iteration should be driven by a specific modeling activity, implying that activity ordering in iterative development is a legitimate and important question. As we show in the next section of this chapter, existing development methodologies differ in a prescribed order of performing these activities: some recommend to start with identifying conceptual classes and continue with developing use cases, using the identified classes or objects, while others suggest to start with developing use cases and continue with building a class diagram based on the concepts appearing in the use cases.

Methodologies starting with creating a domain model by building a class diagram argue that the initial class diagram maps the problem domain and allows describing the functional requirements within a well-defined context. The entities in the class diagram serve as an essential glossary for describing the functional requirements and, since it is an abstraction of the part of the real world relevant for the system, it only rarely changes and can serve as a solid basis for other future systems as well. On the other hand, methodologies starting with creating use cases argue that the classes should be based on the functional requirements, and thus should be elicited from them. One reason for this argument is that creating a domain model before learning the functional requirements can lead to a class diagram that include entities that are out of the system's scope.

We expect that creating a domain model prior to defining the functional requirements with use cases should yield better results, i.e. better class diagrams and use cases. This is because objects are more "tangible" than use cases; analysts can identify and describe more easily the objects they are dealing with and their attributes than the functions or use cases of the developed system. Use cases are not "tangible" and may be vague, since different users may define the expected system functionality in different terms. Repeating Dobing & Parsons (2000), the roles and values of use cases are unclear and debatable. Of course, conceptual data modeling is not trivial either; it is not always clear what is an object, how to classify objects into classes, what are the attributes and the relationships, etc. - but still the task of data modeling is more structured and less complex compared to the task of defining and describing use cases. Besides, the analyst has to create just one class diagram for the system rather than many use cases. While in domain modeling the analyst concentrates only on the data-related aspects, in use-case modeling, the analyst actually deals at the same time with more aspects. Use cases are not merely about functions; they are also about data, user-system interaction and the process logic. Because of the above, it seems to us that starting the analysis process with the more simple and structured task should be more efficient (in terms of time) and effective (in terms of quality of the analysis products). Not only that the first product (the conceptual data model) will be good, it will ease the creation of the following one (the uses cases) since creating use cases based on already defined classes reduces the complexity of the task. In the view of the above, we also expect that analysts would prefer working in that order, i.e. first create a class diagram and then use cases.

The above expectations and assumptions can be supported by both theory and previous experimental work. Shoval & Kabeli (2005) have studied the same issue in the context of the FOOM methodology. According to their experiment, analysts who start the analysis process with data modeling produce better class diagrams than those who start the process with functional modeling. They also found that analysts prefer working in this order. The current study can be viewed as an extension of Shoval & Kabeli (2005) to UML-based software development.

One theory supporting our expectations and assumptions about the better order of analysis activities is the ontological framework proposed by Wand and Weber, known as BWW (Bunge, Wand & Weber); see, for example (Wand & Weber, 1995; Weber, 2003). According to that framework, the world consists of things that possess properties; this fits the "data-oriented" approach, particularly in the context of a class diagram. More complex notions such as processes (functions) are built up from the foundations of things and properties; this fits the "functional orientation" of use cases. If we accept that a key purpose of modeling is to represent aspects of the real world, then it follows from the above framework that it is appropriate to start the modeling process by dealing with the basic "things and properties", i.e. data modeling, and based on that continue with the modeling of the more complex constructs, i.e. functional modeling.

Another support for our assumptions comes from cognitive theory, specifically from COGEVAL (Rockwell & Bajaj, 2005), a propositional framework for the evaluation of conceptual models. COGEVAL tries to explain the quality and readability (comprehensibility) of models using various cognitive theories. Proposition 1 of COGEVAL relies on Capacity theories, which assume that there is a general limit on a person's capacity to perform mental tasks, and that people have considerable control over allocation of their capacities for the task at hand. Based on these theories, for a given set of requirements, the proposition is that it will be easier to create correct models that support chunking of requirements. This proposition supports our belief that it will be easier to create a data model as it deals only with data structure constructs.

Proposition 3 of COGEVAL is based on Semantic Organization theories which indicate how items are organized in long-term memory (LTM). Spreading activation theory depicts concept nodes by relationship links, with stronger links having shorter lengths. The modeling process can be considered as taking concepts and relationships in the users LTM and capturing them in a model. Based on this proposition, data modeling methods whose constructs allow for creation of concept-node and relationship-arc models, as a class diagram, will offer greater modeling effectiveness than other modeling methods, such as use cases, whose constructs are not organized as concept nodes and relationships.

Proposition 4 of COGEVAL is based on theories on levels-of-processing and long-term memory, which divides cognitive processes into three levels of increasing depth: structural, phonemic and semantic. According to that theory, a modeling method whose constructs require only structural processing will offer greater modeling effectiveness and efficiency compare to a modeling method that requires a greater amount of semantic processing. It follows from this proposition that data modeling, which involves only structural processing, offers greater effectiveness and efficiency than functional modeling with use cases, which requires more semantic processing. From this, we deduct that it would be better to begin the analysis with data modeling where the analyst is concerned only with structural modeling, creating only a class diagram; then, using the already defined classes, he/she will create the use cases.

Based on the above discussion, previous results and the theoretical frameworks, we expect that it is better to start with data modeling. However, as we show in the survey of next section, there are in fact different methodologies, which advocate different orders of activities. Therefore, for the sake of this study, we hypothesize (in the third section) that there is no difference in the quality of the analysis models when created in either order. Similarly, we hypothesize that there is no difference in the analysts' preference of the order of activities. These hypotheses are tested in a comparative experiment that is described in the fourth section. The fifth section presents the experimental results, whereas the sixth section discusses future trends and provides the concluding remarks.

BACKGROUD

Order of Analysis in System Development Methodologies

In the mid 70's, as software development became more and more complex, the first methodologies for analysis and design of information systems were introduced. One purpose of those methodologies was to reduce development complexity by separating the analysis phase, where the users' needs are defined, from the design phase, where we decide how the system will meet those needs. Over the years, the way of performing the analysis evolved, as new ideas and concepts have been introduced.

In the early methodologies, such as DeMarco (1978), the emphasis of analysis was on describing the functional requirements, by conducting a functional decomposition of the systems using data flow diagrams (DFDs) or similar techniques. Later on, with the introduction of conceptual data models, many methodologies included also data modeling in the analysis phase (e.g. Yourdon, 1989). Nevertheless, functional analysis was still the primary activity and the data analysis was only secondary to it.

In the early 90's, with the emergence of the object-oriented (OO) approach to software development, new methodologies were introduced to support the analysis and design of information systems according to this concept. Some of these methodologies (e.g. Bailin, 1989; Coad & Yourdon, 1991; Meyer, 1998; Rumbaugh et al., 1991; Shlaer & Mellor, 1992) suggested a different analysis concept: object-driven, instead of functional-driven, i.e. the analysis emphasis is on finding the domain objects, while in the design phase the designer identifies the services which these objects ought to provide and assigns responsibilities to them. These objects are not necessarily the ones from which the system is eventually built, but the principal entities from the problem domain. Furthermore, the analysis class diagram usually does not include functionality. Since this domain object model and the conceptual data diagram are much alike, the analysis tasks eventually remained akin, whereas the order of performing them was inverted. However, there were still methodologies that kept with the functional-driven approach, where a functional analysis is performed first and then the object model is derived from it, whether directly or through a conceptual data model. See for example Alabiso (1988), Jacobson et al. (1992), and Rubin & Goldberg (1992).

As part of the new OO methodologies, many methods and tools were introduced to describe different products during the development of an OO system. Thus, aiming at solving the problems raised by the vast amount of methods and tools, UML (Unified Modeling Language) was adopted by OMG (Object Management Group) as its standard for modeling OO systems. UML consists of a set of models and notations for describing various products during the lifecycle of an OO system, and is in use today by many leading methodologies.

UML's main tools for describing the user/functional requirements and the object model of a system are use cases and class diagram, respectively. A class diagram is an enhanced variation of common tools that have been used in many forms and semantics over the years (notably Entity-Relationship model). A use case is a piece of the system's functionality, presented as an interaction between the system and an entity external to it called "actor", for the purpose of achieving a goal of that actor. Its details are described in a semi-structured format. Relationships between use cases are visualized in a use case diagram, where the system appears as a "black box", meaning that the internal structure of the system and its internal operations are not described. That is why this model is most appropriate to use during the analysis phase. Appendix A presents two illustrative diagrams: a class diagram (Figure 2) and a use case diagram (Figure 3) of a certain system, and provides two examples of use case descriptions .

Being merely a modeling language, UML defines no development process. However, many methodologies that use the UML models/tools have been developed. Despite many variations between different UML-based methodologies, in most of them the analysis phase comprises two main activities: data modeling, i.e., creating a class diagram to describe the application domain; and functional modeling, i.e., creating use case diagrams and descriptions to describe the functional requirements of the system. UML-based methodologies which adopt use cases as the requirements description tool are usually "use case-driven", meaning that the entire development process is derived by describing, realizing, and developing use case scenarios.

The Rational Unified Process – UP (Jacobson, Booch & Rumbaugh, 1999; RUP) is one of the most documented and influential use case-driven methodologies. It provides a wide and general framework for systems development, and as such offers guidelines with many optional variations. The first analysis activity according to UP is creating use case scenarios, whilst an initial class diagram is only created in the next phase called Use Case Analysis . Since UP is a most commonly used methodology, it is common in industry that users spend a great deal of effort on conducting use case analysis aimed at identifying the business requirements, while class diagrams are seen as more closely associated with system design and implementation - therefore often delayed until a reasonable use case analysis is performed.

Larman (2002) supports the UP idea that the requirements analysis starts with creating use cases. The iterative process suggests starting with an initial use case model, stating the names of the use cases in the systems, and describing only the most important and risky ones; then continuing with analysis-design-development iterations. In the analysis phase of each iteration, the use cases are specified and a semantic class diagram called Domain Model is created. The concepts for the domain model are identified from the nouns in the use case descriptions.

Another example for a methodology that starts with functional modeling is Insfran et al. (2002). It starts by specifying high-level functional requirements, which are then systematically decomposed into a more detailed specification that constitutes the conceptual schema of the system. Insfran et al. define a requirements model (RM) that captures the functional and usage aspects, and a requirements analysis process that translates those requirements into a conceptual schema specification. The RM includes a functional refinement tree, i.e. a hierarchical decomposition of the business functions, which are used for building use case specifications. The use cases are used as basis for the conceptual modeling phase, which is carried out according to OO-Method (Pastor et al., 2001). The major activities at this stage are to define an object model, a dynamic model, and a functional model. This process is based on the analysis of each use case aimed at identifying classes and allocating their responsibilities. It utilizes sequence diagrams, which are created for each use case, where each sequence diagram includes the classes that participate in the use case.

Brown's process (Brown, 2002) starts with creating a context diagram, very similar to the one used with hierarchical DFDs. This diagram presents the system as a big square surrounded by the external entities communicating with it, including users and other systems. The next phase is creating use cases to describe the functional requirements, from which the classes are extracted later on.

But there are also UML-based methodologies that start the analysis process with data modeling. One example is UMM (Hofreiter et. al, 2006), an enhancement of UP specifically intended to model business processes and support the development of electronic data interchange (EDI) systems. Unlike the previously presented methodologies, UMM starts with a phase called Business Modeling in which the first step is performing Domain Analysis. Identifying the "top-level entities" in the problem domain is done using Business Operations Map (BOM), a tool that presents a hierarchy of Business Areas, Process Areas and Business Processes. Subsequently, use cases are created to describe the functional requirements. Though not using class diagrams UMM precedes the functional analysis with a mapping of the problem domain.

Catalysis (D'Souza & Wills, 1998) describes a similar process, suggesting creating a Problem Domain Model, which is a diagram presenting the main entities in the problem domain and their relationships, before writing the use cases.

ICONIX (Rosenberg & Kendall, 2001) suggest starting the analysis by creating a class diagram describing the real world entities and concepts in the problem domain, using the name Domain Model for this preliminary class diagram. According to the authors, the general class diagram, which describes the domain and not a specific solution, is an important basis and a glossary for creating use cases that describe the functional requirements. In fact, ICONIX is the only UML-based process we found that actually discusses the issue of analysis order and explains why it is better to create a domain model before describing the functional requirements.

Song et al. (2005) present a taxonomic class modeling (TCM) methodology for object-oriented analysis of business applications. In their approach, the first step is to develop a domain class diagram representing the important business activities at the analysis level. In later stages, design classes and then implementation classes, are added. The domain class diagram is created by examining various source documents, namely problem statements. Their methodology helps discovering classes from nouns, verb phrases and "hidden classes" not explicitly stated in the problem statement. The methodology synthesizes several class modeling techniques under one framework, integrating the noun analysis method, class categories, English sentence structure, checklists, and other heuristic rules for modeling.

While some methodologies advocate starting the analysis with functional modeling, and other – with class modeling, Maciaszek (2005) claims that there is no particular order for creating the use cases and class models, as these two activities are used concurrently and drive each another in successive development iterations. At the same time, he emphasizes the "leading role" of use cases in the software lifecycle. Thus Maciaszek (2005) also suggests that one of the analysis activities should drive the software development. The objective of our study is to address this specific issue and try to determine whether there is a preferable order of these activities.

Previous Research about Analysis Activities

Dobing & Parsons (2000) studied the role of use cases in UML and identified several problems with both the application and the theoretical underpinnings of use cases. In their opinion, the roles and values of use cases are unclear and debatable. Moreover, they claim that the process of moving forward, from use cases to classes, is neither universally accepted, even among use case adherents, nor does it appear to be clearly defined or articulated. Their statement, however, is not based on any empirical evidence.

The above argument is partially supported by Siau & Lee (2004) who examined the values of use case diagrams in interpreting requirements when used in conjunction with a class diagram. Their subjects were 31 university student volunteers who had completed at least one object-oriented UML course. In their study, they found out that the interpretation of a sequence combination of use cases and class diagrams had no effect on the problem domain understanding. They assert that, given there is no significant difference between the sequences of interpretations, the order of using or constructing the diagrams during the requirements analysis may not be important, and perhaps both diagrams may need to be constructed concurrently and modified interactively. It should be noted that Siau & Lee (2004) only examined user comprehension of the diagrams, not the quality of their construction, and that they considered only use case diagrams, but not the use case descriptions.

Shoval & Kabeli (2005) is the only experimental research we are aware of which deals explicitly with the order of analysis activities. They describe an experiment made to compare two orders of analysis using the Functional and Object-Oriented Methodology - FOOM (Shoval & Kabeli, 2001). The subjects received a requirements document, and were asked to perform the two analysis activities according to that methodology: to create a class diagram to model the data requirements, and to create OO-DFDs to model the functional requirements. The subjects were divided into two groups, each performing the same tasks in a different order. The results of that experiment revealed that starting the analysis with creating a class diagram leads to better class diagrams; yet no significant differences were obtained regarding the quality of the functional models. In addition, the subjects were asked about the better order of analysis activities; they preferred starting with creating a class diagram rather than OO-DFDs. This paper extends the study of Shoval & Kabeli (2005) to use-case based functional modeling.

RESEARCH GOAL AND HYPOTHESES

The goal of this study is to determine whether there is a preferred order for performing the analysis activities when class diagrams and use cases are used. With this goal in mind, we are interested to investigate the following research hypotheses:

H1: The class diagram quality is affected by the analysis order.

H2: The class diagram quality is affected by the order of presenting the requirements.

H3: The use case quality is affected by the analysis order.

H4: The use case quality is affected by the order of presenting the requirements.

H5: The analysts prefer a specific order of analysis activities.

However, statistical hypothesis testing is aimed at testing the so-called null hypotheses about the population of interest (the system analysts in our case), which usually state that the collected observations are the result of pure chance. Thus, the above research hypotheses H1-H5 can be considered as alternative hypotheses, while our null hypotheses, to be rejected or retained by our experiments, can be formulated as follows:

H0a: The class diagram quality is independent of the analysis order.

H0b: The class diagram quality is independent of the order of presenting the requirements.

H0c: The use case quality is independent of the analysis order.

H0d: The use case quality is independent of the order of presenting the requirements.

H0e: The analysts are indifferent with respect to the order of analysis activities.

THE EXPERIMENTAL DESIGN

To examine whether there is a preferred order for performing the analysis activities, we carried out a comparative experiment. In an ideal experiment we would compare the quality of the final systems produced by independent development teams each using a different order of analysis. However, since this is not practical, we can evaluate the analyst performance in doing the analysis itself, assuming that the analysis models are the foundations for design and implementation, and their quality is therefore critical to the success of the entire system. In order to simulate the analysis process, we provided the subjects with a requirements document describing the various requirements of the system, for which each of them was asked to create use cases and a class diagram. The quality of each analysis model was evaluated using the pre-determined grading schemes presented in sub-section 4.5 below.

The Research Model

Most experiments aimed at evaluating analysis products refer to data models only, and use a variation of Jenkins' model for evaluating user performance (Jenkins, 1982) that identifies the effect of three factors and the interaction between them: the data model being used, the task characteristics, and the human characteristics. A wide review of these studies is provided by Topi & Ramesh (2002), who suggest an enhancement of Jenkins' model. Based on these models, Figure 1 describes the research model of our experiment.

Figure 1. The research model

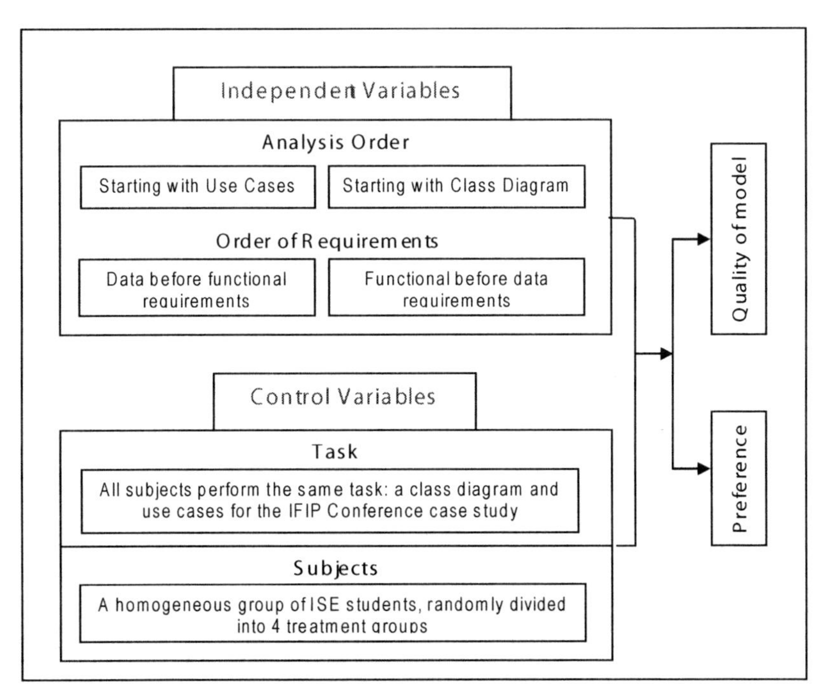

Table 1. Research groups

Group	Analysis Order	Order of Requirements
Group A1	1. Class Diagram 2. Use Cases	1. Data requirements 2. Functional requirements
Group A2	1. Class Diagram 2. Use Cases	1. Functional requirements 2. Data requirements
Group B1	1. Use Cases 2. Class Diagram	1. Data requirements 2. Functional requirements
Group B2	1. Use Cases 2. Class Diagram	1. Functional requirements 2. Data requirements

Independent Variables

Our main interest is the order of performing the two main analysis activities: creating a class diagram to describe the problem domain and writing use cases to describe the functional requirements. The analysis order is therefore our main independent variable. It has two possible values: starting with a class diagram and proceeding with the use cases, or vise versa.

In reality, analysts interact with the users, elicit their requirements, and based on that create the analysis models of the developed system. However, in an experimental setting we create an artificial environment: instead of a real problem and real users who express their needs and interact with the analysts, we prepared a requirements document, which is a narrative description of the user needs; the analysts (the experiment subjects) are supposed to create the analysis models based on that written document without any interactions with the actual users. Of course, such a requirements document should include both data-related and functional-related requirements. This may raise a problem since the order of presentation of the two types of requirements may affect the outcome, i.e. the quality of the analysis models created by the analysts. To explore the possible effect of the order of requirements presentation on the quality of models created by analysts, we prepared two versions of the requirements document: one version presenting the data-related requirements first and then the functional requirements; and another version presenting the same requirements in an opposite order. Putting together these two independent variables, we obtained four treatment groups, as shown in Table 1.

Control Variables

Two control variables are identified in the model: the task and the subjects. As the experiment task, we chose to use the IFIP Conference case study (Mathiassen et al, 2000) that was also used in Shoval & Kabeli (2005). This problem is of manageable size and can be solved easily. Furthermore, using the same case study as Shoval & Kabeli (2005) would strengthen the validity of the comparison between the results of the two experiments. Appendix A presents the class diagram and two use cases from the IFIP Conference case study solution.

The subjects were senior undergraduate students of Information Systems Engineering. We performed the experiment as a mid-term exam in the OO Analysis and Design course. Before the exam, the subjects learned the OO analysis approach and UML, including use cases and class diagrams. Having a homogeneous group of subjects, i.e., students in the same class who took the same courses and were trained by the same instructor, allowed us controlling potential biases

Table 2. Subjects' experience

Type of Experience	Number of Subjects			
	Study only	Less than 6 months	Less than a year	More than a year
Analysis	101	7	3	3
UML Analysis	112	1	0	1
Programming	97	8	3	6
OO Programming	105	4	1	4

such as differences in individual characteristics, analysis skills, and task-related experience. We verified this homogeneity using a questionnaire in which the subjects were asked to describe their analysis experience and skills. The results of this questionnaire are presented in Table 2. As can be seen, nearly all subjects had no practical experience in analysis, UML and OO programming. Besides all this, the subjects have been randomly assigned to the four treatment groups.

In addition, to control the training variable and to direct the subjects toward a unified style of describing use cases, an extra hour and a half tutorial was conducted in the class, which included teaching Alistair Cockburn's (2001) use case writing guidelines and solving a sample exercise.

In order to verify the tasks clarity and to estimate the time needed to complete each, a preliminary experiment, with participation of three experienced analysts, was conducted. They were asked to perform the same tasks and then to provide their comments. Following this preliminary experiment and the subjects' feedback, a few minor changes were made to the requirements document to make it clearer and some of the requirements were omitted to allow completing the tasks within two hours. Furthermore, we decided to hand the subjects of the real experiment a solution for one of the use cases as an example for the expected way of describing the use cases. To motivate the subjects to perform the tasks to the best of their ability, their grades in the experiment (mid-term exam) were considered as a part of the final course grade.

Table 3. Grading scheme of the class diagram

Element	Error	Error points
Class	Missing class	-6
	Superfluous class	-2
	Incorrect class type	-1
Attribute	Missing attribute or attribute in the wrong class	-2
	Superfluous attribute	-1
Relationship	Missing relationship	-4
	Erroneous relationship	-3
	Superfluous relationship	-3
	Incorrect relationship type	-2
Relationship multiplicity	Missing or incorrect multiplicity	-1
Inheritance	Missing inheritance	-6
	Ordinary relationship instead of inheritance	-2
	Superfluous inheritance	-2

Dependent Variables

The main dependent variable we used to evaluate the analysts performance is the quality of the created models. The model quality was measured using a grading scheme that represents the correctness of the analysis artifacts vs. the benchmark solution. The grading scheme for each model is described below.

In addition, we asked the subjects about their subjective preferences regarding the order of analysis activities, using a 7-point ordinal scale.

The Grading Schemes

The quality of the analysis models was measured using grading schemes that determined the number of points to deduct for each error type in every model. The grading schemes we created were based on the ideas from Batra, Hoffer & Bostrom (1990) and used in other studies to evaluate conceptual data models including class diagrams (e.g., Kim & March, 1995; Shoval & Shiran, 1997; and Shoval & Kabeli, 2005). Table 3 presents the grading scheme for the class diagram.

Since class diagram is a well-defined tool with a strict structure and syntax, mapping the possible errors in it was straightforward. Use cases, on the other hand, are less structured and described using free text. Mapping the possible errors in the use case descriptions required first defining the components that the analysts are expected to describe and that we would like to measure in terms of quality. Assisted by the use of case error mapping in Anda & Sjoberg (2002), which is also based on Cockburn's approach (Cockburn, 2001), we identified the following three components:

- Actor: the external entity owning the goal of executing the use case
- User goal: the topmost goal addressed by the use case; each use case has to deliver a user goal
- Sub-goal: each user goal is a collection of sub-goals - steps in accomplishing that goal

After identifying the components, as in the class diagram, we mapped the possible errors in each component, and determined the error points. Table 4 presents the grading scheme for use cases.

In addition to these "semantic errors" (Batra, 1993) we identified the following "syntactic errors" and assigned them altogether six error points: irrational solution, untidiness and lack of logical order, unclear script, and inconsistent description.

RESULTS

The experiment was conducted in a controlled environment and in an exam format. The exam

Table 4. Grading scheme of the use cases

Element	Error	Error points
Actor	Incorrect actor	-4
	Inconsistent actor	-2
User goal	Missing goal	-10
	Goal appears in title only	-6
	Goal described as a part of other goal	-4
	Superfluous goal	-2
Sub-goal	Missing sub-goal	-3
	Superfluous or erroneous sub-goal	-2

was taken by 121 undergraduate students; it was planned to take two hours, but an extension of half hour was granted to allow the subjects to complete their tasks.

We were interested to investigate the effect caused by the two independent variables and their interaction on the dependent variable Quality (grade). Since grade is a continuous variable, and each independent factor has a discrete number of levels, the suitable statistical test is two-way analysis of variance (Montgomery et al., 2001). The analysts' preferences were tested using Wilcoxon test, a non-parametric test that allows testing results from an ordinal scale without imposing any other restrictions on the data samples.

Quality of Models

The two analysis models are of different nature and require different grading schemes to evaluate, which makes them incomparable. We hence compared the quality (grade) of each model separately. The null hypothesis for the statistical tests is that there is no difference between the values of the dependent variable (model quality) for the different levels of the independent variables tested.

Table 5 presents the grading results of the class diagrams. The first column presents the factor (independent variable) whose effect on the dependent variable (class diagram grade) is examined; the second column presents the values of the independent variables; the third (N) is the number of subjects in the group; the fourth is the mean grade of the class diagram; the fifth is the F statistic of the independent variable; the sixth is the p-value; and the last column indicates if the difference between the results is statistically significant at the $\alpha = 0.05$ significance level (i.e., cases where the null hypothesis can be rejected).

The first row presents the effect of the main independent variable: analysis order on the class diagram grades. As can be seen, the grades are significantly higher when starting the analysis with class diagram (73.63) compared to when starting with use cases (70.25).

The second row presents the effect caused by the secondary independent variable: order of requirements in the requirements document. As can be seen, no significant difference is found between the two orders. It means that the order in which the requirements are presented to the analysts does not affect the quality of the resulting class diagrams.

Table 5. Class diagram grades

Factor	Factor value	N	Mean grade (%)	F	p-value	Significance in favor of
Analysis Order	Class Diagram → Use Cases	57	**73.63**	4.386	.038	Starting with class diagram
	Use Cases → Class Diagram	64	**70.25**			
Order of Requirements	Data → Functional	61	71.43	.178	.674	-
	Functional → Data	60	72.27			
Interaction: Analysis Order × Order of Requirements	Class Diagram → Use Cases Data → Functional	27	73.04	.081	.776	-
	Class Diagram → Use Cases Functional → Data	30	74.17			
	Use Cases → Class Diagram Data → Functional	34	70.15			
	Use Cases → Class Diagram Functional → Data	30	70.37			

Table 6. Use Case grades

Factor	Factor value	N	Mean grade (%)	F	p-value	Significance in favor of
Analysis Order	Class Diagram → Use Cases	57	63.72	.192	.662	-
	Use Cases → Class Diagram	64	65.03			
Order of Requirements	Data → Functional	61	66.62	3.186	.077	-
	Functional → Data	60	62.17			
Interaction: Analysis Order × Order of Requirements	Class Diagram → Use Cases Data → Functional	27	65.19	.398	.529	-
	Class Diagram → Use Cases Functional → Data	30	62.40			
	Use Cases → Class Diagram Data → Functional	34	67.76			
	Use Cases → Class Diagram Functional → Data	30	61.93			

The third row presents the effect caused by the interaction between the analysis order and the order of requirements, showing no significant effect of this interaction on the grades of the class diagram. This means that the effect of one of the independent factors on the dependent variable is not affected by the value of the other independent variable.

It should be noted that the quality results of the data model are consistent with the results obtained in Shoval & Kabeli (2005), where OO-DFDs (rather than use cases) were used to model the functional requirements.

Table 6 presents the quality results of the use cases. As can be seen there are no significant differences between the two analysis orders. As for the effect of the requirements order, again, there is no significant difference, and no significant effect caused by the interaction between the analysis order and the requirements order. It should be noted that the quality results of the use case-based functional model are consistent with the results obtained in Shoval & Kabeli (2005) for the OO-DFD functional model.

Since the p-value for the order of requirements factor (.077) is close to 0.05 (α) and no significant interaction was found between the two independent factors, we performed a one-tailed t-test with the same p-value for the use case grades. In this test, the alternative hypothesis was that use case grades are higher when data requirements are presented before functional requirements and vice versa (while the interaction between the order of requirements and analysis order is not tested). The test showed that the difference between the two requirements orders is significant, namely: there is an advantage if data requirements are presented before functional requirements. This is also visible in the "Interaction" row of Table 6: when the analysis starts with a class diagram and data requirements are presented first, the average grade is 65.19 compared to 62.40 when the analysis is performed in the same order but functional requirements are presented first. Similarly, when the analysis starts with use cases the average grades are 67.76 and 61.93, respectively.

Looking at the grades of the class diagrams (Table 5) and the use cases (Table 6), we see that the use case grades were, on average, lower than those of the class diagrams. This may be explained by several factors: A) Different grading schemes: it is possible that because of the grading schemes and error points we adopted, more points were deducted due to errors in uses cases than in class diagram. B) Task complexity: as discussed earlier, it is possible that use case modeling is more complex than class modeling, disregarding the order they are worked out. C) Task clarity: it is possible that the functional requirements were less clear than the data requirements. D) Model

Table 7. Analysts' Preferences

The order in which the subjects worked	N[1]	Mean preference	Standard deviation
Class Diagram → Use Cases	22	2.91	1.54
Use Cases → Class Diagram	18	2.61	1.82
All together	**40**	**2.78**	**1.66**

formality: As already discussed, a class diagram is well structured and has a clear and simple syntax, while a use case description is less structured. A lack of well-defined syntax increases the frequency of errors caused by misunderstanding of the required task (Shneiderman, 1978). At any rate, as previously indicated, we did not attempt to combine or compare the results of the two models; we only compared the differences between the results within each model. Therefore, the above differences in the grades across the different models do not have any impact on the conclusions of this study.

Analysts' Preferences

After the experiment each subject was asked to express to what degree he/she believes that the order of analysis used is good/appropriate using a 1-7 point scale, where 1 means total preference to start with a class diagram, 4 means indifference, and 7 means total preference to start with use cases . Table 7 presents the results, for each group of subjects and for all together.

The results show that the subjects definitely believe that it is better first to create a class diagram and then use cases (mean preference of all is 2.78; much closer to 1 than to 7). It is interesting to see that the subjects who started the analysis with use cases showed even stronger preference to start with creating a class diagram (2.61 compared to 2.91), though this difference was not found statistically significant. The preference towards an order of analysis starting with a class diagram matches both our expectation regarding the analysts' preferences and the results obtained in an earlier experiment with OO-DFD functional models (Shoval & Kabeli, 2005).

SUMMARY AND FURTHER RESEARCH

The principal purpose of this research was to compare two interchangeable orders of performing the main analysis tasks in a use case-based approach: creating a class diagram to model the problem domain and creating use cases to describe the functional requirements of the system. Different existing use case-based development methodologies recommend using different analysis orders, usually without explaining the reason behind this recommendation; some do not refer to the issue at all. The potential effect of the analysis order on the quality of the resulting analysis models, and thus on the quality of the entire software system, is the reason why studying this effect is so important.

The results of our experiment reveal that starting the analysis by creating a class diagram leads to a better class diagram. On the other hand, we did not find a significant effect of the analysis order on the quality of the use cases. Thus, starting the analysis process with class diagrams should result in a better overall quality of the analysis models. For instructors in UML courses, our results suggest that the class diagrams should be taught before use cases as the class diagrams should drive the analysis activities in a real-world setting.

We also tested the effect of the order of the user requirements presentation on the results, by preparing two different versions of the requirements document. We found that the use case grades are higher at the 7.7% significance level when the data requirements are presented first in the requirements document. However, no significant effect of the requirements order was found on the class diagram grades. Using a subjective questionnaire after the experiment we found that the subjects believe it is better to begin the analysis by creating a class diagram.

Interestingly, most of the results we obtained in this experiment are consistent with those obtained in the earlier experiment (Shoval & Kabeli, 2005) in which a variation of the same requirements document was used, but utilizing a somewhat different class diagram notation, and OO-DFDs instead of use cases to model the functional requirements. It appears that the conclusions with respect to the preferred order of analysis activities hold irrespective of the analysis methodology.

As other controlled experiments, which compare methods and models in a laboratory setting, this one too has limitations threatening its external validity. For a discussion on common limitations of such experiments, see Topi & Ramesh (2002). An obvious limitation is that we used a relatively small and simple problem while in reality problems are much bigger and more complex; we cannot be sure how size and complexity of a system would affect the results with respect to the order of analysis activities.

Another limitation which hampers the external validity of the results is that they are based on one case study, the IFIP Conference system, which may represent only data-driven (or MIS) information systems. We cannot be sure that the results are also valid for other types (e.g. real-time systems).

Of course, there is the limitation that we used students with almost no industrial experience as surrogates for analysts. As we know, this limitation is common to almost all experimental work published in the area (Topi & Ramesh, 2002; Sjøberg et al., 2005). We cannot predict if and how the cumulative experience of analysts might affect the preferred order of analysis activities.

A potential limitation of the experiment is its grading schemes. Some subjectivity may exist in the weights of error points given in Tables 3 and 4. We determined the weights based on our assessment of the importance of each error type. In doing so we followed earlier studies who also adopted subjective grading schemes to assess quality of models. For example, Batra, Hoffer & Bostrom (1990) categorized errors based on severity and distinguished between minor, medium, and major error types. Similarly, Meyer (1998) distinguished between minor and major errors and assigned penalty points to each error type. They based their values on the computed overall performance of their subjects using a formula that considered the number of errors and the penalty points, similar to what we did in this study. The potential problem with such grading schemes is that the subjective weights (error points) assigned to error types may affect the overall results. The problem is that there are no objective weights and grading schemes for different methods or models. This issue deserves a separate research.

Being a "laboratory" experiment, we used a requirements document to represent the real world and the users' needs; we actually forced a one-way modeling process, where the analyst/subject reads a given requirements document and creates from it the analysis models to the best of his/her understanding. This is not really an analysis task. In reality, there are real users who have real needs, and the analysis process involves a lot of interaction between analysts and users. Although we may assume that such interaction would affect the quality of the resulting models, the question of which is the better order of activities is still valid. As already discussed, in spite of being aware of the interactive nature of the analysis process, still different methodologies prescribe certain orders of activities without even questioning if the prescribed order is good. Even if we agree that this study does not necessarily simulate a real analysis process, it at least proposes a good strategy to create analysis models in cases where user requirements are already given in the form of requirements documents. Moreover, it suggests a good strategy to teach and train UML techniques.

For further research, we suggest to repeat the experiment using several case studies of different size/complexity and from different domains, beyond data-driven systems, to see if the preferred order of analysis activities is affected by problem size/complexity and domain. It is especially interesting to see the results when simulating the analysis process similar to the real-world analysis, where the analysts have to elicit the requirements rather than working with pre-defined requirement documents. Another point is to conduct controlled experiments with subjects who are experienced analysts rather than undergraduate students.

REFERENCES

Alabiso, B. (1988). Transformation of data flow analysis models to object-oriented design. *Proceeding of OOPSLA '88 Conference*, San Diego, CA, 335-353.

Anda, B., & Sjoberg, D. (2002). Towards an inspection technique for use case models. *Proceeding of the 14th International Conference on Software Engineering and Knowledge Engineering (SEKE '02),* Ischia, Italy, 127-134.

Bailin, S. (1989). An object-oriented requirements specification method. *Communication of the ACM, 32*(5), 608-623.

Batra, D. (1993). A framework for studying human error behavior in conceptual database modeling. *Information & Management, 24*, 121-131.

Batra, D., Hoffer, J., & Bostrom, R. (1990). Comparing representations with the Relational and Extended Entity Relationship model. *Communications of the ACM, 33*, 126-139.

Brown, D. (2002). *An Introduction to Object-Oriented Analysis* (2nd Edition): Wiley.

Coad, O., & Yourdon, E. (1991). *Object-Oriented Design.* Englewood Cliffs, NJ: Prentice Hall.

Cockburn, A. (2001). *Writing Effective Use Cases.* Addison-Wesley.

DeMarco, T. (1978). *Structured Analysis and System Specifications*: Yourdon Press.

Dobing, B., & Parsons, J. (2000). Understanding the role of use cases in UML: A review and research agenda. *Journal of Database Management, 11*(4), 28-36.

D'Souza, D., & Wills, A. (1998). *Objects, Components, and Frameworks with UML: The Catalysis Approach.* Addison-Wesley.

Hofreiter, B., Huemer, C., Liegl, P., Schuster, R., & Zapletal, M. (2006). *UN/CEFACT'S modeling methodology (UMM): A UML profile for B2B e-commerce.* Retrieved March, 2008, from http://dme.researchstudio.at/publications/2006/

Insfran, E., Pastor, O., & Wieringa, R. (2002). Requirements engineering-based conceptual modeling. *Requirements Engineering, 7*, 61-72.

Jacobson, I., Booch, G., & Rumbaugh, L. (1999). *The Unified Software Development Process.* Addison-Wesley.

Jacobson, I., Christerson, M., Jonsson, P., & Overgaard, G. (1992). *Object-Oriented Software Engineering: A Use Case Driven Approach.* Addison Wesley.

Jenkins, A. (1982). *MIS Decision Variables and Decision Making Performance.* Ann Arbor, MI: UMI Research Press.

Kim, Y., & March, S. (1995). Comparing data modeling formalisms. *Communications of the ACM, 38*(6), 103-115.

Larman, C. (2002). *Applying UML and Patterns: An Introduction to Object-Oriented Analysis and*

Design, and the Unified Process (2nd Edition). Prentice Hall.

Maciaszek, L. (2005). *Requirements Analysis and System Design: Developing Information Systems with UML.* 2nd Edition: Addison-Wesley.

Mathiassen, L., Munk-Madsen, A., Nielsen, P., & Stage, J. (2000). *Object Oriented Analysis and Design.* Marko Publishing, Alborg, Denmark.

Meyer, B. (1998). *Object Oriented Software Construction.* Prentice Hall.

Montgomery, C., Runger, C., & Hubele, F. (2001). *Engineering Statistics* (2nd Edition). Wiley.

Pastor, O., Gomez, J., Insfran, E., & Pelechano, V. (2001). The OO-Method approach for information systems modeling: From object-oriented conceptual models to automated programming. *Information Systems, 26*(7), 507-534.

Rockwell, S., & Bajaj, A. (2005). COGEVAL: Applying cognitive theories to evaluate conceptual models. In K. Siau (Ed.), *Advanced Topics in Databases Research.* Idea Group, Hershey, PA, 255-282.

Rosenberg, D., & Kendall, S. (2001). *Applied Use Case-Driven Object Modeling.* Addison-Wesley.

Rubin, K., & Goldberg, A. (1992). Object Behavior Analysis. *Communications of the ACM,* 35(9), 48-62.

Rumbaugh, J., Blaha, M., Premerlani, W., Eddy, F., & Lorensen, W. (1991). *Object Oriented Modeling and Design.* Englewood Cliffs, NJ: Prentice Hall.

RUP - Rational Unified Process. Retrieved March, 2008 from http://www.e-learningcenter.com/rup.htm

Shlaer, S., & Mellor, S. (1992). *Object Lifecycles: Modeling the World in States*: Prentice Hall.

Shneiderman, B. (1978). Improving the human factor aspects of database interactions. *ACM Transactions on Database Systems, 3*(4), 417-439.

Shoval, P., & Kabeli, J. (2001). FOOM: functional- and object-oriented analysis and design of information systems: An integrated methodology. *Journal of Database Management, 12*(1), 15-25.

Shoval, P., & Kabeli, J. (2005). Data modeling or functional analysis: which comes next? - an experimental comparison using FOOM methodology. *Communications of the AIS, 16*, 827-843.

Shoval, P., & Shiran, S. (1997). Entity-relationship and object-oriented data modeling - an experimental comparison of design quality. *Data & Knowledge Engineering, 21*, 297-315.

Siau, K., & Lee, L. (2004). Are use case and class diagrams complementary in requirements analysis? an experimental study on use case and class diagrams in UML. *Requirements Engineering, 9*, 229-237.

Sjøberg, D., Hannay, J., Hansen, O., Kampenes, V., Karahasanovi, A., Liborg, N., & Rekdal, A. (2005). A survey of controlled experiments in software engineering. *IEEE Transactions on Software Engineering, 31*(9), 733-753.

Song, I., Yano, K., Trujillo, J., & Lujan-Mora, S. (2005). A taxonomic class modeling methodology for object-oriented analysis. In K. Siau (Ed.), *Advanced Topics in Database Research, 4*. 216-240. Hershey: Idea Group.

Topi, H., & Ramesh, V. (2002). Human factors research on data modeling: a review of prior research, an extended framework and future research directions. *Journal of Database Management, 13*(2), 188-217.

Wand, Y., & Weber, R. (1995). On the deep structure of information systems. *Journal of Information Systems, 5*, 203-223.

Weber, R. (2003). Conceptual modeling and ontology: possibilities and pitfalls. *Journal of Database management, 14*(3), 1-20.

Yourdon, E. (1989). *Modern Structured Analysis*: Prentice Hall.

ENDNOTES

[1] Due to space limitation, we do not explain the diagrams and the use case descriptions.

[2] UP defines a preliminary optional phase called Business Use Case Model, which includes creating a class diagram named Domain Model that describes the main entities in the problem domain. However, it does not suggest any relationship between the domain model, the use cases, and the analysis class diagram.

[3] OO-DFD, Object-Oriented DFD, is a variation of DFD that is used in FOOM. OO-DFDs include classes rather than data-stores in traditional DFD.

[4] Due to a possible ethical problem that participants assigned in one group would score less than the other because of difference in models, we computed the average grade of participants in the "best" group and compensated members in the other groups accordingly.

[5] There might be some subjectivity in the above grading schemes, in particular in the number of error points. This limitation will be elaborated in the Summary section. At any rate, note that we applied a specific grading scheme for each model separately, but we did not combine or compare the results across the two models.

[6] The importance of this finding on the preferred order of requirements is limited, however, because in reality, the users' requirements are not presented in a specific order that can be controlled, as in the experiment, but rather the data and functional orders are intermingled. Note also that we did not find an equivalent result in Table 5 for the quality of the data models. In that case, the mean grades in the two orders of requirements were almost the same.

[7] Recall that each participant performed the tasks according to one order only, so he/she could only express his subjective preference based on his/her order of task performance.

[8] In this case, not all participants replied to this question, only 40.

APPENDIX A. CLASS DIAGRAM AND USE CASES OF THE IFIP CONFERENCE CASE STUDY

Figure 2 presents the class diagram of the IFIP Conference system. Figure 3 presents the use case diagram of system. Below are descriptions of two use cases of the system: UC1: Create Session; and UC2: Register to a Conference.

UC1: Create Session

Goal: The Program Committee (PC) Chairman creates a session according to the conference subjects.
Primary Actor: PC Chairman
Pre-Condition: A preliminary conference program exists.
Post-Condition: The new session's details are stored in the system.
Main Success Scenario:

1. The PC Chairman requests to create a new conference session.

Steps 2-4 can be performed in any order:

2. The PC Chairman enters the session's details: name, subject, date, start time, end time and conference hall.
3. The PC Chairman marks the session as a Panel and enters its description.
4. The PC Chairman enters the resources needed for the session.
5. The system saves the new session's details.

Extensions:

3a. The PC Chairman marks the session as a Lecture.
3a1. The PC Chairman enters the lecture's abstract.
3b. The PC Chairman marks the session as an Article Presentation.

UC3: Register to a Conference

Goal: An Attendee registers to a conference.
Primary Actor: Attendee
Pre Condition: A preliminary conference program exists.
Post Condition: The attendee's details are stored in the system and he is registered to the conference.
Main Success Scenario:

1. The attendee requests to register for a conference.
2. The attendee enters his details.
3. The system saves the attendee's details.

Extensions:

2a. the attendees requests to apply for a function in the conference:
2a1. the attendee enters the details of the function he is interested in (Lecturer, Panelist or Panel Chairman), his qualifications and interest areas.
2a2. the system sets the attendee's status to "Suggested".

Figure 2. Class diagram of the IFIP Conference System

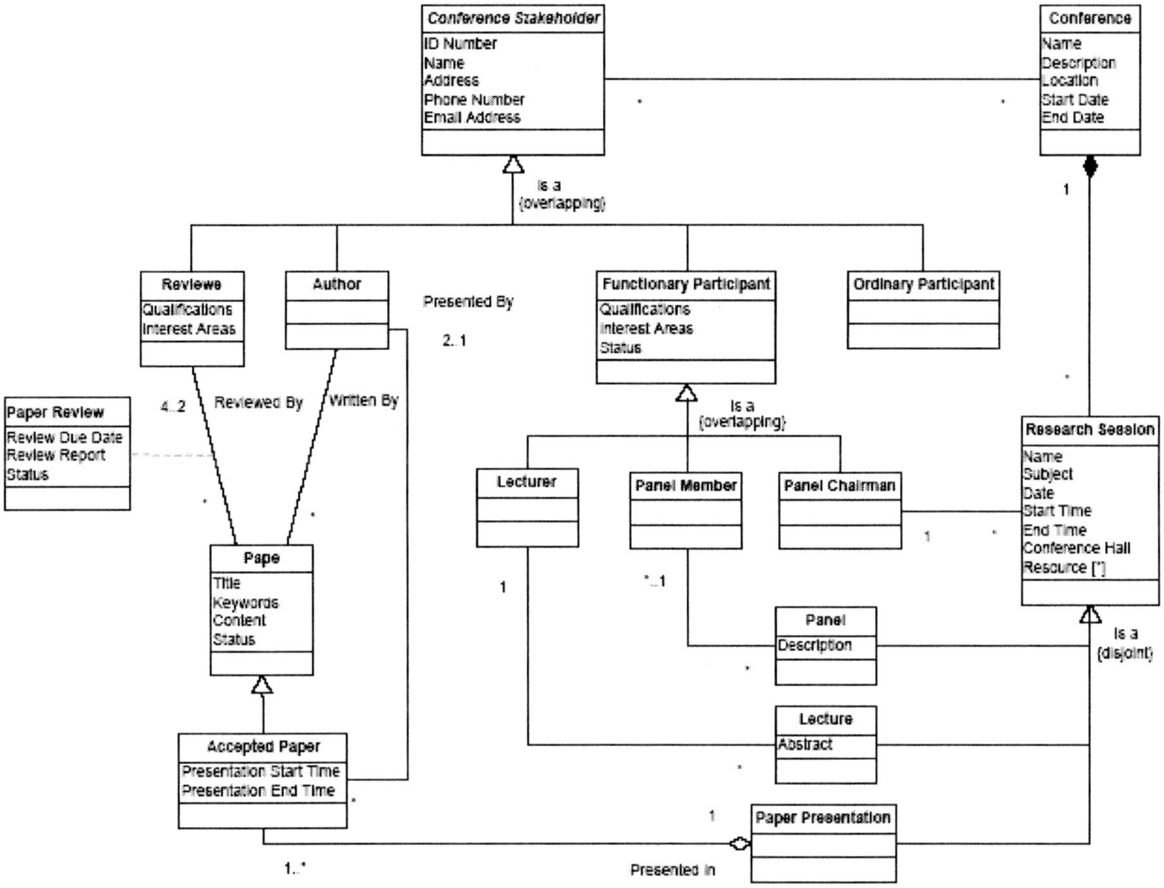

Figure 3. Use case diagram of the IFIP Conference System

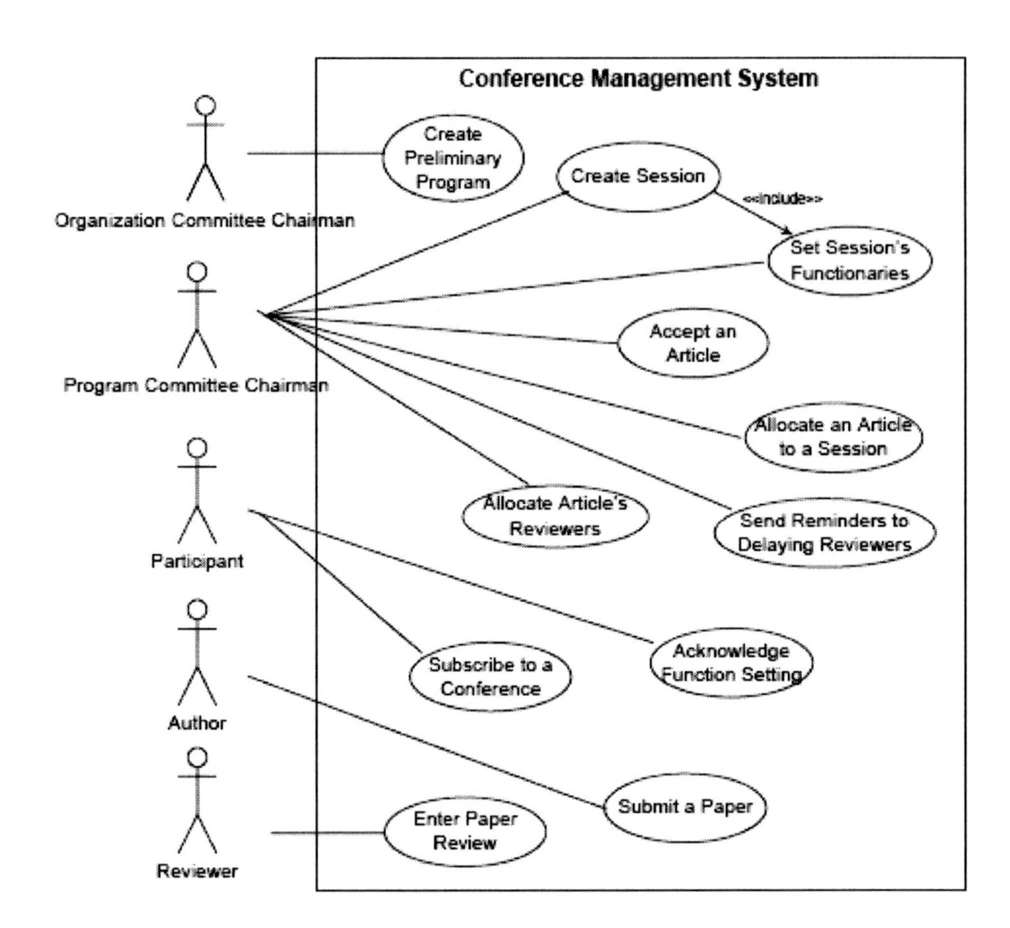

Section III
Modeling Frameworks, Architectures, and Applications

Chapter VIII
OntoFrame:
An Ontological Framework for Method Engineering

Mauri Leppänen
University of Jyväskylä, Finland

ABSTRACT

A large number of strategies, approaches, meta models, techniques and procedures have been suggested to support method engineering (ME). Most of these artifacts, here called the ME artifacts, have been constructed, in an inductive manner, synthesizing ME practice and existing ISD methods without any theory-driven conceptual foundation. Also those ME artifacts which have some conceptual groundwork have been anchored on foundations that only partly cover ME. This chapter presents an ontological framework, called OntoFrame, which can be used as a coherent conceptual foundation for the construction, analysis and comparison of ME artifacts. Due to its largeness, the authors here describe its modular structure composed of multiple ontologies. For each ontology, they highlight its purpose, subdomains, and theoretical foundations. The authors also outline the approaches and process by which OntoFrame has been constructed and deploy OntoFrame to make a comparative analysis of existing conceptual artifacts.

INTRODUCTION

Method engineering (ME) means actions by which an information systems development (ISD) method is developed, and later customized and configured to fit the needs of an organization or an ISD project. ME is far from trivial in practice. In the first place, the ISD methods are abstract things with divergent semantic and pragmatic meanings. The former implies that conceptions of what the ISD methods should contain may vary substantially (Fitzgerald et al., 2002; Hirschheim et al., 1995; Iivari et al., 2001; Graham et al., 1997; Heym et al., 1992; Avison et al., 1995; Leppänen 2005). The latter suggests that views of roles, both technical and political, which the

ISD methods play in ISD may be quite different (Chang et al., 2002; Fitzgerald et al., 2002; Wastell, 1996). The existing methods also differ from one another in their fundamental assumptions and approaches (Fitzgerald et al., 2002; Iivari et al., 2001). Second, it is often difficult to characterize the target ISD situation in a way which makes it possible to conduct a proper selection from and a suitable adaptation in existing methods for an organization or a project (Aydin, 2007). Third, it is frequently unclear which kind of strategies (i.e. from "scratch", integration, adaptation) and processes should be applied at each stage of the engineering of an ISD method. Fourth, most of the method engineering (ME) situations suffer from the lack of time and other resources, causing demands for carrying out ME actions in a straightforward and efficient manner.

A large array of ME strategies and approaches (e.g. Kelly 2007; Kumar et al., 1992; Oie, 1995; Plihon et al., 1998; Ralyte et al., 2003; Rolland et al., 1996), meta models (e.g. Graham et al., 1997; Harmsen, 1997; Heym et al., 1992; Kelly et al., 1996; OMG, 2005; Prakash, 1999; Venable, 1993), ME techniques (e.g. Kinnunen et al., 1996; Kornyshova et al., 2007; Leppänen, 2000; Punter et al., 1996; Saeki, 2003) and ME procedures (e.g. Harmsen, 1997; Karlsson et al., 2004; Nuseibeh et al., 1996; Song, 1997) have been suggested to support method engineering. These *ME artifacts*, as we call them here, sustain, however, several kinds of shortcomings and deficiencies (Leppänen, 2005). One of the major limitations in them is the lack of a uniform and consistent conceptual foundation. Most of the ME artifacts have been derived, in an inductive manner, from ME practice and existing ISD methods without any theory-based conceptual ground. Also those ME artifacts that have a well-defined underpinning have been anchored on foundations that only partly cover the ME domain.

ME is a very multifaceted domain. It concerns not only ME activities, ME deliverables, ME tools, ME actors and organizational units, but, through its main outcome, an ISD method, also ISD activities, ISD deliverables, ISD actors, ISD tools, etc. Furthermore, ME involves indirectly, through information system (IS) models and their implementations, the IS contexts as well as those contexts that utilize information services provided by the ISs. Thus, in constructing an ME artifact it is necessary to anchor it on a coherent conceptualization that covers ME, ISD and IS, as well as the ISD and ME methods. In ontology engineering literature (e.g. Gruber, 1993) a specification of the conceptualization of a domain is commonly called an *ontology*. Hence, what we need here is a coherent set of ontologies which cover all the aforementioned sub-domains of ME.

The purpose of this chapter is to suggest an ontological framework, called OntoFrame, which serves as a coherent conceptual foundation for the construction, analysis and comparison of ME artifacts. OntoFrame is composed of multiple ontologies that together embrace all the sub-domains of ME. It has been constructed by searching for "universal" theoretic constructs in the literature (the *deductive* approach), by analyzing existing frameworks, reference models and meta models (the *inductive* approach), and by deriving more specific ontologies from generic ontologies above them in the framework (the *top-down* approach, Uschold et al., 1996). The construction work has been directed by the goals stated in terms of extensiveness, modularity, consistency, coherence, clarity, naturalness, generativeness, theory basis and applicability. The ontological framework is quite large, comprising 16 individual ontologies (Leppänen, 2005). Here, we are only able to describe its overall structure and outline the ontologies on a general level (Section 2). We also discuss the theoretical background and approaches followed in engineering it (Section 3). In addition, we demonstrate the applicability of OntoFrame by deploying it in a comparative analysis of relevant works (Section 4). The chapter ends with the discussion and conclusions (Section 5).

ONTOFRAME

A *conceptual framework* is a kind of intellectual structure that helps us determine which phenomena are meaningful and which are not. OntoFrame, as an ontological framework, aims to provide concepts and constructs by which we can conceive, understand, structure and represent phenomena relevant to method engineering. Deriving from two disciplines, ontology engineering (e.g. Burton-Jones et al., 2005; Gruber, 1995; Ruiz et al., 2004) and conceptual modeling (e.g. Falkenberg et al., 1998; Krogstie, 1995), the following goals have been set for OntoFrame. To advance communication between people, the framework should be clear and natural. To balance different needs for specificity and generality, OntoFrame should be composed of ontologies that are located at different levels of generality. To build on some more stable and solid ground, the main building blocks of OntoFrame should be driven from relevant theories. To enable extensions and still maintain coherence and consistence, OntoFrame should be modular and generative. To cover the relevant phenomena of ME, OntoFrame should also be extensive. And last but not least, OntoFrame should be applicable in the construction, analysis and comparison of ME artifacts.

In the following, we first present an overall structure of OntoFrame, and then describe each of its parts in more detail.

An Overall Structure

The ontological framework is composed of four main parts. The first part, the *core ontology*, provides basic concepts and constructs to conceive and structure human conceptions of reality and the use of language in general. The second part, the *contextual ontologies*, conceptualizes information processing as contexts or parts thereof, on some processing layer, from some perspective, and as being modeled on some model level. The third part, the *layer-specific ontologies*, has been specialized from those above in the framework to cover the sub-domains of IS, ISD and ME with more special concepts and constructs. The fourth part, the *method ontologies*, conceptualizes the nature, structure and contents of the ISD method and the ME method. The main parts and the ontologies included in them are described in Figure 1. The heads of the arrows point to the ontologies which the other ontologies have been derived from. In the following, we describe each of these main parts and ontologies in terms of purposes, sub-domains, and theoretical foundations.

Core Ontology

The purpose of the core ontology is to provide the basic concepts and constructs for conceiving, understanding, structuring and representing fundamentals of reality. It comprises seven ontologies: generic ontology, semiotic ontology, intension/extension ontology, language ontology, state transition ontology, abstraction ontology, and UoD ontology. Figure 2 presents the ontologies and the most fundamental concepts in the generic ontology and the semiotic ontology, as well as, how the ontologies are related to one another. Each of these ontologies has the scope, purpose and role of its own. The *generic ontology*, rooted on the constructivist position (Falkenberg et al., 1998), provides the most generic concepts from which all the other concepts have been specialized (see the arrows in Figure 2). It is the top ontology (Guarino, 1998), or the foundational ontology (Guizzardi et al., 2001), in our framework. The most elementary concept is 'thing', meaning any phenomenon in the "objective" or subjective reality. The *semiotic ontology* defines concepts that are needed to recognize the semiotic roles of and relationships between the things. The main concepts, adopted from semiotics (Ogden et al., 1923), are 'concept', 'sign', and 'referent'. The *intension/extension ontology* serves as a conceptual mechanism to specialize the notion of concept and defines its semantic meaning (Hautamäki, 1986).

Figure 1. An overall structure of OntoFrame

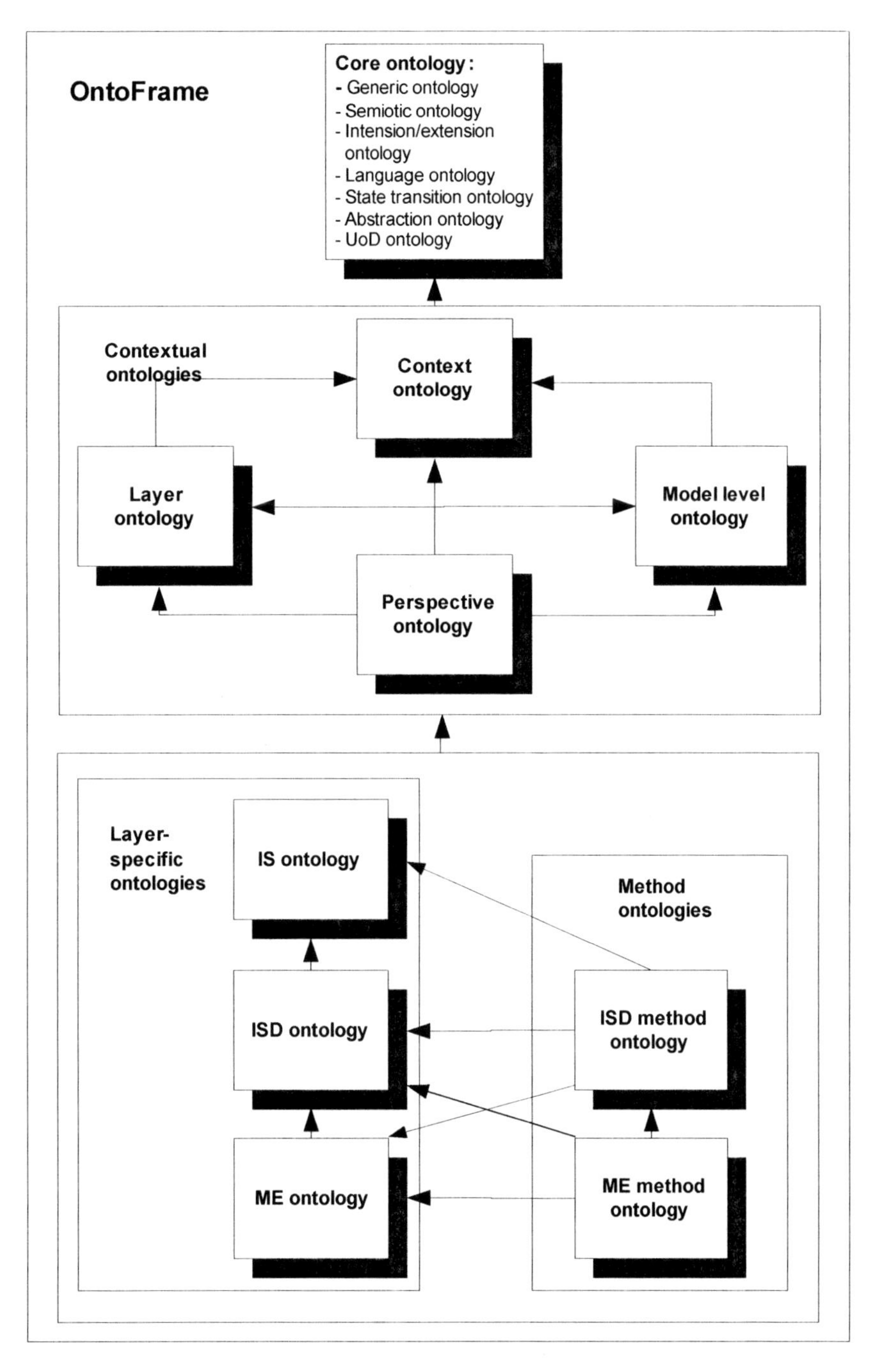

The notions of intension and extension enable to differentiate between 'basic concept', 'derived concept', 'abstract concept', 'concrete concept', 'instance concept' and 'type concept'.

The *language ontology* provides concepts for defining the syntax and semantics of a language. Based on linguistics (e.g. Morris, 1938) and some works in the IS field (e.g. Falkenberg et al., 1998; Krogstie, 1995; ter Hofstede et al., 1997), it defines

Figure 2. An overall structure of the core ontology

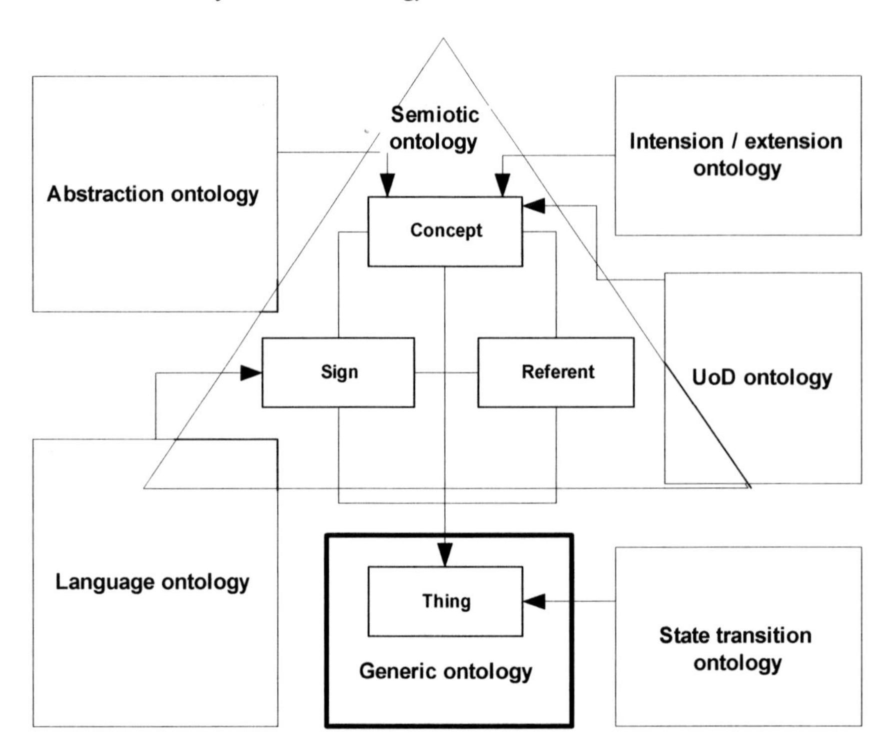

concepts such as 'language', 'alphabet', 'symbol', and 'expression'. The *state transition ontology* is composed of concepts and constructs for the recognition of dynamic phenomena in reality in terms of states, state transitions, and events (cf. Mesarovic et al., 1970; Falkenberg et al., 1998). The *abstraction ontology* serves concepts and constructs for abstraction by classification, generalization, aggregation, and grouping. Deriving from the intension/extension ontology, it also distinguishes between the first order abstraction and the second order abstraction (or the predicate abstraction). It is based on the philosophy of science and abstraction theories by e.g. Motschnig-Pitrik et al. (1995), Mylopoulos (1998), Motschnig-Pitrik et al. (1999), Goldstein et al. (1999), Wand et al. (1999), Opdahl et al. (2001), and Henderson-Sellers et al. (1999). The *UoD* (Universe of Discourse) *ontology* is composed of consolidated concepts through which reality can be conceived from a selected viewpoint. These concepts are 'UoD state', 'UoD behavior' and 'UoD evolution'.

Contextual Ontologies

The contextual ontologies help us recognize, understand, structure and represent phenomena related to information processing (a) as some contexts, (b) on some processing layer, (c) from some perspective, and (d) as being modeled on some model level. These ontologies are: context ontology, layer ontology, perspective ontology, and model level ontology. Figure 3 represents the four ontologies as orthogonal to one another. In the middle of the figure there are phenomena that are related to an information processing system (IPS), as well as its utilizing system (US) and object system (OS). The *utilizing system* is a system that exploits information provided by the IPS. The *object system* means a system about which the IPS collects, stores, processes and disseminates information for the US.

The *context ontology* defines seven related contextual domains, called the purpose domain (P), the actor domain (Ar), the action domain (An),

Figure 3. Contextual ontologies

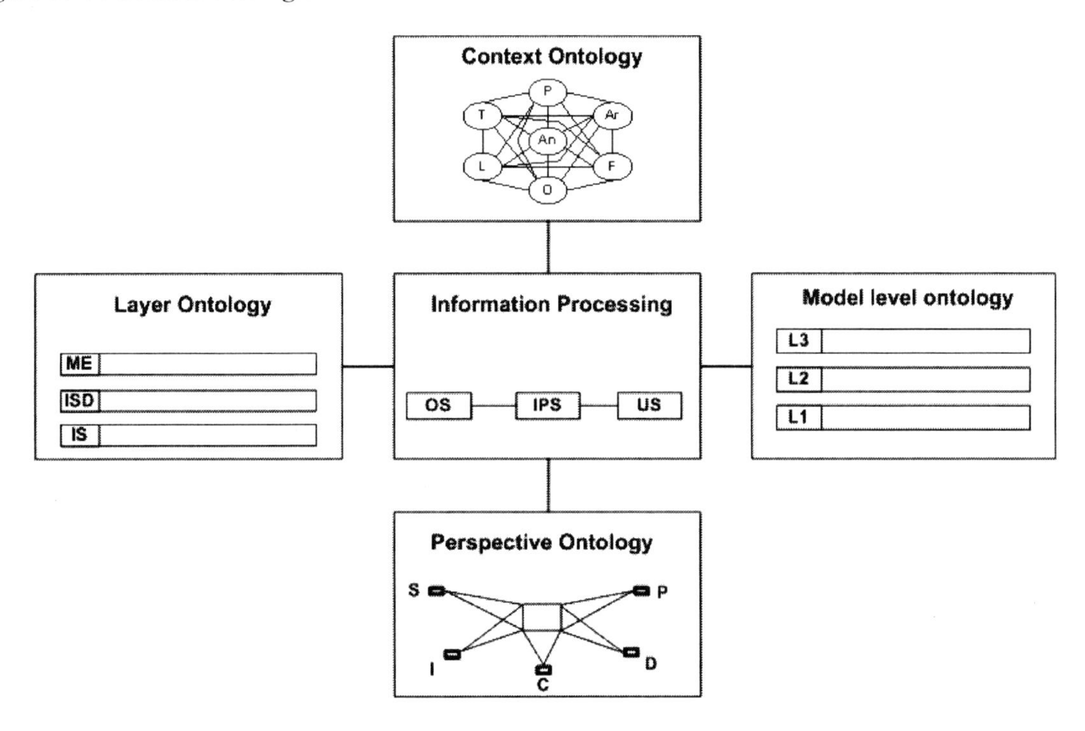

the object domain (O), the facility domain (F), the location domain (L), and the time domain (T). The purpose domain consists of concepts which refer to goals, motives, or intentions of someone or something. The actor domain encompasses concepts which concern individuals, groups, positions, roles, and organizations. The concepts of the action domain refer to functions, activities, tasks, or operations carried out in the context. The object domain comprises concepts which refer to something which an action is targeted to. The facility domain addresses means, whether tools or resources, by which something can be done or is done. The location domain is composed of concepts which refer to parts of space occupied by someone or something. The concepts of the time domain can be used to refer to temporal aspects in the context. For each domain, the most essential concepts and constructs have been defined. The arrangement of the domains is illuminated by the contextual scheme, called the *Seven S's scheme* (Leppänen, 2005), as follows: "For Some purpose, Somebody does Something for Someone with Some means, Sometimes and Somewhere". The ontology is rooted in case grammar (Fillmore, 1968), pragmatics (Levinson, 1983), and activity theory (Engeström, 1987). They concern sentence context, conversation context, and action context, correspondingly.

The *layer ontology* helps us structure and relate, on a general level, phenomena of information processing and its development at three layers, namely information systems (IS), information systems development (ISD) and method engineering (ME). This ontology is based on systems theory (Mesarovic et al., 1970) and some works in information systems science (e.g. Gasser, 1986; Falkenberg et al., 1992; Stamper, 1996; Hirschheim et al., 1995).

The *perspective ontology* provides a set of well-defined perspectives to focus on and structure the conceptions of contextual phenomena. The perspectives are: systelogical (S), infological (I), conceptual (C), datalogical (D), and physical

(P) perspectives. According to the *systelogical perspective,* the IS is considered in relation to its utilizing system. Applying the *infological perspective*, the IS is seen as a functional structure of information processing actions and informational objects, independent from any representational and implementational features. The *conceptual perspective* guides us to consider the IS through the semantic contents of information it processes. From the *datalogical perspective* the IS is viewed, through representation-specific concepts, as a system, in which actors work with facilities to process data. The *physical perspective* ties datalogical concepts and constructs to a particular organizational and technological environment. The perspective ontology is based on systems theory (e.g. Mesarovic et al., 1970), semiotics (Peirce, 1955), and some seminal works of IS researchers, such as Langefors et al. (1975), Welke (1977) and Iivari (1989).

With the *model level ontology*, one is able to create, specify and present models about reality. The kernel of this ontology is a hierarchy composed of the levels of instance models (L0), type models (L1), meta models (L2), and meta meta models (L3). The ontology is based on Dietz (1987), van Gigch (1991), ter Hofstede et al. (1997), and Falkenberg et al. (1998), many of which have their roots in linguistics and philosophy of science.

Layer-Specific Ontologies

The third main part of OntoFrame is called the layer-specific ontologies. While the layer ontology gives the basic structures to distinguish between and relating the information processing layers, the layer-specific ontologies elaborate the conceptualizations of IS, ISD and ME. These ontologies are the IS ontology, the ISD ontology and the ME ontology, correspondingly. Each of them has been specified through the concepts and constructs of the contextual domains and the perspectives defined above.

The *IS ontology* helps us conceive, understand, structure and represent phenomena in the IS, its object system and utilizing system. As the IS is a context, it is conceived through the concepts derived from the contextual domains. To give an example, Figure 4 presents the basic concepts of the IS actor domain. The IS ontology has been derived from the context ontology, and by integrating constructs from multiple works (e.g. Loucopoulos et al., 1998; Lin et al., 1998; Kavakli et al., 1999; Mesarovic et al., 1970; Thayer, 1987; Simon, 1960; Herbst, 1995; Stamper, 1978; Lee, 1983; Katz, 1990; Barros, 1991; Borgo et al., 1996; Bittner et al., 2002; Allen, 1984).

The *ISD ontology* provides concepts for the conceptualization of ISD. Figure 5 represents the basic concepts of the ISD actor domain. Besides having derived from more generic ontologies in the framework, it has been built by selecting, abstracting, modifying and integrating concepts from a large array of IS and ISD literature (e.g. Pohl, 1994; Cysneiros et al., 2001; Acuna et al., 2004; Baskerville, 1989; Checkland, 1988; Thayer, 1987; Kruchten, 2000; Heym et al., 1992; Sol, 1992; Falkenberg et al., 1998; Goldkuhl et al., 1993; Ramesh et al., 2001).

The *ME ontology* covers contextual phenomena in method engineering. It is also organized through the seven domains and the five perspectives. The ontology has been built on works such as Kumar et al. (1992), Rolland et al. (1999), Ralyte et al. (2003), Saeki (1998), Harmsen (1997), Brinkkemper (1990), Graham et al. (1997), Tolvanen (1998), Gupta et al. (2001), and Rossi et al. (2004).

Method Ontologies

The fourth main part of OntoFrame is the method ontologies, comprising the ISD method ontology and the ME method ontology. These ontologies provide concepts and constructs to conceive, understand, and represent the nature, structure and contents of the methods. The contents of the

Figure 4. IS actor domain

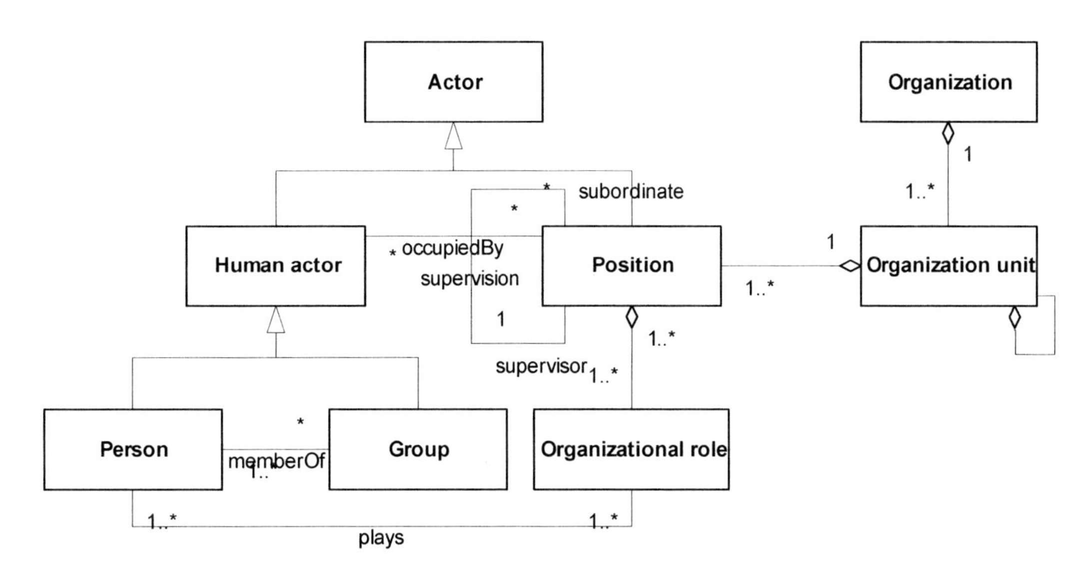

Figure 5. ISD actor domain (Leppänen, 2007c p. 8)

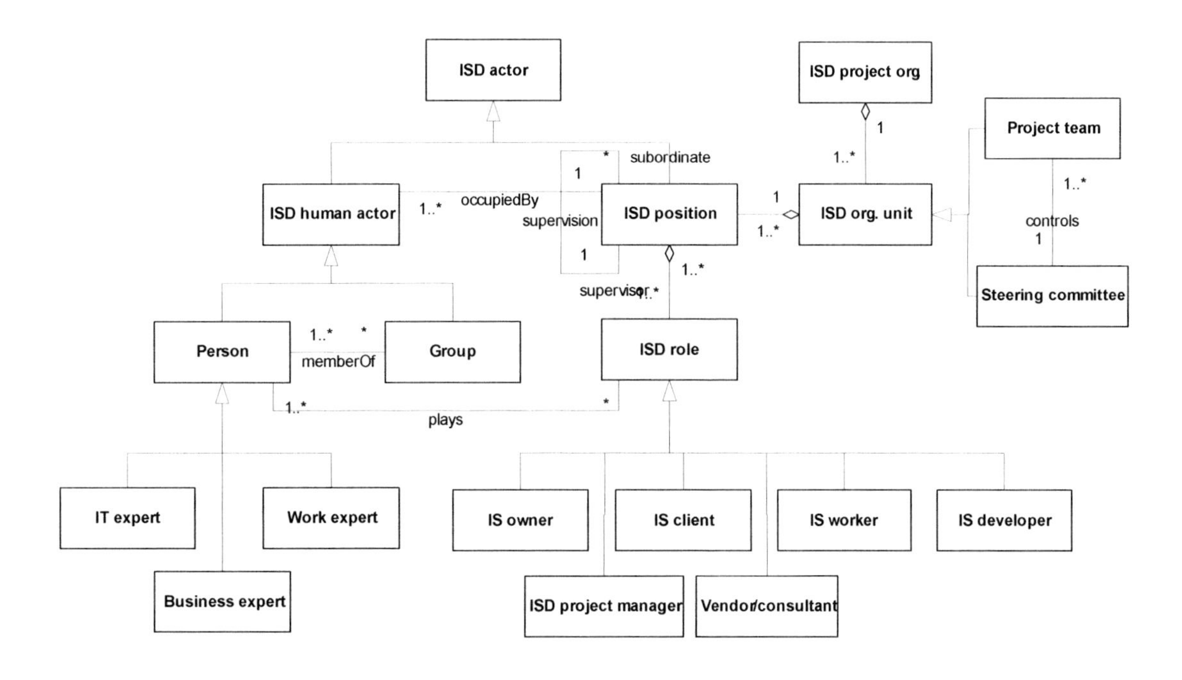

methods are conceptualized by the ISD ontology and the ME ontology, correspondingly. The method ontologies have been structured in accordance with the semantic ladder (Stamper, 1996) and derived from a number of frameworks of ISD methods (Fitzgerald et al., 2002; Hirschheim et al., 1995; Iivari et al., 2001; Zhang et al., 2001; Harmsen, 1997; Rossi et al., 2004; Graham et al., 1997; Heym et al., 1992; Lyytinen, 1986; Avison et al., 1995; Tolvanen, 1998). To address the most essential facets of the method, we have defined seven views, called the *methodical views*: the generic view, the historical view, the application view, the contents view, the presentation view, the physical view, and the structural view. The generic view provides an overview and general understanding of the method. The historical view and the application view "contextualize" the method into the past contexts in which the method has been engineered and applied, and into the future contexts in which the method is to be customized and deployed, correspondingly. The contents view reveals, through the concepts of the ISD/ME ontology, the conceptual contents of the method. The presentation view is used to consider the method as a set of expressions presented in some language. From the physical view the method is considered as a materialized artifact. The structural view decomposes the method into a modular structure of various descriptive and prescriptive parts (Leppänen, 2005).

Summary

OntoFrame provides a holistic view of the subdomains involved in ME. The rationale behind seeing OntoFrame as a composition of various ontologies is based on generativeness, modularity and contextuality. Generativeness facilitates one to select an appropriate level of specificity on which phenomena in reality are to be conceived. Modularity helps us manage complexity inherently resulting from a myriad of concepts and constructs. Contextuality enables the use of concepts and constructs for capturing deeper meanings of single phenomena through relating them to other phenomena in the context(s). Generating specific concepts from generic concepts by specialization advances the coherence and consistence of the framework.

THE APPROACHES AND PROCESS OF ENGINEERING ONTOFRAME

Ontology engineering means categorizing, naming and relating phenomena in reality in an explicit way. There are two sources of ontological categories (Sowa, 2000): observation and reasoning. Observation provides knowledge of the physical world, and reasoning makes sense of observation by generating a framework of abstraction. OntoFrame is based on the extensive reasoning from the large literature on universal theories such as semiotics, linguistics and systems theory, and works related more specifically to IS, ISD and ME.

There are two approaches to deriving from the literature in ontology engineering. In the *inductive approach,* source material is collected from single instance-level artifacts (e.g. ontologies, frameworks, and methods) to abstract a more generic one. In the *deductive approach*, some universal-like theoretic constructs are first selected and then deployed as an underlying groundwork for an ontology. We applied both of these approaches. First, in building the core ontology we made a thorough analysis of generic frameworks and ontologies (e.g. Bunge, 1977; Chisholm, 1996; Lenat et al., 1990; Sowa, 2000; Wand, 1988; Wand et al., 1995; Falkenberg et al., 1998; Guizzardi et al., 2002) and derived the ontology from them by selection, integration, and customization. In contrast, in engineering the context ontology we first searched for disciplines and theories that address meanings in sentence contexts (Fillmore, 1968), conversation contexts (Levinson, 1983) and action contexts (Engeström, 1987), and derived the fundamental categorization of concepts into seven contextual domains. After that we enriched the contents and structure of each domain with constructs from existing artifacts. The fundamental structures in the perspective ontology, the model level ontology and the layer ontology were also inferred from the relevant theories (e.g. systems theory, semiotics, linguistics). For the rest of OntoFrame we mostly applied the deductive approach to generate lower-level ontologies from higher-level ontologies. In this process the existing literature was heavily utilized, in order to complete and customize the derived concepts and constructs for the concerned sub-domains.

Many of the conceptual frameworks in the ISD literature have been constructed applying the inductive approach (e.g. Harmsen, 1997; Saeki et al., 1993; Song et al., 1992). Harmsen (1997), for instance, built his MDM model (Methodology Data Model) by deriving from existing classifications and frameworks, resulting in a large set of IS-specific and ISD-specific concepts that were justified through their source artifacts. A drawback of this kind of approach is that it does not encourage bringing forward novel insights. In contrast, the BWW model (Wand, 1989; Wand et al., 1995) has been anchored in Bunge's ontology (Bunge, 1977). Through this deductive approach the model pursues "universality" of concepts and constructs. In the approach such as this there is, however, a risk that the theories originally crafted for different domains may not cover the whole range of phenomena in the concerned domain. We have tried to overcome these problems and risks by applying both of the approaches. Theory-based constructs provided an underpinning that was tested, enhanced and elaborated by the inductive derivation from current artifacts. The use of the theories advanced not only the soundness of the framework but also innovations.

Another way to characterize our process of engineering OntoFrame is to use the categorization of the approaches into top-down, bottom-up and mixed approaches (Uschold et al., 1996; Noy et al., 2001). Our process mainly proceeded in a top-down manner through the following stages: (1) building the core ontology, (2) deriving the contextual ontologies, (3) establishing the layer-specific ontologies, and (4) deriving the method ontologies (see Figure 1). From the main strategies available for ontology engineering we applied the integration strategy whenever possible. In this way we could import existing knowledge from those sub-fields in which views and concepts are relatively stable and fit our main premises. Adaptation was carried out when needed.

For each of the ontologies in the framework we applied, in an iterative manner, an ontology engineering procedure with the following steps (cf. Uschold et al., 1996): (a) determine the purpose and domain of an ontology, (b) consider reusing existing artifacts, (c) conceptualize, (d) formalize (i.e. present in a graphical model), (e) evaluate, and (f) document.

Our ontological framework is aimed at a means of communication between human beings, not for the use of computers. It comprises a large set of concepts and constructs, and most of them are highly abstract. For these reasons, it is important to present the framework in a concise yet understandable form. We deploy two presentation forms, informal and semi-formal (Leppänen 2005). The concepts are defined in a vocabulary presented in English. In addition, each of the ontologies is expressed in a graphical model. From a large set of semi-formal languages we selected a graphical language (see arguments by Guizzardi et al. (2001)), and preferred the UML language to special ontology representation languages (e.g. CLEO, LINGO and DAML+OIL) because of its large and rapidly expanding user community, intrinsic mechanism for defining extensions, and largely available computer-support. Figures 4 and 5 show examples of the graphical models presented in UML.

COMPARATIVE ANALYSIS OF RELATED WORKS

In this section we portray a big picture of related works by briefly describing and comparing them with OntoFrame. First, we define the criteria for the selection of the artifacts and specify the features to be considered. Then, we report on the findings from the analysis carried out on two levels of detail.

The literature suggests hundreds of frameworks, meta-models, frames of reference and ontologies, shortly *artifacts*, for IS, ISD, ISD methods and ME. Many of them are not ontological but provide, for instance, just lists of features for characterizing ISD methods, or bring forward taxonomies for approaches and viewpoints (e.g. Olle et al. 1983; Olle et al., 1986). We deployed four criteria in the selection of artifacts: purpose, coverage, familiarity, and specificity. An artifact should support the description of the sub-domains, and/or the analysis and construction of other artifacts in the sub-domains (*Purpose*). An artifact should cover as many relevant sub-domains as possible (*Coverage*). An artifact has been published in a recognized journal or in the proceedings of a recognized conference (*Familiarity*). Concepts and constructs within an artifact are defined in a specific manner (*Specificity*), preferably in a formal or semi-formal manner. We ended up selecting 15 artifacts which comprise three models (Harmsen, 1997; Heym et al., 1992; Wand, 1988), nine frameworks (Essink, 1988; Falkenberg et al., 1998; Heym et al., 1992; Iivari, 1989; Iivari, 1990; Olle et al., 1988; Song et al., 1992; Sowa et al., 1992; van Swede et al., 1993) and four meta-models (Freeman et al., 1994; Grosz et al., 1997; Gupta et al., 2001; Saeki et al., 1993). We accepted Harmsen's dissertation work (Harmsen, 1997) because parts of it have been published in journal articles (e.g. Brinkkemper et al., 1999).

The comparative analysis is carried out in two parts. In the first part, we consider the purposes, sub-domains and theoretical bases of the artifacts. In the second part, we analyze, in more detail, the scope and emphases of the artifacts by using OntoFrame as the frame of reference. The results of the first part are summarized in Table 1. The artifacts are classified into four groups based on which sub-domains they primarily address. The groups are: 1) comprehensive artifacts, 2) IS domain-based artifacts, 3) ISD domain-based/ISD method-based artifacts, and 4) ME domain-based artifacts. In the following, we shortly describe and analyze the artifacts within each of the groups.

There is only one artifact, the MDM (the Methodology Data Model) (Harmsen 1997), which can be considered to be comprehensive. It addresses IS, ISD and ISD method and, to some degree, ME. The IS domain-based artifacts (Essink, 1988; Freeman et al., 1994; Iivari, 1989; Olle et al., 1988; Sowa et al., 1992; van Swede et al., 1993; Wand, 1988) provide concepts and constructs that facilitate the recognition, understanding and representation of structural and dynamic features of IS and its utilizing system. All of them, except the BWW model (Wand, 1988; Wand et al., 1995), also provide some perspectives to classify and structure the features according to pre-defined viewpoints.

The third group is composed of those artifacts, which offer concepts and constructs for the ISD domain (Essink, 1988; Iivari, 1990; Grosz et al., 1997; Saeki et al., 1993) and / or for the ISD method (Heym et al., 1992; Gupta et al., 2001; Song et al., 1992). Iivari (1990) and Grosz et al. (1997) mainly apply a process view, whereas Heym et al. (1992), Gupta et al. (2001) and Song et al. (1992) take a broader perspective. The artifacts in this group are mostly intended for the description, assessment, comparison and integration of ISD methods or techniques.

There are only a few artifacts (Grosz et al., 1997; Harmsen, 1997) that describe structural and dynamic features of the ME domain. Grosz et al. (1997) apply a process meta-model for structuring "meta-way-of-working" in the meta-processes of the ME. Harmsen (1997) defines a few conceptual structures for the ME domain.

As seen in Table 1, only some of the artifacts have been explicitly founded on some theories. The Frisco framework (Falkenberg et al., 1998) is based on semiotics, systems theory, cognitive science and organization theory. Iivari's (1989, 1990) frameworks have been rooted in systems theory, organization theory, semiotics, and information economics. The BWW model (Wand, 1988; Wand et al., 1995) has been derived from philosophy of science (Bunge, 1977) and systems theory. In the process meta-model Grosz et al., 1997) we can see some structures of decision

Table 1. Purposes, sub-domains and theoretical bases of the analyzed artifacts

Name/ Reference	Purpose	Sub-domains	Theoretical basis
MADIS framework (Essink, 1986, 1988)	"aimed at providing capability for matching available methods and techniques to particular problem classes". ."aimed at providing means to integrate elements of different methods" (Essink 1986 p. 354)	IS, ISD	none
Frisco-framework (Falkenberg et al., 1998)	"To provide an ordering and transformation framework allowing to relate many different IS modeling approaches to each other" (ibid p. 1]	IS	semiotics, linguistics, cogn. science, systems science, org. science
Organizational metamodel (Freeman et al., 1994)	"to represent all aspects of an information system, necessary for system understanding and software maintenance at four levels of abstraction" (ibid p. 283)	IS	none
Process meta-model (Grosz et al., 1997)	"an overview of the process theory for modeling and engineering the RE process" (ibid p. 115)	ISD, ME	(Decision making theory)
MDM (Harmsen ,1997)	to describe parts of ISD methods, thus supporting method engineering	IS, ISD, ISD method ME	none
Framework and ASDM (Heym et al., 1992)	a framework for describing ISD, and a semantic data model (ASDM) for describing ISD methods (ibid p. 215-216)	ISD, ISD method	none
Conceptual framework (Iivari, 1989)	to facilitate "the systematic recognition, comparison and synthesis of different perspectives on the concept of an information system" (ibid p. 323)	US, IS	systems theory, org. theory, semiotics,
Hierarchical spiral framework (Iivari, 1990)	"a hierarchical spiral framework for IS and SW development including evolution dynamics, main-phase dynamics, learning dynamics related to each other" (ibid p. 451)	ISD	socio-cybernetics, information economics
Framework for understanding (Olle et al. 1988)	"a framework for tacking systems planning, analysis and design, into which many existing methodologies can be fitted" (ibid p. vi)	US, IS	none
Decision-oriented meta-model (Gupta et al., 2001)	a decision-oriented meta-model to be used for instantiating a method representation	ISD, ISD method	none
Meta-model (Saeki et al., 1993)	"a meta-model for representing software specification and design methods" (ibid p. 149)	ISD	none
Framework (Song et al., 1992)	"A framework for aiding the understanding and handling the complexity of methods integration and thus making integration more systematic" (ibid p. 116)	ISD, ISD method	none
ISA framework (Sowa et al., 1992)	to provide "a taxonomy for relating the concepts that describe the real world to the systems that describe an information system and its implementation" (ibid p. 590)	US, IS	none
Framework (van Swede et al., 1993)	"for classifying ISD modelling techniques, to assess the modelling capacity of development methods, and as a checklist for project leaders to construct project scenarios" (ibid p. 546)	IS	none
BWW model (Wand, 1988; Wand et al., 1995)	"is aimed at to be used as an ontology to define the concepts that should be represented by a modelling language, that is the semantics of the language" (Wand, Monarchi et al., 1995 p. 287)	IS	philosophy of science, systems theory,

making theories although they are not mentioned as the theoretical basis.

In the second part of the analysis, we give the grades between 0 and 5 to indicate the degree to which the concepts and constructs in each artifact correspond, in terms of scope and comprehensiveness, to the concepts and constructs in our ontologies. The grades have the following meanings: empty space = not considered, 1 = considered slightly, 2 = considered fairly, 3 = considered equally, 4 = considered more completely, 5 = considered "fully". Note that the comparison is carried out in proposition to our framework, not to the other parts of the artifact, neither to the other artifacts, and the grades given are naturally indicative, not to be taken exactly. The summary of the results from the second part of the analysis is presented in Appendix.

Only five artifacts provide concepts and constructs for some part of the core ontology. The most comprehensive artifact in this respect is the Frisco Framework (Falkenberg et al., 1998), which addresses all seven ontologies, although abstraction is only slightly discussed. The BWW model (Wand, 1988; Wand et al., 1995) provides a rather large set of fundamental concepts and constructs and offers a particularly deep consideration of phenomena related to states and state transitions. In contrast, it overlooks issues related to semiotics and language. The third artifact addressing the core ontology is Iivari's (1989) framework, which is rather strong in defining concepts and constructs within the state transition ontology and the UoD ontology but does not consider the other parts of the core ontology. Harmsen (1997) and Gupta et al. (2001) suggest only some fundamental concepts for the core ontology.

All the artifacts, except Gupta et al. (2001) and the BWW model (Wand, 1988; Wand et al., 1995), provide concepts for at least some of the four contextual ontologies (i.e. the context ontology, the layer ontology, the perspective ontology, the model level ontology). As expected, the most comprehensive treatment is given in the artifacts, which are focused on the IS domain. The context ontology and the perspective ontology are most largely addressed in Iivari (1989) and Sowa et al. (1992). The other artifacts defining concepts and constructs of these ontologies are the frameworks of Essink (1988) and Olle et al. (1988). Falkenberg et al. (1998) and Harmsen (1997) define a large set of concepts and constructs related to the context ontology but nothing for the perspective ontology. The two other contextual ontologies, the model level ontology and the layer ontology, are addressed in Essink (1988), Grosz et al. (1997), Harmsen (1997) and Heym et al. (1992), but on a rather general level.

Concepts and constructs related to the ISD ontology or the ISD method ontology are defined on a detailed level in Heym et al. (1992), while the artifacts of Harmsen (1997) and Gupta et al. (2001) remain on a coarse level. The other artifacts addressing these ontologies, although only slightly, are Grosz et al. (2001), Iivari (1990), Olle et al. (1988), Saeki et al. (1993), and Song et al. (1992). The only artifacts that extend to the ME ontology are the process meta-model by Grosz et al. (2001) and the MDM (Harmsen, 1997). Their definitions for concepts and constructs cover, however, only a small part of the ME domain.

To summarize, there is none among the analyzed artifacts that would come close to OntoFrame as regard the coverage of the sub-domains relevant to ME. Most commonly the artifacts conceptualize either IS, ISD or ISD method. A more detailed analysis of the artifacts in Leppänen (2005) reveals that the artifacts have adopted quite a narrow scope of even those sub-domains they address. Most of the artifacts also lack a theoretical basis. We have, however, to point out that most of the artifacts have been published in journals or conference proceedings for which the space allowed is quite limited. Thus, it is not fare to expect these artifacts to be as comprehensive as those presented in dissertation theses or the like. There are also some researchers (e.g. Falkenberg, Iivari, Jarke, Sølvberg, Rolland, Ralyté, Brinkkemper) and research groups (e.g. Wand & Weber), which have contributed to more sub-domains in separate articles. But because these collections of articles have been published in long periods, they do not necessarily form a unified and coherent whole. We did not analyze these collections of contributions.

We see three alternative strategies to engineer a comprehensive and coherent conceptual foundation for ME. The first option is to select one of those existing artifacts that are strong in the "core domain" and the IS domain, and extend it to embrace the relevant phenomena in the ISD and ME domains as well. Good candidates for that kind of artifact are the Frisco framework (Falkenberg et al., 1998), Iivari's framework (Iivari, 1989) and the BWW model (Wand, 1988; Wand et al., 1995). The second option is to choose one of those artifacts that are strong in the ISD and ISD method domains, and to build for that, in a bottom up manner, a solid and coherent groundwork. The MDM by Harmsen (1997) seems to be the most suitable one for this purpose, although it should also be extended to cover missing concepts and constructs within the ME domain. The third option is to engineer a new artifact, first by building a sound and theory-driven underpinning and then by specializing its concepts and constructs for the relevant sub-domains. Existing artifacts, or parts thereof, are integrated into the body of the artifact when possible. This is the strategy we have adopted. To our view, this is the best way to ensure that all the ontologies, from top to bottom, share the same basic assumptions and views. Deriving specific concepts from fundamental concepts on the "bottom levels" also helps us guarantee that the concepts within all sixteen ontologies are consistent and coherent.

DISCUSSION AND CONCLUSION

In this chapter we have described the overall structure of the large ontological framework, called Ontoframe, which serves as a conceptual foundation for the construction, analysis and comparison of ME artifacts. We have also brought out the purposes, sub-domains and theoretical bases of the ontologies contained in OntoFrame, as well as the approaches and process by which OntoFrame has been constructed. Furthermore, we have deployed OntoFrame in a comparative analysis of related artifacts.

To build a comprehensive conceptual foundation for the ME domain is quite challenging. This was experienced, although on a much smaller scale, by the Frisco group (Falkenberg et al., 1998) in building a conceptual foundation for the IS domain. Difficulties were multiplied in our study due to the much larger scope. We established OntoFrame mainly in the top-down fashion by first building the core ontology and then deriving IS-specific, ISD-specific, ME-specific and method-specific parts from that. We started from generic conceptual constructs and principles, rather than making a synthesis of existing methods and ME artifacts.

OntoFrame shows, what is a necessary and purposeful topology of concepts and constructs that together define the nature, structure and contents of the ISD methods, as well as the type and structure of those contexts in which the ISD methods are engineered and deployed. As far as we know, there is no other presentation that would cover such a large spectrum of sub-domains, on such a detailed level, as OntoFrame does. We have intentionally aspired after this kind of holistic view in order to avoid the fragmentation of views and conceptions that is typical of most of the research in our field. The holistic view enables the recognition, comparison and integration of current artifacts that have been built upon more limited foundations and views. We have also acknowledged conceptions and propositions presented in the past decades. Many researchers, such as Brodie, Chen, Codd, Ein-Dor, Falkenberg, Gorry, Guttag, Hoare, Iivari, Kent, Kerola, Langefors, Mumford,

Sisk, Smith & Smith, Stamper, Welke, Yourdon, and Zisman, just to mention some of them, have presented, already in the 1970's, seminal ideas that are still worthy of respecting and comparing to more recent ones.

OntoFrame is of benefit to both research and practice. With the ontologies contained in OntoFrame it is possible to achieve a better understanding of the contextual phenomena in IS, ISD and ME. OntoFrame provides a reference background for scientists and professionals, thus enabling them to express themselves about matters in the concerned sub-domains in a structured way, and based on that, to analyze, compare and construct ME artifacts. The comprehensive and unified framework establishes bridges between various approaches, disciplines and views across three decades. OntoFrame also provides teachers and students with a large collection of models of sub-domains and a comprehensive vocabulary with clear definitions. OntoFrame is very large. It is not our intention that the whole range of ontologies is applied in every case. Contrary to that, relevant parts of it could be selected depending on the problem at hand.

Assessment of a large ontological framework such as OntoFrame is difficult. Due to the space limit, we discuss it here on a general level in terms of the goals set up for OntoFrame. Coherence, generativeness, modularity, and balance between specificity and generality have been basically advanced by applying the top-down approach by which concepts and constructs of ontologies on lower levels have been derived by specialization from those on higher levels. Extensiveness has been aspired at by anchoring the skeleton of the framework in universal theories. Achievement of coherence and consistency has been aided by cross-checking the definitions in the vocabulary and using graphical models to represent the ontologies. Clarity, naturalness and applicability are goals that should be validated, preferably through empirical tests.

Validation of applicability should, in the first place, involve each of the ontologies in the framework individually. We have responded to this demand with the discussion of validity of individual ontologies in separate articles (i.e. the context ontology (Leppänen, 2007d), the ISD ontology (Leppänen, 2007c), the abstraction ontology (Leppänen, 2007b), the perspective ontology (Leppänen 2007e), the model level ontology (Leppänen, 2006a) and the ISD method ontology (Leppänen, 2007a)). Second, validation should concern the framework as the whole. Some work for that has also been done. The applicability of OntoFrame was demonstrated in the analysis of ME artifacts in Leppänen (2006b). In Leppänen (2005) we deployed OntoFrame to construct methodical support for ME in the form of a methodical skeleton. In Leppänen et al. (2007) OntoFrame was used to structure contingency factors for the engineering of an enterprise architecture method. In Hirvonen et al. (2007) the methodical skeleton was used to analyze the criteria applied in the selection of enterprise architecture method for the Finnish Government. We also deployed the methodical skeleton to make guidelines for the adaptation of the enterprise architecture method for the Finnish Government. Experience from this effort will be reported in the near future. In this chapter, we used OntoFrame to analyze existing conceptual foundations. Although some evidence of the applicability of individual ontologies, as well as of the whole framework has been got, more experience from the use of OntoFrame in different kinds of ME contexts is definitely needed.

REFERENCES

Acuna, S., & Juristo, N. (2004). Assigning people to roles in software projects. *Software – Practice and Experience, 34*(7), 675-696.

Allen, J. (1984). Towards a general theory of action and time. *Artificial Intelligence, 23*(2), 123-154.

Avison, D., & Fitzgerald, G. (1995). *Information systems development: methodologies, techniques and tools*. 2nd ed., London: McGraw-Hill.

Aydin, M. (2007). Examining key notions for method adaptation. In J. Ralytè, S. Brinkkemper & B. Henderson-Sellers (Eds.), *Situational Method Engineering: Fundamentals and Experiences* (pp. 49-63). Boston: Springer.

Barros, O. (1991). Modeling and evaluation of alternatives in information systems. *Information Systems, 16*(5), 537-558.

Baskerville, R. (1989). Logical controls specification: an approach to information systems security. In H. Klein & K. Kumar (Eds.), *Systems Development for Human Progress* (pp. 241-255). Amsterdam: Elsevier Science.

Bittner, T., & Stell, J.G. (2002). Vagueness and Rough Location. *Geoinformatica, 6*(2), 99-121.

Booch, G., Rumbaugh, J., & Jacobson, I. (1999). *The Unified Modeling Language – User guide*. Reading: Addison-Wesley.

Borgo, S., Guarino, N., & Masolo, C. (1996). Towards an ontological theory of physical objects. In *IMACS-IEEE/SMC Conference on Computational Engineering in Systems Applications.(CESA'96)*, Symposium of Modelling, Analysis and Simulation (pp. 535-540). Lille, France.

Brachman, R. (1983). What IS-A is and isn't: An analysis of taxonomic links of semantic networks. *IEEE Computer, 16*(10), 30-36.

Brinkkemper, S. (1990). *Formalization of information systems modeling*. Dissertation Thesis, University of Nijmegen, Amsterdam: Thesis Publishers.

Brinkkemper, S., Saeki, M., & Harmsen, F. (1999). Meta-modelling based assembly techniques for situational method engineering. *Information Systems, 24*(3), 209-228.

Bunge, M. (1977). *Treatise on basic philosophy, Vol. 3: Ontology I: The furniture of the world.* Dortrecht: D. Reidel Publishing Company.

Burton-Jones, A., Storey, V., Sugumaran, V., & Ahluwalia, P. (2005). A semiotic metric suite for assessing the quality of ontologies. *Data & Knowledge Engineering, 55*(1), 84-102.

Chang, L.-H., Lin, T.-C., & Wu, S. (2002). The study of information system development (ISD) process from the perspective of power development stage and organizational politics. In *35th Hawaii International Conference on Systems Sciences.*

Checkland, P. (1988). Information systems and system thinking: time to unite? *International Journal of Information Management, 8*(4), 239-248.

Chisholm, R. (1996). *A realistic theory of categories – an essay on ontology*. 1st edition, Cambridge: Cambridge University Press.

Cysneiros, L., Leite, J., & Neto, J. (2001). A framework for integrating non-functional requirements into conceptual models. *Requirements Engineering, 6*(2), 97-115.

Dietz, J. (1987). *Modelling and specification of information systems* (In Dutch: Modelleren en specificeren van informatiesystemen). Dissertation Thesis, Technical University of Eindhove, The Netherlands.

Engeström, Y. (1987). *Learning by expanding: an activity theoretical approach to developmental research.* Helsinki: Orienta-Konsultit.

Essink, L. (1988). A conceptual framework for information systems development methodologies. In H. J. Bullinger et al. (Eds.), *Information Technology for Organizational Systems* (pp. 354-362). Amsterdam: Elsevier Science.

Essink, L. (1986). A modeling approach to information system development. In T. Olle, H.

Sol & A. Verrijn-Stuart (Eds.), *Proc. of the IFIP WG 8.1 Working Conf. on Comparative Review of Information Systems Design Methodologies: Improving the Practice* (pp. 55-86). Amsterdam: Elsevier Science.

Falkenberg, E., Hesse, W., Lindgreen, P., Nilsson, B., Oei, J. L. H., Rolland, C., Stamper, R., van Asche, F., Verrijn-Stuart, A., & Voss, K. (1998). *A framework of information system concepts, The FRISCO Report* (Web edition), IFIP.

Falkenberg, E., Oei, J., & Proper, H. (1992). A conceptual framework for evolving information systems. In H. Sol & R. Crosslin (Eds.), *Dynamic Modelling of Information Systems II* (pp. 353-375). Amsterdam: Elsevier Science.

Fillmore, C. (1968). The case for case. In E. Bach & R. T. Harms (Eds.) *Universals in Linguistic Theory.* New York: Holt, Rinehart and Winston, 1-88.

Fitzgerald, B., Russo, N., & Stolterman, E. (2002). *Information systems development – methods in action.* London: McGraw Hill.

Freeman, M., & Layzell, P. (1994). A meta-model of information systems to support reverse engineering. *Information and Software Technology, 36*(5), 283-294.

Gasser, L. (1986). The integration of computing and routine work. *ACM Trans. on Office Information Systems, 4*(3), 205-225.

Gigch van, J. (1991). *System design modeling and metamodeling.* New York: Plenum Press.

Goldkuhl, G., & Cronholm, S. (1993). *Customizable CASE environments: a framework for design and evaluation.* Institutionen for Datavetenskap, Universitetet och Tekniska Högskolan, Linköping, Research Report.

Goldstein, R., & Storey, V. (1999). Data abstraction: Why and how? *Data & Knowledge Engineering, 29*(3), 293-311.

Graham, I., Henderson-Sellers, B., & Younessi, H. (1997). *The OPEN process specification.* Reading: Addison-Wesley.

Grosz, G., Rolland, C., Schwer, S., Souveyet, C., Plihon, V., Si-Said, S., Achour, C., & Gnaho, C. (1997). Modelling and engineering the requirements engineering process: an overview of the NATURE approach. *Requirements Engineering, 2*(2), 115-131.

Gruber, T. (1993). A translation approach to portable ontology specification. *Knowledge Acquisition, 5*(2), 119-220.

Gruber, T. (1995). Towards principles for the design of ontologies used for knowledge sharing. *International Journal of Human-Computer Studies, 43*(5/6), 907-928.

Guarino, N. (1998). Formal ontology and information systems. In N. Guarino (Ed.), *Formal Ontology in Information Systems (FOIS'98)* (pp. 3-15). Amsterdam: IOS Press.

Guizzardi, G., Falbo, R., & Filho, J. (2001). From domain ontologies to object-oriented frameworks. In G. Stumme, A. Maedche & S. Staab (Eds.), *Workshop on Ontologies (ONTO'2001)*, 1-14.

Guizzardi, G., Herre, H., & Wagner, G. (2002). On the general ontological foundations of conceptual modeling. In S. Spaccapietra, S. March & Y. Kambayashi (Eds.), *Conceptual Modeling – ER 2002* (pp. 65-78). LNCS 2503, Berlin: Springer.

Gupta, D., & Prakash, N. (2001). Engineering methods from method requirements specifications. *Requirements Engineering, 6*(3), 135-160.

Harmsen, F. 1997. *Situational method engineering.* Dissertation Thesis, University of Twente, Moret Ernst & Young Management Consultants, The Netherlands.

Hautamäki, A. (1986). *Points of views and their logical analysis.* Helsinki: Acta Philosophica Fennica, Vol. 41.

Henderson-Sellers, B., & Barbier, F. (1999). What is this thing called aggregation? In R. Mitchell, A. C. Wills, J. Bosch & B. MeyerProc. (Eds.), *TOOLS EUROPE'99* (pp. 236-250). MD: IEEE Computer Society Press.

Herbst, H. (1995). A meta-model for business rules in systems analysis. In J. Iivari, K. Lyytinen & M. Rossi (Eds.), *Advanced Information Systems Engineering* (pp. 186-199). LNCS 932, Berlin: Springer.

Heym, M., & Österle, H. (1992). A reference model for information systems development. In K. Kendall, K. Lyytinen & J. DeGross (Eds.), *The Impacts on Computer Supported Technologies on Information Systems Development* (pp. 215-240). Amsterdam: Elsevier Science.

Hirschheim, R., Klein, H., & Lyytinen, K. (1995). *Information systems development – conceptual and philosophical foundations.* Cambridge: Cambridge University Press.

Hirvonen, A., Pulkkinen, M., & Valtonen, K. (2007). Selection criteria for enterprise architecture methods. In Proc. of the European Conference on Information Management and Evaluation (pp. 227-236). Reading, UK: Academic Conference International.

Hofstede ter, A., & Proper, H. (1998). How to formalize it? Formalization principles for information system development methods. *Information and Software Technology, 40*(10), 519-540.

Hofstede ter, A., & Verhoef, T. (1997). On the feasibility of situational method engineering. *Information Systems, 22*(6/7), 401-422.

Iivari, J. (1989). Levels of abstraction as a conceptual framework for an information system. In E. Falkenberg & P. Lindgren (Eds.), *Information System Concepts: An In-Depth Analysis* (pp. 323-352). Amsterdam: Elsevier Science.

Iivari, J. (1990). Hierarchical spiral model for information system and software development. Part 2: Design process. *Information and Software Technology, 32*(7), 450-458.

Iivari, J., Hirschheim, R., & Klein, H. (2001). A dynamic framework for classifying information systems development methodologies and approaches. *Journal of Management Information Systems, 17*(3), 179-218.

Karlsson, F., & Ågerfalk, P. (2004). Method configuration: adapting to situational characteristics while creating reusable assets. *Information and Software Technology,* 46(9), 619-633.

Katz, R. (1990). Toward a unified framework for version modeling in engineering databases. *ACM Surveys, 22*(4), 375-408.

Kavakli, V., & Loucopoulos, P. (1999). Goal-driven business process analysis application in electricity deregulation. *Information Systems, 24*(3), 187-207.

Kelly, S. (2007). Domain-specific modeling: The killer app for method engineering? In J. Ralytè, S. Brinkkemper & B. Henderson-Sellers (Eds.), *Situational Method Engineering: Fundamentals and Experiences* (pp. 1-5). Boston: Springer.

Kelly, S., Lyytinen, K., & Rossi, M. (1996). MetaEdit+: a fully configurable multi-user and multi-tool CASE and CAME environment. In Y. Vassiliou & J. Mylopoulos (Eds.), *Advanced Information Systems Engineering (CAiSE'96)* (pp. 1-21). Berlin: Springer.

Kinnunen, K., & Leppänen, M. (1996). O/A matrix and a technique for methodology engineering. *Journal of Systems and Software,* 33(2), 141-152.

Kornyshova, E., Deneckere, R., & Salinesi, C. (2007). Method chunks selection by multicriteria methods: An extension of the assembly-based approach. In J. Ralytè, S. Brinkkemper & B. Henderson-Sellers (Eds.), *Situational Method Engineering: Fundamentals and Experiences* (pp. 64-78). Boston: Springer.

Krogstie, J. (1995). *Conceptual modeling for computerized information systems support in organizations.* Dissertation Thesis, NTH, University of Trondheim, Norway.

Kruchten, P. (2000). *The Rational Unified Process: An introduction.* Reading: Addison-Wesley.

Kumar, K., & Welke, R. (1992). Methodology engineering: a proposal for situation specific methodology construction. In W. Kottermann & J. Senn (Eds.), *Challenges and Strategies for Research in Systems Development* (pp. 257-269). Chichester: John Wiley & Sons.

Langefors, B., & Sundgren, B. (1975). *Information systems architecture.* New York: Petrocelli.

Lee, R. (1983). Epistemological aspects of knowledge-based decision support systems. In H. Sol (Ed.), *Processes and Tools for Decision Support Systems* (pp. 25-36). Amsterdam: Elsevier Science.

Lenat, D., & Guha, R. (1990). *Building large knowledge-based systems.* Reading: Addison-Wesley.

Leppänen, M. (2000). Toward a method engineering (ME) method with an emphasis on the consistency of ISD methods. In K. Siau (Ed.), *Evaluation of Modeling Methods in Systems Analysis and Design (EMMSAD'00),* Stockholm: Sweden.

Leppänen, M. (2005). *An Ontological Framework and a Methodical Skeleton for Method Engineering,* Dissertation Thesis, Jyväskylä Studies in Computing 52, University of Jyväskylä, Finland. Available at: http://dissertations.jyu.fi/studcomp/9513921867.pdf

Leppänen, M. (2006a). An integrated framework for meta modeling. In Y. Manolopoulos, J. Pokomy, & T. Sellis (Eds.), *Advances in Databases and Information Systems (ADBIS'2006)* (pp. 141-154). LNCS 4152, Berlin: Springer-Verlag.

Leppänen, M. (2006b). Conceptual evaluation of methods for engineering situational ISD methods. *Software Process: Improvement and Practice, 11*(5), 539-555.

Leppänen, M. (2007a). A Contextual method integration. In G. Magyar, G. Knapp, W. Wojtkowski, W.G. Wojtkowski, & J. Zupancic (Eds.), *Information Systems Development – New Methods and Practices for the Networked Society (ISD 2006)* (pp. 89-102). Vol. 2, Berlin: Springer-Verlag.

Leppänen, M. (2007b). Towards an abstraction ontology. In M. Duzi, H. Jaakkola, Y Kiyoki, & H. Kangassalo (Eds.), *Information Modelling and Knowledge Bases XVIII, Frontiers in Artificial Intelligence and Applications* (pp. 166-185). The Netherlands: IOS Press.

Leppänen, M. (2007c). Towards an ontology for information systems development – A contextual approach. In K. Siau (Ed.), *Contemporary Issues in Database Design and Information Systems Development* (pp. 1-36). New York: IGI Publishing.

Leppänen M. (2007d). A context-based enterprise ontology. In W. Abramowicz (Ed.), *Business Information Systems (BIS 2007)* (pp. 273-286). LNCS 4439, Berlin: Springer.

Leppänen, M. (2007e). IS ontology and IS perspectives. In H. Jaakkola, Y. Kiyoki, & T. Tokuda (Eds.), *Information Modelling and Knowledge Bases XIX, Frontiers in Artificial Intelligence and Application.* The Netherlands: IOS Press (in print).

Leppänen, M., Valtonen, K., & Pulkkinen, M. (2007). Towards a contingency framework for engineering an enterprise architecture planning method. In *Proc. of the 30th Information Systems Research Seminar in Scandinavia (IRIS 2007),* Tampere, Finland.

Levinson, S. (1983). *Pragmatics.* London: Cambridge University Press.

Lin, C.-Y., & Ho, C.-S. (1999). Generating domain-specific methodical knowledge for requirements analysis based on methodology ontology. *Information Sciences*, 114(1-4), 127-164.

Loucopoulos, P., Kavakli, V., Prekas, N., Rolland, C., Grosz, G., & Nurcan, S. (1998). *Using the EKD approach: the modelling component*. ELEKTRA – Project No. 22927, ESPRIT Programme 7.1.

Lyytinen, K. (1986). *Information systems development as social action: framework and critical implications*. Jyväskylä Studies in Computer Science, Economics, and Statistics, No. 8, University of Jyväskylä, Finland, Dissertation Thesis.

Mesarovic, M., Macko, D., & Takahara, Y. (1970). *Theory of hierarchical, multilevel, systems*. New York: Academic Press.

Morris, C. W. (1938). Foundations of the theory of signs. In O. Neurath, R. Carnap & C. Morris (Eds.) *International Encyclopedia of Unified Science*. Chicago: University of Chicago Press, 77-138.

Motschnig-Pitrik, R., & Kaasboll, J. (1999). Part-whole relationship categories and their application in object-oriented analysis. *IEEE Trans. on Knowledge and Data Engineering,* 11(5), 779-797.

Motschnig-Pitrik, R., & Storey, V. (1995). Modelling of set membership: the notion and the issues. *Data & Knowledge Engineering, 16*(2), 147-185.

Mylopoulos, J. (1998). Information modelling in the time of the revolution. *Information Systems,* 23(3/4), 127-155.

Noy, N., & McGuinness, D. (2001). *Ontology development 101: a guide to creating your first ontology.* Stanford Knowledge Systems Laboratory Technical Report KSL-01-05 and Stanford Medical Informatics Technical Report SMI-2001-0880. Retrieved June 3, 2004 from http://smi-web.stanford.edu/pubs/SMI_Abstracts/SMI-2001-880.html

Nuseibeh, B., Finkelstein, A., & Kramer. J. (1996). Method engineering for multi-perspective software development. *Information and Software Technology, 38*(4), 267-274.

Ogden, C., & Richards, I. (1923). *The meaning of meaning.* London: Kegan Paul.

Oei, J. (1995). A meta model transformation approach towards harmonization in information system modeling. In E. Falkenberg, W. Hesse & A. Olive (Eds.), *Information System Concepts – Towards a Consolidation of Views* (pp. 106-127). London: Chapman & Hall.

Olle, T., Sol, H., & Tully, C. (Eds.) (1983) *Proc of the IFIP WG8.1 Working Conf. on Feature Analysis of Information Systems Design Methodologies.* Amsterdam: Elsevier Science.

Olle, T., Sol, H., & Verrijn-Stuart, A. (Eds.) (1986) *Proc. of the IFIP WG8.1 Working Conf. on Comparative Review of Information Systems Design Methodologies: Improving the Practice.* Amsterdam: Elsevier Science.

Olle, T., Hagelstein, J., MacDonald, I., Rolland, C., Sol, H., van Assche, F., & Verrijn-Stuart, A. (1988). *Information Systems Methodologies – A Framework for Understanding.* 2nd edition. Reading: Addison-Wesley.

OMG. (2005). *Software process engineering metamodel specification*, Version 1.1, January 2005. Available at URL: < http://www.omg.org/technology/ documents/formal/ spem.htm>.

Opdahl, A., & Henderson-Sellers, B. (2001). Grounding the OML metamodel in ontology. *The Journal of Systems and Software,* 57(2), 119-143.

Plihon, V., Ralyté, J., Benjamen, A., Maiden, N., Sutcliffe, A., Dubois, E., & Heymans, P. (1998). A re-use-oriented approach for the construction of scenario based methods. In *Software Process (ICSP'98),* Chicago, Illinois, 14-17.

Peirce, C. (1955). *Philosophical writings of Peirce*, edited by J. Buchle. New York: Dover.

Pohl, K. (1994). The three dimensions of requirements engineering: a framework and its application. *Information Systems* 19(3), 243-258.

Prakash, N. (1999). On method statics and dynamics. *Information Systems,* 24(8), 613-637.

Punter, T., & Lemmen, K. (1996). The MEMA-model: towards a new approach for methods engineering. *Journal of Information and Software Technology,* 38(4), 295-305.

Ralyte, J., Deneckere, R., & Rolland, C. (2003). Towards a generic model for situational method engineering. In J. Eder & M. Missikoff (Eds.), *Advanced Information Systems Engineering (CAiSE'03)*(pp. 95-110). LNCS 2681, Berlin: Springer-Verlag.

Ramesh, B., & Jarke, M. (2001). Towards reference models for requirements traceability. *IEEE Trans. on Software Engineering, 27*(1), 58-93.

Rolland, C., & Prakash, N. (1996). A proposal for context-specific method engineering. In S. Brinkkemper, K. Lyytinen & R. Welke (Eds.), *Method Engineering: Principles of Method Construction and Tool Support* (pp. 191-208). London: Chapman & Hall.

Rolland, C., Prakash, N., & Benjamen, A. (1999). A multi-model view of process modeling. *Requirements Engineering, 4*(4), 169-187.

Rossi, M., Lyytinen, K., Ramesh, B., & Tolvanen, J.-P. (2005). Managing evolutionary method engineering by method rationale. *Journal of the Association of Information Systems (JAIS), 5*(9), 356-391.

Ruiz, F., Vizcaino, A., Piattini, M., & Garcia, F. (2004). An ontology for the management of software maintenance projects. *International Journal of Software Engineering and Knowledge Engineering, 14*(3), 323-349.

Saeki, M. (1998). A meta-model for method integration. *Information and Software Technology, 39*(14), 925-932.

Saeki, M. (2003). Embedding metrics into information systems development methods: an application of method engineering technique. In J. Eder & M. Missikiff (Eds.), *Advanced Information Systems Engineering (CAiSE 2003)* (pp. 374-389). LNCS 2681, Springer-Verlag: Berlin.

Saeki, M., Iguchi, K., Wen-yin, K., & Shinokara, M. (1993). A meta-model for representing software specification & design methods. In N. Prakash, C. Rolland & B. Pernici (Eds.), *Information Systems Development Process* (pp. 149-166). Amsterdam: Elsevier Science.

Simon, H. (1960). *The new science of management decisions.* New York: Harper & Row.

Sol, H. (1992). Information systems development: a problem solving approach. In W. Cotterman & J. Senn (Eds.), *Challenges and Strategies for Research in Systems Development* (pp. 151-161). Chichester: John Wiley & Sons.

Song X. (1997). Systematic integration of design methods. *IEEE Software, 14*(2), 107-117.

Song, X., & Osterweil, L. (1992). Towards objective, systematic design-method comparison. *IEEE Software, 9*(3), 43-53.

Sowa, J. (2000). *Knowledge representation – logical, philosophical, and computational foundations.* Pacific Grove, CA: Brooks/Cole.

Sowa, J., & Zachman, J. (1992). Extending and formalizing the framework for information system architecture. *IBM Systems Journal, 31*(3), 590-616.

Stamper, R. (1978). *Towards a semantic model for the analysis of legislation.* Research Report L17, London School of Economics.

Stamper R. (1996). Signs, information, norms and information systems. In B. Holmqvist, P. Andersen, H. Klein & R. Posner (Eds.) *Signs are Work: Semiosis and Information Processing in Organisations* (pp. 349-392). Berlin: De Gruyter.

Swede van, V., & van Vliet, J. (1993). A flexible framework for contingent information systems modeling. *Information and Software Technology, 35*(9), 530-548.

Thayer, R. (1987). Software engineering project management – a top-down view. In R. Thayer (Ed.), *Tutorial: Software Engineering Project Management* (pp. 15-56). IEEE Computer Society Press.

Tolvanen, J.-P. (1998). *Incremental method engineering with modeling tools – Theoretical principles and empirical evidence.* Dissertation Thesis, Jyväskylä Studies in Computer Science, Economics and Statistics, No. 47, University of Jyväskylä, Finland.

Uschold, M., & Gruninger, M. (1996). Ontologies: principles, methods and applications. *Knowledge Engineering Review, 11*(2), 93-155.

Venable, J. (1993). CoCoA: *A conceptual data modeling approach for complex problem domains.* Dissertation Thesis, State University of New York, Binghampton, USA.

Wand, Y. (1988). An ontological foundation for information systems design theory. In B. Pernici & A. Verrijn-Stuart (Eds.), *Office Information Systems: The Design Process* (pp. 201-222). Amsterdam: Elsevier Science.

Wand, Y., Monarchi, D., Parson, J., & Woo, C. (1995). Theoretical foundations for conceptual modeling in information systems development. *Decision Support Systems, 15*(4), 285-304.

Wand, Y., Storey, V., & Weber, R. (1999). An ontological analysis of the relationship construct in conceptual modeling. *ACM Trans. on Database Systems,* 24(4), 494-528.

Wand, Y., & Weber, R. (1995). On the deep structure of information systems. *Information Systems Journal,* 5(3), 203-223.

Wastell, D. (1996). The fetish of technique: methodology as a social defense. *Information Systems Journal,* 6(1), 25-40.

Welke, R. (1977). Current information system analysis and design approaches: framework, overview, comments and conclusions for large – complex information system education. In R. Buckingham (Ed.), *Education and Large Information Systems* (pp. 149-166). Amsterdam: Elsevier Science.

Zhang, Z., & Lyytinen, K. (2001). A framework for component reuse in a metamodelling-based software development. *Requirements Engineering,* 6(2), 116-131.

APPENDIX

Table 2. Scopes and emphases of the analyzed artifacts compared to the ontologies in OntoFrame

Ontologies in OntoFrame	[A]	[B]	[C]	[D]	[E]	[F]	[G]	[H]	[I]	[J]	[K]	[L]	[M]	[N]	[O]
Generic ontology		3			1	1									3
Semiotic ontology		3													
Intension/ extension ontology		2													
Language ontology		4													
State transition ontology		4						2							5
UoD ontology		2						2							1
Abstraction ontology		1				1									1
Context ontology	2	2	1			2		3		2	1	1	3	1	
Layer ontology	1			1		2	1								
Perspective ontology	2		2					3	1	2			2	2	
Model level ontology	1	2		1		2	1								
ISD ontology				2	1	1	3		1	1	1				
ISD method ontology					2	2	3					1			
ME ontology				1		1									

Legend:

[A] = Essink (1988)
[B] = Falkenberg et al. (1998)
[C] = Freeman et al. (1994)
[D] = Grosz et al. (1997)
[E] = Gupta et al. (2001)
[F] = Harmsen (1997)
[G] = Heym et al. (1992)
[H] = Iivari (1989)
[I] = Iivari (1990)
[J] = Olle et al. (1988)
[K] = Saeki et al. (1993)
[L] = Song et al. (1992)
[M] = Sowa et al. (1992)
[N] = van Swede et al. (1993)
[O] = Wand (1988), Wand et al. (1995)

Chapter IX
Concepts and Strategies for Quality of Modeling

Patrick van Bommel
Radboud University Nijmegen, The Netherlands

Stijn Hoppenbrouwers
Radboud University Nijmegen, The Netherlands

Erik Proper
Capgemini, The Netherlands

Jeroen Roelofs
BliXem Internet Services, Nijmegen, The Netherlands

ABSTRACT

A process-oriented framework (QoMo) is presented that aims to further the study of analysis and support of processes for modeling. The framework is strongly goal-oriented, and expressed largely by means of formal rules. The concepts in the framework are partly derived from the SEQUAL framework for quality of modelling. A number of modelling goal categories is discussed in view of SEQUAL/QoMo, as well as a formal approach to the description of strategies to help achieve those goals. Finally, a prototype implementation of the framework is presented as an illustration and proof of concept.

INTRODUCTION

This chapter aims to contribute to the area of conceptual modeling quality assessment and improvement, in particular by providing some fundamental concepts concerning the quality of the *process* of modeling, and for structured description of *ways of achieving* quality models. Though operationalization of the concepts and strategies is still limited in this version of the framework, an initial application has been realized and is discussed.

There is a clear link between the work presented and the field of Situational Method Engineering. In particular, the basic idea of combining (patterns of) language related aspects of methods with process related aspects is commonplace in method engineering (see for example Mirbel and Ralyté, 2006; Ralyté et al., 2007). We believe the specific contribution of the current chapter lies in its formal, rule-based nature, and a strong emphasis on combinations of rather specific modeling goals. Also, we focus only on modeling, whereas method engineering in general also covers other activities in systems engineering. Finally, we choose a relatively fine-grained view on the activity of modeling, whereas method engineering generally deals with process aspects only at the level of clearly distinguishable *phases* (i.e. has a more course-grained view, which is not to say that such a view is not a very useful one in its own right).

We first present a process-oriented 'Quality of Modeling' framework (QoMo), which for a large part is derived from the established SEQUAL framework for quality of models. QoMo is based on knowledge state transitions, the cost of the activities bringing such transitions about, and a goal structure for activities-for-modeling. Such goals are directly linked to concepts of SEQUAL.

We then proceed in two steps. In the first, generic step (section 5: "a generic rule-based meta-model for methods and strategies") we consider the underlying generic structure of strategies for modeling. We discuss how QoMo's goals for modeling can be linked to a rule-based way of describing processes for modeling. Such process descriptions hinge on *strategy frames* and *strategy descriptions*, which may be used descriptively (for studying/analyzing real instances of processes) as well as prescriptively (for the guiding of modeling processes). We present a set of concepts for describing quality-oriented strategies.

In the second, implementation step (section 6: "Implementing goals and strategies in a concrete workflow language") we consider an example implementation involving a concrete operational workflow language. We present results of a case study in which a specialized version of our generic framework is applied to the description of an elementary method for requirements modeling, as taught in the 2nd year of an Information Science Bachelor's curriculum. We discuss and exemplify how concepts from the generic framework were used, and in some cases how they were amended to fit the task at hand.

BACKGROUND

Interest in frameworks for quality and assessment of conceptual models has been gradually increasing for a number of years. A generic overview and discussion can be found in (Moody, 2006). A key framework for analysis of the quality of conceptual models is the SEQUAL framework (Krogstie et al., 2006; Krogstie, 2002; Krogstie and Jorgesen, 2002) . This framework takes a semiotics-based view on modeling which is compatible with our own (Hoppenbrouwers et al., 2005a). It is more than a quality framework for models as such, in that it includes not just the model but also the knowledge of the modelers, the domain modeled, the modeling languages, agreement between modelers, etc.; it bases quality assessment on relations between such model-related items, i.e. respects the broader context of the model.

As argued in (Hoppenbrouwers et al., 2005b), in addition to analysis of the quality of models, the *process* of which such models are a product should also be taken into account. We briefly summarize the main arguments here:

1. Though some have written about detailed stages in and aspects of "Ways of Working" in modeling, i.e. its process or procedure (for example, see Halpin, 2001), the detailed "how" behind the activity of creating models is still mostly art rather than science. There

is, therefore, a purely scientific interest in improving our understanding of the operational details of modeling processes.

2. In addition, such a study should enable us to find ways of improving the modeling process (for example, its quality, efficiency, or effectiveness; from a more methodological angle: reproducibility, stability, and precision; also traceability, and so on).
3. Indeed, some aspects of quality, it seems, can be better achieved through a good modeling process than by just imposing requirements on the end product and introducing a feedback cycle (iteration). This holds in particular (though not exclusively) for matters of validation and grounding in a socio-political context.
4. A score of more practical arguments follow from the ones above. For example, for improvement of case tool design, a more process-oriented approach seems promising.

As mentioned, models as such are not the only product of a modeling process. The related knowledge development, agreements, etc. are arguably as important. Hence, our process-oriented view suggests that we take aboard such additional *model items*, and quality concepts related to them. The latest SEQUAL version explicitly allows for this approach, though it does not make explicit use of a meta-concept such as "model item".

The importance of the modeling process in view of model quality is commonly confirmed in the literature in general (Poels et al, 2003; Nelson and Monarchi, 2007). If we want to evaluate a modeling process, we can take (at the least) the following four different points of view:

1. Measure the success of the process in fulfilling its goals. This boils down to using the current SEQUAL framework as-is, and directly link such an analysis of the model items to the success of the process. However, one might also analyze *intermediate steps* in the process against intermediate or sub-goals set. By evaluating partial or intermediary products (possibly in view of a prioritization of goals), the process may be *steered* along the way. Also, the steps described will eventually become so small that *essential* modeling goals/activities can perhaps be identified (steps in the detailed thinking process underlying modeling), opening up the black box of the modeling process.
2. Calculate the cost-benefit ratio: achievements set against the cost. Such cost again can hold for the process as a whole, but also for specific parts.
3. We can look at the process and the working environment as an information system. This then is a 2nd order information system: an IS that serves to develop (bring forth) information systems. The information system underlying the modeling process (probably but not necessarily including IT support) can be evaluated in a way similar to evaluation of information systems in general. In particular, aspects like usability and actability but also traceability and even portability are relevant here.
4. Views 1-3 concern operational evaluations of particular process instantiations. At a higher level of abstraction, we can also look at properties and control aspects of a process in terms of, for example, repeatability, optimization, etc. (Chrissis et al., 2006).

In this paper, we focus on 1 and 2. View 3 depends very much on implementation and support of some specific process (in particular, tooling), which is outside the grasp of our current study. View 4 is essential in the long run yet stands mostly apart from the discussion in this paper, and will not be elaborated on any further now. Admittedly, many fundamental aspects of process quality (process control) are expected to be covered by 4, rather than 1 and 2. However, 1 and 2 do provide the concepts we direly need for applying 4 in any concrete way. This paper, therefore, is arguably a SEQUAL-based 'step up' towards analysis at the level of viewpoint 4.

Our main contribution in this chapter twofold: 1. a preliminary version of a framework for Quality of Modeling (QoMo), which not only takes into account the products of modeling but also *processes*. Analogous to SEQUAL, QoMo is not yet an operational method for model-oriented quality assessment and management. The framework is expected to evolve as our knowledge of dealing with quality of modeling processes increases. 2. We provide a set of concepts for capturing the basics of strategies for realizing the QoMo goal. The concepts are made operational for the first time in context of a case implementation.

THE SEQUAL FRAMEWORK: OVERVIEW

Since we cannot, nor wish to, present an elaborate overview or discussion of the SEQUAL framework, we will provide a short summary of its key concepts, as based on its latest substantial update (Krogstie et al., 2006). However, we take into account all SEQUAL concepts, including those not explicitly mentioned in that update. We have rephrased some of the definitions or added some explanations/interpretations of our own, yet made sure we did not stray from the intentions of the SEQUAL framework.

SEQUAL Model Items:

- **G**: goals of modeling (normally organizationally defined).
- **L**: language extension; set of all statements that are syntactically correct in the modeling languages used.
- **D**: the domain; the set of all statements that can be stated about the situation at hand.
- **D^O** the optimal domain; the situation the organization would or should have wanted –useful for comparison with the actual domain D in order to make quality judgments.
- **M**: the externalized model; the set of all statements in someone's model of part of the perceived reality written in a language.
- **K_s**: the relevant knowledge of the set of stakeholders involved in modeling (i.e. of the audience at large).
- **K_m**: a subset of K_s; the knowledge of only those stakeholders actively involved in modeling.
- **K^N**: knowledge need; the knowledge needed by the organization to perform its tasks. Used for comparison with K_s in order to pass quality judgments.
- **I**: the social actor interpretation, that is, the set of all statements that the audience thinks that an externalized model consists of.
- **T**: the technical actor interpretation, that is, the statements in the model as 'interpreted' by the different modeling tools.

SEQUAL Quality Definitions

- **Physical quality**: how the model is physically represented and available to stakeholders; a matter of *medium*.
- **Empirical quality**: how the model comes across in terms of *cognitive ergonomics*, e.g. layout for graphs and readability indexes for text.
- **Syntactic quality:** conformity to the syntax of the modeling language, involving L.
- **Semantic quality**: how well M reflects K_s.
- **Ideal descriptive semantic quality**: Validity: $M/D=\varnothing$; Completeness: $D/M=\varnothing$.
- **Ideal prescriptive semantic quality**: Validity: $M/D^O=\varnothing$; Completeness: $D^O/M=\varnothing$.
- **Domain quality**: how well the domain fits some desired situation: D compared with D^O.

- **Quality of socio-cognitive interpretation**: how an individual or group interprets the model, i.e. how I matches M, in view of how M was intended to be interpreted by one or more of its modelers.
- **Quality of technical interpretation**: similarly, how a tool or group of tools interprets the model, i.e. how T matches M.
- **Pragmatic quality -actability**: how the model, or the act of modeling, influences the actability of the organization. Note that this enables description of the effect of the modeling process even in case the model as such is discarded.
- **Pragmatic quality -learning**: how the modeling effort and/or the model as such contribute to organizational learning.
- **Knowledge quality**: how well actual knowledge K_s matches knowledge need K^N.
- **Social quality**: the level of agreement about the model among stakeholders (individuals or groups) about the statements of M.

PRODUCT GOALS AND PROCESS GOALS

There has been some preliminary process oriented work related to the use of SEQUAL (e.g. going back to Sindre and Krogstie, 1995). Our current proposal concerns a strongly goal-oriented approach. The model items and qualities of the SEQUAL framework can be used as an abstract basis for expressing product quality of a model process, and alternatively to specify goals for model quality. Note that during a modeling process, each model item (for example the model (M), stakeholder knowledge (K_s), or the state of agreement about M (Covered by "Social quality" in SEQUAL) may change, in principle, at any step in the process. The SEQUAL framework can therefore be used not only for expressing how well a process as a whole has done in terms of achieving its goals, but also which specific steps or series of steps in the process have contributed to specific advances in achieving specific (sub)goals. We can thus directly use SEQUAL concepts for expressing *product goals*: both *product end-goals* and *intermediary product goals*.

In addition, it should be possible to link product goals and sub-goals, and the level of achievement in view of these goals, to the notion of *benefit*, and weigh this against its *cost*. Note that cost (for example in terms of time or money) is something that can be especially well calculated in view of a work process and the people and resources involved in it. This entails that, as is common in process modeling, account should be taken of the tasks or actions involved as well as the people (typically, *roles*) performing them. Both the cost of the entire process and, again, the cost of steps/parts in the process can then be calculated. This entails that the cost-benefit ratio for an entire process, or parts of it, can be calculated. As argued earlier, this is a useful way of evaluating a modeling process.

In (van Bommel et al., 2006) we presented an initial list of *modeling goals* (slightly amended here) of which the relation to SEQUAL will be made clear. The goals are, or course, generically covered by G in SEQUAL, but they also relate to most of the other SEQUAL concepts.

Usage goals (including actability and knowledge goals): These stand apart from our "modeling goals": they represent the *why* of modeling; the modeling goals represent the *what*. In SEQUAL, the usage goals are covered primarily by the Pragmatic qualities (both learning and actability) and, related to the former, Knowledge quality. The overall cost-benefit ratio will mostly relate to Usage goals, but optimizing (aspects of) modeling methods in an operational sense requires us to look at the other goals, the "modeling goals":

Creation goals (list of model items/deliverables): This relates to what we might generalize as "required deliverables": M, in a very broad sense (i.e. also including textual documents etc.). If made explicit, K_s and/or K_m are to be included here. Creation goals are primarily related to the SEQUAL notions of *completeness* (as part of "Ideal descriptive/prescriptive semantic quality") and *validity* as defined under "Ideal descriptive/prescriptive semantic quality". Note that "Completeness" in an operational sense would in fact be defined as $K_s/M = \varnothing$ ((Krogstie et al., 2006) has it as $M/D = \varnothing$). Validity would then be $M/K_s = \varnothing$. There is a complication, however, because some definitions of validity also strongly involve Social Quality (see Validation goals below), linking validation with levels of agreement with (parts of) the model by specific actors. We observe that SEQUAL allows us to differentiate between these two notions of validity, and yet combine them.

Validation goals: These are related to Social Quality: the level and nature of agreement between stakeholders concerning the model. As discussed, our analysis allows us to differentiate between two common notions of "validity": one now falling under Creation Goals, one (the one related to Social Quality) under Validity Goals.

Argumentation goals: In some cases, arguments concerning particular model items may be required. Though a weak link with Social Quality can be suggested here, it seems that this type of modeling goals is not as of yet explicitly covered by SEQUAL. Argumentation goals arguably are an extension of Validation goals.

Grammar goals: Language (L) related: concerns syntactic quality.

Interpretation goals: Related to quality of socio-cognitive interpretation, and possibly also to technical interpretation. The latter may be covered by Grammar goals if the language and its use are fully formal and therefore present no interpretation challenges whatsoever. Note that Interpretation Goals may be seen as a refinement of Validation Goals.

Abstraction goals: This is as of yet a somewhat obscure category. It boils down to the question: does the model (or parts of it) strike the right level of abstraction? This seems to be a crucial matter, but also one that is terribly hard to operationalize. There seems to be a link with Semantic quality (and it is a different link than the one covered by Creation Goals), but its precise nature is yet unclear to us. Quite probably, Abstraction Goals are sub-goals of Usage Goals, related to the utilitarian relevance of various levels of and details in models.

ACHIEVING GOALS BY MEANS OF STRATEGIES

Given usage goals, modeling goals, and a modeling environment, strategies can be formulated to best execute the modeling process. In other words:

Usage goal + Modeling goal + Modeling environment ⇒ Modeling strategy

Moving to concepts that are more specifically related to actual modeling processes, we will now briefly present an approach to describing the detailed steps in modeling processes. This approach can be used either descriptively or prescriptively.

It is customary in run-of-the-mill method description to view procedures or processes in terms of fairly simple flows or "steps". In view of the many entwined goals and sub-goals at play in modeling, and the different sorts of equally entwined actions taken to achieve these, it seems more helpful at a fundamental level to let go of (work)flow in the traditional sense as a metaphor for modeling procedures. Instead, we advocate a rule-based approach in which an analysis of the states of all relevant model items, in combination with a set of rules describing strategies, leads to "run-time" decisions on what to do next. Importantly, the framework is capable of capturing *ad hoc* activities of modeling as well as tightly pre-structured ones. In order to achieve this, we define constraints on states and activities, i.e. not necessarily fully specified goals and strategies. As for workflow-like procedures: these are likely to play a role at some level, as reference procedures. However, we believe that while such procedures may be valid instruments for supporting a modeling process, they are not so well suited to capture or reflect actual, individual modeling processes.

Our view of methods and the processes they govern takes shape in the use of the following concepts. These concepts are based on some initial studies of modeling processes (for example, Hoppenbrouwers et al., 2005b).

Results of steps in a method are *states*. States are typically described by means of some logic, or a semi-natural verbalization thereof. *Goals* are a sub-type of states, and typically describe desired states-of-affairs in the project/domain; *situations* are also a sub-type of state, that describe states-of-affairs that contribute towards achieving their related goal(s). Importantly, situations are not mere negative versions of a goal they are linked with: they are positively stated descriptions of a state of affairs that is a way point on the way to achieving a goal. If no helpful state-of-affairs can be identified, the situation is "void": no particular precondition state is assumed.

States cannot be labeled situations or goals in an absolute sense; such labeling is linked to the use of some state in relation to some *transition(s)* and some other state(s). In other words, a state that is a situation in view of some goal can also be a goal in view of another situation. Combinations of states and transitions can be graphically depicted by means of plain directed graphs. States can be described at type level as well as at instance level. This means that a state description may be an abstract description of one or more aspects of actual (instantiated) states. Situation descriptions as well as goal descriptions are typically used as abstract patterns to be matched against factual (instance level) descriptions of the method domain as the operational process takes place.

The combination of, minimally, two states (a situation and a goal) linked by a transition, we call a strategy *context*[a]. Complex strategy contexts are allowed; they are composed of three or more states, and two or more transitions. A *strategy* is a course of action associated with one or more strategy contexts. We distinguish three types of strategy in our framework for method description:

Ad hoc strategies. In ad hoc strategies, no concrete modeling guidelines are available. We know what we have (input) and we have an indication of what we should produce (output), but we do not know how we should produce this.

Guided strategies. In guided strategies, a concrete guided description of the actions to be performed is available. This description can for example have the form of (a) a piece of text in natural language, (b) a piece of text in some controlled language such as a subset of natural language (c) an algorithm written in pseudo code or (d) a graphical notation (for example, a workflow).

Composed strategies. A composed strategy consists of more (sub) strategies. The nesting structure of composed strategies may be cyclic, as long as eventual termination by means of ad hoc or guided strategies is guaranteed.

Some very basic notions of *temporality* are also part of our framework. A strategy (complex or not) may be defined as preceding another strategy: "strategy s occurs before strategy t". This may be captured in a (logical) rule constraining the occurrence in time of s and t. It allows for other strategies to be executed between execution of s and of t. In addition, a stricter notion of immediate order is used: if a strategy t is defined as occurring immediately after strategy s, this means that no strategy may be executed between the execution of s and t. For the moment, these simple temporal notions are all we need for our purposes.

The basic concepts being in place, we need to refine those generic concepts for more specialized categories of strategy. For this, we use the goals discussed in the previous section. Note that the distinction in goal types is based strictly on the terms in which their states are described, and that the various strategy/goal types can be combined and even tightly interwoven in the fabric of a method model. For example, validation strategies can be combined with creation strategies at a low level; alternatively, validation may be seen as an activity separated in time from creation. To our regret, in the remainder of this chapter we only have space to consider examples for two goal types: grammar goals and creation goals. We emphasize, however, our conviction that the rule-based approach to method modeling can be successfully used beyond the detailed study and guidance of modeling in some particular modeling language.

A GENERIC RULE-BASED META-MODEL FOR METHODS AND STRATEGIES

We now first focus on generic descriptions of composed modeling strategies. Our examples in this section concern grammar strategies related to a technique for formal conceptual modeling called Object Role Modeling (Halpin, 2001). We then proceed with the introduction of the descriptive concepts of strategy space and strategy frame, and we elaborate on these in order to achieve further refinement.

Basic Description of (Grammar) Strategies

In this subsection we focus on basic grammar-related strategy descriptions. Trivial though the differences between grammar strategies may seem to the casual observer, they have given rise to considerable debate among experienced modelers, and even schisms between schools of conceptual modeling that are in principle closely related. In addition, novice modelers often feel uncertain about ways to proceed in modeling. Where to start? Where to go from there? What to aim for? Similar questions are raised in any modeling situation, for any modeling language. In practice, several main ORM grammar strategies have been observed to co-exist. Several styles are in fact used in the literature. We consider some styles which are in part reflected in ORM-related literature and practice. Here we first consider an "object-type-driven" modeling strategy. This strategy starts with the introduction of "object types":

Strategy description 1 *Object-type-driven strategy (basic).*

a. Provide object types.
b. Provide fact types.
c. Provide facts and constraints.
d. Provide objects.

Thus, strategy description 1 focuses on grammar goals, where a (partial) modeling procedure is initiated by the specification of object types, followed by fact types. Next, fact instances and constraints are specified. The modeling procedure is concluded by providing the necessary object instances. Modeling of other styles of object-type driven strategies is of course also possible, for example if the requirement is posed that constraints and object instances are specified before fact instances. However, such an approach goes beyond existing methodological descriptions of ORM modeling. Note that we do not claim that the "object-type-driven strategy" is superior (or inferior, for that matter) to other strategies; finding out which strategy works best in which situation concerns a research question not addressed here.

We make a distinction between model-items and model-item-kinds. Items (instance level) reflect the model-in-progress, whereas item kinds (type level) stem from the method (in this case, the modeling language) used for describing the domain. Examples of items in our grammar strategy are: 'John', 'person', 'John works for the personnel department', and 'persons work

Figure 1. Example strategy space p_1

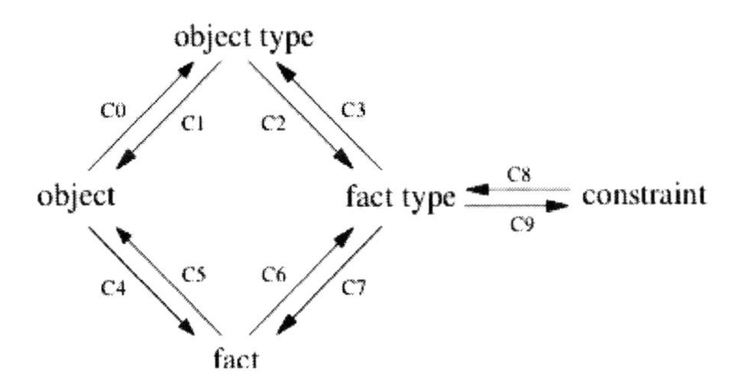

for departments'. The corresponding item kinds are: 'object', 'object type', 'fact', and 'fact type', respectively.

Strategy Space

In order to facilitate effective definition of composed modeling strategies we define the notion of strategy space, which is a light-weight representation of the meta-model of a modeling language. For other approaches to meta-modeling, see for example (Kensche et al., 2005; OMG, 2001; Saeki, 1995). A strategy space P is defined as a structure $P = \langle S,T \rangle$ consisting of states S and transitions T. States are associated with item kinds. We assume $T \subseteq S \times S$ and a transition $\langle x,y \rangle$ has $x \neq y$. A strategy space provides the context for setting up strategies, but also for refining and using strategies.

Example 1.

As an example, figure 1 presents the strategy space p_1. This space contains the following states: object, fact, object type, fact type, and constraint. It contains the transitions c0 , . . . , c9 , where for instance c4 = <object,fact>. In section the section below we show how strategy description 1 is embedded within the space p1. When drawing strategy spaces, a transition is drawn as an arrow. When writing S and T we assume that the corresponding space P is clear from the context. If this is not the case, we write SP and TP or $S(P)$ and $T(P)$.

Strategy Frame

Many different composed modeling strategies can be described in the context of a single strategy space. In order to embed the various strategies within a single strategy space, we define the notion of **strategy frame**. A strategy frame F in the context of a space P is a spanning subgraph of P:

- A frame contains all states, so $SF = SP$.
- A frame contains some of the transitions, so $TF \subseteq TP$.

The set of all frames of a strategy space P is denoted as frames(P). Note that a frame is not necessarily connected.

Example 2.

As an example, the left hand diagram of figure 2 presents the strategy frame f1 $\in$ frames(p1). The frame f1 is used as underlying structure of strategy description 1. In the strategy space p1 the frame f1 focuses on an object-type-driven strategy by first specifying object types, followed by fact types via transition c2. Then constraints are specified via transition c9 and facts and objects are specified via transitions c7 and c5.

Figure 2. Example strategy frames **f1 and f2**

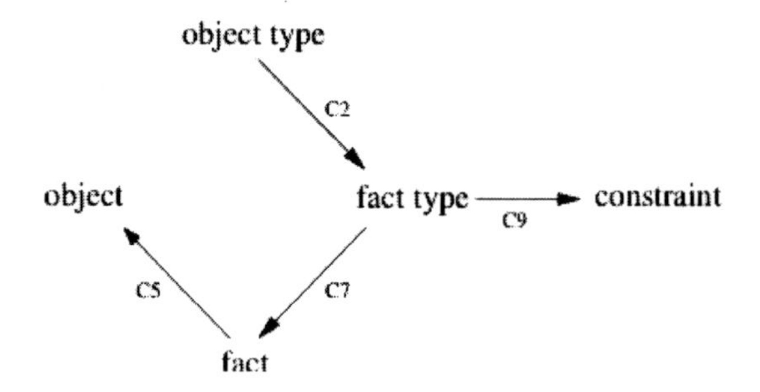

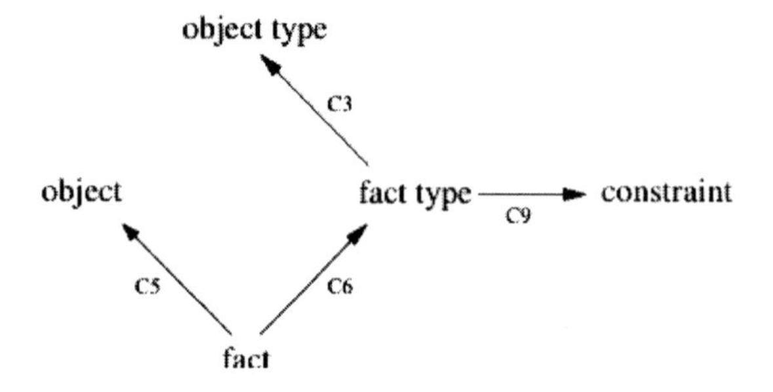

In basic grammar strategy descriptions we usually have a breadth-first traversal of the underlying strategy frame. Refinements of this default approach are discussed later.

According to the definitions of strategy space and strategy frame, each strategy space is a strategy frame as well, or P ∈ frames(*P*). Also, each strategy frame can be used as a new strategy space consisting of subframes. This enables the systematic treatment of sub-strategies.

Example 3.

If the frame f1 in figure 2 were used as a new strategy space, we would have several possible subframes. We have, however, only a single connected subframe in this case, which is the frame f1 ∈ frames(f1).

The Rules Underlying a Strategy Frame

We now provide an example of the rules underlying strategy frame f1 . The relevant goal descriptions for strategy frame f1 are the following[b].

G1 There is at least one FactType

This is an instance-level goal. The rest of the goals are type-level; they express syntactic requirements as dictated by the ORM meta-model (note that we use a mere selection from the complete set of rules that define the metamodel). The other goal-rules relevant to f1 are:

G2 Each FactType is populated by at least one Fact

G3 Each ObjectType belongs to at least one FactType

G4 Each FactType is constrained by zero or more Constraints

G5 Each Fact has at least one Object

Next, we define the one situation that is relevant to f1:

S1 There is at least one ObjectType

So we assume that one or more ObjectTypes have already been identified (presumably as a goal of another strategy) and that these ObjectTypes are used as input for f1. Next, a series of contexts is defined, consisting of states (either situations or goals) linked by a transition. Note that these related contexts are a 1:1 reflection of f1. The states are selected from the rules above. The transitions are verbalized as "SHOULD LEAD TO".

S1 There is at least one ObjectType

C2 SHOULD LEAD TO

G3 Each ObjectType belongs to at least one FactType

G3 Each ObjectType belongs to at least one FactType

C7 SHOULD LEAD TO

G2 Each FactType is populated by at least one Fact

G2 Each FactType is populated by at least one Fact

C5 SHOULD LEAD TO

G5 Each Fact has at least one Object

G3 Each ObjectType belongs to at least one FactType

C9 SHOULD LEAD TO

G4 Each FactType is constrained by zero or more Constraints

So far, our definition does not include temporal ordering. Adding this, we get:

C2 occurs before C7
C2 occurs before C9
C7 occurs before C5
C5 occurs immetiatelyAfter C7

These temporal rules will be elaborated on below. The transitions are to be linked to further strategies (ad hoc, guided, or composed, as explained), which suggest how each particular transition is to be achieved.

This concludes our example. For reasons of space, we will not define other strategy frames at rule level, as the same descriptive principles hold across all examples.

Another Composed Modeling Strategy

Next we consider a fact-driven strategy. In this strategy, is is mandatory that facts are introduced first. Suppose we have the following basic strategy description:

Strategy description 2 *Fact-driven strategy (basic).*

a. *Provide facts.*
b. *Provide fact types and objects.*
c. *Provide object types and constraints.*

Note that strategy description 2 is indeed fact-driven rather than fact-type-driven. A fact-type-driven strategy would require the specification of fact types prior to the specification of fact instances. Comparison of strategies is considered in more detail in section 6.6 "comparison of modeling strategies").

Example 4.

The right-hand side of figure 2 presents the strategy frame f2 $\in$ frames(pl). The frame f2 is used as underlying structure of strategy description 2. In the strategy space pl the frame f2 focuses on the specification of facts including their types via transition c6 and their objects via transition c5, followed by the specification of object types via transition c3 and constraints via transition c9.

Comparison of Modeling Strategies

In order to compare composed modeling strategies, we examine the differences between their underlying frames. We first consider the reversal of individual transitions by the reversal operator rev. Let $x = \langle y,z \rangle \in T$ be a transition. Then the effect of rev_x is that $\langle y,z \rangle$ is removed and a new transition $\langle z,y \rangle$ is added.

Example 5.

As an example we compare the frames of object-type-driven and fact-driven strategies. We see that these frames are quite similar. No transitions are added or deleted, and some transitions are reversed while others are not. Using the reversal operator we thus may have:

$f1 = \text{rev}_{c6}(\text{rev}_{c3}(f2))$

In the above example only a selection of individual transitions has been reversed. Next we consider dual modeling strategies. For a given strategy frame $x \in \text{frames}(P)$, the dual frame $\text{dual}(x)$ is obtained by reversal of all transitions. Note that the dual frame is not necessarily a frame of the same strategy space. Only if each transition in a strategy space is accompanied by its reversal, the dual of a frame is again a frame in that same space:

$P = \text{dual}(P) \Leftrightarrow \forall x \in \text{frames}(P)\ [\text{dual}(x) \in \text{frames}(P)]$

The above is a basic property of frame duality. Note that the dual frame is only one way of deriving an entire strategy with completely different organization. Another way to derive an entirely different strategy is based on the notion of complement strategy. For a given frame $x \in \text{frames}(P)$ the complement frame $\text{compl}(x,P)$ contains all states, and exactly those transitions from P which are absent in the original frame x.

Example 6.

In figure 3 we see two other example strategy frames f3 and f4 . In the left hand diagram, the frame f3 expresses a dual view on object-type-driven strategies, since $f3 = \text{dual}(f1)$. In the right-hand side, the frame f4 expresses a complement view on fact-driven strategies, because $f4 = \text{compl}(f2)$.

The frames f3 and f4 reflect several interesting properties of strategy frames. These will be considered in later sections. We now first express the basic property that a complement frame is again a frame of the same space:

$\forall\, x \in \textbf{frames(P)}\ [\textbf{compl}(x, P) \in \textbf{frames}(P)]$

Strategy Refinement

Here we discuss the refinement of basic strategy descriptions. Consider the following refined object-type-driven strategy:

Strategy description 3 *Object-type-driven strategy (refinement).*

a. Provide object types.

b. Provide fact types including constraints.

c. Provide facts including objects.

Figure 3. Example strategy frames **f3 and f4**

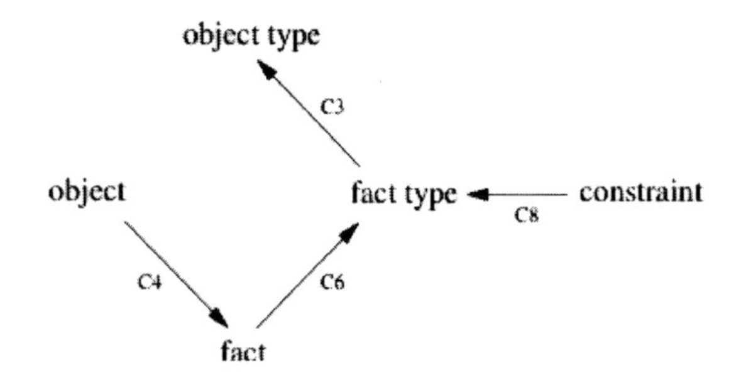

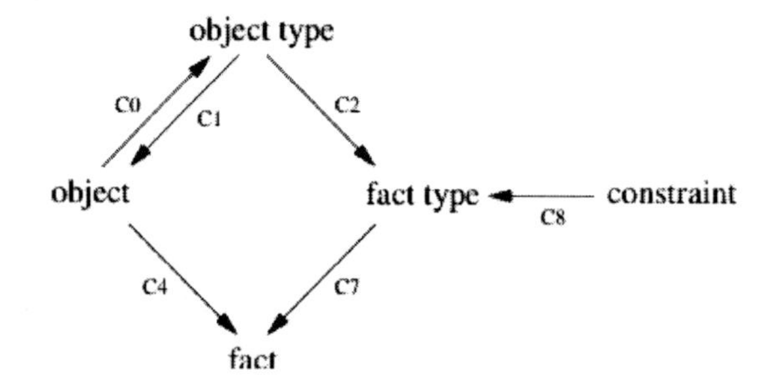

In the above description, we use a refinement of the breadth-first approach that was assumed in strategy description 1. To be able to express such refinements, the transitions in a strategy frame will be ordered.

We let $\prec$ be an ordering relation on transitions. For two transitions $x, y \in T$, the intention of $x \prec y$ is that transition x is handled prior to transition y.

Example 7.

The transition from the basic strategy description 1 to the refined strategy description 3 is obtained by the additional ordering requirement c9 $\prec$ c7 in frame f1.

Next we consider the following refined fact-driven strategy:

Strategy description 4 *Fact-driven strategy (refinement).*

a. Provide facts including objects.
b. Provide fact types including object types.
c. Provide constraints.

In terms of the ordering of transitions, the above refinement is expressed as follows.

Example 8.

The transition from the basic strategy description 2 to the refined strategy description 4 is obtained by the additional ordering requirements c5 $\prec$ c6 and c3 $\prec$ c9 in frame f2.

Besides the $\prec$ constraint, we also have a stronger temporal constraint. This stronger constraint expresses that one transition must be handled immediately after another transition. Note that more temporal constraints may be embedded within our framework, for example notions occurring in workflow specifications. At this moment, these constraints are not needed for our purposes, though.

IMPLEMENTING GOALS AND STRATEGIES IN A CONCRETE WORKFLOW LANGUAGE

In this section, we show how the framework presented thus far has been used in the implementation of a reference method for requirements modelling as taught in the 2nd year Requirements Engineering course of the BSc Information Science curriculum at Radboud University Nijmegen. Please note that the method as such is not subject to discussion in this paper, just the way of describing it. This section is based on work by Jeroen Roelofs (Roelofs, 2007). The original work focused strictly on strategy description; in this paper, some examples of related goal specification are added. The strategy description was implemented as a simple but effective web-based hypertext document that allows "clicking your way through various layers and sub-strategies" in the model (see below).

Case Study and Example: Requirements Modeling Course Method

The main goal behind the modelling of strategies of the case method was to provide a semi-formal, clear structuring and representation thereof that was *usable for reference purposes*. This means that the rule-based nature of the framework was played down, in favour of a clear and usable representation. A crucial step (and a deviation of the initial framework) was taken by replacing the plain directed graphs used so far by workflow-style models in the formal YAWL language (Yet Another Workflow Language: van der Aalst and ter Hofstede, 2005). Below we show the basic concepts of YAWL (graphically expressed), which were quite sufficient for our purposes. We trust the reader will require no further explanation.

The main (top) context of the method is depicted in the following schema:

Figure 4. Basic graphic concepts of YAWL

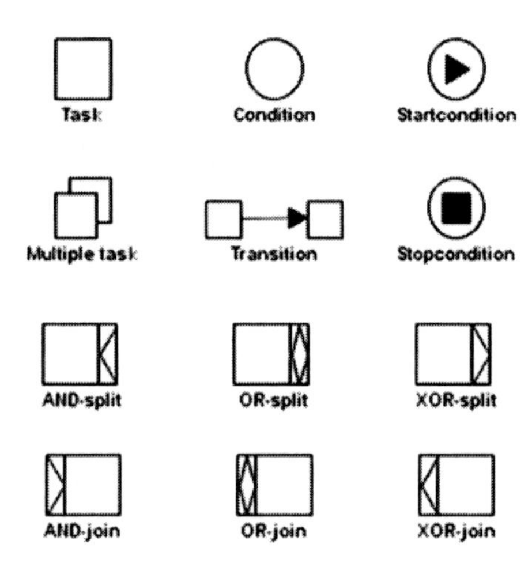

Note that in Figure 5, the square-based YAWL symbols correspond to the QoMo strategy framework in that they represent states (goals/situations). The actual strategies match the arrows between states: the actions to be taken to effectuate the transitions between states. In other words, the diagram is a very concise way of representing a strategy *context*. A useful operational addition to the framework is the use of conditions (circles) for choosing a goal (in going from "Use Case Survey" to either "Scenario" or "Use Case"): this was explicitly part of the existing method and possible in YAWL, and therefore gratefully taken aboard.

All arrows in the diagram have been labelled with *activity names* (which are another addition to the framework). Underlying the activities, there are strategies, which in turn consist of one or more *steps* (another addition). The complete strategy description of the activity "create requirements model" which is graphically captured by the top context (fig. 5) is the following:

1. **Create problem statement**
2. **Create use case survey**
3. **Create use case based on use case survey AND create scenario based on use case**
3. **Create scenario based on use case survey AND create use case based on scenario**
4. **Create domain model based on use case**
4. **Create domain model based on scenario**
5. **Create terminological definition**
6. **Create business rule**
7. **Create integrated domain model**

All steps listed are represented in boldface, which indicates they have underlying *composed* strategies (which implies that each step is linked to a further activity which is in turn linked to an underlying strategy). Concretely, this means that in the hypertext version of the description, all steps are clickable and reveal a new strategy context for each deeper activity. For example, if **"Create domain model based on use case"** is clicked, a new (rather smaller) context diagram

Figure 5. The top strategy context "Create a Requirements Model"

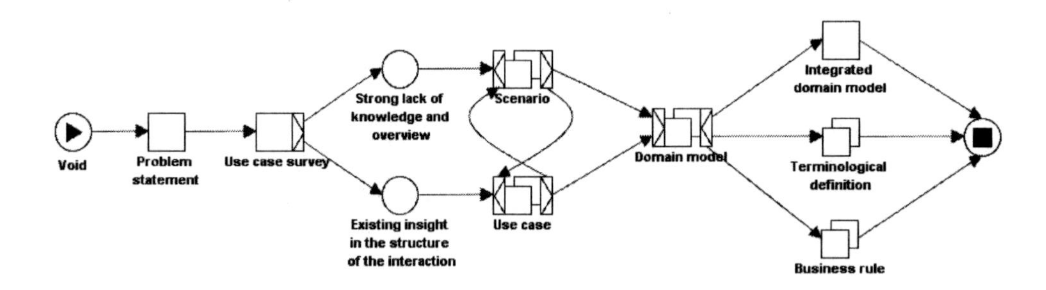

in YAWL is shown, with further refinement of what steps to take (strategy description). We will get back to this particular strategy, but before we do this, some explanation is in order concerning the irregular numbering of steps above. The occasional repetition of numbers (3. 3. and 4. 4.) serves to match the textual description with the YAWL diagram: the XOR split and AND-join in figure 5. In addition, the two possible combinations of steps before the AND-join needed to be combined using an "AND" operator, but note that the activities linked by AND are separately clickable.

Let us now return to the "**Create domain model based on use case**" strategy. It concerns the creation of a "Domain Model" (ORM) based on a "Use Case", which (roughly in line with previous examples) boils down to a basic description of steps in making an ORM diagram based on the interaction between user and system that is described stepwise in the use case (please note the participants in the course are familiar with ORM modelling and therefore need only a sketchy reference process description). The related strategy context is a fragment of the one in the top context:

Apart from this context, the underlying strategy is shown:

1. **Identify relevant type concepts in use case**
2. **Create fact types**
3. <u>Create example population</u>
   - Make sure the example population is consistent with the related scenario(s)
4. Make constraints complete

Steps one and two, represented in boldface, by way of more activities refer to more compositional strategies, so they are clickable and each have an underlying strategy. Activities 3. and 4. are represented differently, respectively signifying a *guided strategy* (underlined and with additional bulleted remark) and an *ad hoc strategy* (normal representation). A guided strategy is a strategy of which a description of some sort is available that helps execute it. In the example, this guidance is quite minimal: simply the advice to "Make sure the example population is consistent with the related scenario(s)". In view of our general framework, this guidance could have been anything, e.g. a complex process description or even an instruction video, but crucially it would not be part of the compositional structure. In context of our case method, we found that a few bulleted remarks did nicely.

There still is the ad hoc strategy linked to the activity name "Make constraints complete" (step 4.). It simply leaves the execution of the activity entirely up to the executor. As explained, it is an "empty strategy" –which is by no means a useless concept because it entails an explicit decision to allow/force the executing actor to make up her own mind about the way they achieve the (sub)goal.

In addition to the strategy context diagrams and the textual strategy descriptions, the hypertext description provided a conceptual diagram (in ORM) for each strategy, giving additional and crucial insights in concepts mentioned in the strategy and relations between them. The ORM diagram complementing the "create domain model based on use case" strategy is given in figure 7[c]. In context of the case, the inclusion of such a diagram had the immediate purpose of clarifying and elaborating on the main concepts used in the strategy description. In a wider context, and

Figure 6. another strategy context – "create domain model based on use case"

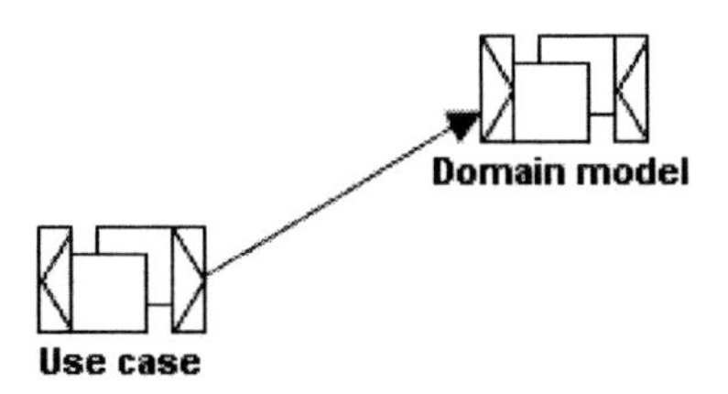

Figure 7. ORM diagram complementing the strategy description

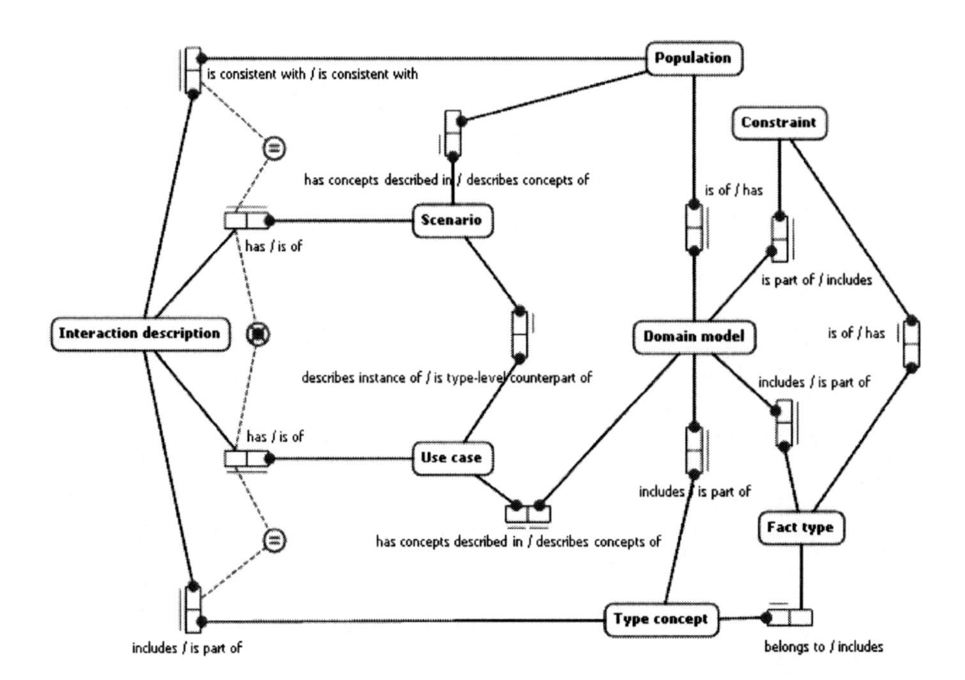

more in line with the more ambitious goals of the general strategy framework, the ORM diagram provides an excellent basis for the creation of formal rules capturing creation goals. We will discuss an extension to that wider context in the next section.

Goal and Procedure Rules Added to the Case

The case strategy description as worked out in detail by Roelofs (2007) stops at providing a workable, well-structured description of the interlinked strategies and concepts of a specific method. Though has been found useful in education, the main aim of creating the description was to test the QoMo strategy framework. However, it could in principle also be a basis for further reaching tool design involving intelligent, rule-based support combining classical model checking and dynamic workflow-like guidance. In order to achieve this, indeed we would need to formalize the goals and process rules of the strategy descriptions to get rules of the kind suggested in (van Bommel et al., 2006) and in section 6.4 of this paper ("the rules underlying a strategy frame"). We will go as far as giving semi-formal verbalizations of the rules.

Fortunately, such rules (closely related to FOL descriptions) are already partly available even in the case example: they can be derived from, or at least based on, the ORM diagrams complementing the strategy descriptions, and the YAWL diagrams that represent the strategy contexts. For example, fig. 6 corresponds to a (minimal) strategy frame as shown in figure 2. The corresponding goals is:

G1 There is at least one Domain Model

This is an instance-level goal. Next, we define the one situation that is relevant to the example strategy "create domain model based on use case":

S1 There is at least one Use Case

So we assume that one or more Use Cases have already been identified (presumably as a goal of another strategy) and that these are used as input for the strategy "create domain model based on use case". We now can weave a rule-based definition combining G1, S1, and various C-rules that correspond to the transitions captured in the strategy description. The key rules raising demands that correspond to steps in the strategy are represented in boldface.

S1 There is at least one Use Case (situation)

C1 SHOULD LEAD TO

G1 There is at least one Domain Model (main goal)

G1.1 Each Use case has concepts described in exactly one Domain model.

G1.2 Each Domain model describes concepts in exactly one Use case.

C2 SHOULD LEAD TO

G2.1 It is possible that more than one Type concept is part of the same Domain model and that more than one Domain model includes the same Type concept.

G2.2 Each Type concept, Domain model combination occurs at most once in the population of Type concept is part of Domain model.

G2.3 Each Type concept is part of some Domain model.

G2.4 Each Domain model includes some Type concept.

G2.5 Each Type concept that is part of an Interaction description of a Use case that has its concepts described by a Domain model should also be part of that Domain model (goal underlying step 1).[d]

C3 SHOULD LEAD TO

G3.1 It is possible that more than one Domain model includes the same Fact type and that more than one Fact type is part of the same Domain model.

G3.2 Each Fact type, Domain model combination occurs at most once in the population of Domain model includes Fact type.

G3.3 Each Domain model includes some Fact type.

G3.4 Each Fact type is part of some Domain model.

G3.5 Each Fact type that is part of a Domain model should include one or more Type concepts that are part of that same Domain model (goal underlying step 2).

C4 SHOULD LEAD TO

G4 Each Fact type of a Domain Model is populated by one or more Facts of the Population of that Domain Model.[e] (goal underlying step 3)

C4 SHOULD LEAD TO

G5.1 Each Scenario describes concepts of exactly one Population.

G5.2 Each Population has concepts described in some Scenario.

G5.3 It is possible that the same Population has concepts described in more than one Scenario.

G5.4 Each Fact that is part of a Population which describes concepts of a Scenario should include at least one Instance concept that is included in that Scenario. (goal underlying the note with step 3)

C5 SHOULD LEAD TO

G6.1 Each Fact type has some Constraint. (goal underlying step 4)

G6.2 Each Constraint is of exactly one Fact type.

G6.3 It is possible that the same Fact type has more than one Constraint.

Note that further restrictions could be imposed on G6.1, demanding explicitly that the constraints applying to a fact type should correspond to the population related to that fact type, and so on. This constraint is left out because it is also missing in the informal strategy description (step 4).

So far, our definition does not include temporal ordering. The following orderings are applied in the C-rules:

C1 no restriction

This reflects the achievement of the main goal, which lies outside the temporal scope of the strategy realizing it. For the rest, rather unspectacularly:

C2 occurs before C3

C3 occurs before C4

C4 occurs before C5

For a somewhat more interesting example of temporal factors, consider the XOR-split and AND-join in fig. 5. (splitting at "use case survey" and joining at "domain model"). Obviously, such split-join constructions involve ordering of transitions:

C1 occurs before C2

C1 occurs before C3

C2 occurs before C3 (under condition Y) XOR C3 occurs before C2 (under condition Z)

C2 AND C3 occur before C6

These expressions of rules covering YAWL semantics are rough indications; a technical matching with actual YAWL concepts should in fact be performed, but this is outside the current scope.

Finally, note that in the implementation, fulfillment of the main goal, "create domain model from use case", is achieved even if the domain model is not finished. However, the unfinished status of the domain model would lead to a number of "ToDo" items. This emphasizes that the strategy is a *initial creation* strategy (bringing some item into existence), which next entails the possibility that a number of further steps have to be taken iteratively (triggered by validity and completeness checks based on, for example, G-rules), hence not necessarily in a foreseeable order.

Findings Resulting from the Implementation

The implementation led to the construction of a specific meta-model reflecting the key concepts used in that implementation (figure 8). We will finish this section by presenting the most interesting findings in the implementation with respect to the generic framework, at the hand of fig. 8.

Sources and Products

The specific flavour of the implementation led to the introduction of the concepts *source*, *product*, *intermediate product*, *raw material*, and *void*. They were needed to operationalize the only goal/strategy category explicitly used in the implementation: creation goals. Situations (which are state descriptions) took the shape of concrete entities (documents) that typically followed each other up in a straightforward order: void or raw material input leading to products, possibly after first leading to intermediary products. Clearly, these concepts classified the items created; such classification emerged as helpful from the discussions that were part of the implementation process.

Use of YAWL Concepts

YAWL concepts (and their graphical representations) were introduced to capture strategy context, while a simple textual description format was used to capture the stepwise strategy description. The YAWL concepts were very helpful in creating easy-to-read context descriptions. In addition, they

Figure 8. Meta-model derived from strategy description case

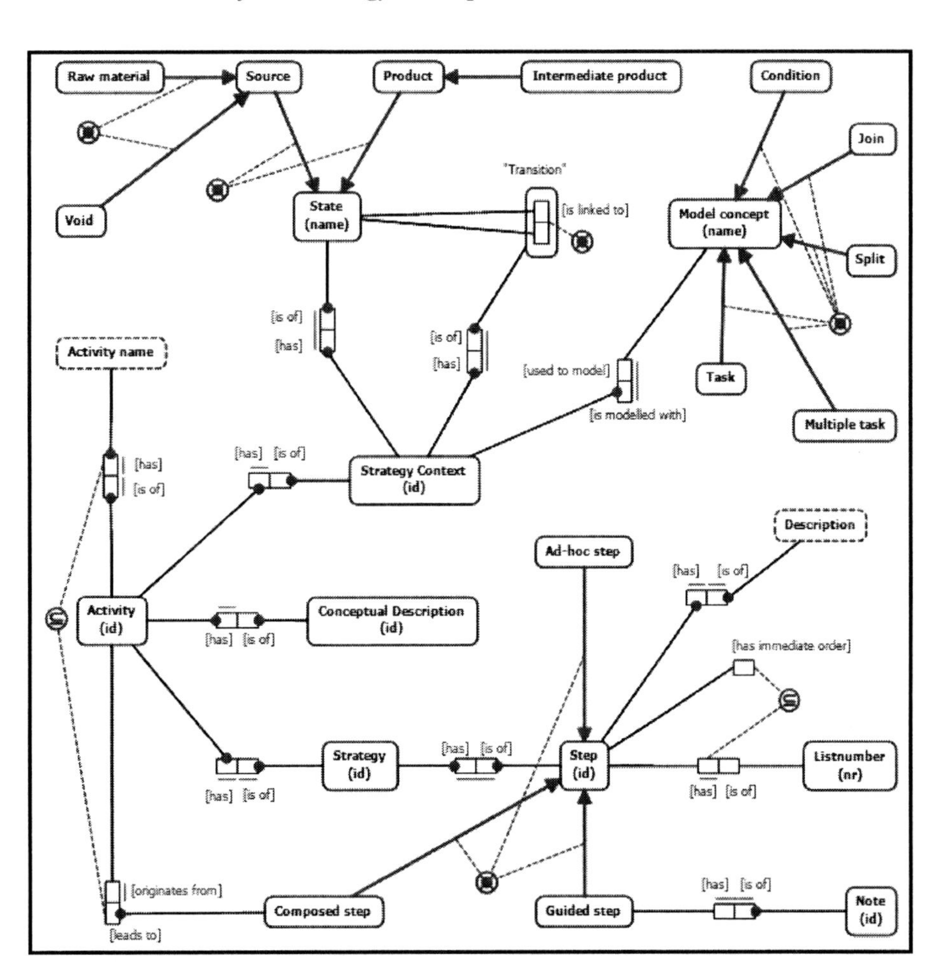

helped in operationalizing the concepts required to capture the workflow-like transitions between states (i.e. between creation situations/goals). Whether YAWL diagrams would be equally useful in describing contexts for other types of goal (for example, validation goals or argumentation goals) remains to be explored.

In addition, YAWL concepts can be used as a basis for formal rule definition capturing the recommended order of steps. The formal underpinnings of YAWL would be extra helpful in case of automated (rule-based) implementation of the strategies, which was still lacking in the prototype (for more on this, see "further research": the "modelling agenda generator").

Activity Descriptions, Names, and Steps

A simple but crucial refinement needed to operationalize the general framework was the introduction of the "activity" and "activity name" concepts. These allowed for the successful implementation of the recursive linking of *activities* to *strategy contexts* to *strategies* to *strategy steps* to further *(sub)activities*, and so on. We expect this amendment to be useful at the generic level, and henceforth we will include it in the main framework.

"Immediate" Concept Not Used

The "immediate" concept was in principle included in the case study implementation but in the end was not used. We still believe it may be required in some strategy descriptions. The ordering in the creation strategies in the case is basic step-by-step. In more complex, dynamic setups, the availability of both immediate and non-immediate ordering still seems useful. However, admittedly the actual usefulness of the "immediate / non-immediate" distinction still awaits practical proof.

Lessons Learned from the Case Study

Apart from he conceptual findings discussed in the previous section, some other lessons were learned though the case study:

- Syntax-like rules can be successfully applied beyond actual modelling language syntax (which amounts to classic model-checking based on grammar goals) into the realm of more generic "creation goals" which may concern various sorts of artefacts within a method.
- The case has shed some light on the fundamental distinction between creation and iteration in dealing with creation goals. While iteration is essentially unpredictable and thus can only receive some ordering (if any at all) through rule-based calculations based on rules and state descriptions, for initial creation people do very much like a plain, useful stepwise description of "what to do": a reference process. Only after initial creation, the far less obvious iteration stage is entered. Also, if a robust rule-based mechanism for guiding method steps is in place, participants may choose to ignore the recommendations of the reference process. This can be compared by the workings of a navigation computer that recalculates a route if a wrong turn has been made.

CONCLUSION AND FURTHER RESEARCH

This chapter set out to present a plausible link between the SEQUAL approach to model *product* quality and our emerging QoMo approach to *process* quality in modeling, and to provide basic concepts and strategies to describe processes aiming for achievement of QoMo goals. We did not aim to, nor did, present a full-fledged framework for describing and analyzing modeling processes, but a basic set of concepts underlying the design of a framework for capturing and analyzing 2nd order information systems was put forward.

We started out describing the outline of the QoMo framework, based on knowledge state transitions, and a goal structure for activities-for-modeling. Such goals were then directly linked to the SEQUAL framework's main concepts for expressing aspects of model items and its various notions of quality, based on model items. This resulted in an abstract but reasonably comprehensive set of main modeling process goal types, rooted in a semiotic view of modeling. We then presented a case implementation of how such goals can be linked to a rule-based way of describing strategies for modeling, involving refinements of the framework. We added concrete examples of rules describing goals and strategies, based on the case implementation.

These process descriptions hinge on strategy descriptions. Such strategies may be used descriptively, for studying/analyzing real instances of processes, as well as prescriptively, for the guiding of modeling processes. Descriptive utility of the preliminary framework is crucial for the quality/evaluation angle on processes-for-modeling. Study and control of a process requires concrete concepts describing what happens in it, after which more abstract process analysis (efficiency, cost/benefit, levels of risk and control) may then follow. Means for such an analysis were not discussed in this paper: this most certainly amounts to future work.

Besides continuing development and operationalization of the QoMo strategy and goal framework for quality modeling by applying it to new and more complex cases, we need to push forward now to implementations that actively support our rule-based approach. An initial implementation, using Prolog and a standard SQL database, is in fact available, but has not been sufficiently tested and documented yet to report on here. This "modeling agenda generator" dynamically generates ToDo lists (with ordered ToDo items if C-rules apply) based on the model states as recorded in the repository. We will finish and expand this prototype, testing it not only in a technical sense but also its usability as a system for supporting real specification and modeling processes. In the longer term, we hope to deploy similar automated devices in CASE-tool like environments that go beyond the mere model or rule editors available today, and introduce advanced process-oriented support and guidance to modelers as required in view of their preferences, needs, experience, competencies, and goals.

REFERENCES

Van der Aalst, W. And ter Hostede, A. (2005): YAWL: Yet Another Workflow Language. *Information systems, 30*(4), 245-275.

Bommel, P., van, S. J. B. A. Hoppenbrouwers, H. A. (Erik) Proper, & Th.P. van der Weide (2006). Exploring Modeling Strategies in a Meta-modeling Context. In R. Meersman, Z. Tari, and P. Herrero, (Eds.), *On the Move to Meaningful Internet Systems 2006: OTM 2006 Workshops*, 4278 of *Lecture Notes in Computer Science*, (pp. 1128-1137), Berlin, Germany, EU, October/November 2006. Springer.

Chrissis, M. B., Konrad, M., & Shrum, S. (2006). *CMMI: Guidelines for Process Integration and Product Improvement*, Second Edition. Addison-Wesley.

Halpin, T. A. (2001). *Information Modeling and Relational Databases, From Conceptual Analysis to Logical Design*. Morgan Kaufmann, San Mateo, California, USA, 2001.

Hoppenbrouwers, S.J.B.A., H.A. (Erik) Proper, & Th.P. van der Weide (2005a). A Fundamental View on the Process of Conceptual Modeling. In *Conceptual Modeling - ER 2005 - 24 International*

Conference on Conceptual Modeling, 3716 of *Lecture Notes in Computer Science*, pages 128-143, June 2005.

Hoppenbrouwers, S.J.B.A., H.A. (Erik) Proper, and Th.P. van der Weide (2005b). Towards explicit strategies for modeling. In T.A. Halpin, K. Siau, and J. Krogstie, editors, *Proceedings of the Workshop on Evaluating Modeling Methods for Systems Analysis and Design* (EMMSAD'05), held in conjunction with the 17th Conference on Advanced Information Systems 2005 (CAiSE 2005), pages 485-492, Porto, Portugal, EU, 2005. FEUP, Porto, Portugal, EU.

Hoppenbrouwers, S.J.B.A., H.A. (Erik) Proper, and Th.P. van der Weide (2005c). Fact Calculus: Using ORM and Lisa–D to Reason About Domains. In R. Meersman, Z. Tari, and P. Herrero, editors, *On the Move to Meaningful Internet Systems 2005: OTM Workshops* – OTM Confederated International Workshops and Posters, AWeSOMe, CAMS, GADA, MIOS+INTEROP, ORM, PhDS, SeBGIS, SWWS, and WOSE 2005, Agia Napa, Cyprus, EU, volume 3762 of Lecture Notes in Computer Science, pages 720–729, Berlin, Germany, October/November 2005: Springer–Verlag.

Kensche, D., Quix, C., Chatti, M. A., & Jarke, M. (2005). GeRoMe: A Generic Role Based Metamodel for Model Management. In R. Meersman, Z. Tari, and P. Herrero, editors, *On the Move to Meaningful Internet Systems 2005: CoopIS, DOA, and ODBASE – OTM Confederated International Conferences, CoopIS, DOA, and ODBASE 2005*, Proceedings, Part II, Agia Napa, Cyprus, EU, volume 3761 of Lecture Notes in Computer Science, pages 1206–1224. Springer–Verlag, October/November 2005.

Krogstie, J. (2002). A Semiotic Approach to Quality in Requirements Specifications. In L. Kecheng, R.J. Clarke, P.B. Andersen, R.K. Stamper, and E.-S. Abou-Zeid, editors, *Proceedings of the IFIP TC8 / WG8.1 Working Conference on Organizational Semiotics: Evolving a Science of Information Systems*, pages 231-250, Deventer, The Netherlands, EU, 2002. Kluwer.

Krogstie, J., & Jorgensen H. D. (2002). Quality of Interactive Models. In M. Genero, Grandi. F., W.-J. van den Heuvel, J. Krogstie, K. Lyytinen, H.C. Mayr, J. Nelson, A. Olivé, M. Piattine, G. Poels, J. Roddick, K. Siau, M. Yoshikawa, and E.S.K. Yu, editors, *21st International Conference on Conceptual Modeling (ER 2002)*, volume 2503 of *Lecture Notes in Computer Science*, pages 351-363, Berlin, Germany, EU, 2002. Springer.

Krogstie, J., Sindre, G., & Jorgensen, H. (2006). Process models representing knowledge for action: a revised quality framework. *European Journal of Information Systems*, 15, 91-102.

Mirbel, I., & J. Ralyté (2006). Situational Method Engineering: combining assembly-based and roadmap-driven approaches. *Requirements Engineering*, 11, 58-78.

Moody, D. L. (2006). Theoretical and practical issues in evaluating the quality of conceptual models: current state and future directions. *Data and Knowledge Engineering*, (55), 243-276.

Nelson, H. J., & Monarchi, D. E. (2007). Ensuring the Quality of Conceptual Representations. In: *Software Quality Journal*, *15*, 213-233. Springer.

Object Management Group OMG (2001): *Common Warehouse Metamodel (CWM) metamodel, version 1.0*, Februari 2001.

Poels, G., Nelson, J., Genero, M., & Piattini, M. (2003): "Quality in Conceptual Modeling – New Research Directions". In: A. Olivé (Eds.): *ER 2003* Ws, LNCS 2784, pp. 243-250. Springer.

Ralyté, J., Brinkkemper, S., & Henderson-Sellers, B. (Eds.), (2007). *Situational Method Engineering: Fundamentals and Experiences.* Proceedings of the IFIP WG 8.1 Working Conference, 12-14 September 2007, Geneva, Switzerland. Series:

IFIP International Federation for Information Processing , Vol. 244.

Roelofs, J. (2007). *Specificatie van Strategieën voor Requirement Engineering.* Master's thesis, Radboud University Nijmegen. In Dutch.

Saeki, M. (1995). Object–Oriented Meta Modelling. In M. P. Papazoglou, (Ed,), *Proceedings of the OOER'95, 14th International Object–Oriented and Entity–Relationship Model ling Conference*, Gold Coast, Queensland, Australia, volume 1021 of Lecture Notes in Computer Science, pages 250–259, Berlin, Germany, EU, December 1995. Springer.

Sindre, G., & Krogstie, J. (1995). Process heuristics to achieve requirements specification of feasible quality. In *Second International Workshop on Requirements Engineering: Foundations for Software Quality* (REFSQ'95), Jyväskylä, Finland.

ENDNOTES

[a] We use a broad definition of "strategy". This term corresponds closely to what, for example, (Mirbel and Ralyté, 2006) call "guidelines". We use our own terminology for the moment because our current discussion has a relatively generic focus on method modeling, but this is not to say we reject Mirbel and Ralyté's conceptual distinctions as such. We acknowledge that proper alignment of terminology will require due attention in the near future.

[b] For sake of readability, we use simple natural language statements for state/rule description, in ORM verbalization style. Each of these descriptions can be easily represented in logic.

[c] The ORM diagrams in this paper were produced by means of the NORMA case tool developed by Terry Halpin and his co-workers at Neumont University: http://sourceforge.net/projects/orm.

[d] In expressing this complex rule, we use a controlled language called Object Role Calculus: see (Hoppenbrouwers et al., 2005c)

[e] Rules G4 and G5.4 refer "Facts" and "Instance concepts", which are not included in figure 7 but in the ORM diagram (not presented in this paper) supporting a different strategy, namely "create domain model based on scenario". In the implementation, populations are defined as included in a domain model.

Chapter X
Service Oriented Architecture: A Research Review from the Software and Applications Perspective

John Erickson
University of Nebraska-Omaha, USA

Keng Siau
University of Nebraska-Omaha, USA

ABSTRACT

This chapter presents the basic ideas underlying Service Oriented Architecture as well as a brief overview of current research into the phenomena also known as SOA. SOA is defined, and principal components of one proposed SOA framework are discussed. The more relevant historical background behind the move toward SOA is presented, including SOA antecedents such as Web Services, SOAP, and CORBA, and enabling technologies such as XML and EJB. A basis for understanding SOA is presented, based on Krafzig, Banke, and Slama's (2005) three-level hierarchical perspective. The common SOA components including UDDI, Application Programming Interface, Service Bus, Service Contract, Interface, Implementation, Data, and Business Logic are also presented. Finally, relevant research in four categories is presented, including implementation strategies, patterns and blueprints, tool development, standards proposals, or modifications (including middleware), and ontological or meta-model development or modification.

INTRODUCTION

Service Oriented Architecture (SOA), Service-Oriented Computing, and Web Services, along with some of their underlying middleware realization schemas such as SOAP, UDDI, XML, CORBA, and many other ideas or approaches to cutting edge information system architectures, have become the buzzwords of the day for many in the business world. In many ways the entire area of "Service" is also somewhat hazy from a definitional perspective in the Information Technology (IT), and Information Systems (IS) communities as well. While the situation among Enterprise Architecture, Service Oriented Enterprises, and Services is arguably more developed and stable, there is nevertheless a lacking in cohesiveness that seems to plague the entire Service area. It has proven quite difficult, perhaps nearly impossible to pick up any relatively current practitioner publication without encountering an article focusing on at least one of the above topics. A recent library database search using key words Service Oriented Architecture, Web Services, and SOA resulted in 800-plus returns. Further investigation revealed that roughly 25 of those 800 articles were sourced in research journals while the other (still roughly 800) articles were all from more practitioner-oriented sources.

When it comes to adopting and implementing SOA applications, it appears that businesses are doing it at astounding rates. Of course what they are actually doing, even though they may say that their efforts represent a move toward Service Oriented Architecture, may not match anyone else's definition of SOA but their own. Further, how can SOA be defined, and how can we define the benefits of moving toward such architectures? It seems that there is little agreement among practitioners and researchers alike as to a standard definition of SOA.

Worse still, a growing number of practitioners are now beginning to question the business return of some of the approaches. For example, Dorman (2007), Havenstein, (2007), Ricadela (2006), and Trembly (2007) indicate that there is doubt emerging as to the real value of SOA for adopting businesses and organizations. Perhaps the question of Return on Investment (ROI) should not be that surprising since it sometimes seems that each organization has its own definition of what SOA really is.

Finally, an entire genre of research has emerged relatively recently that focuses on the architecture of the enterprise itself in conjunction with service orientation. It is not the intent of this chapter to examine or discuss the enterprise architecture element, but to focus on the software and application side of the service issue.

This paper attempts to reach for a more clear understanding of what SOA really is from the software perspective, and proposes some possible areas of research into SOA that could help clear up some of the definitional confusion, which could in turn and if conducted, help lead to better understanding of ROI as it relates to SOA. Section 1 consists of the introduction, while Section 2 provides existing definitions of Service Oriented Architecture (SOA), Web Services, and Section 3 details some of the related and underlying technologies and protocols. Section 4 combines the various definitions of SOA into a more coherent form, while Section 5 reviews current SOA research. Finally, Section 6 will conclude the paper with recommendations for future research efforts.

BACKGROUND, HISTORY AND DEFINITIONS OF SERVICE ORIENTED ARCHITECTURE

A minimum of nine formal definitions of SOA exist as of this writing, from sources such as the OASIS Group, the Open Group, XML.com, javaworld.com, the Object Management Group (OMG), W3C (World Wide Web Consortium), Webopedia, TechEncyclopedia, WhatIs.com, and Webopedia.org. In addition, many other definitions put forth by numerous industry experts, such as those from IBM, further cloud the issue, and worse yet, still other formal definitions might also exist. In other words, the concept of "Service Oriented Architecture" appears in many ways to be a virtually content free description of an IT-based architecture. It is not our intent here to add yet another definition to this already crowded arena of definitions, but to try to cull the common, base meanings from the various distinct definitions.

Prior to about 2003, the term Service Oriented Architecture was not in general use for the most part, according to Wikipedia (2007). However, since that time, SOA has exploded nearly everywhere in the business and technology world. SOA appears to derive or develop in many cases from more basic Web Services. These services can include enabling technologies such as SOAP (Simple Object Access Protocol), CORBA (Common Object Request Broker Architecture), EJB (Enterprise Java Beans), DCOM (Distributed Component Object model), and even SIP (Session Initiated Protocol) among many others; services may also include other middleware created with XML (Sulkin, 2007).

Service Oriented Architecture Definitions

The Open Group defines SOA as "an architectural style that supports service orientation" (Open Group, 2007). The definition goes on to also include descriptions of architectural style, service orientation, service, and salient features (of SOA). The OASIS Group defines SOA as "...a paradigm for organizing and utilizing distributed capabilities that may be under the control of different ownership domains" (OASIS, 2006). The OASIS definition includes what they call a "reference model" in which the details of the definition are expanded and formalized. The Object Management Group (OMG) defines SOA as "...an architectural style for a community of providers and consumers of services to achieve mutual value" (OMG, 2007). OMG adds that SOA allows technical independence among the community members, specifies the standards that the (community) members must agree to adhere to, provides business and process value to the (community) members, and "allows for a variety of technologies to facilitate (community) interactions" (OMG, 2007).

W3C defines SOA as "a form of distributed systems architecture that is typically characterized by...a logical view, a message orientation, a description orientation, granularity and platform neutrality." W3C adds details describing what they mean by logical view, message and description orientations, granularity and platform neutrality. XML.com defines SOA as follows: "SOA is an architectural style whose goal is to achieve loose coupling among interacting software agents. A service is a unit of work done by a service provider to achieve desired end results for a service consumer. Both provider and consumer are roles played by software agents on behalf of their owners (XML.com, 2007).

The javaworld.com SOA definition, composed by Raghu Kodali, is as follows, "Service-oriented architecture (SOA) is an evolution of distributed computing based on the request/reply design paradigm for synchronous and asynchronous applications" (javaworld, 2005). Kodali also goes on to describe four characteristics of SOA. First the self-describing interfaces composed in XML, using WSDL (Web Services Description Language) for the self-description. Second, XML schema called XSD should be used for messaging. Third, a UDDI-based (Universal Description, Definition, and Integration) registry maintains a list of the services provided. Finally, each service must maintain a level of quality defined for it via a QoS (Quality of Service) security requirement.

Finally, IBM proposes that SOA "describes a style of architecture that treats software components as a set of services." Further, they insist that business needs should "drive definition" of the services, and that the value proposition be centered around the re-usability and flexibility of the defined services IBM (UNL-IBM System i Global Innovation Hub, 2007).

SERVICE ORIENTED ARCHITECTURE COMPONENTS

We begin the SOA discussion with an overview of SOA provided by Krafzig, Banke, and Slama (2005). They proposed a three-level hierarchical perspective on SOA, in which level 1includes the Application Front End, the Service, the Service Repository, and the Service Bus. Accordingly, only the Service child has children, consisting of Contract, Implementation, and Interface. Finally, the last level of the proposed hierarchy is composed of Business Logic and Data, children of Implementation. The next sub-sections will discuss the general ideas of the elements included in the hierarchy proposed by Krafzig, Banke, and Slama (2005) described previously. This is not to recommend adoption of the hierarchy and description as the final description of SOA, but rather as a framework for discussing the meaning of SOA for the balance of this paper.

Application Front End

This part of SOA comprises a source code interface, and in SOA terminology, it is referred to as the Application Programming Interface (API). In accordance with most commonly accepted design principles, the underlying service requests, brokerage (negotiation), and provision should be transparent to the end-user.

Service Repository

The Service Repository could be thought of as the library of services offered by a particular SOA. This would likely consist of an internal system that describes the services, and provides the means in the user interface to call a particular service. UDDI (Universal Description, Discovery and Integration) could be seen as a realization of the Service Repository idea. UDDI is a global registry that allows businesses to list themselves on the Internet. UDDI is platform independent and XML based. The point of UDDI is for businesses to list the (web or SOA-type) services they provided, so that other companies searching for such services can more easily locate and arrange to use them.

Service Bus

The Service Bus (SB), more commonly referred to as the Enterprise Service Bus (ESB), provides a transportation pathway between the data and the end-user application interface. Using an ESB does not necessarily mean a SOA implementation, but ESB or some sort of SB use is almost always part of a SOA deployment. According to Hicks (2007), Oracle's idea of an ESB includes multiple protocols that "separate integration concerns from applications and logic." What this means is that ESBs have now become commercialized, and can be licensed for use much like other UDDI-based services. So, companies searching for ESB solutions as part of a SOA effort now have multiple choices and do not necessarily have to re-create the wheel by building their own ESB.

Common Services

It seems apparent from many of the SOA definitions that many of the technologies included in an SOA definition, and by default SOA implementations, are established and conventional protocols. To better understand the services provided in many SOA definitions, a brief explanation of some of the more commonly used underlying technologies is provided. A particular service may or may not be explicitly Web based, but in the end it matters little, since the architecture provides that the service should be transparently designed, implemented and provided. The general consensus from most involved in Web Services is that the services are meant to be modular. This means that no single document encompasses all of them, and further, that the specifications are multiple and (more or less) dynamic. What results is a small number of core specifications. Those core services can be enhanced or supported by other services as "the circumstances and choice of technology dictate" (Web Service, 2007).

XML (Extensible Markup Language) allows users to define and specify the tags used to capture and exchange data, typically between distinct and usually incompatible systems from different companies or organizations. This means that XML is a good example of middleware; it also means that XML enables Web Services. XML was one of the initial drivers that provided the ability to conduct e-business for many businesses in the Internet era. XML cannot really be considered a service, but as the language used to write many of the Web Services or Service stack protocols.

SOAP (Simple Object Access Protocol), like all protocols, consists of a set list of instructions detailing the action(s) to be taken in a given circumstance. SOAP is designed to call, access, and execute objects. The original SOAP protocol was typically for communications between computers, and usually involved XML-based messages. SOAP and its underlying XML programming, comprised one of the first Web Service communication stacks. One of the original Web Services that SOAP provided was called Remote Procedure Call (RPC), which allowed a remote computer to call a procedure from another computer or network. More recently, SOAP has taken on a somewhat modified meaning, so that the acronym now means Service Oriented Architecture Protocol. In both cases, what SOAP does is to use existing communications protocols to provide its services. The more common early SOAP protocol contracts included XML applications written for http, https, and smtp, among others. It should be apparent from these, that many early SOAP implementations involved e-commerce or e-business applications, which means that the concern at the time many were first developed was to move sales and other data collected in Web portals to back-end data stores.

CORBA is an OMG developed standard that allows different software components that are usually written in different languages and installed on different computers to work together (OMG, 2007). CORBA was developed in the early 1990s, and while not overtly SOA at the time, it actually performs many of the functions in a SOA, using an IIOP (Internet Inter-Orb Protocol) based service stack.

EJB is a component, typically situated on the server that "encapsulates the business logic of an application" (Wikipedia, 2007). EJB enables creation of modular enterprise (and other) applications. The intent of EJB is to facilitate creation of middleware that acts as a go-between tying front end applications to back end applications or data sources.

SIP (Session Initiated Protocol) is a signaling protocol designed for use in telecommunications, at the application layer. It has generally become one of the primary protocols used in VoIP, H.323 and other communications standards. SIP can be seen as a primary provider of Web Services for Internet-based voice communications such as VoIP (Voice over Internet Protocol) (Sulkin, 2007).

Contract (Services)

Components of a service contract typically include primary and secondary elements. The primary elements consist of the Header, Functional requirements, and Non-functional requirements. Sub-elements for the header consist of the Name, Version, Owner, RACI, and Type. Under Functional requirements are Functional (requirement description), Service operations, and Invocation. Non-function requirements include Security constraints, QoS (Quality of Service), Transactional (is the service part of a larger transaction), Service Level Agreement, and Process (Wikipedia (SOA), 2007). The contract generally includes metadata about itself; who owns it, and how it is brokered, bound, and executed.

Interface

At this level of service provision, the interface referred to is a segment of code that connects the service with the data and/or business logic (process). The Interface describes how data will be moved into and out of the data source by the Service, and must be designed to comply with the physical (Data, data structures, etc.) and process (Business Logic) requirements of the existing and/or legacy system.

Implementation

The Implementation specifies the Contract and Interface to be used for each Service requested, and contains the direct pathway into the Data and Business Logic.

Architecture

The Service component of SOA has been discussed, though admittedly at a high level. However the Architecture component has not yet been addressed and it will be helpful to speak briefly about the Architecture segment of SOA. Architecture in general refers to the art (or science) behind the design and building of structures. Alternatively, an architecture may refer to a method or style of a building, or a computer system. So, if SOA is taken literally as a description of its function, it could be taken to mean a structured way of organizing or arranging the services in a business or organization.

OVERALL SOA FRAMEWORKS AND STANDARDS CONSOLIDATIONS

It is apparent from the existing definitions and models that Service Oriented Architecture is commonly seen as an architecture, or way of assembling, building, or composing the information technology infrastructure of a business or organization. As such, SOA is not a technology in itself; rather it is a way of structuring or arranging other technologies to accomplish a number of other tasks. This naturally leads to the problem of a multiplicity of definitions of SOA, since many relatively similar structural arrangements of Services are possible. Many of the definitions also indicate that the arrangement and relationships between modules should be loosely coupled rather than tightly coupled. This allows for customization of Services based on need, rather than some pre-determined structure, but the downside is that it also leads toward a plethora of definitions and approaches to SOA implementation.

Some of the common features that it seems sensible to include in a formal definition of SOA would relate to a common framework, such as that specified by Krafzig, Banke, and Slama (2005), or one of the other standards bodies. In other words, a framework would include metadata describing the various important features of SOA, how those features can be arranged, and the libraries or location of services that allow adopting organizations to arrange bindings or contracts between themselves and the service provider, independent of whether the service provider is internal or external. Several of the standards bodies have taken a stance creating or calling for a metamodel, at least in some form, among them the Open Group, the OASIS Group, the OMG, W3C, and to a lesser extent industry related bodies such as javaworld.com, xml.com, IBM, and Oracle.

That UDDI has become a very well known structured repository for services and service components, speaks to the universality of the library or centralized database of services. However, more standardization efforts will be necessary to enhance the interoperability of UDDI.

It also appears, especially with the industry definitions of SOA, that the contracts, bindings, interfaces, service bus, and other implementation related portions of SOA are important elements to be considered when attempting an overall definition of SOA. This unfortunately could easily represent a stumbling block in garnering consensus on a definition of SOA since each of these companies has invested significant time, human, and other likely resources toward development of their specific pieces of the SOA pie. Each company has invested heavily and thus will likely be less willing to risk that investment and any potential return and customer lock-in, in order to simply agree on standards. We can see a similar occurrence of this type of behavior in the current format war going on for the high definition DVD market. Similarly, if the standards bodies have political or industry leanings, agreement on a common SOA definition and standards could be difficult to achieve.

Another more recent development comes from Alkesh Shah and Paul Kalin (2007). They proposed that organizations adopting SOA follow a specific path based on an analysis of business challenges, including SOA business drivers and IT barriers. This lead them to speculate that a specific adoption model be used to guide the SOA implementation process. They indicated that an ad-hoc SOA model is better where the benefits of new services are specific to each individual service, where the technologies may be inconsistently applied (different implementations for the same service in different projects), where services cannot be reused, and where the increases in technological complexity translate into decreased system response times. Shah and Kalin ended with a call for a strategy or program-based SOA adoption model that is situational (2007).

LITERATURE REVIEW

SOA research is quite limited currently. The existing body of work can be classified into several distinct categories. The first type of research examines and recommends implementation strategies related to SOA. These efforts typically include exploratory or recommendation-type research that proposes various means to approach SOA implementation, and may or may not include patterns or blueprints. A second type of research into SOA involves proprietary or non-proprietary tool development. These investigations may or may not include proprietary industry software, but most of these research efforts also propose use of patterns or blueprints, and may overlap somewhat with the first research category. A third type of research examines the existing SOA standards, or proposes new standards designed to solve a problem or shortcoming identified the existing standards. Also included in this category are proposals for middleware development, support or modification in support of SOA efforts. Finally, the last research category represents a higher level approach to SOA research, and proposes new meta-models or changes to existing meta-models. Most importantly for this research, no research effort that we are aware of has attempted to assess SOA critical success factors, at least in metrical or financial terms. Table 1 depicts a classification of past SOA research into the above categories.

SOA Implementation Strategies; Patterns and Blueprints

Stal (2006) advocated using architectural patterns and blueprints (software engineering patterns) as a means to enable or foster efficient deployment of SOA. He supported loose coupling of services in

Table 1. SOA research classification

	Primary Focus of SOA Research			
Author(s)	**Implementation Strategies; Patterns and Blueprints**	**Tool Development**	**Standards Proposals or Modifications (including Middleware)**	**Ontological or Meta-Model Development or Modification**
Stahl	Y		Y	Y
Kim and Lim	Y		Y	
Shan and Hua	Y		Y	
Schmidt, et al.	Y		Y	
Crawford, et al.	Y		Y	
Brown, et al.	Y	Y		
Ferguson and Stockton	Y	Y		
De Paw, et al.	Y	Y		
Jones		Y	Y	
Chen, et al.				Y
Borker, et al.			Y	
Duke, et al.				Y
Zhang		Y	Y	
Malloy, et al.			Y	
Verheeke, et al.			Y	
Hutchinson, et al.	Y			
Li, et al.	Y			

a registry or library to the extent that he thought that removing the services' dependency on the registry's or provider's distinct location would benefit the deployment of SOA. Stal maintained that this would eliminate, or at least minimize a layer in the SOA framework. He also proposed a more tightly defined and controlled integration of middleware using XML or similar tools. Basically, Stal suggested a metamodel and pattern approach to defining SOA, but did not suggest what the research might accomplish, or how the research into SOA would be framed. Additionally, since Stahl suggested changes to the SOA meta-model and also proposed changes to SOA standards, his research fits into at least 3 of the arbitrary classifications identified previously. Kim and Lim (2007) also proposed a distinct means to implementing SOA, using in this instance, Business Process Management, in addition to a variant of the SOA framework specifically dealing with the telecommunications industry. Similar to Stal (2006), Kim and Lim (2007) did not propose empirical research into SOA, but rather focused on implementation and standards in a specific industry.

Shan and Hua (2006) followed a now common form in that they proposed an SOA approach for the Internet banking industry. They also compiled a list of patterns that have proven to be successful for other online service industries. However, the models they used and end with are very detailed regarding how SOA should be implemented for first online companies in general, and then Internet banking specifically. This again does not propose or frame specific research but rather suggest an implementation approach and a structure for SOA.

The Enterprise Service Bus (ESB) is explained in detail, but from a general perspective, rather than a company specific approach in Schmidt, Hutchison, Lambros and Phippen's (2005) expository. The article is informative regarding ESB implementation and design patterns, but it is not research oriented.

Crawford *et al.* (2005) proposed a different way to structure SOA and they called it on-demand SOA. They essentially proposed an even greater loose coupling of services and their connecting elements, than other perspective of SOA, and that would allow much more flexibility for the adopting organizations and the end-users.

Hutchinson, Henzel, and Thwaits (2006) described a case in which an SOA based system was deployed for a library extension collaboration project. Much of the case details the SOA approach itself, and explains the experiences of the project developers and implementers. They noted that while the SOA architecture could be expected to reduce the operational maintenance costs overall, the way the system was specified and delivered in this particular case might require more work from IT to keep some services, such as flash players, etc. up to date. While it does not specifically mention it in the article, perhaps a more loosely coupled architecture might alleviate some of those operational maintenance costs.

Li *et al.* (2007), proposed a methodology to migrate the functionality of legacy systems to Web Services or SOA architecture. They used a case study to investigate the efficacy of their proposed methodology, finding that while it was possible to make such a migration from legacy to SOA (or Web Services), the changes that it required from the organization were considerable, and that some process re-engineering would likely be necessary.

Tool Development

Brown *et al.* (2005) presented an industry oriented perspective on the SOA puzzle. They suggested an approach to service orientation using the proprietary IBM Rational platform. Their recommendations follow similar paths as previous researches, but are also filtered through the IBM Rational lens. The article is primarily illustrative in nature, suggesting how to best implement SOA using IBM Rational tools. In a similar vein, Ferguson and Stockton (2005), also detail IBM's programming model and product architecture.

De Pauw *et al.* (2005) described the benefits of Web Services Navigator, a proprietary tool created to provide a better visualization of SOA and Web Services in a loosely coupled architecture. The tool can help with design pattern, business logic, and business process analysis, and thus, help with SOA architecture design and implementation.

Jones (2005) suggested that SOA, Service, and Web Service standards were "on the way" and provided a list of existing tools, such as UML and/or the Rational Unified Process that could aid the SOA (or service) design process. But, he also advocated the push toward formal definitions of such SOA basics as services, to the end of providing a more coherent and cohesive structure that he thought would enhance the ability of developers and adopters to understand and deploy SOA.

Standards Proposals or Modifications (including Middleware Proposals)

Borker *et al.* (2006) suggested a way of handling XML-based data in an SOA or service environment. Their idea involved the use of queryable and unqueryable data and would necessarily also involve XML formatted data. This represents empirically based research into a part of SOA, namely the underlying services.

Zhang (2004) explored the connection between web services and business process management, and described the modular nature of the service (and Web Service) perspective. He detailed the software industry's approach to Web Services, and provided evidence that standards development would quickly mature, beginning in 2005. He maintained that once standards were agreed upon, a connection to Business Process Management would be easier to sell to business. Zhang's efforts go a bit further than others, because he did not stop at this point, but developed a prototype e-procurement system that composed external service to operate.

Malloy *et al.* (2006) developed an extension to WSDL (Web Services Description Language). They insisted that Web Services' specifications were "...typically informal and not well-defined," and proposed what they called an intermediate step between requiring more formal and rigorous service specifications and the informal nature of the existing service specifications. They accomplish this balance by extending the WSDL to include support for application argument and return that would help automate and expand the ability of services to operate in multiple environments. They provided an example of how their WSDL extension could allow a single service to function successfully in different applications using multiple zip code formats (5 versus 9 digits, and hyphens versus no hyphens).

Verheeke, Vanderperren, and Jonckers (2006) proposed and developed a middleware level that they called Web Services Management Layer (WSML). They saw the primary advantage of their approach in that it provided a reusable framework. They further believed that the use of their framework would enable "...dynamic integration, selection, composition, and client-side management of Web Services in client applications" (p. 49). They were aware that their approach could cause some problems in a distributed system, since implementation of it resulted in a centralized architecture.

Ontological or Meta-Model Development or Modification

Chen, Zhou and Zhang proposed an ontologically based perspective on SOA, Web Services and Knowledge Management (2006). They attempted, with some success, to integrate two separate research streams into one. They presented a solution to show that semantic and syntactic based knowledge representations could both be depicted with a comprehensive ontology that also described web service composition. While their framework represents a step toward automated (web) service composition, more research is still needed.

Duke, Davies and Richardson (2005) recommended and provided details on using the Semantic Web to organize an organization's approach to SOA and Web Service orientation. They suggested that combining the Semantic Web and SOA into what they called Semantic SOA would provide benefits to adopting organizations. Then, they further proposed an ontological model of the semantic SOA, attempting essentially to create a meta-meta-model of SOA, using their experience with the telecommunications industry as a case example.

FUTURE RESEARCH DIRECTIONS RELATED TO SOA

Assuming that the definitions and standards can reach a point of balance or agreement in the short term, what are some possible directions for SOA research? Possibilities in several areas seem likely to produce useful results. For example, in the broad categories of managerial, technical, and organizational effects, SOA could, and should be examined as to its efficacy and impact.

As noted earlier, the overall success of a SOA implementation is an area more and more managers are beginning to question. The measures of success might be managerial, technical or organizational in nature, and research into these could be accomplished via some of the usual activities, such as capturing ROI, IRR, NPV, and other financially based estimation techniques. Other techniques might be necessary to capture some of the more intangible benefits of SOA.

Nah, Islam, and Tan (2007) proposed a framework and critical success factors for estimating the success of ERP implementations. They empirically assessed a variety of implementation success factors including top management support, project team competence, and interdepartmental cooperation, among many others. While the study answered a number of important questions regarding ERP implementations, the issue of assessing intangibles in terms of success factors remains a problem, and not only for ERP type implementations but also for other system types as well, and more critically for SOA, since the SOA approach can be seen as an alternative in many ways to ERP.

Langdon (2006) noted that while many economically-based studies indicate that IT projects add value at the macro level, little has been done to assess how value is added at the more micro or individual project level. Specifically, Langdon proposed and evaluated a research model that included (IS) integration and (IS) flexibility as capabilities that could lead to IT business value. Of course, flexibility and integration are only two components of a larger IT capabilities structure, but the study indicates that the first steps have been taken to study intangibles in the context of IT systems development project.

Two intangibles in the IT success factor context are the oft-cited agility or nimbleness of a company or organization. An entire genre of systems development has emerged based on the principle of agility. However, there is little empirical evidence supporting the value added from such development approaches (Erickson, Lyytinen, and Siau, 2005). Since a growing number of SOA installations are constructed as ad-hoc, which is in a basic sense agile, we propose that in environments where agility and nimbleness are important, so in turn are SOA and SOC important.

On the technical side of the SOA puzzle, how does implementing a SOA project change the Total Cost of Ownership (TCO) and IT infrastructures for a company? How does delivery of (IT) services change, and does that impact the average cost per transaction for a company? Some of these questions could be addressed in a well designed and executed research project into assess the technical impacts of SOA.

Finally, another area of interest involving SOA and Web Services adoption is the cultural and structural impacts on the organization or business. A number of articles note the importance of those elements, but little has been accomplished in terms of research specifically connecting SOA or Web Services with cultural and structural changes in organizations. Do the Standard Operating Procedures change post SOA implementation? Does SOA impact organizational structure? Is HR adequately addressing training issues? It seems that the organizational impact of SOA has received very little attention, and could benefit from research in a number of these areas.

From the literature collected it appears that only a few efforts could be said to be empirical research. A majority of the research efforts involved creating tools or language extensions that would increase the interoperability of services, while other research proposed standards modifications. Many of the remaining articles published proposed new tools or use of existing proprietary tools, described an approach to SOA from specific perspectives, or proposed model or meta-model changes. A limited number of case studies detailing SOA, Web Services or Service deployments or implementation efforts provide experience reports on how best to implement such systems.

As far as we can determine virtually no research has been formally done regarding the benefits and drawbacks of SOA or Web Services. Two problems with this are likely to revolve around the nebulous nature of SOA and Web Services in terms of the widely varying definition, and the emerging standards issue. An effort to identify SOA and Web Services metrics would help to get research into this area started.

A variety of standards bodies are working separately toward formal definitions including meta models, and a number of SOA vendors, among them some of the very large and established software industry players, have emerged. While the effort toward standardization is direly needed and commendable, a more collaborative approach would, in our opinion, benefit the industry and implementing companies and organizations as well. The seeming result of the rather haphazard approach to SOA appears to indicate that an increasing number of implementing organizations are finding it difficult to assess the cost benefit of the entire services approach. Research efforts at this point appear to be in a similar state of disarray. Until a more coherent picture of SOA emerges, its image is likely to remain slightly out of focus, and research in the area is likely to remain somewhat unfocussed as a result.

REFERENCES

Borker, V., Carey, M., Mangtani, N., McKinney, D., Patel, R., & Thatte, S. (2006). XML Data Services. *International Journal of Web Services Research, 3*(1), 85-95.

Brown, A., Delbaere, M., Eeles, P., Johnston, S., & Weaver, R. (2005). Realizing Service Oriented Solutions with the IBM Rational Software Development Platform. IBM Systems Journal, *44*(4), 727-752.

Chen, Y., Zhou, L., & Zhang, D. (2006). Ontology-Supported Web Service Composition: An Approach to Service-Oriented Knowledge Management in Corporate Financial Services. *Journal of Database Management, 17*(1), 67-84.

Crawford, C., Bate, G., Cherbakov, L., Holly, K., & Tsocanos, C. (2005). Toward an on Demand Service Architecture. *IBM Systems Journal, 44*(1), 81-107.

De Pauw, Lei, M., Pring, E., & Villard, L. (2005). Web Services Navigator: Visualizing the Execution of Web Services. *IBM Systems Journal, 44*(4), 821-845.

Dorman, A. (2007). FrankenSOA. *Network Computing, 18*(12), 41-51.

Duke, A., Davies, J., & Richardson, M. (2005). Enabling a Scalable Service oriented Architecture with Semantic Web Services. *BT Technology Journal, 23*(3), 191-201.

Erickson, J., Lyytinen, K., & Siau, K. (2005). Agile Modeling, Agile Software Development, and Extreme Programming: The State of Research. *Journal of Database Research, 16(*4), 80-89.

Erickson, J., & Siau, K. (2008). Web Services, Service Oriented Computing, and Service Oriented Architecture: Separating Hype from Reality. *Journal of Database Management, 19*(3), 42-54.

Ferguson, D., & Stockton, M. (2005). Service oriented Architecture: Programming Model and Product Architecture. *IBM Systems Journal, 44*(4),753-780.

Havenstein, H. (2006). Measuring SOA Performance is a Complex Art. *Computer World, 40*(2), 6.

Hicks, B. (2007). *Oracle Enterprise Service Bus: The Foundation for Service Oriented Architecture*. Retrieved10-18-2007 from: http://www.oracle.com/global/ap/openworld/ppt_download/middleware_oracle%20enterprise%20service%20bus%20foundation_250.pdf.

Hutchinson, B., Henzel, J., & Thwaits, A. (2006). Using Web Services to Promote Library-Extension Collaboration. *Library Hi Tech, 24*(1), 126-141.

Jones, S. (2005). Toward an Acceptable Definition of Service. *IEEE Software*. May/June. (pp. 87-93).

Kim, J., & Lim, K. (2007). An Approach to Service Oriented Architecture Using Web Service and BPM in the Telcom OSS domain. *Internet Research, 17*(1), 99-107.

Krafzig, D., Banke, K., & Slama, D. (2005). *SOA Elements*. Prentice Hall. Retrieved on 10-02-2007 from: http://en.wikipedia.org/wiki/Image:SOA_Elements.png

Langdon, C. (2007). Designing Information Systems to Create business Value: A Theoretical Conceptualization of the Role of Flexibility and Integration. *Journal of Database Management, 17*(3),1-18.

Li, S., Huang, S., Yen, D., & Chang, C. (2007). Migrating Legacy Information Systems to Web Services Architecture. *Journal of Database Management, 18*(4),1-25.

Malloy, B., Kraft, N., Hallstrom, J., & Voas, J. (2006). Improving the Predictable Assembly of Service Oriented Architectures. *IEEE Software*. March/April. pp. 12-15.

Nah, F., Islam, Z., & Tan, M. (2007). Empirical Assessment of Facotrs Influencing Success of Enterprise Resource Planning Implementations. *Journal of Database Management, 18*(4), 26-50.

OMG (Object Management Group). (2007). Retrieved on 9-25-2007 from: http://colab.cim3.net/cgi-bin/wiki.pl?OMGSoaGlossary#nid34QI.

OASIS (Organization for the Advancement of Structured Information Standards). (2006). Retrieved on 9-25-2007 from: http://www.oasis-open.org/committees/tc_home.php?wg_abbrev=soa-rm

Open Group. (2007). Retrieved on 9-25-2007 from http://opengroup.org/projects/soa/doc.tpl?gdid=10632.

Ricadela, A. (2006). The Dark Side of SOA. *Information Week.* September 4th. (pp. 54-58).

Schmidt, M., Hutchison, B., Lambros, P., & Phippen, R. (2005). Enterprise Service Bus: Making Service Oriented Architecture Real. *IBM Systems Journal, 44*(4), 781-797.

Shah, A., & Kalin, P. (2007). SOA Adoption Models: Ad-hoc versus Program-based. *SOA Magazine*. July 6.

Shan, T., & Hua, W. (2006). Service Oriented solution Framework for Internet Banking. *Internet Journal of Web Services Research, 3*(1), 29-48.

Stal, M. (2006). Using Architectural Patterns and Blueprints for Service oriented Archtiecture. *IEE Software*. March/April. (pp. 54-61).

Sulkin, A. (2007). SOA and Enterprise Voice Communications. *Business Communications Review, 37*(8), 32-34.

Trembly, A. (2007). SOA: Savior or Snake Oil? National Underwriter Life & Health, 111(27), 50.

UNL-IBM System i Global Innovation Hub. (2007). *Making SOA Relevant for Business.* Retrieved on 10-09-2007 from http://cba.unl.edu/outreach/unl-ibm/documents/SOA_Relevant_Business.pdf

Verheeke, B., Vanderperren, W., & Jonckers, V. (2006). Unraveling Crosscutting Concerns in Web Services Middleware. *IEEE Software.* January/February. pp. 42-50.

W3C (World Wide Web Consortium). (2007). Retrieved on 9-25-2007 from: http://colab.cim3.net/cgi-bin/wiki.pl?WwwCSoaGlossary#nid34R0.

Web Service. (2007). Retrieved on 10-18-2007 from: http://en.wikipedia.org/wiki/Web_service on.

Wikipedia (EJB). (2007). Retrieved on 10-12-2007 from: http://en.wikipedia.org/wiki/Ejbhttp://en.wikipedia.org/wiki/Ejb.

Wikipedia (SOA). (2007). Retrieved on 9-25-2007 from: http://en.wikipedia.org/wiki/Service-oriented_architecture#SOA_definitions.

XML.com. (2007). Retrieved on 9-25-2007 from: http://www.xml.com/pub/a/ws/2003/09/30/soa.html.

Zhang, D. (2004). Web Services Composition for Process Management in e-Business. *Journal of Computer Information Systems, 45*(2), 83-91.

Chapter XI
Designing Web Information Systems for a Framework-Based Construction

Vítor Estêvão Silva Souza
Universidade Federal do Espírito Santo, Brazil

Ricardo de Almeida Falbo
Universidade Federal do Espírito Santo, Brazil

Giancarlo Guizzardi
Universidade Federal do Espírito Santo, Brazil

ABSTRACT

In the Web Engineering area, many methods and frameworks to support Web Information Systems (WISs) development have already been proposed. Particularly, the use of frameworks and container-based architectures is state-of-the-practice. In this chapter, we present a method for designing framework-based WISs called FrameWeb, which defines a standard architecture for framework-based WISs and a modeling language that extends UML to build diagrams that specifically depict framework-related components. Considering that the Semantic Web has been gaining momentum in the last few years, we also propose an extension to FrameWeb, called S-FrameWeb, that aims to support the development of Semantic WISs.

INTRODUCTION

The World Wide Web (also referred to as WWW or simply Web) was created as a means to publish documents and make them available to people in many different geographical locations. However, the advent of the Common Gateway Interface (CGI), in 1993, allowed for authors to publish software instead of documents and for visitors to execute them, producing dynamic results.

The evolution of Web development technology and the emergence of high-level languages (such as PHP, ASP, JSP, etc.) and platforms (such as Microsoft .NET and Java Enterprise Edition) allowed for more complex applications to be built on the Web. Soon enough, a handful of large B2C (business-to-consumer, such as online stores) and B2B (business-to-business, such as supply chain management systems) applications were being deployed on the Internet.

Thus, the concept of Web Applications (WebApps) was born. WebApps consist of a set of Web pages or components that interact with the visitor, providing, storing and processing information. WebApps can be informational, interactive, transactional, workflow-based, collaborative work environments, online communities, marketplaces or web portals (Ginige & Murugesan, 2001).

In this chapter, however, we focus on a specific class of Web Applications, called Web-based Information Systems (WISs). WISs are just like traditional information systems, although deployed over the Internet or on an Intranet. These systems are usually data-centric and more focused on functionality rather than content and presentation. Examples are online stores, cooperative environments, and enterprise management systems, among many others.

Although many Software Engineering principles have long been established before the creation of the Web, first-generation WebApps were constructed in an ad-hoc manner, with little or no concern for them. However, with the increase of complexity of the WebApps, which is especially true for WISs, the adoption of methodologies and software processes to support the development team becomes crucial.

Thus, a new discipline and research field was born. Web Engineering (or WebE) can be defined as "the establishment and use of engineering principles and disciplined approaches to the development, deployment and maintenance of Web-based Applications" (Murugesan et al., 1999, p. 2). Pressman (2005) complements this definition stating that WebE borrows many conventional Software Engineering fundamental concepts and principles and, in addition, incorporates specialized process models, software engineering methods adapted to the characteristics of this kind of application and a set of enabling technologies.

In this field, a lot of methods and modeling languages have been proposed. Some well known works are WebML (Ceri et al., 2000), WAE (Conallen, 2002), OOWS (Fons et al., 2003), UWE (Koch et al., 2000), and OOHDM (Schwabe & Rossi, 1998), among others.

Parallel to the academic research, the industry and the developer community have also proposed new technologies to provide a solid Web infrastructure for applications to be built upon, such as frameworks and container-based architectures. Using them we can improve productivity at the coding phase by reusing software that has already been coded, tested and documented by third parties. As their use becomes state-of-the-practice, methods that focus on them during software design could provide a smoother transition from models to source code.

This has motivated us to develop a WebE design method that focuses on frameworks. The Framework-based Design Method for Web Engineering (FrameWeb) (Souza & Falbo, 2007) proposes a basic architecture for developing WebApps and a UML profile for a set of design models that brings concepts used by some categories of frameworks, which are applied in container-based architectures as well.

Meanwhile, many researches have been directed to the construction of what is being considered the future of the WWW: the Semantic Web. Coined by Berners-Lee et al. (2001), the term represents an evolution of the current WWW, referred by some as the "Syntactic Web". In the latter, information is presented in a way that is accessible only to human beings, whereas in the former data is presented both in human-readable and machine-processable formats, in order to promote the development of software agents that would help users carry out their tasks on the Web.

However, for Berners-Lee's vision to become a reality, Web authors and developers must add semantic annotations to their Web Applications. This is neither an easy nor a small task and support from tools and methods is needed. Thus, an extension of FrameWeb was proposed. The Semantic FrameWeb (S-FrameWeb) (Souza et al., 2007) incorporates into the method activities and guidelines that drive the developer in the definition of the semantics of the WISs, resulting in a "Semantic Web-enabled" application.

The objective of this chapter is to discuss the current research state regarding Web Engineering (methods and modeling languages), frameworks and the Semantic Web, and to present FrameWeb, and its extension S-FrameWeb, as a new method based on best practices for the development of Web-based Information Systems. We close the chapter by presenting future trends and opportunities for this research.

BACKGROUND

Web Engineering was born from the need to apply Software Engineering principles to the construction of WebApps, adapting them to the application's size, complexity and non-functional requirements. A lot of research on methods and modeling languages has already been conducted, providing an extensive background for new research.

Meanwhile, companies and independent developers create frameworks and propose container-based architectures to promote reuse and improve productivity while maintaining good design principles. Furthermore, research on the Semantic Web has been pointing out some directions on what the Web may become in the future.

This section discusses the current state-of-the-art and state-of-the-practice on Web Engineering, modeling languages for WebE, frameworks for development of WebApps and the Semantic Web.

Web Engineering

Web Engineering (or WebE) uses scientific, engineering, and management principles and systematic approaches to successfully develop, deploy, and maintain high-quality WebApps (Murugesan et al., 1999).

As with conventional software engineering, a WebE process starts with the identification of the business needs, followed by project planning. Next, requirements are detailed and modeled, taking into account the analysis and design perspective. Then the application is coded using tools specialized for the Web. Finally, the system is tested and delivered to end-users (Pressman, 2005).

Considering that, in general, the platform in which the system will run is not taken into account before the design phase of the software process, developing a WebApp would be just like developing any other application up to that phase. However, many differences between Web Engineering and Conventional Software Engineering have been identified by researchers and practitioners (Ahmad et al., 2005), such as sensitivity to content, short time frames for delivery, continuous evolution, focus on aesthetics, etc (Pressman, 2005).

This has motivated researchers to propose different methods, modeling languages and frameworks for Web Engineering. The amount of propositions is quite vast, demonstrating that academics and practitioners have not yet elected a standard concerning Web development. In this subsection we briefly present some methods, while the following subsections focus on modeling languages and frameworks.

Web Application Extension (WAE) (Conallen, 2002) defines an iterative and incremental software process, centered on use cases and based on the Rational Unified Process (Krutchen, 2000) and the ICONIX Unified Process (Rosenberg & Scott, 1999). It proposes activities such as business analysis, project planning, configuration management and an iterative process that includes the usual software development cycle from requirement gathering to deployment.

OOWS (Object Oriented Web Solution) (Fons et al., 2003) is an extension of the OO-Method (Pastor et al., 2001) for WebApp specification and development. It divides the software process in two main steps: conceptual modeling and solution development. In the conceptual modeling step, the system specification is obtained by using conceptual models. For that, OOWS introduces new models for representing navigational and presentational characteristics of web applications. In the solution development step, the target platform is determined, and a specific architectural style is chosen. Then, a set of correspondences between abstraction primitives and the elements that implement each tier of the architectural style are applied in order to automatically obtain the final system (Pastor et al., 2003).

The UML-based Web Engineering (UWE) (Koch et al., 2000) is a development process for Web applications with focus on systematic design, personalization and semi-automatic generation. It is an object-oriented, iterative and incremental approach based on the Unified Modeling Language (UML) and the Unified Software Development Process (Jacobson et al., 1999). The notation used for design is a "lightweight" UML profile. The process is composed by requirement analysis, conceptual navigation and presentation design, supplemented with task and deployment modeling and visualization of Web scenarios (Koch & Kraus, 2002).

Lee & Shirani (2004) propose a component-based methodology for WebApp development, which is divided in two major parts: component requirements analysis and component specifications. Analysis begins identifying the required component functions and is followed by a comparison with the functions available in existing components. The component specification phase has three activities: rendering specification, integration specification and interface specification.

The Ariadne Development Method (Díaz et al., 2004) proposes a systematic, flexible, integrative and platform-independent process for specification and evaluation of WebApps and hypermedia systems. This process is composed of three phases: conceptual design, detailed design and evaluation. Each phase is further subdivided into activities, which in turn defines sets of work products to be built.

Díaz et al. (2004, p. 650) also define the hypermedia paradigm as one that "relies on the idea of organizing information as a net of interrelated nodes that can be freely browsed by users selecting links and making use of other advanced navigation tools, such as indexes or maps". We consider hypermedia methods quite different than methods for the development of WISs, as they focus on content and navigational structures instead of functionality and seem to be better suitable for information-driven WebApps.

Although hypermedia development methods are not on our focus, it is worthwhile to cite OOHDM (Object Oriented Hypermedia Design Method) (Schwabe & Rossi, 1998), a well-known method that is representative of hypermedia methods. It was born from the need to represent hypermedia structures such as links, text-based interfaces and navigation, and more recently has also been applied to Web development. For instance, an extension of this method, called OOHDM-

Java2 (Jacyntho et al., 2002), was proposed, which consists of a component-based architecture and an implementation framework for the construction of complex WebApps based on modular architectures (e.g. Java EE). The OOHDM process is divided into five steps: requirements gathering, conceptual design, navigational design, abstract interface design and implementation.

During our research we have also found several other methodological approaches that target specific contexts or scenarios, such as:

- The Business Process-Based Methodology (BPBM) (Arch-int & Batanov, 2003), which blends advantages of the structured and object-oriented paradigms for identifying and designing business components. The central idea of business component modeling is reusability of elementary units, which are business activities. An elementary unit that represents an atomic changeable business process can be implemented with a portable set of Web-based software components;
- The Internet Commerce Development Methodology (ICDM) (Standing, 2002), which is focused on the development of B2C e-commerce applications, emphasizing not only technical aspects, but also strategic, business and managerial aspects.

Some of the methods presented above also propose a modeling language that better suits its purposes, such as WAE and UWE . In the next subsection, some of them are briefly presented.

Modeling Languages for Web Engineering

Modeling languages define notations to be used on the creation of abstract models to solve problems. The Unified Modeling Language (UML) (Booch et al., 2005), for instance, is a modeling language that defines on its metamodel standardized notations for different kinds of models, such as class diagrams and use case diagrams. However, UML does not define when and to which purpose each model should be used.

Hence, methodologies usually present their own modeling language or, as is most commonly seen, use and extend UML, defining a UML Profile. For this purpose, UML includes extension mechanisms, such as stereotypes (definition of a new model element based on an existing one), tagged values (attachment of arbitrary textual information to elements using label/value pairs) and constraints (semantic specification for an existing element, sometimes using a formal language).

Based on these extension mechanisms, Conallen (2002) proposed the Web Application Extensions (WAE), which extends UML to provide Web-specific constructs for modeling WebApps. WAE also advocates the construction of a new model, the User Experience (UX) Model, which defines guidelines for layout and navigation modeling from requirements specification through design. Models, like the navigation diagram, the class diagram and the component diagram (the last two specific for the web tier), use WAE to represent Web components such as screens, server pages, client pages, forms, links and many more.

The UML-based Web Engineering (UWE) (Koch et al., 2000) also defines a UML profile. Based on class and association elements, it defines new elements to describe Web concepts, such as navigation, indexes, guided tours, queries, menus and many others.

Another method that defines a modeling language based on UML is OOWS (Fons et al., 2003). For the construction of the navigational model, UML packages represent navigational contexts and form a directed graph where the arcs denote pre-defined valid navigational paths. Each context is further modeled using a class diagram to show the navigational classes that form them.

Not all modeling languages are UML-based. WebML (Ceri et al., 2000) is an example. It allows developers to model WebApp's functionalities in a high level of abstraction, without committing to any architecture in particular. WebML is based on XML, but uses intuitive graphical representations that can easily be supported by a CASE tool. Its XML form is ideal for automatic generation of source code, producing Web applications automatically from the models.

Methods and modeling languages aid developers mostly during analysis and design of information systems. However, one can also find tools that focus on the implementation phase. In the next subsection, we discuss frameworks that have been extensively used for the development of WISs.

Frameworks for Web Development

WISs have very similar architectural infrastructure. Consequently, after the first systems started to be built, several frameworks that generalize this infrastructure were developed to be reused in future projects. In this context, a framework is viewed as a code artifact that provides ready-to-use components that can be reused via configuration, composition or inheritance. When combined, these frameworks allow large-scale n-tier WISs to be constructed with less coding effort.

Putting together several of these frameworks can produce what we call a container-based architecture. A container is a system that manages objects that have a well-defined life cycle. A container for distributed applications, such as the applications servers for the Java Enterprise Edition (Shannon, 2003), manage objects and offer services such as persistence, transaction management, remoting, directory services, etc.

The use of these frameworks or container-based architectures has a considerable impact in the development of a WIS. Since it is possible to find many frameworks for the exact same task, we categorized them according their objectives into the following classes:

- Front Controller (or MVC) frameworks;
- Decorator frameworks;
- Object/Relational Mapping frameworks;
- Dependency Injection frameworks;
- Aspect-Oriented Programming frameworks;

Figure 1. General architecture of a Front Contoller framework

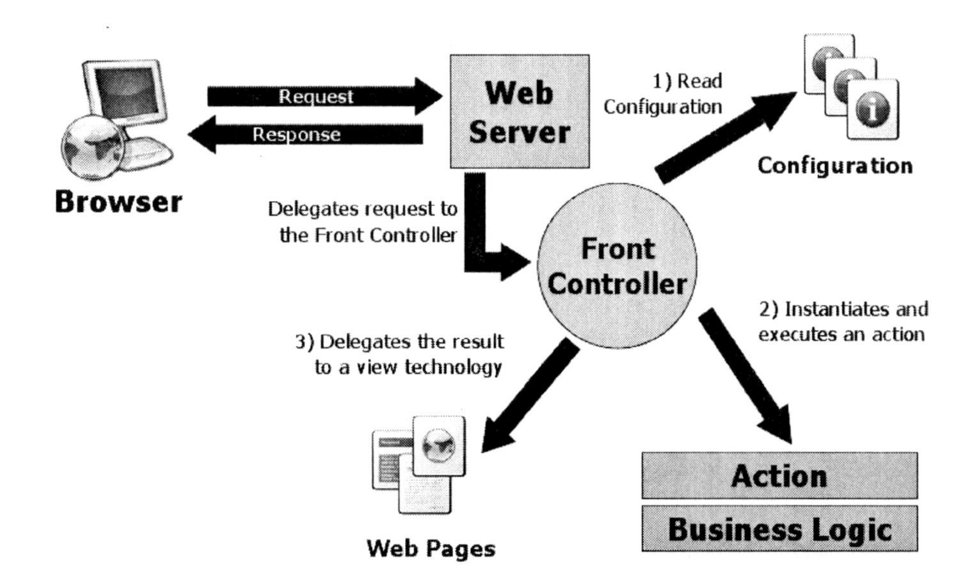

- Authentication & Authorization frameworks.

Front Controller Frameworks

MVC stands for Model-View-Controller (Gamma et al., 1994). It is a software architecture that was developed by the Xerox PARC for the Smalltalk language in 1979 (Reenskaug, 1979) and has found great acceptance by Web developers. When applied to the Web, the MVC architecture is adapted and receives the name "Front Controller" (Alur et al., 2003, p.166). Both terms are used indistinguishably by Web developers.

The Front Controller architecture is depicted in Figure 1. When structured in this architecture, a WebApp manages all requests from clients using an object known as Front Controller. Based on a customizable configuration, this object decides which class will respond to the current request (the action class). Then, following the Command design pattern (Gamma et al., 1994), it instantiates an object of that class and delegates the control to it, expecting some kind of response after its execution. Based on that response, the controller decides the appropriate view to present as result, such as a web page, a report, a file download, among other possibilities.

One of these possibilities is using a template engine that defines a template language that is usually more suitable for the view layer than the usual dynamic web technology (such as JSP, ASP or PHP). The template language is usually simpler, making it possible for Web Designers without specific programming skills to build them. Also, they tend to help developers not to break the MVC

Figure 2. The process of decoration of websites

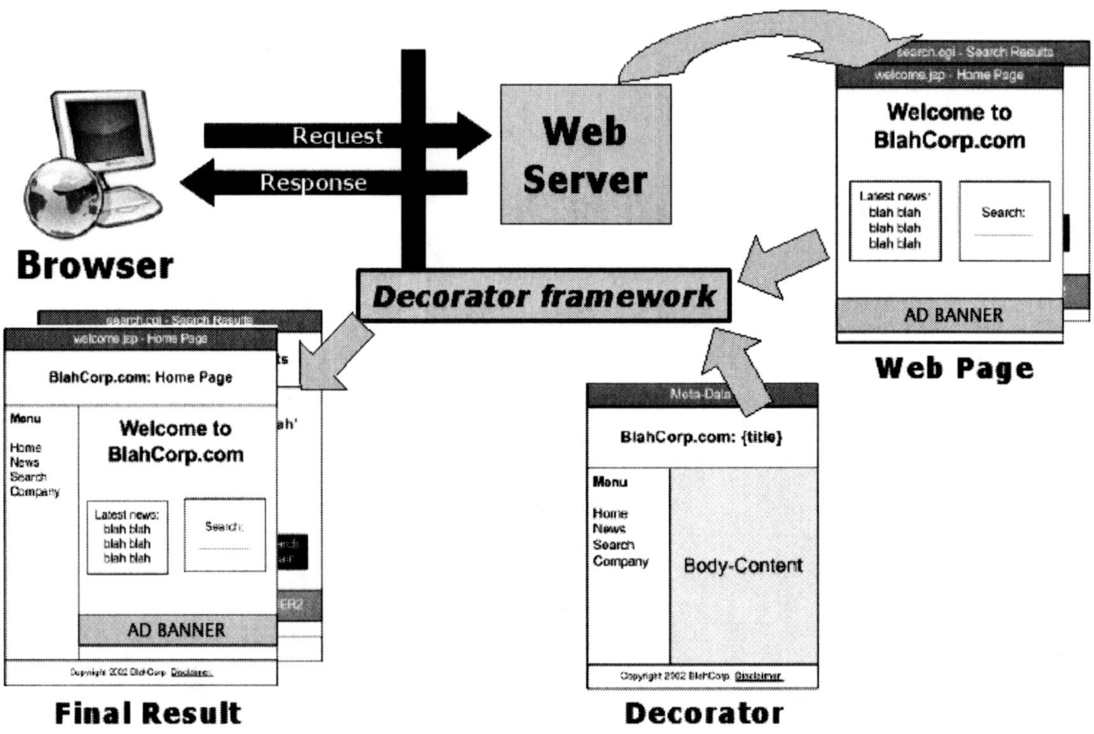

architecture by restricting what can be done in the template language (e.g. can not directly connect to a database from a template).

MVC Frameworks usually provide the front controller, a super-class or interface for action classes, several result types and a well defined syntax for the configuration file. The template engine is a separate tool, but the framework usually provides integration to it. Note that on n-tier applications, this framework belongs to the Web tier and should delegate business and persistence tasks to components on appropriate tiers.

Only for the Java platform, for instance, there are more than 50 MVC frameworks. Some of the most popular are Struts[1], Spring MVC[2] and Tapestry[3].

Decorator Frameworks

Decorator frameworks automate the otherwise tedious task of making every web page of the site have the same layout, meaning: header, footer, navigational bar, color schemes and other graphical layout elements produced by a Web design team. Figure 2 shows how a decorator framework works.

They work like the Decorator design pattern (Gamma et al., 1994), providing a class that intercepts requests and wraps their responses with an appropriate layout before it is returned to the client. It also provides dynamic selection of decorators, making it easy to create alternate layouts, such as a "print version" of the page. Examples of this kind of framework are SiteMesh[4] and Tiles[5].

Object/Relational Mapping Frameworks

Relational Database Management Systems (RDBMS) have long been the *de facto* standard for data storage. Because of its theoretical foundations (relational algebra) and strong industry, even object oriented applications use it for object

Figure 3. Persistence of objects using an ORM framework

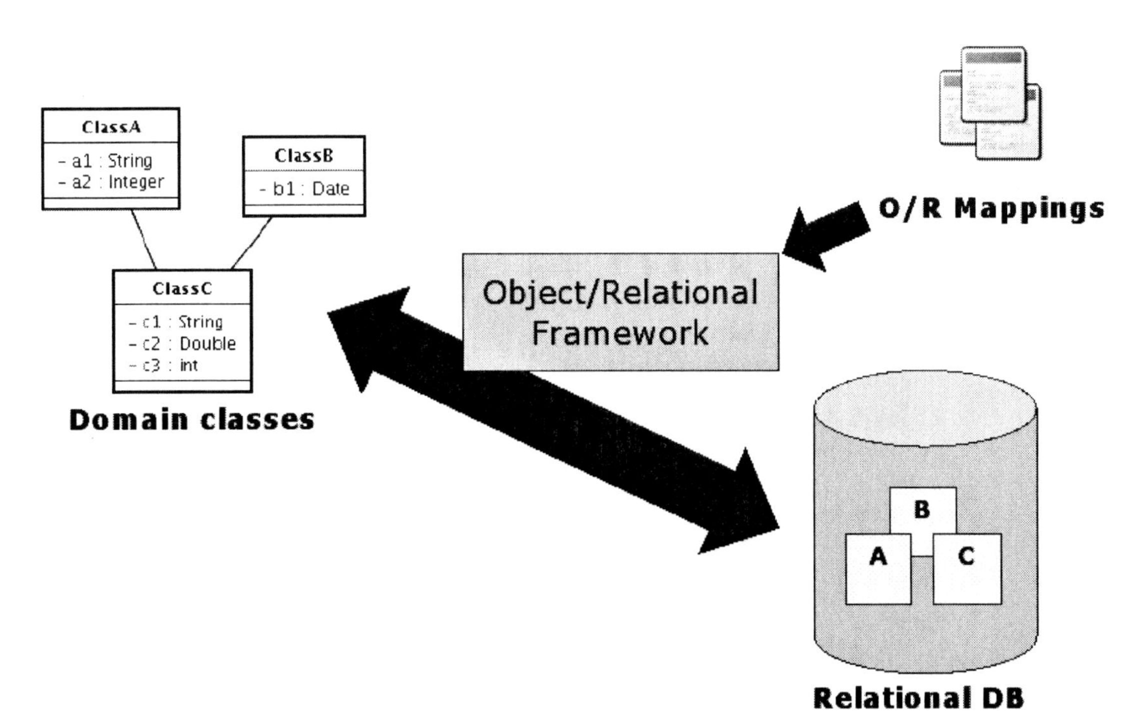

persistence, giving rise to a "paradigm mismatch" (Bauer & King, 2004): tables, rows, projection and other relational concepts are quite different from a graph of interconnected objects and the messages they exchange.

Among the many options to deal with this problem, there is the Object/Relational Mapping (ORM) approach, shown in Figure 3, which is the automatic and transparent persistence of objects to tables of a RDBMS using meta-data that describe the mapping between both worlds (Bauer & King, 2004). Instead of assembling a string with the SQL command, the developer provides mapping meta-data for the classes and call simpler commands, such as `save()`, `delete()` or `retrieveById()`. An object-oriented query language can also be used for more complex retrievals.

The use of ORM frameworks is not restricted to Web applications and has been in use for quite some time now in all kinds of software. The most popular Java ORM framework is Hibernate[6]. Other well-known frameworks are Java Data Objects[7], Apache Object Relational Bridge[8] and Oracle Toplink[9].

Dependency Injection Frameworks

Object-oriented applications are usually built in tiers, each of which having a separate responsibility. According to Fowler (2007), when we create classes that depend on objects of other classes to perform a certain task, it is preferred that the dependent class is related only to the interface of its dependencies, and not to a specific implementation of that service.

Creational design patterns, such as Factory Method, Abstract Factory and Builder (Gamma et al., 1994), help implementing this good practice in programming, known today as "programming to interfaces, not implementations" (Schmidt, 2007). For instance, if a service class depends on a data access class, it does not need to know *how* the data access class will perform its duty, but only *what* it will do and what method should be called for the job to be done.

Figure 4. Dependency injection using a framework

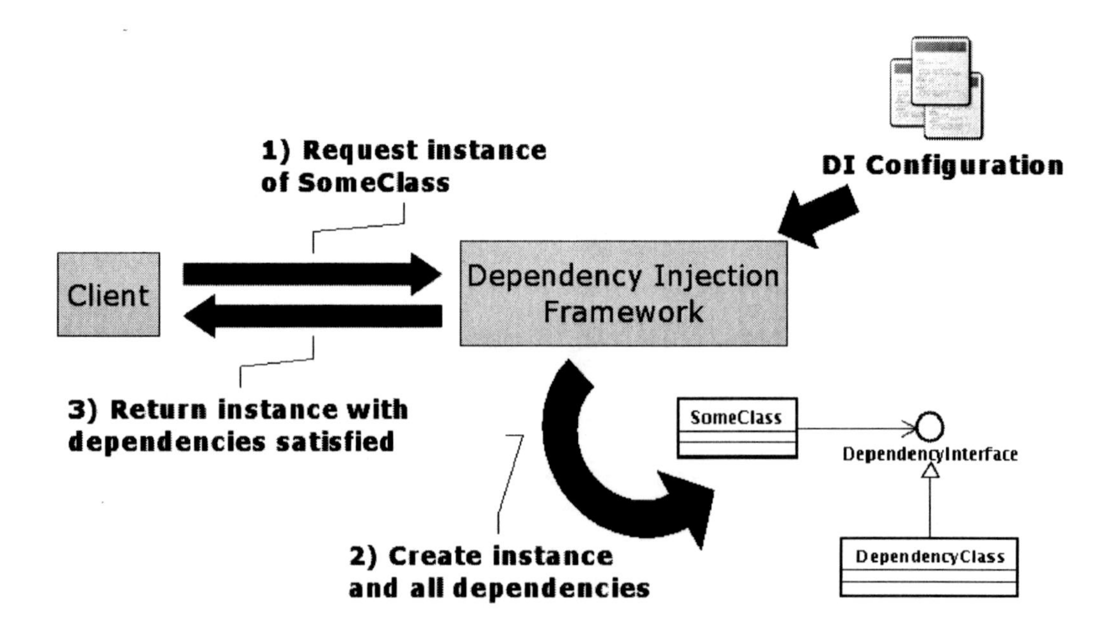

Figure 5. example of application of AOP using an AOP runtime framework

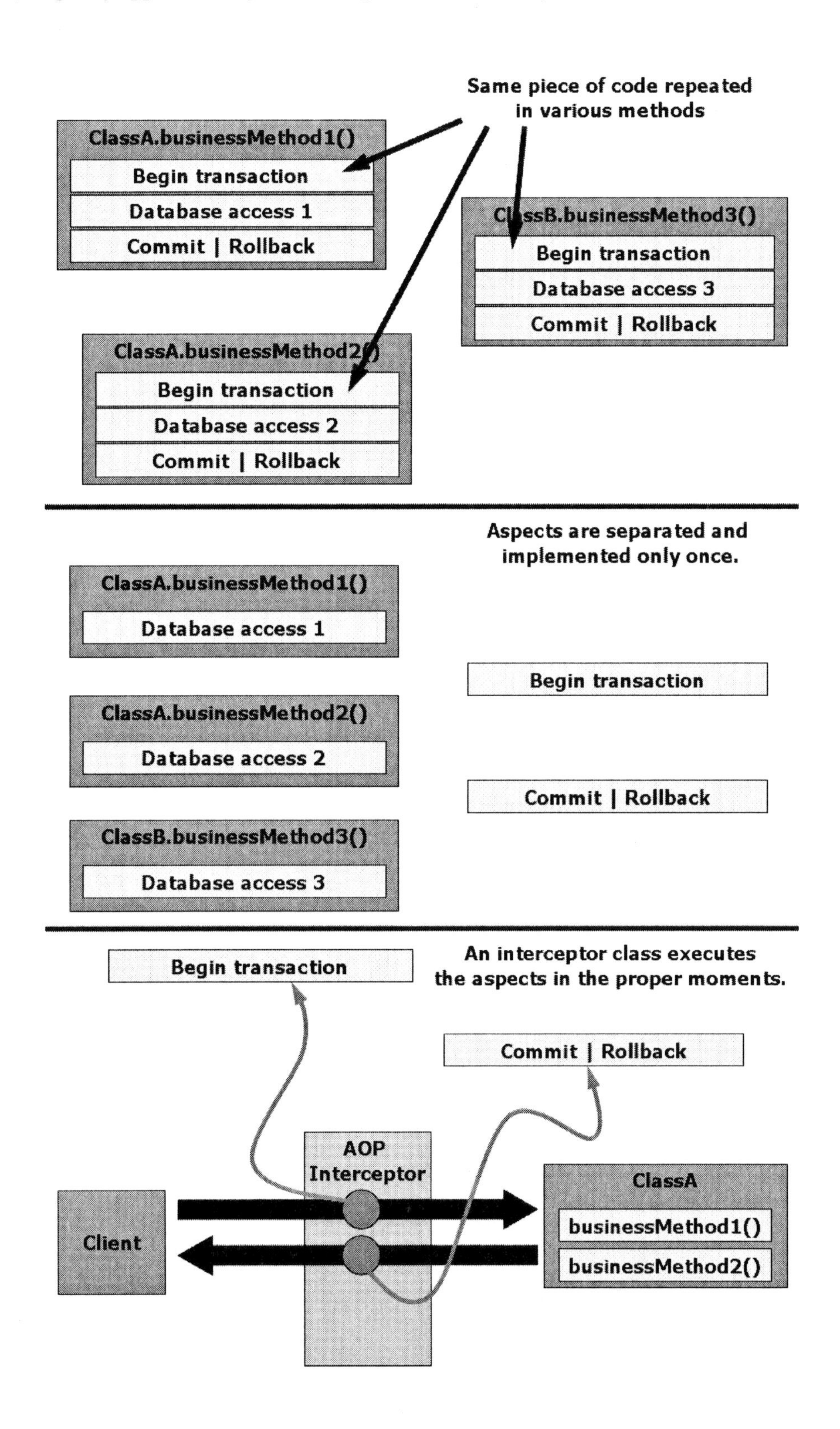

Dependency Injection (DI) frameworks allows the developer to program to interfaces and specify the concrete dependencies as meta-data in a configuration file. When a certain object is obtained from the DI framework, all of its dependencies are automatically injected and satisfied. An abstract example is shown in Figure 4: when the client asks for an instance of `SomeClass`, the DI framework first satisfies `SomeClass`' dependencies and delivers the object with all dependencies fulfilled – in the example, an instance of `DependencyClass`.

These frameworks are also known as Inversion of Control (IoC) frameworks, since the control (who creates the objects) is removed from the dependent classes and given to the framework. As well as ORM frameworks, DI frameworks are not used exclusively for WebApps, although they tend to integrate more seamlessly with applications that run inside containers, just like a WebApp runs inside a Web server. Lots of frameworks provide this service, including Spring Framework, PicoContainer[10], Apache Hivemind[11], etc.

Aspect-Oriented Programming Frameworks

The Aspect-Oriented paradigm is based on the concept of separation of concerns: the idea is to separate different concerns of a system to be treated separately, thus reducing the complexity of development, evolution and integration of software (Resende & Silva, 2005). Although it concerns the whole development process, its biggest influence is at the coding phase, with Aspect Oriented Programming (AOP).

Once a *cross-cutting concern* is identified (e.g.: logging, transaction management), instead of repeating similar code in different points, the functionality can be implemented in a single place, becoming an *aspect*. Then, the different places where that aspect should be applied are identified (these are called *pointcuts*) and, before the code is executed, a process called *weaving* is conducted to automatically spread the aspect all over the code.

The weaving can be conducted by an AOP framework during runtime or by an AOP compiler during compilation time. Many infrastructure concerns that are usual in Web applications are good candidates for this separation, making AOP frameworks very popular. One example, depicted in Figure 5, is that of transaction management. An AOP framework can make all business methods transactional with few configuration steps, avoiding the effort of repeatedly implementing the same logic in all of them.

Some well-known AOP frameworks for the Java platform are AspectJ[12], Spring Framework and JBoss AOP[13].

Authentication & Authorization Frameworks

Another common concern of Web information systems is that of guaranteeing the security of the information. This is usually done by two different procedures: authentication (verifying if an access key is valid to access the application) and authorization (verifying the level of access of the authenticated user and what she is allowed to do).

Being such an important task, frameworks were created to guarantee its proper execution. They can be configured to support many different "auth" methods, using, as usual, meta-data and configuration files. Some well-known auth frameworks for the Java platform are Acegi Security for Spring[14], Apache Cocoon Authentication[15] and the Java Authentication and Authorization Services[16].

In spite of frameworks being much used, there is no Web Engineering method that explores their use in the design phase of the software process. To fill this gap, we proposed FrameWeb, a Framework-based Design Method for Web Engineering (Souza & Falbo, 2007), which is presented later in this chapter.

The Semantic Web

The Semantic Web is being proposed as an evolution of the current WWW, in which information is provided both in human-readable and computer-processable formats, in order to allow for the semi-automation of many tasks that are conducted on the Web.

In order for the software agents to reason with the information on the Web (reasoning meaning that the agents are able to understand it and take sensible actions according to a predefined goal), web pages have to be presented also in a machine-readable form. The most usual way for this is annotating the pages using formal knowledge representation structures, such as ontologies.

An ontology is an engineering artifact used to describe a certain reality, plus a set of explicit assumptions regarding the intended meaning of its vocabulary words (Guarino, 1998). Along with ontology representation languages such as OWL (W3C, 2007a), they are able to describe information from a website in formal structures with well-defined inference procedures that allow software agents to perform tasks such as consistency checking, to establish relationships between terms and to systematically classify and infer information from explicitly defined data in this structure.

Designing an ontology is not a straightforward task. There are many methodologies for their construction (Gomez-Perez et al., 2005) and attention has to be given to the selection of concepts, their properties, relationships and constraints. However, after the ontology is built, the annotation of static Web pages with languages such as OWL becomes a simple task, especially with the aid of tools, such as OILEd[17] and Protégé[18].

However, several websites have their Web pages dynamically generated by software retrieving information from data repositories (such as relational databases) during runtime. Since these pages cannot be manually annotated prior to their presentation to the visitor, another approach has to be taken. Two approaches that have been proposed are dynamic annotation and semantic Web services.

The former works by recognizing whether the request belongs to a human or a software agent, generating the proper response depending on the client: in the first case, a HTML human-readable Web page; in the second, a document written in an ontology specification language containing meta-data about the information that would be displayed in the HTML version. Although the solution seems appropriate, many aspects still need to be addressed, such as: how are the agents supposed to find the Web page? How will they know the correct way to interact with it? For instance, how will they know how to fill in an input form to submit to a specific request?

The latter approach is based on Web Services, which are software systems designed to support interoperable machine-to-machine interaction over a network (W3C, 2007b). Web Services provide a nice way for software agents to interact with other systems, requesting services and processing their results. If semantic information is added to the services, they could become interpretable by software agents. Meta-data about the service are written in a markup language, describing its properties and capacities, the interface for its execution, its requirements and the consequences of its use (McIlraith et al., 2001). Many tasks are expected to be automated with this, including service discovery, invocation, interoperation, selection, composition and monitoring (Narayanan & McIlraith, 2002).

As the research on the Semantic Web progresses, methods are proposed to guide developers on building "Semantic Web-enabled" applications. An example of this is the Semantic Hypermedia Design Method (SHDM) (Lima & Schwabe, 2003). Based on OOHDM (Schwabe & Rossi, 1998), SHDM is a comprehensive model-driven approach for the design of Semantic WebApps.

SHDM's process is divided in 5 activities. In the first step, Requirements Gathering, requirements are gathered in the form of scenarios, user interaction diagrams and design patterns. The next phase, Conceptual Design, produces a UML-based conceptual model, which is enriched with navigational constructs in the Navigational Design phase. The last two activities are Abstract Interface Design and Implementation.

Some Considerations

In our research, we haven't found a method focused on the use of frameworks for the construction of WISs nor for the development of Semantic Web applications. In the next sections, we present FrameWeb, our proposal for the design of framework-based WISs, and its extension, S-FrameWeb, which incorporate into the method activities and guidelines that drive the developer in the definition of the semantics of the WISs, resulting in a "Semantic Web-enabled" application.

FRAMEWEB

FrameWeb is a design method for the construction of Web-based Information Systems (WISs) based on frameworks. The main motivations for the creation of this method were:

1. The use of frameworks or similar container-based architectures has become the *de facto* standard for the development of distributed applications, especially those based on the Web;
2. There are many propositions in the area of Web Engineering, including methodologies, design methods, modeling languages, frameworks, etc. However, we haven't found one that deals directly with the particularities that are characteristic of the use of frameworks;
3. Using a method that fits directly into the architecture chosen for the implementation promotes a greater agility to the software process, which is something that is desired in most Web projects.

In general, FrameWeb assumes that certain types of frameworks will be used during the implementation, defines a basic architecture for WISs and proposes design models that are closer to the implementation of the system using these frameworks.

Being a design method, it doesn't prescribe a complete software process. However, it suggests the use of a development process that includes the following activities, as presented in Figure 6: requirements elicitation, analysis, design, coding, testing and deployment. For a more systematic usage of the method, it also suggests that, during Requirement Elicitation and Analysis, use case diagrams are used to model requirements and class diagrams are used to represent the conceptual model.

Also, as mentioned earlier, one of the motivations for the creation of FrameWeb is the demand for agility that surrounds Web projects. Thus, although the method brings more agility especially

Figure 6. A simple software process suggested by FrameWeb

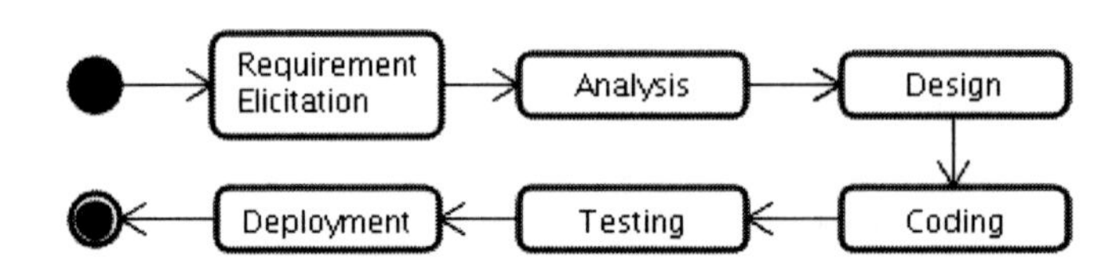

Figure 7. A simplified use case diagram for LabES Portal

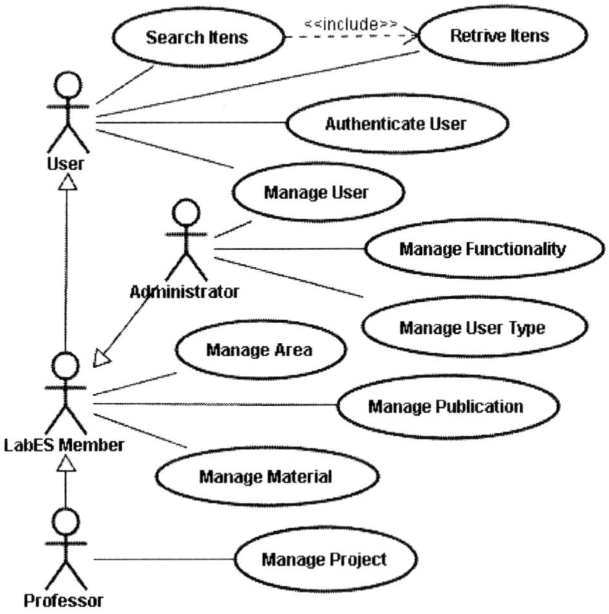

Figure 8. Conceptual model for the User Control module of the LabES Portal

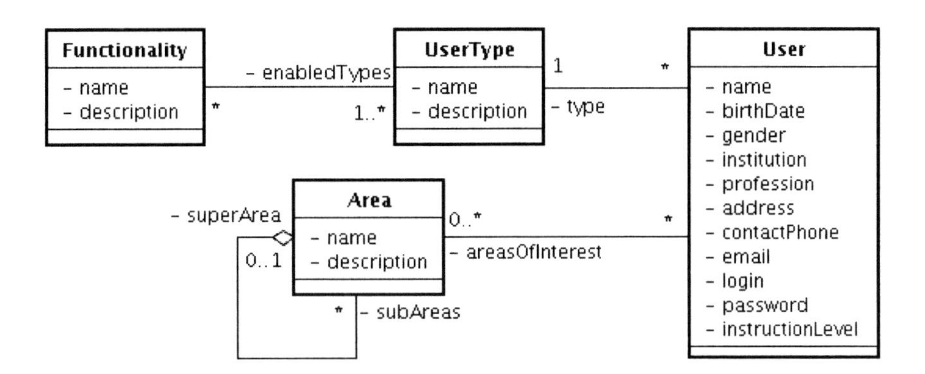

to the design and coding phases, developers are advised to follow principles of agility during requirements analysis, as the ones proposed by Agile Modeling (Ambler & Jeffries, 2002).

The main contributions of the method are for the Design phase: (i) the definition of a basic architecture that divides the system in layers with the purpose of integrating better with the frameworks; (ii) a UML profile for the construction of four different design models that bring the concepts used by the frameworks to the design stage of the software process.

The Coding phase is facilitated by the use of frameworks, especially because design models show components that can be directly related to them. The use of frameworks can also have impacts on Testing and Deployment, but these are yet subject to study and research.

Throughout the next subsections we detail FrameWeb's basic architecture and its UML pro-

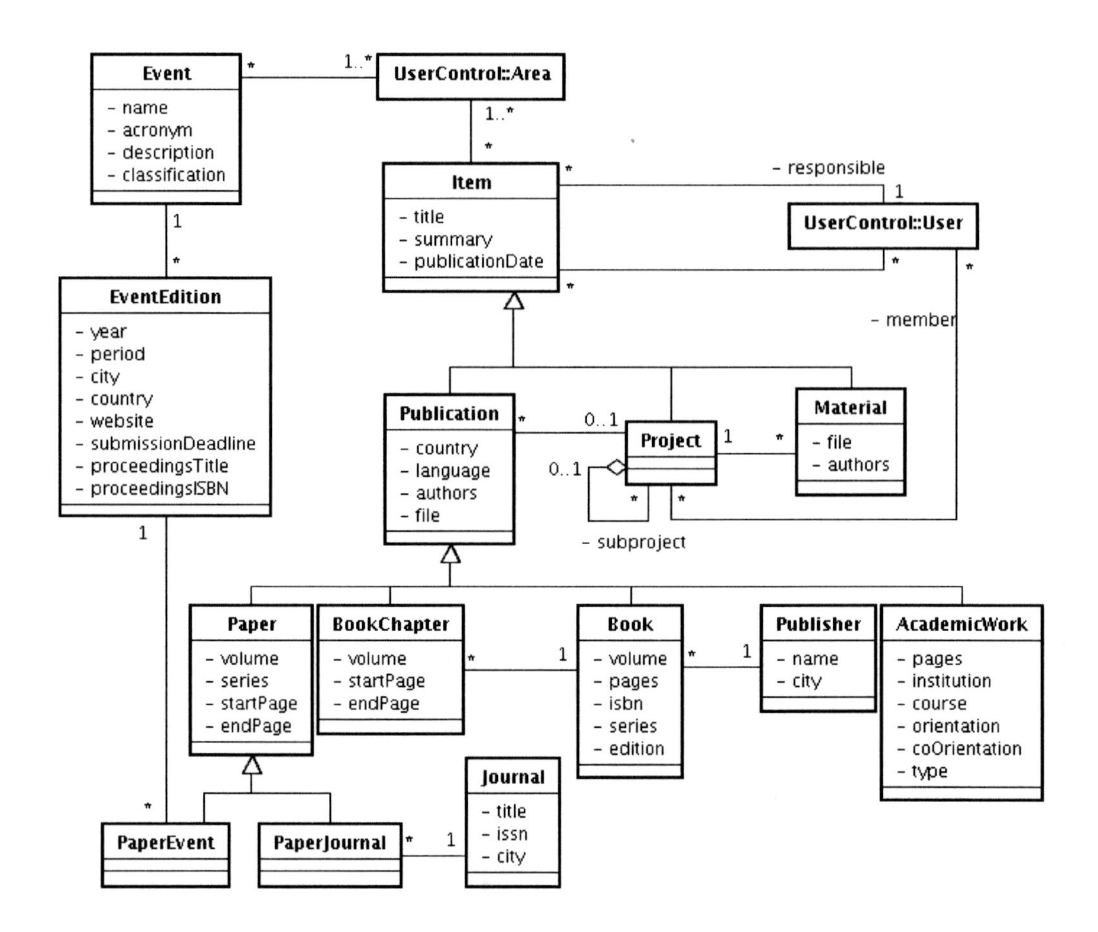

Figure 9. Conceptual model of the Item Control module of the LabES Portal

file. Examples diagrams were taken from the development of a portal for the Software Engineering Lab (LabES) of the Federal University of Espírito Santo State using FrameWeb. Figure 7 shows its use case diagram, simplified for brevity.

The "LabES Portal" was proposed to provide a better interaction with the Software Engineering community. This WIS has a basic set of services providing information about current LabES projects, areas of interest, publications and other material available for download. Figures 8 and 9 show the conceptual models produced during Analysis.

Basically, the portal makes a collection of items available. These items can be organized in projects and subprojects or belong to the lab in general. Publications (papers, books, book chapters and academic works) and generic materials can be published in the portal. Items are also related to users (responsible user, editing users) and areas of interest.

Framework-Based WebApp Architecture

The Design activity, traditionally executed after requirement elicitation and analysis, has as purpose the description of the logical and physical architectures of the system as well as the development of structural and behavioral models built based on the models developed in the previous phases, but that now consider the specific characteristics of the chosen implementation platform.

Figure 10. FrameWeb's basic architecture for WISs

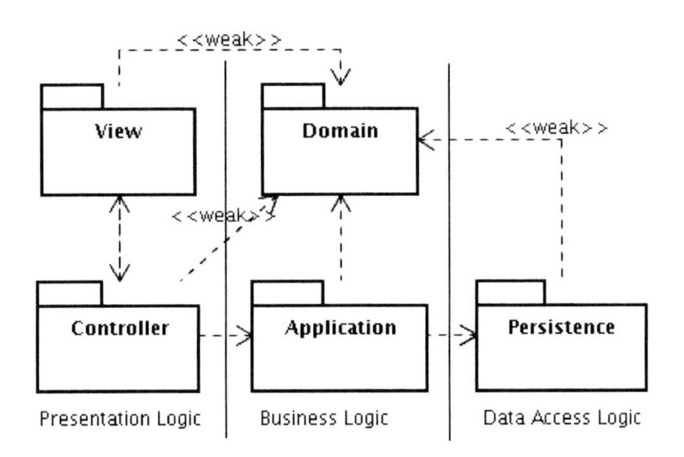

FrameWeb defines a logical architecture for WISs based on the architectural pattern Service Layer (Fowler, 2002, p. 133). As depicted in Figure 10, the system is divided in three main layers: presentation logic, business logic and data access logic.

The first layer concerns the graphical user interfaces. The `View` package contains the Web pages, style sheets, images, client-side scripts, templates and everything else related to the exhibition of information to the user. The `Controller` package encompasses action classes and other files related to the Front Controller framework. These two packages are mutually dependent, since `View` elements send user *stimuli* to `Controller` classes while these process the response using pages, models and other `View` components.

The business logic is implemented in the second layer, divided in two packages: `Domain` and `Application`. The former contains classes that represent concepts of the problem domain identified and modeled by the class diagrams during analysis and refined during design. The latter has the responsibility of implementing the use cases defined in the requirements specification, providing a service layer independent of the user interface. The `Application` classes deal directly with `Domain` objects to implement system functionality and, thus, this dependency is represented in the diagram.

The `Controller` package, on the presentation layer, depends on the `Application` package since it mediates the user access to the system functionalities. User *stimuli* coming from `View` are transformed by the `Controller`'s classes in method calls to classes in the `Application` package. `Controller` and `View` have also dependency relationships with `Domain`, but this is tagged as weak to denote low coupling: `Domain` objects are used only for exhibition of data or as parameters on method invocations between one package and another, i.e., the presentation layer does not have the right to alter domain entities.

The third and last layer regards data access and has only the `Persistence` package. This package is responsible for the storage and retrieval of persistent objects in long-term duration media, such as databases, file systems, naming services, etc. In the case of FrameWeb, it expects the use of an ORM framework through the Data Access Object (DAO) pattern (ALUR et al., 2003, p. 462). The DAO pattern adds an extra abstraction layer, separating the data access logic of the chosen persistence technology in a way that the `Application` classes do not know which ORM framework is being used, allowing for its replacement, if necessary. It also facilitates unit testing, as one can provide mock DAOs for the `Application` classes to be tested alone.

As we can see in Figure 10, the `Application` package depends on the `Persistence` package to retrieve, store and delete domain objects as the result of use case execution. Since the `Persistence` package works with `Domain` objects, a weak dependency is also portrayed in the figure.

This architecture provides a solid base for the construction of WISs based on the types of frameworks presented earlier in this chapter. Each package contains classes or other elements that integrate with these frameworks and, to model all these elements, FrameWeb proposes a modeling language based on the UML, which is presented next.

Modeling Language

During design, besides specifying the system architecture, the artifacts that will be implemented by the programmers on the coding phase should be modeled. Since FrameWeb is based on the frameworks presented earlier, we felt the need for a modeling language that would represent the concepts that are present in these frameworks.

Following the same approach as other modeling languages such as WAE and UWE, FrameWeb uses UML's lightweight extensions to represent typical Web and framework components, creating a UML profile that is used for the construction of four kinds of diagrams, which are presented in the following subsections: domain model, persistence model, navigation model and application model.

Domain Model

The domain model is a UML class diagram that represents domain objects and their persistence mapping to a relational database. This model is used by the programmers to implement the classes of the `Domain` package. FrameWeb suggests its construction in two steps:

1. Adapt the conceptual model produced during the Requirement Analysis phase to FrameWeb's architecture and to the chosen platform of implementation. This requires choosing data types for attributes, defining navigabilities of the associations, promoting attributes to classes (if necessary), etc.;
2. Add persistence mappings.

Persistence mappings are meta-data that allow ORM frameworks to convert objects in memory to tuples in Relational Data Base Management Systems and vice-versa. Mappings are added to the domain model using stereotypes and constraints that guide developers in the configuration of the ORM framework during implementation. Despite the fact that these mappings are more related to persistence than domain, they are shown in this model because the classes that are mapped and their attributes are shown here.

Table 1 describes the possible O/R mappings for the domain model. For each mapping, the table presents the extension mechanism used and what are its possible values or syntax. None of the mappings is mandatory and most of them have sensible defaults, reducing the amount of elements that have to be modeled. The default values are shown in the third column in boldface.

The Domain Model for the User Control module of sLabES Portal is shown in Figure 11. According to the default values, all classes are persistent and class and attribute names are used as table and column names respectively.

As we can see in the diagram, attributes have received mappings such as nullability and size. The `birthDate` attribute was mapped as date-only precision. The recursive association in Area was configured to be sorted naturally (will be implemented in the programming language) and

Table 1. Possible OR mappings for the Domain Model

Mapping	Extension	Possible Values
If the class is persistent, transient or mapped (not persistent itself, but its properties are persistent if another class inherits them)	Class stereotype	**<<persistent>>** <<transient>> <<mapped>>
Name of the table in which objects of a class will be persisted	Class constraint	table=*name* **(default: class' name)**
If an attribute is persistent or transient	Attribute stereotype	**<<persistent>>** <<transient>>
If an attribute can be null when the object is persisted	Attribute constraint	**null** not null
Date/time precision: store only the date, only the time or both (timestamp)	Attribute constraint	precision = (date | time | **timestamp**)
If the attribute is the primary-key of the table	Attribute stereotype	<<id>>
How the ID attribute should be generated: automatically, obtained in a table, use of IDENTITY column, use of SEQUENCE column or none	Attribute constraint	generation = (**auto** | table | identity | sequence | none)
If the attribute represents the versioning column.	Attribute stereotype	<<version>>
If an attribute should be stored in a large object field (e.g.: CLOB, BLOB)	Attribute stereotype	<<lob>>
Name of the column in which an attribute will be persisted	Attribute constraint	column=*name* **(defaults to the attribute's name)**
Size of the column in which an attribute will be persisted	Attribute constraint	size=*value*
If the association should be embedded (instead of having its own table, the associated child class' attributes are placed in the parent's table)	Attribute stereotype	<<embedded>>
Inheritance mapping strategy: one table for each class using UNION, one table for each class using JOIN or single table for the entire hierarchy	Inheritance stereotype	<<union>> <<join>> **<<single-table>>**
Type of collection which implements the association: bag, list, set or map	Association constraint	collection = (bag | list | **set** | map)
Order of an association's collection: natural ordering (implemented in code) or order by columns (ascending or descending)	Association constraint	order = (natural | *column names* [asc | desc])
Cascading of operations through the association: nothing, persists, merges, deletions, refreshs or all	Association constraint	cascade = (**none** | persist | merge | remove | refresh | all)
Association fetching strategy: lazy or eager.	Association constraint	fetch = (lazy | **eager**)

to cascade all operations (e.g. if an area is deleted, all of its subareas are automatically deleted).

None of the classes have ID or version attributes because they are inherited from a utility package, as shown in Figure 12. The `mapped` stereotype indicates that `DomainObjectSupport` and `HibernatePersistentObject` are not persistent entities, but their subclasses, which are entities, inherit not only their attributes but also their O/R mappings. All domain classes in LabES Portal are said to extend `HibernatePersistentObject`, inheriting, thus, the UUID[19], the persistence ID and the version attribute.

The parameters `I` and `V` are generic, allowing for the user to choose the type of ID and version attributes. `HibernateBaseDAO` is a base class for data access objects, described in the persistence model, discussed in the next subsection.

Figure 11. Domain Model for the User Control module of LabES Portal

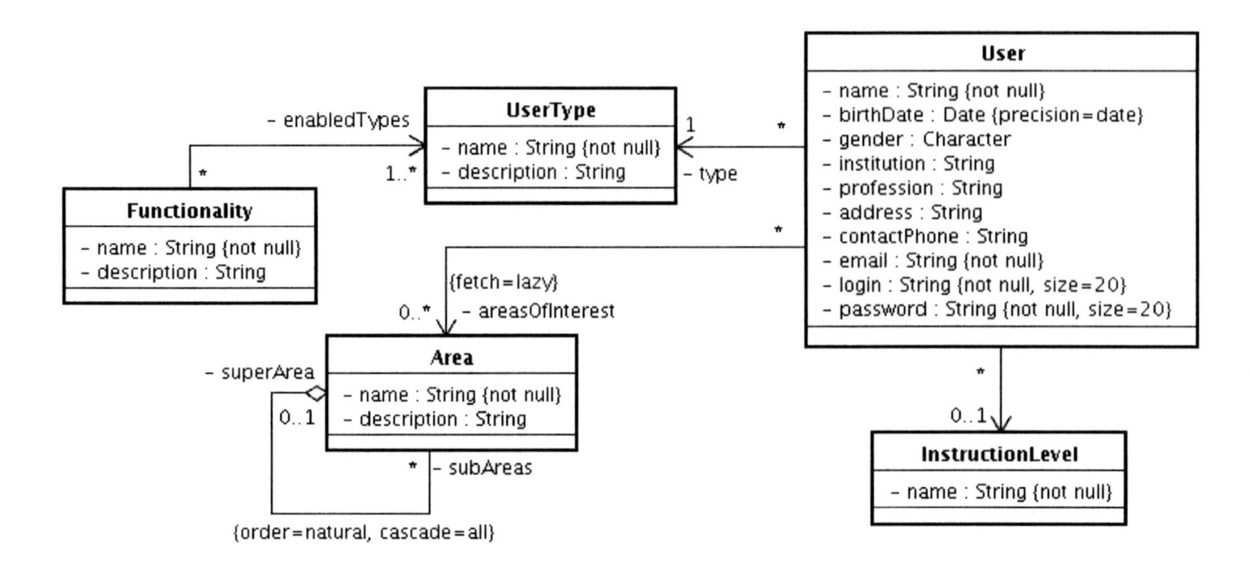

Figure 12. Utility classes for persistence

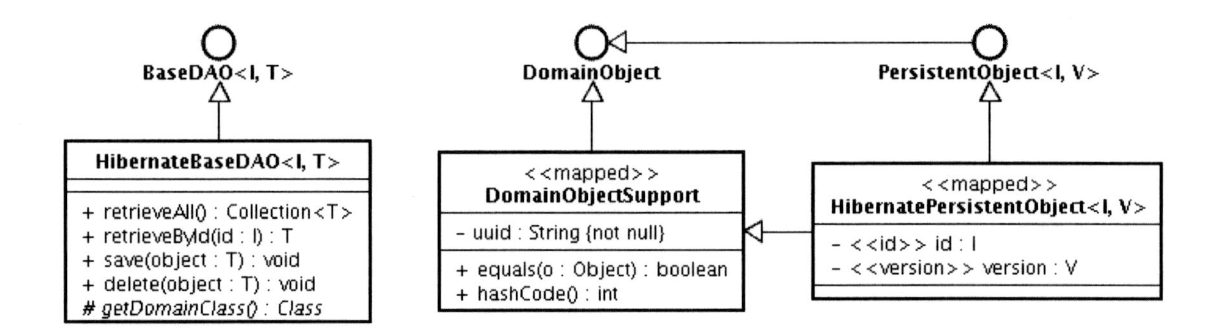

Persistence Model

As mentioned before, FrameWeb indicates the use of the DAO design pattern (ALUR et al., 2003, p. 462) to the construction of the data access layer. Thus, the persistence model is a UML class diagram that represents DAO classes responsible for the persistence of the domain classes. Therefore, it guides the implementation of the classes from the `Persistence` package. FrameWeb suggests three steps for its construction:

1. Model the interface and concrete implementation of the base DAO (an example is shown in Figure 12);
2. Define which domain classes need basic persistence logic and create a DAO interface and implementation for each one;
3. For each DAO, evaluate the need of specific database queries, adding them as operations in their respective DAOs.

The persistence model presents, for each domain class that needs data access logic, an

Figure 13. Persistence model of the User Control module of LabES Portal

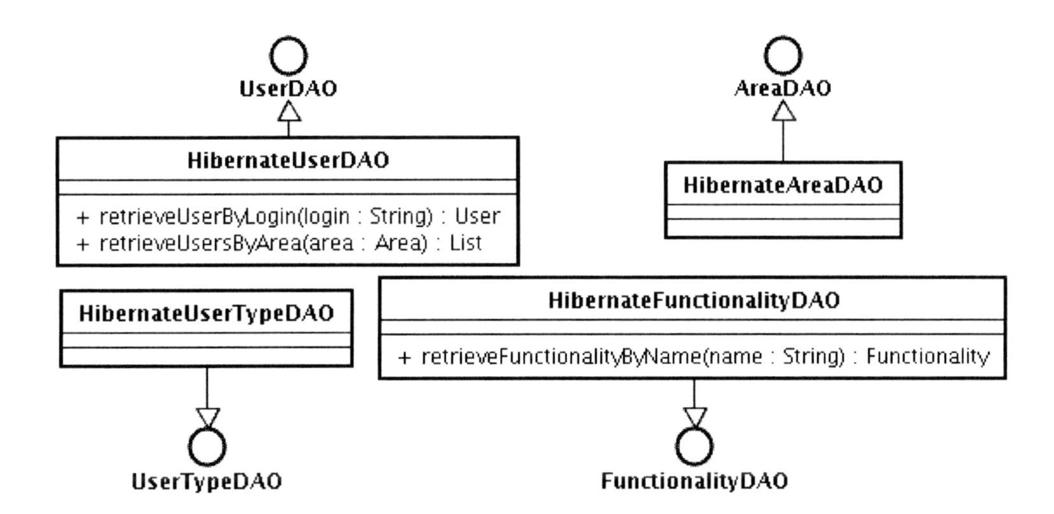

interface and a concrete DAO that implements the interface. The interface has to be unique and defines the persistence methods for a specific domain class. One concrete class is modeled for each persistence technology used.

To avoid repeating in each DAO operations that are common in all of them (e.g.: save, delete, retrieve by ID, etc.), a Base DAO (interface and implementation class) is modeled in a utility package. Automatically all DAO interfaces inherit from the BaseDAO interface and the same happens with concrete implementations, without the need to explicitly state that in the diagram. Also, to avoid repeating methods in the interface and implementations, the designer can choose to display them in one of the two only and it is inferred that all public methods are defined in the interface and implemented in the concrete class.

Figure 12 shows the interface and implementation using Hibernate ORM framework, designed for the LabES Portal project. Both interface and class and declared using generic types, leaving to their subclasses to specify which class is being persisted and what is the type for its ID attribute. The Base DAO defines methods to retrieve all persistent entities of a given class, retrieve an entity given its ID, save and delete an entity. As stated before, all public methods modeled in

Table 2. UML stereotypes used in the navigation model

Stereotype	What it represents
(none)	An action class, to which the Front Controller framework delegates the execution of the action.
<<page>>	A static or dynamic Web page.
<<template>>	A template that is processed by a template engine and is transformed into a Web page.
<<form>>	A HTML form.
<<binary>>	Any binary file that can be retrieved and displayed by the browser (e.g.: images, reports, documents, etc.).

Table 3. Dependency associations between an action class and other elements

From	To	What it represents
Page / template	Action class	A link in the page/template that triggers the execution of the action.
Form	Action class	Form data are sent to the action class when the form is submitted.
Action class	Page / template	The page/template is shown as one of the results of the action class.
Action class	Binary file	A binary file is shown as one of the results of the action class.
Action class	Action class	An action class is executed as result of another. This process is known as "action chaining".

`HibernateBaseDAO` are inferred to be defined in the `BaseDAO` interface.

Figure 13 shows the modeling of four DAOs from the LabES Portal project, for the persistence of the classes in the UserControl module. `AreaDAO` and `UserTypeDAO` are simple, as they inherit all basic operations from the Base DAO and don't need to define any extra ones. The other two define extra operations. For example, UserDAO defines an operation to retrieve all users that have a given area of interest. This is necessary because there is no navigability from Area to User (see Figure 11) and the "Manage Area" use case needs to prevent an area from being deleted if it is associated with any user.

As we can see, the persistence model does not define any UML extensions to represent the concepts that are needed to implement the data access layer, but only some rules that make this modeling simpler and faster.

Navigation Model

The navigation model is a UML class diagram that represents different components that form the presentation layer, such as Web pages, HTML forms and action classes from the Front Controller framework. Table 2 shows the UML stereotypes used by the different elements that can be represented in a navigation model. This model is used by developers to build classes and components of the `View` and `Controller` packages.

For Web pages and templates, the attributes of the classes represent information from the domain that is supposed to be displayed in the page. Dependency relationships between them indicate hyperlinks while composition associations between pages and forms denote the presence of the form in that page.

In HTML forms, attributes represent the form fields and their types follow the HTML standard for types of fields (e.g.: input, checkbox, etc.) or the names of the JSP tags used by the framework (e.g., for Struts[2], textfield, checkbox, checkboxlist, etc.).

The action class is the main component of the model. Its dependency associations show the control flow when an action is executed. Table 3 lists the different meanings of this kind of association, depending on the components that are connected by it. Dependencies that are navigable towards an action class represent method calls, while the others represent results from the action execution.

The attributes of the action class represent input and output parameters relevant to that action. If there is a homonymous attribute in an HTML form being submitted to the action, it means that the data is injected by the framework in the action class (input parameter). Likewise, when one of the result pages/templates show an attribute with the same name of an attribute of the action class, this indicates that the framework makes this information available for the output.

When an action is executed, the framework will execute a default action method or allow/request the explicit definition of which method to execute. In the latter case, the designer must specify which method is being executed using the constraint `{method=method-name}` in the dependency association. The same is true for associations that represent results. Naturally, these methods should be modeled in the diagram.

When modeling action chaining, it's sometimes necessary to indicate the method that was executed in the first action and the one that will be executed in the following. These can be specified with the constraints `outMethod` and `inMethod`.

For dependency associations that represent results there are two other constraints that can be used:

- `{result=result name}` specifies a keyword that represents this control flow, i.e., when the action class returns this keyword as result of the action execution, the framework will follow this flow and show the appropriate result page/template/binary file;
- `{resultType=type name}` determines the type of result, among those supported by the framework. Usually, at least the following types of result are available: `binary` (display a binary file), `chain` (action chaining), `dispatch` (dispatches the request), `redirect` (redirects the request) and `template` (processes a template using a template engine).

The difference between a dispatch and a redirection is that the first makes the action's output parameters available to the view, while the second does not. When a dependency association doesn't specify a type, it means it is a `dispatch`. The default result is defined by the framework.

The designer is free to choose the granularity of the action classes, building one for each use case scenario, one for each use case (encompassing many scenarios), one for multiple use cases, and so forth. Moreover, he/she should decide if it's best to represent many actions in a single diagram or have a separate diagram for each action. Figure 14 is the navigation model for a use case of LabES Portal.

The figure shows that in the initial page of the portal (represented by `web::index`), there is a form where login and password can be filled. When submitted, this data goes to the action class for the execution of the `executeLogin()` method, which would access the business logic layer to perform the use case. If the information filled is correct (`result = success`), the user is taken to `web::home`, which represents the starting page for authenticated users. Otherwise, the user will be taken back to `web::index` (`result = input`), showing once again the login form and an error message.

If the user forgot his/her password, he/she can click on a link in the initial page to go to the `web::remindpassword` page, where his/her login would be informed and sent to the action class. The `executeRemindPassword()` method requests the business logic layer to send the password to the user's email address and informs the user that the message has been sent. To log out, the user clicks on the appropriate link and is redirected back to the initial page.

During the conception of FrameWeb, there has been a discussion on whether the navigation model would be better represented by a sequence diagram, as it could represent better the control flow. Two main reasons led to the choice of the class diagram: (a) it provides a better visualization of the inner elements of action classes, pages and forms; and (b) it models composition between pages and forms with a more appropriate notation. Nonetheless, designers are advised to build sequence diagrams to represent complex flows when they see fit.

Last but not least, FrameWeb suggests four steps for the construction of a navigation model:

Figure 14. Navigation Model for the use case "Autheticate User"

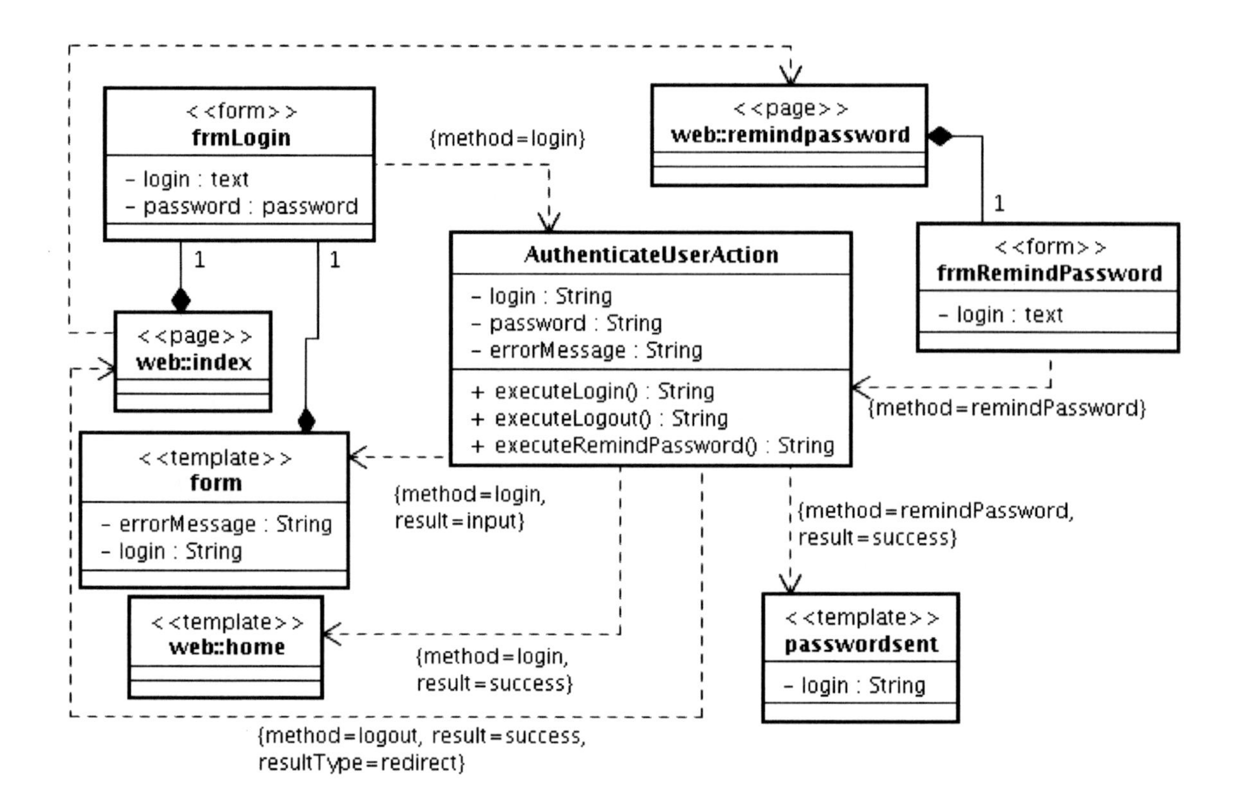

1. Study the use cases modeled during requirements analysis to define the granularity of the action classes (using, preferably, names that can relate the actions to the use cases/scenarios they refer to);
2. Identify how the data gets to the action class, modeling input pages and forms and the appropriate attributes on them and in the action class;
3. Identify what are the possible results and model the output pages/templates/binary files, also adding attributes when appropriate. We suggest that results that come from exceptions should not be modeled to avoid polluting the diagram;
4. Periodically check if the model is getting too complex and consider dividing it into two or more navigation models.

Application Model

The application model is a UML class diagram that represents classes from the `Application` package and their relationship with the `Controller` and `Persistence` packages. Besides guiding the implementation of application classes, this diagram also instructs developers on the configuration of the Dependency Injection framework, which is responsible for managing the dependencies among these three packages.

The granularity of the application classes can be chosen by the developer in the same way as the granularity of the action classes. The application model also shares similarities with the persistence model, as it does not define any UML extension and uses the "programming to interfaces" principle, indicating the modeling of an interface for each application class.

When an application class is modeled, all action classes that depend on it should be displayed in the diagram, with the appropriate namespaces and relationships depicted. Analogously, all DAOs required by the application class to execute the use case should have their interfaces shown in the model, along with the relationship with the application class. Both relationships are represented by directed associations and the multiplicity is not required, as it is always 1.

Figure 15 shows part of an application model of LabES Portal, depicting the classes that implements the "Manage User" and "Authenticate User" use cases and its relationships with controller and persistence components. The methods of the classes represent each scenario of each use case and define the parameters that should be given for them.

Application classes manipulate domain objects and, thus, depend on them. These relationships, however, are not shown in the diagram to avoid increasing the complexity of the model. One can know about these relationships by reading the description of each use case.

FrameWeb suggests four steps for the construction of an application model:

1. Study the use cases modeled during analysis to define the granularity of the application classes (using, preferably, names that can relate the classes to the use cases/scenarios they implement);
2. Add to the interfaces/classes the methods that implement the business logic, giving special attention to the name of the method (as before, with the name of the class), its parameter, the parameters types and its return type;
3. By reading the use case descriptions, identify which DAOs are necessary for each application class and model the associations;
4. Go back to the navigation model (if already built) and identify which action classes depend on which application class and model their associations.

By defining the standard architecture and a UML profile for the construction of these four

Figure 15. Part of an Application Model of the User Control module of LabES Portal

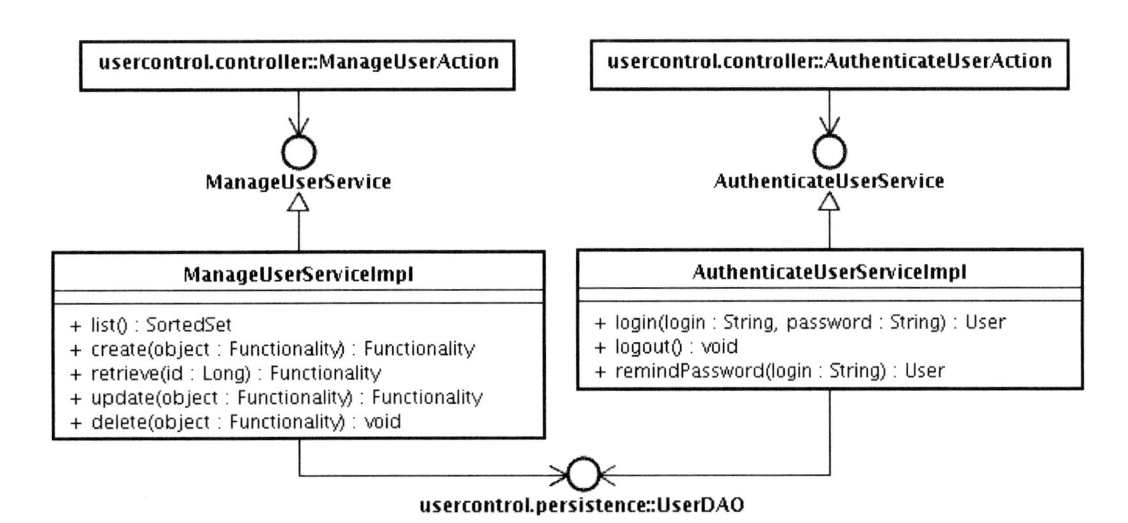

diagrams, FrameWeb provides the appropriate tools for the design of framework-based WISs. To promote the construction of "Semantic Web-enabled" WISs, an extension called S-FrameWeb was proposed and it is presented in the next section.

S-FRAMEWEB

The main goal of S-FrameWeb is to make WISs "Semantic Web-enabled". Being a framework-centered method, the chosen approach is to have the Front Controller framework produce dynamic annotations by identifying if requests come from human or software agents. In the former case, the usual Web page is presented, while in the latter, an OWL document is returned.

To accomplish this, S-FrameWeb extends FrameWeb in the following manners:

- The activity of Domain Analysis should be conducted in the beginning of the project to build an ontology for the domain in which the software is based. If it already exists, it should be reused (and eventually modified);
- Requirement Specification and Analysis go as usual, except for the fact that conceptual models build during Analysis can now be based on the domain ontology built in the previous activity;
- During design, FrameWeb's Domain Model (FDM) receives semantic annotations based on the domain ontology;
- During implementation, the MVC framework has to be extended in order to perform dynamic annotation.

Figure 16 shows the software process suggested by S-FrameWeb while Table 4 summarizes the evolution of the models throughout that software process.

The following subsections go through the suggested software process discussing it in more detail.

Domain Analysis

The first step for bringing a WIS to the Semantic Web is formally describing its domain. As discussed previously in this chapter, this can be achieved by the construction of an ontology. S-

Figure 16. The software process suggested by S-FrameWeb (SOUZA et al., 2007)

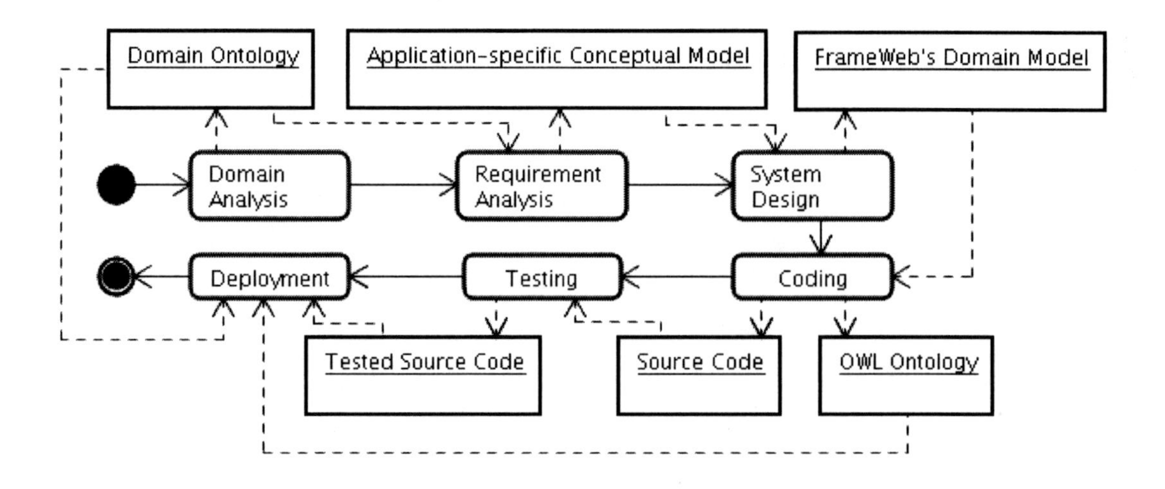

Table 4. Models produced by the software process suggested by S-FrameWeb (SOUZA et al., 2007)

Activity	Artifact	What the model represents
Domain Analysis	Domain Ontology	Concepts from the domain to which the software is being built. Modeled in ODM, but converted to OWL for deployment.
Requirement Analysis	Conceptual Model	Concepts that are specific to the problem being solved. Modeled in ODM.
System Design	FrameWeb's Domain Model (FDM)	Same as above plus OR mappings. Modeled using S-FrameWeb's UML profile.
Coding	OWL code	OWL representation of FDM, without OR mappings.

FrameWeb indicates the inclusion of a Domain Analysis activity in the software process for the development of a domain ontology (we don't use the term "domain model" to avoid confusion with FrameWeb's Domain Model – FDM –, which is a design model).

Domain Analysis is "the activity of identifying the objects and operations of a class of similar systems in a particular problem domain" (Neighbors, 1981; Falbo et al., 2002). When a software is built, the purpose is to solve a problem from a given domain of expertise, such as medicine, sales or car manufacturing. If the domain is analyzed prior to the analysis of the problem, the knowledge that is formalized about the domain can be reused when another problem from the same domain needs a software solution (Falbo et al., 2002).

S-FrameWeb does not impose any specific method for the construction of ontologies. It also doesn't require a specific representation language, but suggests the use of OMG's[20] Ontology Definition Metamodel (ODM) (OMG, 2007), "a language for modeling Semantic Web ontologies in the context of MDA" (Đurić, 2004). ODM defines an ontology UML profile that allows developers to represent ontologies in UML class diagrams.

In the development of the LabES Portal, the SABiO method (Falbo, 2004) was followed, resulting in the construction of an ontology for educational portals that deals with competency questions such as: what are the roles of the people in the educational institution?, what are the areas of interest of these people and the institution?, how is the institution organized?, etc. The ontology was divided into two separate diagrams: one for the general structure of educational protals and another specific for publications. Figure 17 shows the first one.

The domain ontology serves as a basis for the construction of the application's conceptual model (during Requirement Analysis), which should derive some classes and associations from the ontology, adding and modifying elements as needed, concerning the specific problem being solved.

Requirement Specification and Analysis

The activities of Requirement Specification and Analysis should be conducted by the development team using its methodology of preference. S-FrameWeb, like FrameWeb, does not prescribe any methods or languages to this phase of the software process. However, as during Domain Analysis, it suggests the use of ODM for the graphical representation of the conceptual model, as it eases its conversion to FDM and, later on, to code (using OWL).

Figure 18 shows the conceptual model for the User Control module of the LabES Portal.

The stereotype `<<OntClass>>` indicates domain classes, `<<ObjectProperty>>` models associations between domain classes,

Figure 17. Diagram of the structural part of the ontology for educational portals

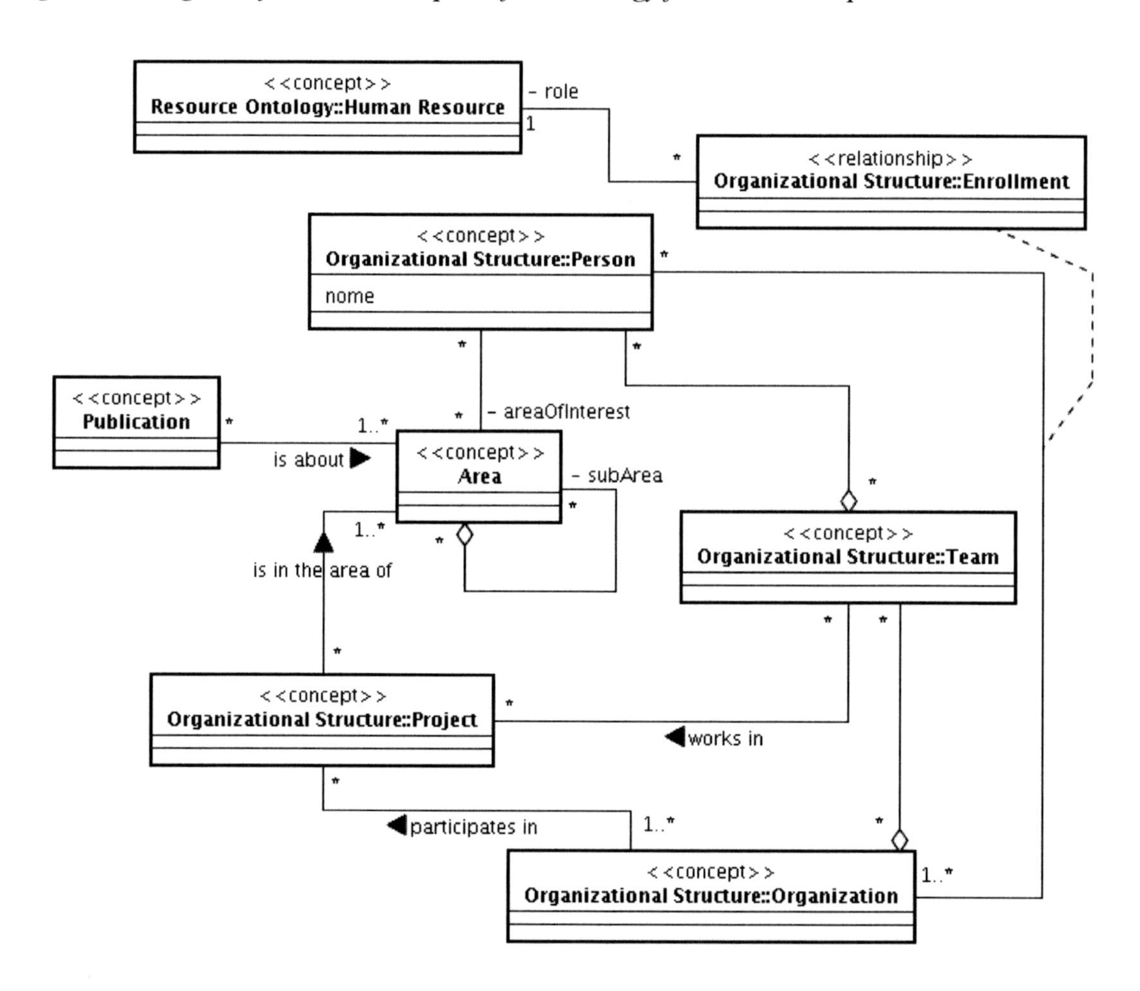

<<DataType>> represents XML data types and <<DatatypeProperty>> models associations between classes and data types.

The reader accustomed with UML conceptual models may notice that associations are represented as classes in ODM. This is because in OWL associations are independent from classes and, for instance, can form their own subsumption hierarchy. This could also happen with attributes, for the same reasons. More on ODM's semantics can be found at (OMG, 2007).

In the cases where there is no need to represent associations or attributes as UML classes, S-FrameWeb suggests the conceptual model is simplified, such as the one shown in Figure 19. Notice that this diagram is very similar to the one in Figure 8.

Design

As discussed before, FrameWeb proposes the creation of four kinds of models during design: domain, persistence, navigation and application models. These models are still used with S-FrameWeb, although the domain model (FDM) should be adapted to a representation more suitable to the purposes of this semantic extension. Therefore, S-FrameWeb suggests a new UML profile for this diagram, mixing the profile defined by ODM with the one proposed by FrameWeb.

Figure 18. The conceptual model for the User Control module of LabES Portal, in ODM

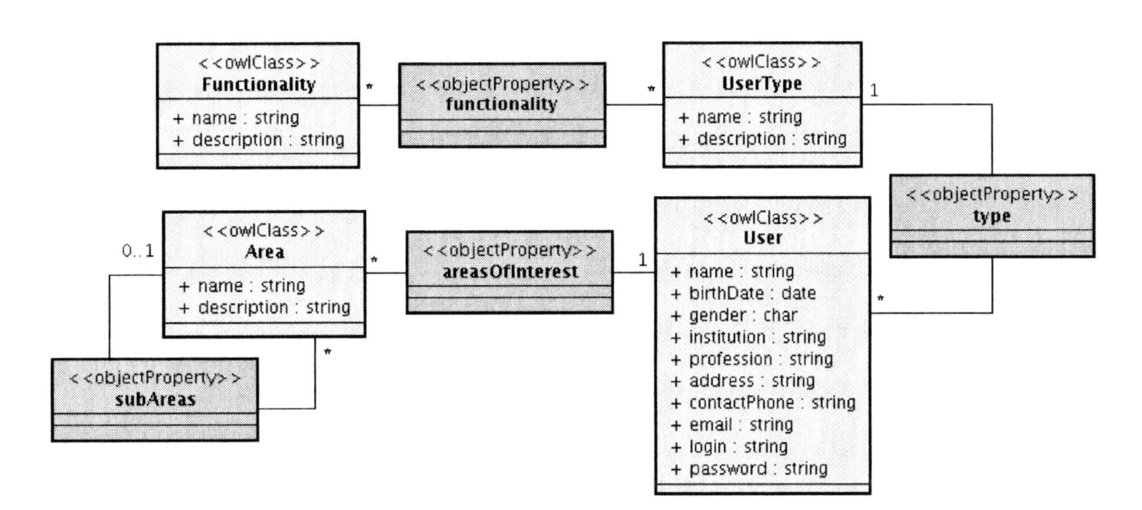

Figure 19. The conceptual model for the User Control module of LabES Portal, in its simplified version

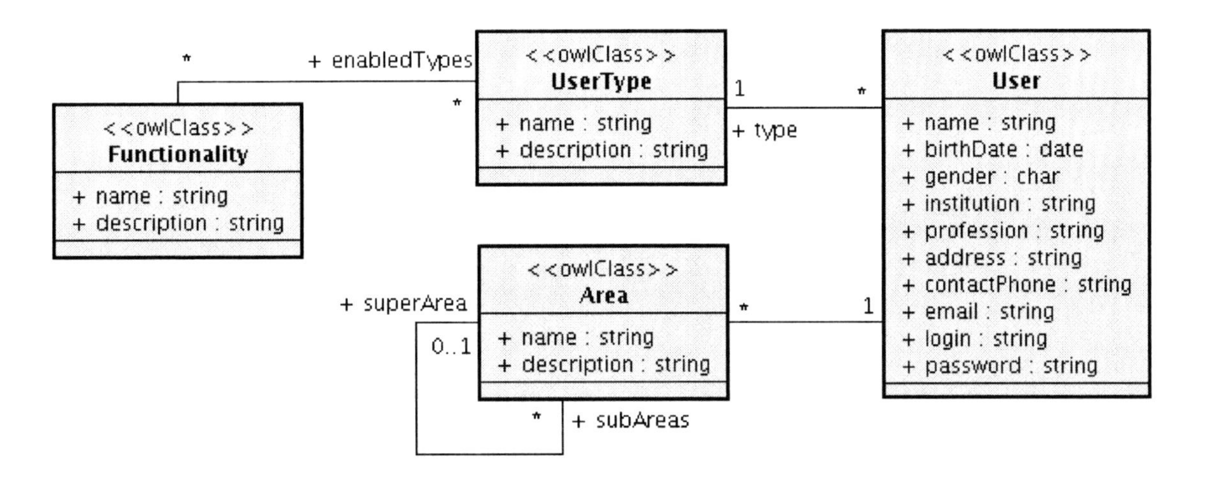

This new modeling language consists basically of the one defined by ODM, with the following adaptations:

1. Specification of association navigabilities for the implementation of the classes;
2. Addition of the O/R mappings for the configuration of the ORM framework;
3. Use of the data types of the implementation platform instead of those defined by the XML Schema Definition (XSD) standard[21];
4. Simplification of ODM's syntax when possible (if not already done previously).

Naturally, the construction of the FDM should be based on the conceptual model already built in previous activities. Figure 20 shows the FDM

Figure 20. S-FrameWeb's Domain Model for the User Control module of LabES Portal

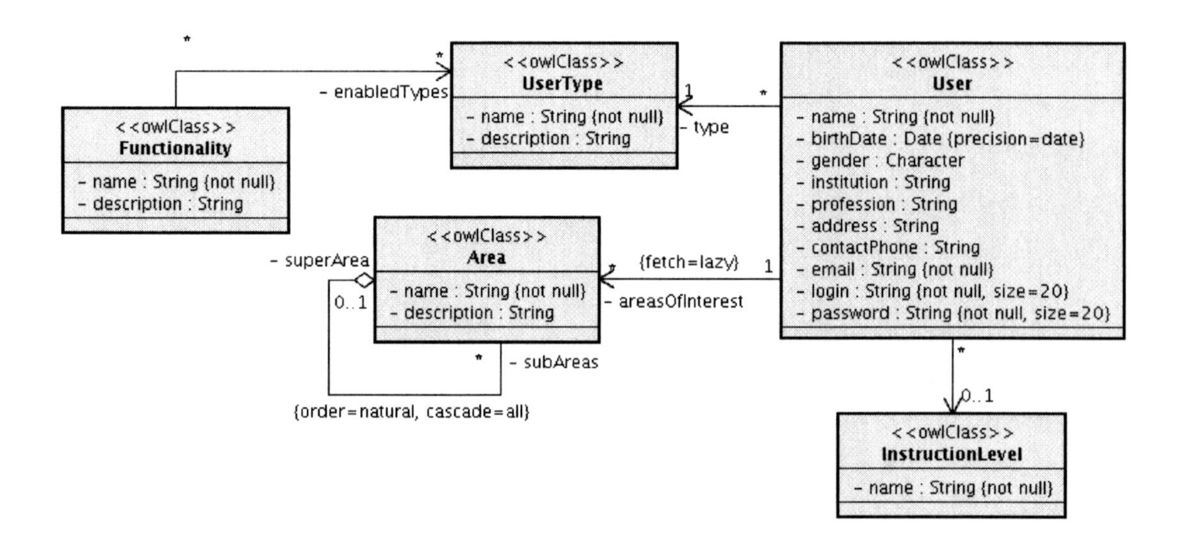

for the LabES Portal. We can see that, based on the simplified version of the conceptual model, association navigabilities were defined, data types were chosen among those of the implementation platform and that some O/R mappings were included. The result is very similar of that of Figure 11, due to the simplifications performed.

The representation of this model in a language that mixes profiles from both ODM and FrameWeb attempts to facilitate the implementation phase, when an OWL file representing the conceptual model should be created and the ORM framework should be configured.

Implementation, Testing and Deployment

During implementation, the classes that, integrated with the frameworks, provide a software solution to the problem at hand are developed. S-FrameWeb adds a new task to this activity: the construction of OWL files representing the domain ontology and the application conceptual model (based on the FDM). As stated before, this task is facilitated by the use of ODM in both models.

The OWL files should be used by the Front Controller framework to implement dynamic annotation on the Web pages. S-FrameWeb proposes an extension to this kind of framework that recognizes when a request comes from a human or from a software agent by analyzing a specific HTTP request parameter (e.g. `owl=true`). In the case of a software agent, the framework should respond with an OWL file that is based on the domain ontology and the conceptual model, and represents the data that would be shown in the human-readable version of the page.

To experiment this approach in practice, a prototype of an extension for the Struts[2] framework was built. Figure 21 shows this extension and how it integrates with the framework. The client's web browser issues a request for an action to the framework. Before the action gets executed, the controller automatically dispatches the request through a stack of interceptors, following the pipes and filters architectural style.

An "OWL Interceptor" was developed and placed as first of the stack. When the request is made, this interceptor verifies the HTTP parameter and, if present, creates a pre-result listener that will deviate successful requests to the "OWL

Result Class", another custom-made component that is responsible for producing this result.

The listing below is an excerpt of an OWL document produced by the search of publications with "FrameWeb" in their names. Publications that are returned by the applications are placed under the <results> tag, while objects associated with them are placed under <instancesList> tag. The association is made using the UUID of each object. See Exhibit A.

Since this result should be based on the application ontology, it was necessary to use an ontology parser. For this purpose, we chose the Jena Ontology API, a framework that provides a programmatic environment for many ontology languages, including OWL. With Jena and Java's reflection mechanisms, the OWL Result Class reads all properties that are made available to the Web page by the action, produces an OWL document containing their information and delivers it to the software agent.

Testing should be conducted in order to check not only the source code, but also the ontologies codified in OWL. In the context of S-FrameWeb, however, this is still open to research and study. Deployment works as the same as other WISs, but should also include the OWL files in a specific place in order to be used by the Front Controller's extension.

FUTURE TRENDS

Web Engineering is a relatively new field of research. New methods, languages and frameworks are proposed to provide practitioners with tools that can facilitate and increase the productivity when developing WebApps.

FrameWeb is a new tool, targeting WISs that have their architecture based on frameworks. By suggesting a standard architecture and bringing

Exhibit A.

```
<results>
        <instance>
                <uuid>2a6304f5-34c9-4356-a1ce-baa1e7b99e04</uuid>
                <areasOfInterest>130c2f70-5a37-4ad3-815b-841922584cd9</areasOfInterest>
                <areasOfInterest>93fcbf36-cdfe-4fd3-be23-a0d6ab3b45e8</areasOfInterest>
                <publishDate>2007-06-20</publishDate>
                <summary>An Application of the S-FrameWeb Method</summary>
                <participants>e5242491-b2be-4a34-b7b4-b0d9b7537517</participants>
        </instance>
</results>
<instancesList>
        <instance>
                <uuid>130c2f70-5a37-4ad3-815b-841922584cd9</uuid>
                <name>Semantic Web</name>
        </instance>
        <instance>
                <uuid>93fcbf36-cdfe-4fd3-be23-a0d6ab3b45e8</uuid>
                <name>WebApps</name>
        </instance>
        <instance>
                <uuid>e5242491-b2be-4a34-b7b4-b0d9b7537517</uuid>
                <name>Vítor Souza</name>
                <birthDate>1981-06-15</birthDate>
                <gender>M</gender>
                <profession>Professor</profession>
                <institution>UFES</institution>
                <type>2e9c5b6e-0d0f-4da0-b99b-24178ca6873a</type>
        </instance>
</instancesList>
```

concepts from the frameworks to the design models, developers can translate models to code more easily and designer have more control on the outcome of the implementation.

FrameWeb was first applied in the development of the Portal of the Software Engineering Lab – LabES. First, developers were trained in general concepts of Web Engineering, in the use of FrameWeb and also in the following frameworks: WebWork2, FreeMarker (template engine), SiteMesh, Hibernate and Spring.

In general, the development went smoothly. The method allowed the developers to deliver the models mostly in time and few deadlines had to be extended. However, some developers had difficulties on capturing the idea of some frameworks, especially the MVC framework. All of them had some experience with the Java platform, but most did not have any experience with Web development.

At the end of the development, the developers were asked to provide feedback on the work done. This feedback can be summarized in the following items:

- Allowing to directly model aspects related to the use of frameworks is the biggest strength of FrameWeb;
- Implementing in Java what was modeled during design was very much facilitated by the clear understanding of the semantics of the four models (domain, persistence, navigation and application);
- The simplicity of the models facilitated the adoption of FrameWeb, except for the navigation model, which added some complexity to the method.

Two other case studies were conducted. The local Java User Group ESJUG[22] modeled a collaborative learning environment called JSchool[23] using FrameWeb for the same set of frameworks used in the LabES Portal project. This helped mature the method in its initial version.

Another case study reimplemented the LabES Portal changing the Front Controller framework. This helped identify some extensions that should be added to FrameWeb in order to cope with some characteristics of different frameworks. For instance, this work suggested the addition of the `<<formBean>>` stereotype for the navigation model to represent how the framework Struts sends data from the web page to the action class. It also reached the conclusion that the navigation model in FrameWeb is somewhat dependent on the instance of Front Controller frameworks used, and not generic as it was assumed before.

More case studies should be conducted to assess the effectiveness of the method and its appropriateness to different instances of frameworks. Many improvements can come from more practical experiences.

The use of framework-based architectures is becoming the standard for implementation of medium-to-large-sized WIS. Taking the Java platform as example, the definition of standards as JavaServer Faces (JSF)[24] for Web development and the new Enterprise JavaBeans (version 3.0)[25] for distributed components reinforce that conclusion. JSF defines a MVC-like architecture, and EJB 3.0 had all of its persistence model reconstructed based on Hibernate ORM framework and also makes heavy use of Dependency Injection.

The research on the Semantic Web points out to the future of the World Wide Web. Methods for the development of WISs should prepare for, or even help build, this new paradigm. S-FrameWeb suggests a software process that facilitates the development of Semantic WISs by automating certain tasks concerning the generation of semantic annotations on dynamic Web pages. Nonetheless, FrameWeb and S-FrameWeb are far from ideal: there are several opportunities to improve the method. Future work may include:

- Further research on the impact of the use of frameworks and FrameWeb on the activity

of Testing. The current work provides no discussion on the subject of testing;
- Proposals on layout and interaction models. Complete methods for the design of WebApps should include models that model aesthetics and usability;
- Conduction of more formal experiments with the method, evaluating more precisely the gains in the productivity of the development team. Currently, only informal experiments have been conducted and conclusions have been reached by requesting developer's opinions;
- Tools could be developed to help create the models or to convert the models to code, automatically implementing much of the infrastructure code and configuration for the most used frameworks available;
- To make FrameWeb's models more generic, the development of an ontology on Web Applications and frameworks to guide the evolution of FrameWeb's modeling language. New concepts brought by new frameworks could be included in the ontology and, thus, taken to the modeling language;
- Continuation on the research on the Semantic Web and in-practice experiments on the construction of a Semantic WIS using S-FrameWeb;
- Deeper discussions on how to tackle specific Semantic Web issues such as: how will agents find the desired web page?, how will they know how to interact with it?, how will they know if a concept "table" refers to a piece of furniture or a systematic arrangement of data usually in rows and columns?, will a top-level ontology be used for all the Internet?
- Evaluation on how to use Semantic Web Services with S-FrameWeb instead of the dynamic page approach and a comparison of both solutions.

CONCLUSION

The amount of propositions in the Web Engineering area, including methods, frameworks and modeling languages, is quite vast, demonstrating that academics and practitioners haven't yet elected a standard when it comes to Web development.

Parallel to this, many frameworks and containers for the implementation of WISs were created, denoting the need for a basic infra-structure that helps on the quick development of reliable software with low future maintenance costs. With several ready-to-use and extensively tested components, frameworks promote reuse and good programming practices.

The large utilization of these frameworks and containers by practitioners and the absence of a design method directed to them has motivated the proposal of FrameWeb, a method based on frameworks for the design of WISs. The current research on the Semantic Web, with many efforts on bringing this idea to reality has impelled us to extend this method and create S-FrameWeb: a method based in frameworks to the construction of semantic WISs.

Given all of the options available, FrameWeb comes in as another one that targets a specific architecture, one based on the use of frameworks. In this case, FrameWeb excels for its agility, because models are directed towards the framework architectures and allow for quick understanding of the implementation. It also doesn't introduce much complexity, allowing organizations to use their own processes up to design with few adaptations, if any. Of all the proposed design models, the navigation model is the only one we consider a little bit complex, making FrameWeb very easy to learn and use.

S-FrameWeb complements FrameWeb, adding activities that promote the construction of Semantic WISs. Given that the Semantic Web vision will not come true unless Web authors add semantic to their websites, S-FrameWeb is a step in that direction, giving directives for WISs developers to follow in order to add Semantic to Web Applications.

REFERENCES

Ahmad, R., Li, Z., & Azam, F. (2005). Web Engineering: A New Emerging Discipline. In *Proceedings of the IEEE Symposium on Emerging Technologies* (pp.445-450). Catania, Italy: IEEE.

Alur, D., Malks, D., & Crupi, J. (2003). *Core J2EE Patterns: Best Practices and Design Strategies, 2nd edition*. Prentice Hall.

Ambler, S., & Jeffries, R. (2002). *Agile Modeling: Effective Practices for Extreme Programming and the Unified Process, 1st edition*. John Wiley & Sons.

Arch-Int, S., Batanov, D. N. (2003). Development of industrial information systems on the Web using business components. *Computers in Industry*, 50 (2), 231-250, Elsevier.

Bauer, C., & King, G. (2004). *Hibernate in Action, 1st edition*. Manning.

Berners-Lee, T., Hendler, J., & Lassila, O. (2001). The Semantic Web. *Scientific American*, 284, 34-43.

Booch, G., Rumbaugh, J., & Jacobson, I. (2005). *The Unified Modeling Language User Guide. 2nd edition*. Addison-Wesley Professional.

Ceri, S., Fraternali, P., & Bongio, A. (2000). Web Modeling Language (WebML): a modeling language for designing Web sites. *Computer Networks*, 33 (1-6), 137-157, Elsevier.

Conallen, J. (2002). *Building Web Applications with UML, 2nd edition*. Addison-Wesley.

Díaz, P., Montero, S., & Aedo, I. (2004). Modelling hypermedia and web applications: the Ariadne Development Method. *Information Systems*, 30 (8), 649-673, Elsevier.

Đurić, D. (2004). MDA-based Ontology Infrastructure. *Computer Science and Information Systems* 1 (1), ComSIS Consortium.

Falbo R. A., Guizzardi, G., & Duarte, K. C. (2002). An Ontological Approach to Domain Engineering. In *Proceedings of the 14th International Conference on Software Engineering and Knowledge Engineering* (pp. 351- 358). Ischia, Italy: Springer.

Falbo, R. A. (2004) Experiences in Using a Method for Building Domain Ontologies. In *Proceedings of the Sixteenth International Conference on Software Engineering and Knowledge Engineering* (pp. 474-477), Banff, Alberta, Canadá.

Fons, J., Valderas, P., Ruiz, M., Rojas, G., & Pastor, O. (2003). OOWS: A Method to Develop Web Applications from Web-Oriented Conceptual Models. In *Proceedings of the International Workshop on Web Oriented Software Technology* (pp. 65-70), Oviedo, Spain.

Fowler, M. (2002). *Patterns of Enterprise Application Architecture*. Addison-Wesley.

Fowler, M. (2007). *Inversion of Control Containers and the Dependency Injection pattern*. Retrieved on November 19, 2001, from http://www.martinfowler.com/articles/injection.html

Gamma, E., Helm, R., Johnson, R., & Vlissides, J. (1994). *Design Patterns: Elements of Reusable Object-Oriented Software*. Addison-Wesley.

Ginige, A., & Murugesan, S. (2001). Web Engineering: An Introduction. *IEEE Multimedia*, 8 (1), 14-18, IEEE.

Gomez-Perez, A., Corcho, O., & Fernandez-Lopez, M. (2005). *Ontological Engineering*. Springer.

Guarino, N. (1998). Formal Ontology and Information Systems. In *Proceedings of the 1st*

International Conference on Formal Ontologies in Information Systems (pp. 3-15), Trento, Italy: IOS Press.

Jacobson I., Booch G. & Rumbaugh J. (1999). *The Unified Software Development Process*, Addison Wesley.

Jacyntho, M. D., Schwabe, D., & Rossi, G. (2002). A Software Architecture for Structuring Complex Web Applications. In *Proceedings of the 11th International World Wide Web Conference, Web Engineering Alternate Track*, Honolulu, EUA: ACM Press.

Koch, N., Baumeister, H., Hennicker, R., & Mandel, L. (2000). Extending UML to Model Navigation and Presentation in Web Applications. In *Proceedings of Modelling Web Applications in the UML Workshop*, York, UK.

Koch, N., & Kraus, A. (2002). The Expressive Power of UML-based Web Engineering. In D. Schwabe, O. Pastor, G. Rossi e L. Olsina (Ed.), *Proceedings of the Second International Workshop on Web-Oriented Software Technology* (pp. 105-119), CYTED.

Krutchen, P. (2000). *The Rational Unified Process: An Introduction, 2nd edition*. Addison-Wesley.

Lee, S. C., & Shirani, A. I. (2004). *A component based methodology for Web application development. Journal of Systems and Software*, 71 (1-2), 177-187, Elsevier.

Lima, F., & Schwabe, D. (2003). Application Modeling for the Semantic Web. In Proceedings of the 1st Latin American Web Conference (pp. 93-102), Santiago, Chile: IEEE-CS Press.

McIlraith, S. A., Son, T. C., & Zeng, H. (2001). Semantic Web Services. *Intelligent Systems*, 16 (2), 46-53, IEEE.

Murugesan, S., Deshpande, Y., Hansen, S., & Ginige, A. (1999). Web Engineering: A New Discipline for Development of Web-based Systems. In *Proceedings of the 1st ICSE Workshop on Web Engineering* (pp. 3-13). Australia: Springer.

Narayanan, S., & McIlraith, S. A. (2002). Simulation, Verification and Automated Composition of Web Services. In *Proceedings of the 11th international conference on World Wide Web* (pp. 77-88), Hawaii, USA: ACM.

Neighbors, J. M. (1981). *Software Construction Using Components*. Ph.D. Thesis. Department of Information and Computer Science, University of California, Irvine.

OMG (2007). *Ontology Definition Metamodel Specification*. Retrieved on January 29, 2007, from http://www.omg.org/cgi-bin/doc?ad/06-05-01.pdf

Pastor O., Gómez, J., Insfrán, E., & Pelechano, V. (2001). The OO-Method Approach for Information Systems Modelling: From Objetct-Oriented Conceptual Modelling to Automated Programming. *Information Systems*, 26 (7), 507-534, Elsevier.

Pressman, R. S. (2005). *Software Engineering: A Practitioner's Approach, 6th edition*. McGraw Hill.

Reenskaug, T. (1979). *THING-MODEL-VIEW-EDITOR, an Example from a planning system*. Xerox PARC Technical Note.

Resende, A., & Silva, C. (2005). *Programação Orientada a Aspectos em Java*. Brasport.

Rosenberg, D., Scott, K. (1999). *Use Case Driven Object Modeling with UML : A Practical Approach*. Addison-Wesley.

Schmidt, D. (2007). *"Programming Principles in Java: Architectures and Interfaces", chapter 9*. Retrieved on January 29, 2007 from http://www.cis.ksu.edu/~schmidt/CIS200/

Schwabe, D., & Rossi, G. (1998). An Object Oriented Approach to Web-Based Application Design. *Theory and Practice of Object Systems* 4 (4), Wiley and Sons.

Shannon, B. (2003). *Java[TM] 2 Platform Enterprise Edition Specification, v1.4.* Sun Microsystems.

Souza, V. E. S., & Falbo, R. A. (2007). FrameWeb - A Framework-based Design Method for Web Engineering. In *Proceedings of the Euro American Conference on Telematics and Information Systems 2007* (pp. 17-24). Faro, Portugal: ACM Press.

Souza, V. E. S., Lourenço, T. W., Falbo, R. A., & Guizzardi, G. (2007). S-FrameWeb – a Framework-based Design Method for Web Engineering with Semantic Web Support. In *Proceedings of Workshops and Doctoral Consortium of the 19th International Conference on Advanced Information Systems Engineering* (pp. 767-778), Trondheim, Norway.

Standing, C. (2002). Methodologies for developing Web applications. *Information and Software Technology*, 44 (3), 151-159, Elsevier.

W3C (2007a). *OWL Web Ontology Language Guide, fev. 2004.* Retrieved on December 17, 2007, from http://www.w3.org/TR/owl-guide/

W3C (2007b). *W3C Glossary and Dictionary.* Retrieved on January 23, 2007, from http://www.w3.org/2003/glossary/

ENDNOTES

1. http://struts.apache.org/2.x/index.html
2. http://www.springframework.org
3. http://tapestry.apache.org
4. http://www.opensymphony.com/sitemesh
5. http://struts.apache.org/struts-tiles
6. http://www.hibernate.org
7. http://java.sun.com/products/jdo
8. http://db.apache.org/ojb/
9. http://www.oracle.com/technology/products/ias/toplink
10. http://www.picocontainer.org
11. http://jakarta.apache.org/hivemind
12. http://www.eclipse.org/aspectj
13. http://labs.jboss.com/portal/jbossaop
14. http://www.acegisecurity.org
15. http://cocoon.apache.org
16. http://java.sun.com/products/jaas
17. http://oiled.man.ac.uk/
18. http://protege.stanford.edu/
19. The relationship between an object's identity in memory and its primary key in the database raises several issues that are discussed in the article "Hibernate, null unsaved value and hashcode: A story of pain and suffering" from Jason Carreira (http://www.jroller.com/page/jcarreira?entry=hibernate_null_unsaved_value_and). The idea of using a Universal Unique Identifier (UUID) was taken from this article.
20. Object Management Group – http://www.omg.org/ontology/
21. The XML Schema standard can be found at http://www.w3.org/XML/Schema. Its data types are described in a specific page, at http://www.w3.org/TR/xmlschema-2.
22. http://esjug.dev.java.net
23. http://jschool.dev.java.net
24. http://jcp.org/en/jsr/detail?id=127
25. http://jcp.org/en/jsr/detail?id=220

Section IV
Selected Readings

Chapter XII
Business Process Simulation: An Alternative Modelling Technique for the Information System Development Process

Tony Elliman
Brunel University, UK

Tally Hatzakis
Brunel University, UK

Alan Serrano
Brunel University, UK

ABSTRACT

This paper discusses the idea that even though information systems development (ISD) approaches have long advocated the use of integrated organisational views, the modelling techniques used have not been adapted accordingly and remain focused on the automated information system (IS) solution. Existing research provides evidence that business process simulation (BPS) can be used at different points in the ISD process to provide better integrated organisational views that aid the design of appropriate IS solutions. Despite this fact, research in this area is not extensive; suggesting that the potential of using BPS for the ISD process is not yet well understood. The paper uses the findings from three different case studies to illustrate the ways BPS has been used at different points in the ISD process. It compares the results against IS modelling techniques, highlighting the advantages and disadvantages that BPS has over the latter. The research necessary to develop appropriate BPS tools and give guidance on their use in the ISD process is discussed.

INTRODUCTION

This article looks at Information Systems Development (ISD) and examines the potential role of simulation techniques within the Information System (IS) developer's toolkit. Since the inception of business data processing in the 1950s, ISD has remained a complex and unreliable process with the research repeatedly reporting high levels of failed projects (Standish Group, 1999).

Early approaches to discipline ISD focused on treating it as a production process and gave rise to the linear, or waterfall, Systems Development Life Cycle (SDLC). This was perceived to have three advantages: (1) it follows a series of specific and sequential phases from the beginning of the project until its end; (2) it advocates the use of techniques and tools to formulate step by step the detailed design and to implement the IS; and (3) it introduces the use of project management tools to control the overall process.

Despite the initial success of the linear SDLC, it did not deliver a dramatic reduction in the project failure rate, and a number of limitations was identified. For example, it is argued that instead of meeting organizational objectives, the traditional or linear SDLC aims to design an IS to help to solve low-level operational tasks (Avison & Fitzgerald, 2003). In addition, it is claimed that the traditional SDLC focuses on automating processes rather than proposing innovative integrated solutions (Rhodes, 1998). It is important to recognize that in parallel with the adoption of more rigorous ISD techniques, there also has been a progressive demand for IS to deal with more complex and wide-ranging business processes.

In trying to address some of these limitations, IS practitioners have proposed a wide range of alternative ISD approaches by emphasizing different aspects of the development process. For instance, some methodologies claim that organizational objectives can be met better by stressing the analysis of the organizational processes. Examples of these are structured analysis and design of IS (STRADIS), SSADM (OGC, 2000), and Yourdon Systems Method (YSM). Others, such as information engineering (IE), claim that organizational goals can be addressed better by placing more emphasis on the analysis of the data. Finally, there are approaches like Merise that consider both processes and data with equal importance (Vessey & Glass, 1998). Most of these approaches stress a scientific or functionalist approach by breaking up a complex system into its constituent parts. However, there are other approaches, like soft systems methodology (SSM) (Checkland & Scholes, 1999), that suggest that the properties of the whole system cannot be explained in terms of the properties of its constituent parts but can be understood better when looked at from a holistic perspective. A key issue is the dichotomy between methodologies, like SSM, that see the human actors and decision makers as part of the system and those that focus on the automated all programmed elements as the system. The former wider view introduces complex sociotechnological issues that are avoided in the latter narrower perspective.

Even though ISD approaches long have advocated the use of integrated organizational views, appropriate modeling techniques have not been adopted, and practice remains focused on the automated IS solution. For example, well-defined IS modeling techniques are available in order to understand the overall function of the system in question, to understand IS data structures, or to model the processes involved in the IS software (see Table 1). There is, however, very little indication of modeling techniques for examining organizational views that explicitly integrate automated software and human activities (Giaglis, Hlupic, Vreede, & Verbraeck, 2005).

In order to address this problem, it is proposed that Business Process Simulation (BPS) can be used at different points in the ISD process in order to better integrate the organizational views and

Table 1. Classification of modeling techniques adapted from Avison and Fitzgerald (2003)

Stage/Aspects addressed	Overall	Data	Process
Strategy	Rich Pictures		
Investigation &Analysis	Rich Pictures Objects Martices Strcuture diagrams Use Cases	Entity Modelling Class Diagrams	Data Flow Diagrams Entity Life Cycle Decision Trees Decision Tables Action Diagrams Root Definitions Conceptual Models (UML)
Logical design	Objects Matrices Structure diagrams	Normalisation Entity Modelling Class Diagrams	Decision Trees Decision Tables Action Diagrams
Implementation	Objects Matrices Structure diagrams	Normalisation	Decision Trees Decision Tables Action Diagrams

thereby to aid the design of appropriate IS solutions. To this end, the article is structured in the following way. In order to illustrate the advantages of using BPS for the ISD process, section 2 describes the underlying principles behind BPS. In order to provide a reference point for this critique, sections 3 through 6 describe the objectives pursued in the main phases of the linear SDLC. In addition, a critique of the modeling techniques used in these phases is provided together with a description of how BPS has been used in ISD projects to address some of the limitations found in the critique. The linear SDLC paradigm, as described by Avison and Fitzgerald (2003), was chosen as the reference point, because it can be seen as a generalization of the variety of IS methodologies available in the field. Arguably, iterative, star, and spiral SDLC models modify rather than escape from this basic linear model. Advocates of specific ISD approaches can refer back to the linear model, and the way BPS is usable in each phase of a linear SDLC can be related to the corresponding phases of particular ISD approaches. Section 7 is a discussion of the implications of this approach and the research needed to establish the use of BPS within the ISD toolkit. Finally, section 8 draws general conclusions from this article and points at future research in the area.

BUSINESS PROCESS SIMULATION

Business Process Simulation (BPS) can be defined as follows:

The process of designing a model of a real system and conducting experiments with this model for the purpose, either of understanding the behaviour of the system or of evaluating various strategies (within the limits imposed by a criterion or set of criteria) for the operation of the system (Shannon, 1975, p. 2).

Simulation can be used to understand the behavior of the existing business system in order to identify problematic tasks and to experiment with alternative scenarios (Hlupic & Robinson, 1998; Vreede, 1998). Business process practitioners long have recognized this advantage and have been using this technique in process innovation projects. In particular, BPS has been used to do the following:

- Evaluate process and information systems (Paul, Hlupic & Giaglis, 1998)
- Allow multidisciplinary teams to understand the system under investigation and to enforce communication among the stakeholders

(Vreede & Verbraeck, 1996; Paul et al., 1998)
- Understand, analyze, and improve business processes (Pegden, Shannon & Sadowski, 1995).
- Provide quantitative information related to the system performance; hence, to make better decisions (Pegden et al., 1995; Sierhuis, Clacey, Seah, Trimble, & Sims, 2003).
- Evaluate different system alternatives (Levas, Boyd, Jain, & Tulskie, 1995; Giaglis, 1999).

Subsequent sections of this article will use this information to show that BPS is a modeling technique that, in principle, can be used to model many of the aspects needed for different stages of the ISD process. In particular, this article concentrates on demonstrating that BPS can be used within the Feasibility Study, System Investigation, System Analysis, and System Design phases of the linear SDLC.

FEASIBILITY STUDY PHASE

The purpose of the feasibility study phase is simply to answer this question: Is this system worth building? A feasibility study will review analysis and design issues in sufficient detail in order to answer this question, but it will not go further. It is, therefore, a preview of the analysis and design process but conducted at low cost within a short timescale. As soon as the question can be answered, the project will go through a full management review, and the decision on whether to make the necessary investment will be made.

Because this phase focuses on capturing general aspects of the present system, the modeling techniques used in this phase are mainly holistic and process-oriented. Rich pictures, root definitions, conceptual models, and cognitive mapping are some of the techniques used in this phase to help to understand the problem situation being investigated by the analysts. Rich pictures are particularly useful as a way to understand the general problem situation at the beginning of the project. Root definitions help the analysts to identify the organizational context with which the system has to deal; in particular, human activity systems. Conceptual models show how the various activities in the human activity system relate to each other.

It can be argued that the aforementioned IS modeling techniques are capable of modeling the information required in this phase with few, if any, limitations. The following section, however, describes the way that BPS can be used to obtain other information that these traditional IS techniques cannot expose.

BPS for the Feasibility Study Phase

The various reasons for using simulation in process innovation projects and the information obtained from simulation models (see the second section) does not differ much from the information collected in a feasibility study. Thus, BPS is a technique that could be used to get most of the results needed for the feasibility study.

The major advantage that BPS has is related to its dynamic properties. Traditional techniques can be used in order to understand the problem situation, to identify the organizational context (people, resources, processes, etc.), and to extract system requirements. However, these models are static in the way that they represent a particular moment during the operation of the system. On the other hand, BPS can be used not only as a graphical representation of the system but also to simulate the operation of the system as it evolves over time (Paul et al., 1998). This feature allows practitioners to gain a better understanding of the behavior of the system, because the analyst can observe the way the system operates without the need to interrupt the organization's operations or the need to be in the organization's premises. The quantitative data provided by simulation runs, such as queuing times, processing time, resource utilization, and so forth, also can be used to complement the qualitative information derived from the graphical interface, providing more information to make better decisions (Pegden et al., 1995; Sierhuis et al., 2003). These metrics also can be used to evaluate different system alternatives (Levas et al., 1995; Giaglis, 1999).

Recent research provides evidence that BPS already has been used successfully in IS projects for similar purposes. For example, Eatock, Paul, and Serrano (2002) used BPS to assess the impact that the insertion of new IT may have on the organizational process. The authors argue that the performance measurements provided by the BPS model helped them to gain a better understanding of the current system. This, in turn, allowed IS practitioners to propose alternative IS solutions that better fit the identified problems. The proposed alternatives also were modeled using BPS so that performance measurements could be compared.

Similar to this research, Giaglis et al. (2004) used BPS to assess the expected operational benefits of Electronic Data Interchange (EDI) in the textile/clothing sector in Greece. The main purpose of the simulation exercise was to provide quantitative measures of the supposed ability of EDI to facilitate inventory reduction in the organizations that use this technology as part of their ordering and logistics processes. The study showed that the process of developing, validating, and using simulation models for the design of BP and IS was a very useful learning exercise for all participants in the study. It generated greater awareness of both the specifications of the proposed system and the conditions of the business operations under which the system can produce the desired results.

SYSTEMS INVESTIGATION PHASE

This phase is an extension of the work performed in the previous phase but in much more detail. This phase usually looks at the following:

- Functional requirements of the existing system and whether these requirements are being achieved
- The requirements of the new system
- Any constraints imposed
- Range of data types and volumes to be processed
- Exception conditions
- Problems of present working methods

The modeling techniques used in this phase are the same as those used in the feasibility phase; namely, rich pictures, root definitions, conceptual models, and cognitive mapping. The major difference is that the models developed in previous phases are elaborated in much more detail. Thus, there is the need to collect detailed information about the system. This phase, therefore, uses other techniques to gather information. Among the most popular ones are the five-fact finding techniques: observation, interviewing, questionnaires, searching records and documentation, and sampling.

The advantages and disadvantages of IS modeling techniques used in this phase were discussed in section 3. In relation to the five-fact finding techniques, the following disadvantages can be listed (Bennett, McRobb, & Farmer, 1999):

- Written documents do not match reality; for instance, company reports can be biased and out of date.
- Lack of access to required people.
- Interviews are time-consuming and can be the most costly form of data gathering.
- The interviewee may be trying to please by saying what he or she thinks the interviewer wants to hear.

- Most people do not like being observed and are likely to behave differently from the way in which they would normally behave.
- Questionnaires are easier to ignore and, hence, suffer from low response rates.
- Good questionnaires are difficult to design.

BPS for the Systems Investigation Phase

The main difference between the investigation phase and the feasibility phase is related to the depth in which the system is analyzed in the former. Thus, the uses of BPS illustrated in the feasibility phase also apply to the systems investigation phase, in which the distinction lies on the depth in which the models are constructed.

Apart from the advantages already described in section 3, Paul and Serrano (2004) provide evidence that BPS also can be used as a requirement-gathering technique. Paul and Serrano (2004) reported that the analysis of BPS models had helped IS analysts to identify IS requirements; in particular, non-functional requirements that were overlooked by traditional IS techniques. Based on the results derived from a case study, the authors reported that in order to reduce the time to complete an order (identified as a system requirement), the system depended on one particular factor: the number of backorders produced by the system. Hence, a non-functional requirement that previously was overlooked and that was derived from the analysis of the BPS model is related to the reduction of backorders. Moreover, the results provided by the simulation model suggested that in order to deliver orders within the period of time set by the organization (24 hours), the system should produce no more than 5% of backorders. This information was obtained because the BPS model produced performance measurements of the whole operational processes, including those supported by the proposed IS solution. In this way, analysts were able to identify system requirements that were related to performance and also to provide specific metrics for those requirements. Therefore, BPS can be used to complement the information derived from traditional gathering techniques.

BPS also can help to overcome some of the limitations found in the five fact-finding techniques. It is argued that a simulation exercise can engage staff in the process, because it presents a dynamic and visual impression of the system or process (Hlupic & Vreede, 2005). By engaging staff, problems related to unambiguous or biased information can be reduced.

SYSTEMS ANALYSIS PHASE

In this phase, the efforts concentrate on understanding the information gathered in the previous phase. It seeks to describe all aspects of the present system, the reasons it was developed as it was, and eventually, proposing alternative solutions for the creation of a new system. The analysis of the present system usually is done by asking the following questions:

- Why do the problems exist?
- Why were certain methods of work adopted?
- Are there alternative methods?

Apart from the modeling techniques used in previous stages, in this phase, analysts count on other modeling techniques in order to capture more specific information. For example, in order to model the data used, produced, and manipulated by the system, data techniques such as entity modeling and class diagrams are used. Similarly, process-oriented techniques, such as Data Flow Diagrams, Entity Life Cycle, Decision Trees, Decision Tables, and Action Diagrams, also are employed as basic techniques for functional decomposition. This is to break down the problem into more and more detail in a disciplined way.

Entity modeling and class diagrams are designed to identify specific issues related to the data that the system uses and manipulates, and they have been proved very reliable to achieve this aim. Thus, little criticism can be made in this respect. This is not the case, however, for process techniques.

Once again, the main disadvantage that traditional IS modeling techniques have is related to their static nature. The main questions posed in this phase, such as identifying the reasons why problems exist and if there are alternative methods of work, are very difficult to answer with static models (Pidd, 1998; Robinson, 1994). IS analysts rely much on their experience and expertise in order to answer such questions, since these techniques mainly are used to portray the analyst perspective, and they rarely provide more information to the analysts to make better decisions.

BPS for the Systems Analysis Phase

BPS has been proved to be an excellent tool for functional decomposition and systems analysis. It has been said that simulation models can be regarded as problem-understanding rather than problem-solving tools (Hlupic & Vreede, 2005). Therefore, BPS can be used to answer these questions: Why do the problems exist? Why were certain methods of work adopted?

A major difference between BPS and traditional IS techniques is that the former is capable of conducting what-if analysis, whereas the latter cannot. Once a BPS model is built and validated, changes to system variables and processes can be done in order to test alternative scenarios. According to Giaglis et al. (2004), there are two main sets of variables to be studied by decision makers: the configuration of the proposed information system (IS functionality) and the organizational arrangement regarding the structures and operations that surround it (business processes). By measuring the performance of the business processes with and without the use of IT, decision makers can collect the quantitative information needed to conduct further investment appraisal and IS design using established methods (Giaglis et al., 2004).

Paul and Serrano (2004) have used BPS to analyze five process solutions for the case study reported in their research. The experiments' results provided more information, such as performance measurements, that helped in the selection of the scenario that better matched the organizational needs. More importantly, prior to the experimentation with BPS models, the scenario that included the use of IT was thought to be the most appropriate one for the organization. The analysis of the simulation results indicated that the scenario that included the insertion of IT did not improve in a significant manner the overall system's performance. It was identified that one of the main problems with the system was due to the way processes were organized rather than the lack of adequate IT infrastructure to support them. Similar to this work, Giaglis et al. (2004) have used BPS to assess different solutions in an IS development project. The main objective of the simulation study was specified to provide a measure of the efficiency gains that could be achieved in inventory control within the textile/clothing value chain. The simulation exercise also aimed to explore the possible benefits of the insertion of EDI in inventory reduction. To this end, the authors developed two simulation models: one to portray the organization's operations as they are and one that included an Electronic Data Interchange (EDI) solution. The results provided evidence that indicated that all inventory levels were reduced after the introduction of EDI. Materials inventory, for example, was reported to be reduced by up to 46% and the product inventory by up to 27%.

SYSTEMS DESIGN PHASE

This stage involves the design of the system. To achieve this aim, analysts use the information gathered during previous phases in order to produce the documentation that portrays the functionality of the new system. Many parts of these documents can be seen in the form of models. Models used in previous phases can be used to derive more detailed models of the way the system will operate. For example, Use Cases is a modeling technique that can be used in the first stages of an ISD process in order to capture the functionality of the system. At the system design phase, the information depicted in Use Cases is used commonly to design collaboration, sequence, and activity or state diagrams. These models provide detailed information about how the system will function at particular points in time. For example, they can provide information about how the system will perform a specific transaction and how it will interact with the user in order to achieve this aim. Traditional IS techniques, however, cannot be used to assess how different workloads may affect the performance of such transactions. More importantly, they cannot be used to assess the impact that this new way of operating will have on the system as a whole.

BPS for the Systems Design Phase

It is argued that misinterpretation of user requirements is one of the main factors that contribute to IS failure (Vessey & Conger, 1994). Therefore, one of the challenges faced by analysts in this phase is to ensure that the functionality proposed for the new system matches user requirements in the best possible way. Because misinterpretation of user requirements may cause significant changes in the system's design, hence adding unexpected time delays and/or expenses, validation of requirements should be done prior the implementation phase. Validation of user requirements frequently is done iteratively throughout previous phases of the ISD process. The techniques used to validate requirements are usually those employed to capture user requirements. For example, Use Case is one modeling technique that commonly is used to capture user requirements. Once requirements are captured and translated into Use Cases, these models are taken to the users in order to validate that their requirements are well represented in such models.

Traditional IS techniques such as use cases, however, cannot provide information on whether the functionality proposed in such models will improve the performance of the system as a whole or will provide predictive metrics of such performance. Use case models cannot provide information related to what could be the benefits of implementing the functionality described, considering the organizational context. In other words, traditional IS techniques cannot be used to answer questions such as the following: What is the performance of the proposed IS functionality? or What impact will the proposed system have on other processes?

When asked to validate requirement models, users typically focus on items of detail rather than on the impact of the system on general working practices. The experience of using the system is not the same as reading about using it. Users will be able to perceive the impact of the system on their individual tasks but not know how these effects combine to change the behavior of the organization as a whole. Long-term systemic impacts often will remain hidden.

Researchers in this area argue that BPS can be used in this phase in order to verify that the functionality proposed for the new system matches global or systemic requirements. Paul and Serrano (2004) and Giaglis et al. (2004) have proposed alternative ways to use BPS in order to simulate the effects that a proposed IS functionality will have on business processes and vice versa. Paul and Serrano (2004) proposed a BPS modeling approach that uses the specifications derived from

IS models, such as use cases, collaboration, and activity diagrams to represent the IS functionality within a BPS model. In this way, analysts can obtain metrics of (1) the performance of the IS as it evolves over time (known as non-functional requirements) and (2) the impact that the functionality proposed by the IS would have on business processes. More importantly, Paul and Serrano (2004) report that the use of BPS models helped analysts to identify flaws in system design and, thus, redesign the proposed IS functionality. With the aid of the BPS model, the authors observed that the IS functionality proposed for their organizational case study would not improve the overall system's performance in a significant manner . They observed that in order to take full advantage of the proposed information system, changes to other processes also were required. This helped them to redesign the system's functionality so it better meets the organizational targets.

Prototyping is a method that long has been used by the IS community to ensure that the proposed IS functionality meets user requirements. Software engineers use the term *prototype* or *prototyping* to reflect a variety of different activities. In this article, we will concentrate on the conventional engineering sense of prototyping. This is the production of a partial system (i.e., interface, key algorithm, etc.) for the purpose of evaluating or selecting an element of the design. Such prototypes are not of adequate quality or sufficiently complete to be regarded as early deliverable versions of the system (Prototype, 2002).

A traditional prototyping process consists of designing and building a scaled-down usable model of the proposed system and then demonstrating the working model to the user with the purpose of obtaining feedback on its suitability and effectiveness. Developers then take the feedback and make corresponding changes in the design. This process is repeated until the users agree that the prototype is satisfactory (Arthur, 1992; Boar, 1984).

There are some cases, however, in which prototypes (all pilot systems) may not be appropriate. Organizational processes and their supporting information system(s) require input from users at different points in time. The time between these points may range from seconds, minutes, or hours to days, months, or even years, depending on the organizational processes. For example, an arbitration process can take more than one year to be completed, having several users input information at different times during this period. Similarly, insurance processes can take months to be completed. Prototype systems need to wait for the processes and related transactions to be completed in real time in order to obtain user feedback. Thus, when processes take long periods of time, prototyping methods cannot provide the desired results within acceptable limits.

Ongoing research in the School of Information Systems and Computing at Brunel University (Elliman & Eatock, 2004) claim that BPS can be used to validate user requirements in cases where long-term processes are involved. The authors propose a modeling approach that combines prototyping with simulation techniques, specifically with BPS. The approach is composed of two main models: a BPS model that simulates the organizational processes and an IS prototype that simulates the functionality of the proposed information system. The business process simulation will model the behavior of actors within or without the organization. It will generate work for the organization and play out the way that actors respond to information from the proposed new IS. Thus, the link between the two components in this prototyping experiment is as follows:

- Signals of events that are recorded by the information system
- Outputs from the information system that change the behavior of actors

Note that the level of implementation required is well below that of a completed information system. For example, the system has no user interfaces or data that affect the state of the information held in such a way so as to change the subsequent behavior of the actors. For example, it is not necessary to work out whether a particular arbitration case requires the use of an expert. In the simulation, one simply can assign a probability to this necessity and ensure that, at random, an appropriate number of cases is tagged as needing an expert witness. The IS implementation simply carries this tag rather than a full set of name, address, and so forth describing the witness. Upon interrogation, the IS can confirm the involvement of the expert and provide the tag value as a sufficient identifier.

Because this approach simulates the interactions among the system's components (i.e., actors, IS, processes), analysts were able to test the way the system would behave without the need to wait long periods of time. Processes that take long periods of time (e.g., months) now were simulated by the BPS model in minutes.

DISCUSSION

Previous sections provide evidence that Business Process Simulation (BPS) is a modeling technique that can be used effectively in different phases of the IS development process paradigm and, more importantly, can be used to overcome some of the limitations identified in traditional IS modeling techniques. To this end the discussion of BPS for the Feasibility, Systems Investigation and Systems Analysis phases of the SDLC (above) has shown that it can provide the information required. More importantly, it provides other information that traditional IS techniques cannot provide, such as performance metrics of the system as it evolves over time.

Although this suggestion of simulation as an ISD technique has a long history, it has not been developed as a routine tool in the analysts' armory. To achieve the potential value set out in this article, two areas of ongoing research are necessary. First, there is the need to develop business process simulation tools and techniques that can be applied rapidly. Second, there is the need to develop awareness and acceptance of the techniques.

The development of a model in the E-Arbitration-T project (Elliman & Eatock, 2004) involved significant technical effort that could have been reduced if appropriate tools were available off the shelf. This project suggested a need for three lines of tool development research. The most important part of a combined information system and business process model is a representation of the IS and its interface to the discrete event simulation of human activity. The point of the IS is to inform the human actors and enable them to change their behaviors appropriately. It is also necessary for the simulated actors to update the IS. Thus, the IS component is unlike any other element in the discrete event model. In order for the simulation to be constructed rapidly and effectively, this component needs to be easily configured and integrated within the model. As described previously, much of a conventional IS need not be constructed, because the simulation requires no Graphical User Interface (GUI) and no long-term data storage. The necessary component only needs to focus on data entity or class identity and some form of entity or system state model. The details of such an IS prototype, however, requires further research in order to create it as a generic component in BPS packages.

The second area of technical development is the need to provide other prebuilt business process elements. Almost all simulation packages provide prebuilt elements for modeling manufacturing systems (i.e., machine tools, stores, conveyors, transport devices). The availability of business elements is less frequent and more basic. Although packages may have elements like call centers, they do not deal with higher levels of knowledge worker behavior (Elliman & Hayman, 1999; Kidd, 1994). Research to formulate and develop these components is also necessary (Elliman, Eatock, & Spencer, 2005). The last area of technical development concerns the generation of work for the simulated business. The demands for information or knowledge services are much more variable than those experienced in general manufacturing. Thus, there is a need to enhance the case or work generation capabilities of most simulation packages in order for them to handle efficiently complex cases of generation. With the increasing use of mass customization and flexible manufacturing improved work generation, tools incidentally may have benefits for manufacturing systems simulation.

These three tool development areas are not independent, and research is needed not only to develop models for each of these tools but also to establish the relationships among them and the different ISD phases. Given the time and cost limits on a feasibility study, the time and cost of setting up current simulation packages could be inappropriate for most IS projects, at least for this phase. BPS practitioners argue, however, that it is possible to create broad-brush models with only limited detail but with enough information to determine whether the synergies exist to deliver the expected benefits or whether the reorganized system contains negative interactions that could undermine the anticipated benefits. Furthermore, the information captured from models developed during the first phase of the SDLC is used frequently to design models for subsequent phases. This suggests that IS analysts could use the simplified version of the BPS model designed in the first phase and gradually modify the level of detail according to the requirements needed for each phase.

In conventional engineering, simulation is an accepted and standard element of design practice. The use of models in wind tunnels or model ships in wave tanks are examples of tried and trusted simulation techniques. Engineers understand the limitations of these models and their relationship to the final product. Similarly, discrete event simulation of a physical production plant is an accepted methodology (Siemer, Taylor, & Elliman, 1995). In ISD, these relationships are less well defined and understood, and thus, there is a reluctance to accept simulation in this context. Further research is needed to refine the techniques and to present them to practitioners. These lines of research are intimately tied up with building a bridge between objective technology and subjective evaluations or perceptions. Developing appropriate guidelines for their use will be important.

The knowledge required to construct adequate BPS models for many elements of the Feasibility Study, System Investigation, and System Analysis phases is relatively simple. IS developers can refer to the simulation steps found in the literature such as those suggested by Banks, Carson, Nelson, & Nicol (2000) or Robinson (1994). However, in order to answer deeper questions about performance measurements of both BPS and IS functionality, particularly in the Design Phase (see section 6.2), developers need significant modifications to the way traditional BPS models are constructed, as described previously.

CONCLUSION

This article argues that BPS models are able to provide the same and more information than traditional IS modeling techniques; thus, they are suitable to address the modeling needs required at different points in the ISD process. Evidence to sustain this argument has been presented in the following ways:

1. BPS has been used successfully in the business process innovation domain to obtain information very similar to that required in different phases of the SDLC paradigm.
2. BPS models already have been used within the IS domain for similar purposes.

The main advantage of BPS over traditional BPS modeling techniques is its ability to simulate the dynamic behavior of the system as it evolves over time. In particular, it provides models that better integrate the dynamics of human activity and the automated IS. It has been discussed that the quantitative metrics provided by BPS models can be used by IS analysts to accomplish the following:

- Better understand the operation of the current system
- Identify possible system bottlenecks
- Evaluate different system alternatives
- Obtain performance measurements of the system's behavior for both processes and IS

To justify these arguments, three case studies that employ BPS in IS projects are used: Paul and Serrano (2004), Giaglis et al. (2004), and Elliman and Eatock (2004).

The evidence presented in this article strongly suggests that BPS models are able to provide more information than traditional IS techniques and that this information can be very useful to design better IS solutions. Thus, the authors of this article advocate the idea that practitioners in this domain routinely should consider the use of BPS as an alternative tool to support different stages of the ISD process. Moreover, the seventh section argues that BPS models can be used to simulate proposed IS functionality and the effect that it may have on the organization as a whole. The development of such models, however, is more complicated than the way traditional BPS models are designed. Thus, further research in this area is needed to improve the BPS toolkit and to demonstrate its effectiveness in various ISD scenarios.

REFERENCES

Arthur, L.J. (1992). *Rapid evolutionary development: Requirements; prototyping and software creation.* Chichester, UK: John Wiley & Sons.

Avison, D.E., & Fitzgerald, G. (2003). *Information systems development: Methodologies, techniques and tools.* London: McGraw-Hill.

Banks, J., Carson, J.S., Nelson, B.L., & Nicol, D.M. (2000). *Discrete-event system simulation.* Upper Saddle River, NJ: Prentice-Hall.

Bennett, S., McRobb, S., & Farmer, R. (1999). *Object-oriented systems analysis and design using UML.* London: McGraw-Hill.

Boar, B.H. (1984). *Applications prototyping: A requirements definition strategy for the 80s.* Chichester, UK: Wiley.

Checkland, P., & Scholes, J. (1999). *Soft systems methodology in action.* Chichester, UK: John Wiley & Sons.

Eatock, J., Paul, R.J., & Serrano, A. (2002). Developing a theory to explain the insights gained concerning information systems and business processes behaviour: The ASSESS-IT project. *Information Systems Frontiers, 4*(3), 303-316.

Elliman, T., & Eatock, J. (2004). Online support for arbitration: Designing software for a flexible business process. *International Journal of Information Technology and Management, 4*(4), 443-460.

Elliman, T., Eatock, J., & Spencer, N. (2005). Modelling knowledge worker behaviour in business process studies. *Journal of Enterprise Information Management, 18*(1), 79-94.

Elliman, T., & Hayman, A. (1999). A comment on Kidd's characterisation of knowledge workers. *Cognition, Technology and Work, 1*(3), 162-168.

Giaglis, G.M. (1999). *Dynamic process modelling for business engineering and information systems* (Doctoral Thesis). London: Brunel University.

Giaglis, G.M., Hlupic, V., Vreede, G.J., & Verbraeck, A. (2005). Synchronous design of business processes and information systems using dynamic process modelling. *Business Process Management Journal, 11*(5), 488-500.

Hlupic, V., & Robinson, S. (1998). Business process modelling and analysis using discrete-event simulation. In *Proceedings of the 1998 Winter Simulation Conference,* Washington, DC (pp. 1363-1369).

Hlupic, V., & Vreede, G.J. (2005). Business process modelling using discrete-event simulation: Current opportunities and future challenges. *International Journal of Simulation & Process Modelling, 1*(1-2), 72-81.

Kidd, A. (1994). The marks are on the knowledge worker. In *Proceedings of the Conference of Human Factors in Computer Systems*, Boston, MA (pp. 186-191).

Levas, A., Boyd, S., Jain, P., & Tulskie, W.A. (1995). The role of modelling and simulation in business process reengineering. In *Proceedings of the Winter Simulation Conferenc*e (pp. 1341-1346).

OGC. (2000). *SSADM Foundation (Office of Government Commerce).* London: The Stationery Office Books.

Paul, R.J., Hlupic, V., & Giaglis, G. (1998). Simulation modeling of business processes. In *Proceedings of the 3rd U.K. Academy of Information Systems Conference,* Lincoln, UK.

Paul, R.J., & Serrano, A. (2004). Collaborative information systems and business process design using simulation. In *Proceedings of the 37th Hawaii International Conference on Systems Sciences,* Big Island, HI, 9.

Pegden, C.D., Shannon, R.E., & Sadowski, R.P. (1995). *Introduction to simulation using SIMAN.* London: McGraw-Hill.

Pidd, M. (1998). *Computer simulation in management science.* Chichester, UK: John Wiley.

Prototype. (2002). *A dictionary of business.* Oxford Reference Online. Retrieved February 27, 2005, from http://www.oxfordreference.com/views/ENTRY.html?subview=Main&entry=t18.e4769

Rhodes, D. (1998). Integration challenge for medium and small companies. In *Proceedings of the 23rd Annual Conference of British Production and Inventory Control Society,* Birmingham, UK (pp. 153-166).

Robinson, S. (1994). *Successful simulation: A practical approach to simulation projects.* Maidenhead, UK: McGraw-Hill.

Shannon, R.E. (1975). *Systems simulation: The art and the science.* Englewood Cliffs, NJ: Prentice Hall.

Siemer, J., Taylor, S.J.E., & Elliman, A.D. (1995). Intelligent tutoring systems for simulation modelling in the manufacturing industry. *International Journal of Manufacturing System Design, 2*(3), 165-175.

Sierhuis, M., Clacey, W.J., Seah, C., Trimble, J.P., & Sims, M.H. (2003). Modeling and simulation for mission operations work system design. *Journal of Management Information Systems, 19*(4), 85-128.

The Standish Group International (1999). *CHAOS: A recipe for success.* West Yarmouth, MA: The Standish Group International.

Vessey, I., & Conger, S.A. (1994). Requirements specification: Learning object, process, and data methodologies. *Communications of the ACM, 37*(5), 102-113.

Vessey, I., & Glass, R. (1998). Strong vs. weak: Approaches to systems development. *Communications of the ACM, 41*(4), 99-102.

Vreede, G.J. (1998). Collaborative business engineering with animated electronic meetings. *Journal of Management Information Systems, 14*(3), 141-164.

Vreede, G.J., & Verbraeck, A. (1996). Animating organizational processes: Insight eases change. *Journal of Simulation Practice and Theory, 4*(3), 245-263.

This work was previously published in Int. Journal of Enterprise Information Systems, Vol. 2, Issue 3, edited by A. Gunasekaran, pp. 43-58, copyright 2006 by IGI Publishing (an imprint of IGI Global).

Chapter XIII
An Agent Based Formal Approach for Modeling and Verifying Integrated Intelligent Information Systems

Leandro Dias da Silva
Federal University of Campina Grande - UFCG, Brazil

Elthon Allex da Silva Oliveira
Federal University of Campina Grande - UFCG, Brazil

Hyggo Almeida
Federal University of Campina Grande - UFCG, Brazil

Angelo Perkusich
Federal University of Campina Grande - UFCG, Brazil

ABSTRACT

In this chapter a formal agent based approach for the modeling and verification of intelligent information systems using Coloured Petri Nets is presented. The use of a formal method allows analysis techniques such as automatic simulation and verification, increasing the confidence on the system behavior. The agent based modelling allows separating distribution, integration and intelligent features of the system, improving model reuse, flexibility and maintenance. As a case study an intelligent information control system for parking meters price is presented.

INTRODUCTION

Intelligent information systems (IIS) have been used in several different domains such as communication, sensor networks, decision-making processes, traffic control, business, and manufacturing systems, among others. Most IIS are distributed because the data is collected, processed, and used at different locations by different entities. These entities are autonomous and communicate among them to perform specific tasks. This scenario of an integrated IIS (IIIS) gives rise to several problems and difficulties to be addressed by the development team. Communication, scheduling, synchronization, databases, workflow, and real-time systems are some of the fields that usually have to be addressed in the development of an IIIS.

The agent-based development has been successfully applied to develop information systems with the characteristics mentioned (Jennings, 2001). Agents are intelligent entities, which are distributed and integrated to other agents. Therefore, developing information systems using the agent abstraction is quite convenient.

On the other hand, there are some features related to IIIS which are not explicitly addressed by the agent approach, such as dependability. Within this context, formal methods have been successfully used to promote confidence on the system behavior. More specifically, the coloured Petri nets (CPN) formal method (Jensen 1992, 1997) is pointed out as suitable for distributed and concurrent systems, which are inherent features of IIIS.

This chapter presents an agent-based approach for formal specification and verification of IIIS using coloured Petri nets. A generic agent-based skeleton CPN model comprising integration and distribution functionalities has been defined. Using this model, it is possible to model and verify agent-based IIIS by only modeling the intelligent functionalities. Distribution, integration and intelligent features are well encapsulated in CPN sub models. Thus, the proposed approach simplifies the IIIS modeling activity and improves its flexibility and maintenance. In order to illustrate the usage of the proposed approach, the modeling and verification of a control parking system is presented.

The remainder of the chapter is organized as follows. In the Background, some related works are discussed and background concepts are presented informally. After, the agent-based formal approach for modeling and verifying IIIS is presented. Next, the modeling and verification of a parking meter control system is presented and some verification results are discussed. In Future Trends, there are some insights and suggestions for future work. In the last section, the chapter is concluded and summarized.

BACKGROUND

Coloured Petri Nets

Coloured Petri nets (CPN) (Jensen 1992, 1997) are a formal method with a mathematical base and a graphical notation for the specification and analysis of systems with characteristics such as concurrent, parallel, distributed, asynchronous, timed, among others. For the mathematical definition, the reader can refer to Jensen (1992, 1997). The graphical notation is a bipartite graph with places, represented as ellipses, and transitions, represented as rectangles. Transitions represent actions and the marking of the places represent the state of the model. A marking of a place at a given moment are the tokens present at that place. A token can be a complex data type in CPN/ML Language (Christensen & Haagh, 1996). Each place has an associated color set that represents the kind of tokens the place can have. The transitions can have guards and code associated to it. Guards are Boolean expression that must be true for the transition to fire. Code can be a function that is executed every time the transition fires. Arcs go from places to transitions and from transitions to places and never from transition to transition or place to place, and can have complicated expressions and function calls associated to it.

For a transition to fire, it is necessary that all input places, that is, places that have arcs that go from the place to the transition, have the number of tokens greater than or equal to the weight of the arc, *w(p,t)*, and the guard of the transition must be true. When these characteristics hold the transition is said to be enabled to fire. An enabled transition can fire at any time and not necessarily immediately. Once a transition fires, it removes *w(pi,t)* from each input place pi, and the output places, that is, places that have arcs from the transition to the place, receive tokens according to the arc expression from the transition to the place: *w(t,p)*.

A CPN model can also have a hierarchy. Two mechanisms—substitution transitions and fusion places — can be used to structure the model in a more organized way. Substitution transitions are transitions that represent another CPN model, called a page. The page where the transition belongs to is called super-page and the page represented by the transition is called sub-page. In the sub-page there are places that can be input, output, or input/output places. These places are associated to input and output places of the transition in the super-page called sockets. Fusion places are places that are physically different but have always the same marking. A change in the marking of one place is reflected in all places that belong to that fusion set. These two mechanisms are just visual, and for simulation and model checking purpose the model is glued together and the hierarchy is considered a flat CPN model. This extension with hierarchy is called hierarchical coloured Petri nets (Jensen 1992, 1997), and there is a tool set to edit and analyze such models called Design/CPN (Jensen et al., 1999).

In Figure 1, a simple HCPN model is illustrated. In this figure, place *P3* is the input socket and place *P4* is the output socket of the substitution transition *T2* in the super-page. The place *P5* in the sub-page is the input place associated to *P3* in the super-page, and the place *P6* is the output place associated to place *P4*, respectively. The inscriptions *Int* closer to each place is the color set of the place and states that the places can have integer tokens. The expression *1`(2)* close to *P2* is the initial marking of this place. That means that in the initial state this place have a marking consisting of one token which value is *2*.

The expression *[i>0]* is the guard for transition *T1*. It is easy to see that at the initial marking *T1* is enabled to fire because there are tokens in the two input places that satisfy the input weight functions *w(P1,T1) = 1* and *w(P2,T2) = 1*, and the guard *[i>0]* as *i=3* in the initial marking. Once this transition fires one token is removed from *P1* and *P2* and one token is added to *P3* which value is the sum between the two input token *i+j*, where *i* and *j* are declared as integers in the declaration node. This node is where all the declarations of colors (the data types), variables, and functions are written.

The sub-page can be an arbitrarily complicated CPN model with its own substitution transitions. Moreover the tokens can have complex data types such as tuples, records, lists, and user-defined types using primitive types.

Model Checking

With simulation, it is possible to investigate the behavior of the model. However, based on this analysis, it is not possible to guarantee that the properties are valid for all possible behavior of the model. Therefore, model checking (Clarke, Grumberg, & Peled, 1999) is used to prove that the desired properties of the model hold for every possible behavior.

The model checking activity consists of three steps. First, a model *M* must be developed. Second, the properties *S* to be verified must be specified in

Figure 1. HCPN model

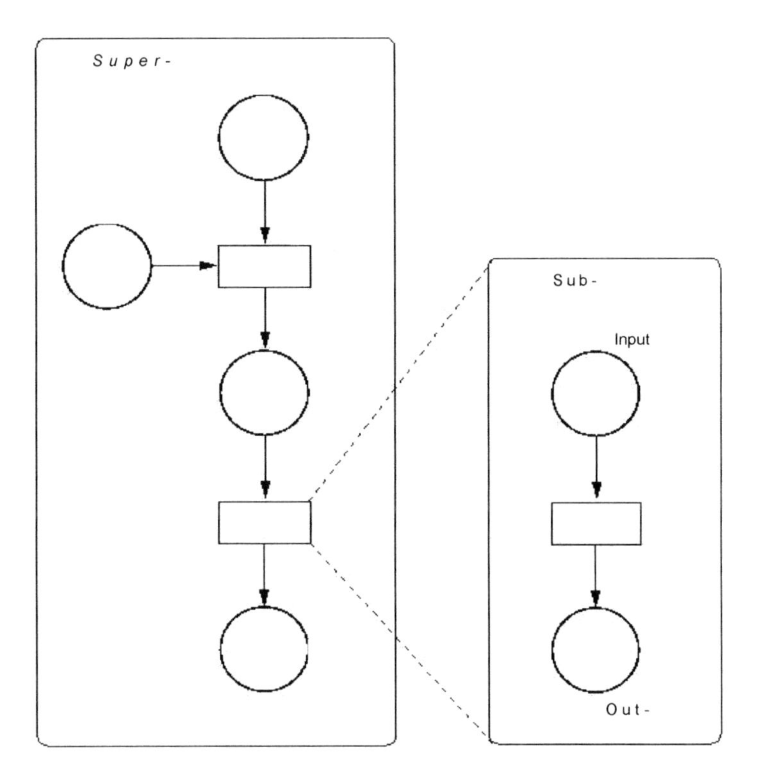

some logic. Third, model checking is performed. Therefore, model checking is used to verify if a given model models the specification: $M \models S$.

To verify the desired properties, through model checking, it is used the Design/CPN tool package is used to generate the state space and its temporal language library named ASK/CTL (Christensen & Mortensen, 1996) is used to perform model checking.

Based on the example shown in Figure 1 and a simplified ASK/CTL syntax, some examples of properties specifications are as follows:

Pa = Marking.P1 1`(_)
Pb = Marking.P6 1`(_)
S = AND (Pa, EV(Pb))

where *Pa* and *Pb* are atomic propositions and *S* is the property specification. *Pa* means that there is a token with any value in places *P1* and *P6*. *S* specifies that *Pa* is true and *Pb* is eventually true. This formula evaluated to true if *Pa* is true and at some point in the future *Pb* becomes true. Otherwise it is false. This example specifies that every token in the initial place always generate a token in the final place of the model. Moreover, it is important to note that this model can be a complicated model and to verify this kind of property can be useful.

Related Work

There are some works related to our approach. In Adam, Atluri, and Huang (1998), for example, it is presented a Petri net-based formal framework for modeling and analyzing workflows. The modeling and analysis activities are used to identify inconsistent dependency specifications among workflow tasks; to test if the workflow terminates in an acceptable state; and to test if it is feasible to execute a workflow with specified temporal constraints, for a given time.

In DEEN (2005), it is presented a multi-agent model of a distributed information system. The objective is to define the agent-based behavior of the distributed system, which must operate correctly and effectively in an error-prone industrial environment, based on a problem-solving model that is also proposed in DEEN (2005).

Another work, presented in Delena and Pratt (2005), focuses on to promote advanced model-based decision support systems by addressing problems such as lack of model reusability. The main motivation for such work is that the existing solutions for decision support systems are only suitable for specific domains and tools. Thus, the idea is to provide a general methodology for decision support, according to the real needs of the decision-makers, in a feasible manner.

Colored Petri nets have also been used to model some previous works such as Almeida, Silva, Perkusich, and Costa (2005) and Almeida, Silva, Silva Oliveira, and Perkusich (2005). In Almeida et al. (2005), a set of guidelines are presented for the modeling and verification of multi-agent systems, focusing on planning activities. In Almeida et al. (2005), a modeling and verification process for component-based systems is presented, with application to embedded control and automation systems.

In this chapter, a formal approach based on Petri nets and multi-agents systems is considered for the specification and analysis of IIIS. This approach is described in the next section.

AGENT-BASED MODELING AND VERIFICATION APPROACH

The Agent Architecture

Many researchers have claimed that the agent-based approach is adequate for complex, distributed, and interaction-centric software development (Jennings, 2001). According to this approach, a system is composed of a society of distributed

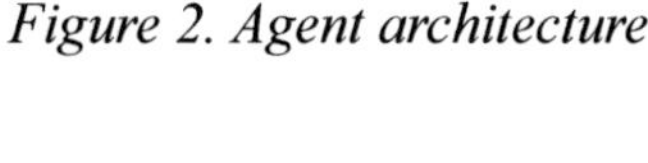

Figure 2. Agent architecture

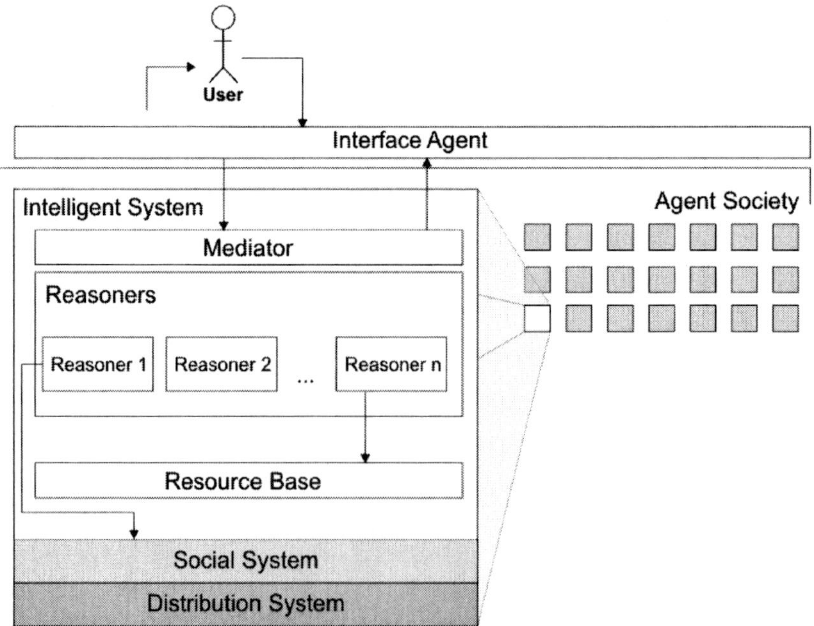

and autonomous agents that interact to reach their goals and provide the system functionalities. Also, agents are intelligent entities. Therefore, developing IIIS using the agent abstraction is quite convenient.

There are several architectures for agent-based systems. In the context of this work, we consider an adaptation of the MATHEMA architecture (Costa, Lopes, &Ferneda, 1995). It has been applied to various domains, such as algebra (Costa et al., 1995), musical harmony (Costa et al., 2002), and Petri nets (Costa, Perkusich, & Figueiredo, 1996), and promotes a clear encapsulation of the integration, distribution, and intelligent system functionalities.

Each agent in our architecture is composed of three systems: intelligent, social, and distribution. The intelligent system implements the functionalities of the system domain, according to the specific IIIS needs: decision support, business

Figure 3. Hierarchy page for the generic CPN model

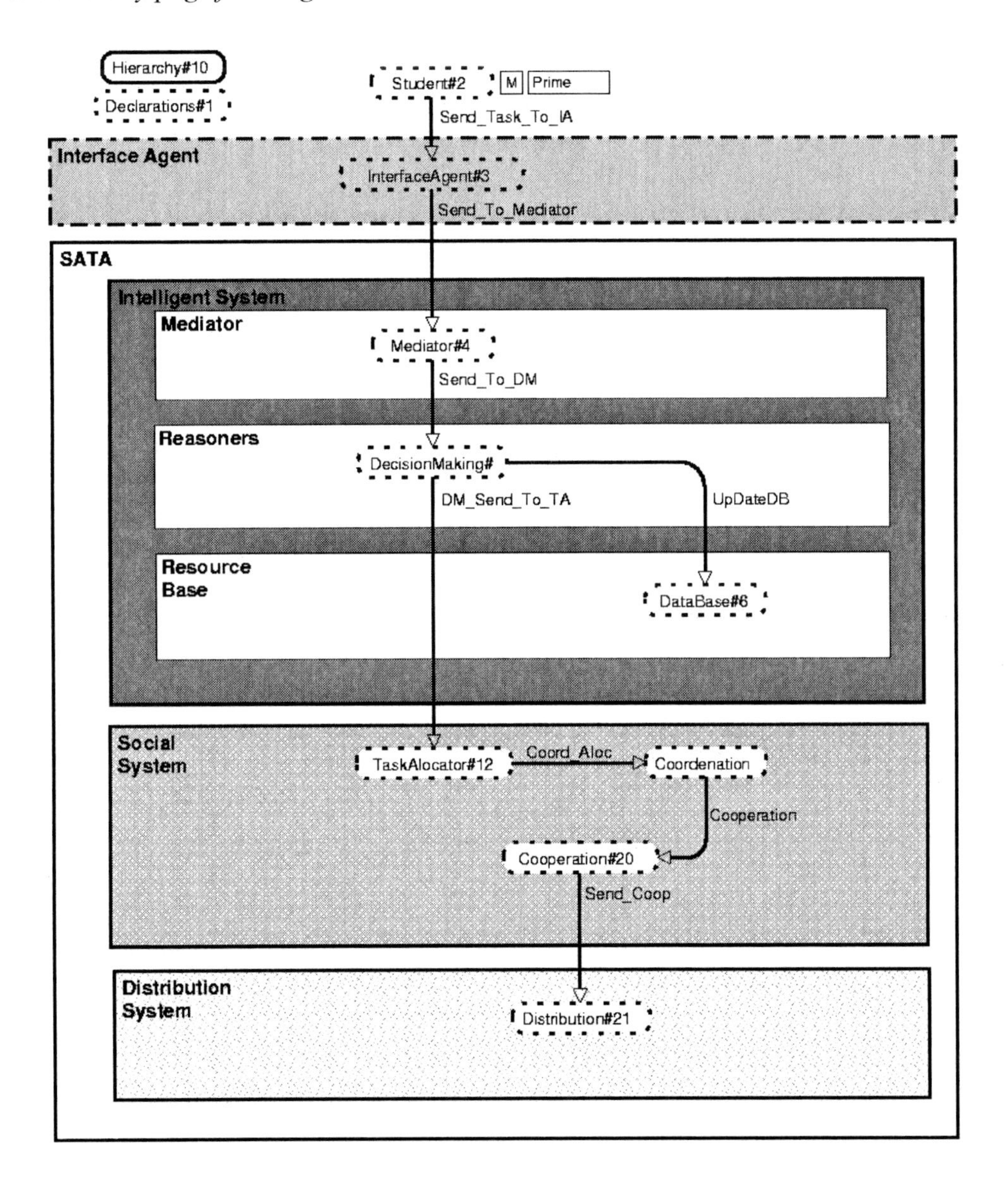

intelligence, problem solving, and so forth. The social system implements the functionalities for agent interaction and integration. The distribution system implements the functionalities to promote the communication among the agents.

Considering the integration and distribution features are inherent of all IIIS, the intelligent system can be seen as the *customizable part* of the agent architecture. Thus, a generic architecture must define customization points where domain-specific IIIS functionalities could be placed to be integrated to the rest of the architecture. For that, the intelligent system is divided into the following entities, which are also illustrated in Figure 2.

- **Mediator:** It implements the interaction mechanisms with the interface agent and thus with the user. It also selects the suitable reasoner in order to respond to the user requests. The interface agent is just a simple agent responsible to interact with the user.
- **Reasoners:** They are the components that implement the domain-specific functionalities of the intelligent tutoring system. As the reasoners could be different for each application, these components are the customizable points of the architecture. For example, if an IIIS requires a decision-making support, a reasoner that implements such support must be integrated to the architecture.
- **Resource base:** It is responsible to store information.

Modeling and Verification Approach

Based on the agent-based generic architecture, we have modeled a generic CPN model for IIIS. The modeling activity has been performed using the design/CPN tool. The hierarchy page for the generic CPN model for IIIS is shown in Figure 3.

As mentioned before, the customizable entities are the reasoners. In Figure 3, the unique reasoner model is in the *decision-making* page. By changing and inserting new reasoners, it is possible to model the intelligent features of a specific IIIS. Since the distribution and integration features are common to all IIIS, we focus here in how to customize the model for the intelligent feature. The detailed models for all pages described in the hierarchy page can be found in Silva, Almeida, and Perkusich (2003).

The reasoners are modeled as CPN *substitution transitions*. In this way, according to the IIIS specific needs, new pages can be modeled and integrated to the model. Also, this makes possible to maintain the separation between intelligent, integration, and distribution features of the IIIS model.

In order to model and verify an IIIS using our approach, the following steps must be accomplished.

1. **Identify the agents:** The agents must be identified according to the IIIS requirements. It can be performed through defined roles involved with the IIIS specific domain and then define an agent or a set of agents to play each role. There are several methodologies for identifying agents based on system requirements, such as Tropos (Giunchiglia, Mylopoulos, & Perini, 2002).
2. **Identify agent reasoners:** When agents are identified, the reasoners can be defined by analyzing the intelligent requirements for each agent or agent role.
3. **Model agent reasoners:** Build a CPN model for each reasoner. Since the distribution and integration features are already modeled, this is the main activity of the IIIS modeling.
4. **Integrate the reasoners models to the generic CPN model:** Integrate the reasoners models to the CPN generic model by defining substitution transitions for each reasoner and linking the mediator, resource base, and social system according to the reasoners needs.

5. **Analyze the whole model:** Use the design/CPN tool to analyze the model and identify potential design errors.

CASE STUDY: CONTROL PARKING SYSTEM

Parking has becoming a serious problem even for small cities. An initial solution is to charge for parking. When it is not a solution anymore, increasing the parking charges and constructing parking structures near busy places are possible solutions.

However, parking structures are sometimes not good because they are expensive and change the view and architecture of the city. Just keep increasing the price is also not a permanent solution because employees and other people cannot pay more money for parking.

An efficient solution is to change the price charged by the parking meters depending on the number of available spaces (Smith, 2005). In this case, 15% is considered the ideal number for specialists. Therefore, every time the number of spaces available goes above, the price can be decreased, and when it goes below 15%, the price can be increased.

The system for parking prices control is composed of sensors, parking meters, and a control system. In Figure 4, the parking scenario with sensors and meters is illustrated.

The sensors send the information regarding its space to the control system that keep track of all information sent by sensors and can decide which parking meter should increase or decrease the price shown.

MODEL

The problem with parking is that people have encountered difficulty in finding available spaces to park their vehicles. As a solution, all the places where people usually park are marked as a space and they are grouped in some parking zones.

A sensor is put in each one of these spaces with the purpose of monitoring its status. It monitors if the space is available or not. A parking meter is used to show the price to be paid. The idea of this solution is that for each parking zone, or group of parking spaces, the sensors send data to

Figure 4. Parking meters

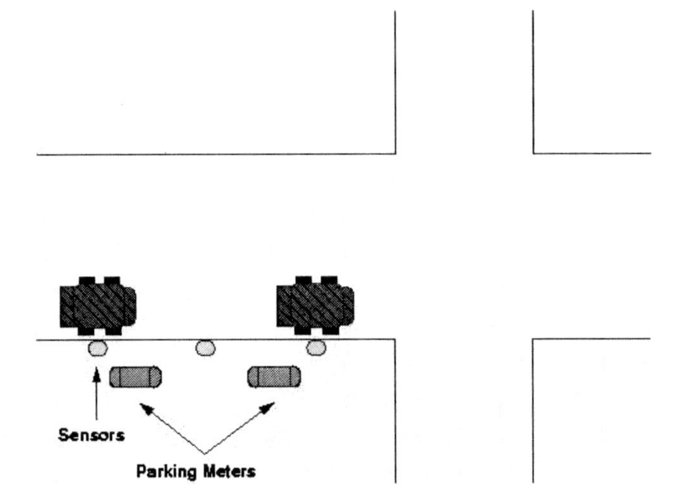

a central computer periodically informing if their respective space are available or not.

The central computer increases or decreases the price of parking according to the ratio of available spaces in each parking zone. In this example, if there are just 15% of available spaces to park in a specific parking zone, the price of vacancies is increased for that parking zone. Otherwise, if there are more than 15% of available spaces, the price is reduced. The act of increasing or reducing the price is done every time the central computer receives data from the sensors. So, the price will be constantly changed depending on the search for spaces to park.

Using the approach described, the steps to develop and analyze this application are:

1. **Identify the agents:** sensors, parking meters, and the decision agent
2. **Identify agent reasoners:** the decision agent
3. **Model agent reasoners:** This model can be seen in Figure 8.
4. **Integrate the reasoners models to the generic CPN model:** This integration is done by the main page as seen in Figure 5, using fusion places, and this whole model is integrated in the model shown in Figure 5 with substitution transitions at the reasoner part.
5. **Analyze the whole model:** This is discussed in the next section.

Figure 5. Main page

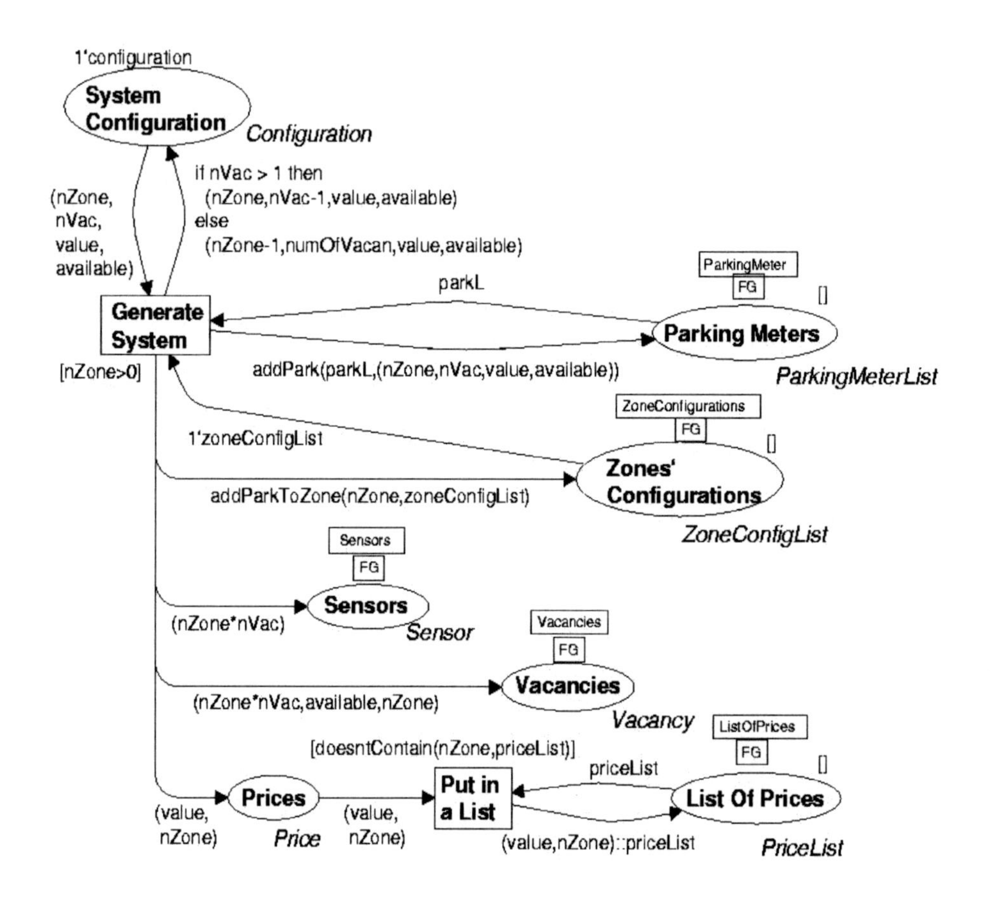

In Figure 5, the page where sensor agents, parking zones, and their vacancies and configurations, the parking meters agents and a list of prices for each parking zone are created is presented.

In the system configuration place, there is a token that has four fields: nZone, nVac, value, and available. The first field determines the number of parking zones to be created. The second means the number of vacancies for each parking zone. The third and fourth ones mean the initial price paid for using the vacancy and the status of the vacancy, respectively.

The *generate system* transition creates *nVac* vacancies for each one of the *nZone* parking zones. The guard *[nZone>0]* guarantees that the process of creation will stop for *nZone* less than or equal to zero.

In Figure 6, the page that describes the sensor behavior is presented. The part of the page surrounded by a dashed box generates a random value that models the status of the vacancy. The value *true* means it is available, *false* means it is not available. The *verify vacancy* transition verifies if the vacancy that corresponds to a sensor, by verifying their identifier values, is available or not. This verified status is sent to the *decision agent* page (central computer) that determines the new price according to the number of available vacancies.

In Figure 7, the page that describes the *decision agent* behavior is presented. It is this agent that controls when the price of a parking zone must be increased or reduced. On the *zones' configuration* place there is a token that has a list of the configuration of parking zones. This configuration has three fields: the parking zone identifier, the number of available vacancies, and the number of unavailable vacancies inside that parking zone. The unique transition of the page, named *update prices*, and the arc expression

Figure 6. Sensor agent page

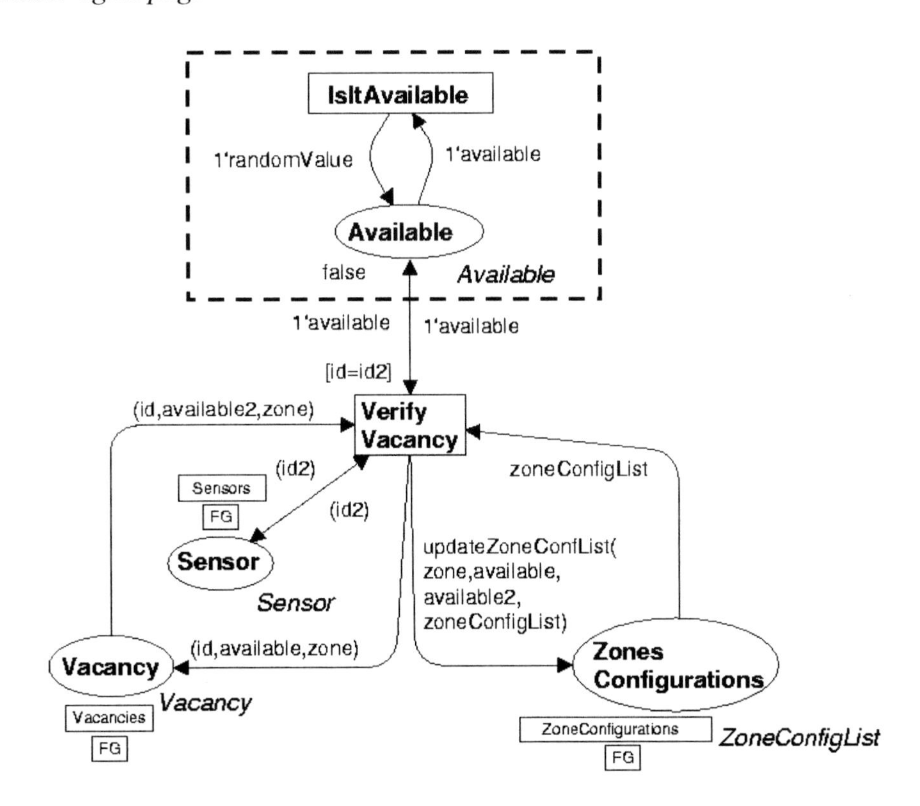

(*updatePrices(zoneConfigList,proceList)*) verify if the configuration of each parking zone is according to the specified threshold. As mentioned, if there are less than 15% of available vacancies in a parking zone, the price is increased, otherwise the price is decreased.

There is also a place named *list of prices*. The prices of all parking zones are on it in a list form. The transition updates this list and send a list with the newest prices, by the arc expression *nPriceList@getNewPrices(zoneConfigList,proceList)*, to all the parking meters of the parking zones that had their prices changed.

In the Figure 8, the *parking meter agent* page is presented. This agent receives a message from the *decision agent* with the new price list of the parking zones. If this list is not empty (see the *update prices* transition guard), the list that contains all the vacancies prices is updated with the new prices. So, it means that each one of the parking meters has changed its screen showing the new price. All this is done by the *update prices* transition.

It is important to note that any number of sensors and parking meters can be present in the system, and they can enter and leave the system without compromising the rest of the system. Also, the business strategy can change and the price can be controlled for a different number of available parking spaces instead of 15%. This change is also transparent for sensors and parking meters. Moreover, other agents can be considered such as a management agent for reporting and auditing purposes.

This model-based approach promotes a better organization and validation of the system as well as a good way to deal with changes in the system or requirements. Every time some change is incorporated to the requirement, the part of the model related to it can be changed and the analysis can be performed again to ensure that the changes respect the desired system behavior.

ANALYSIS

Simulation

Many simulations were done and some things were observed: every time, for each zone it was created *numOfVac* vacancies; for each vacancy, it was created a parking meter and a sensor; and it was always created a list of prices, each one for each vacancy and a list of zone configurations, each one for each parking zone.

By simulating the model, it was detected that when a parking zone price changes, the price of the other parking zone vacancy does not change. And if a parking zone price is changed, maybe this new price does not appear in the parking meter visor because of a newer price. In other words, suppose the price is US$1. And it is changed to US$2. Before this new price is passed to the parking meter, it is changed back to $1, so the first change will not be passed to the users of the vacancies.

A weird feature was detected—the surrounded box of the sensor agent page that monitors if a vacancy is available or not looks preferring true values. Therefore, the model checking technique has to be applied in order to verify all possible behaviors.

Model Checking

Simulation cannot guarantee that all the possible execution traces of the model were explored. Therefore, it is necessary to generate a state space of the CPN model, also known as occurrence graph. In this kind of graph, each one of the nodes represents a possible state of the model.

With the state space in hands, it is used a temporal logic language to specify the desired properties, in this case it is used a library called ASK/CTL (Christensen & Mortensen, 1996). The specified properties were: (a) every time a sen-

Figure 7. Decision agent page

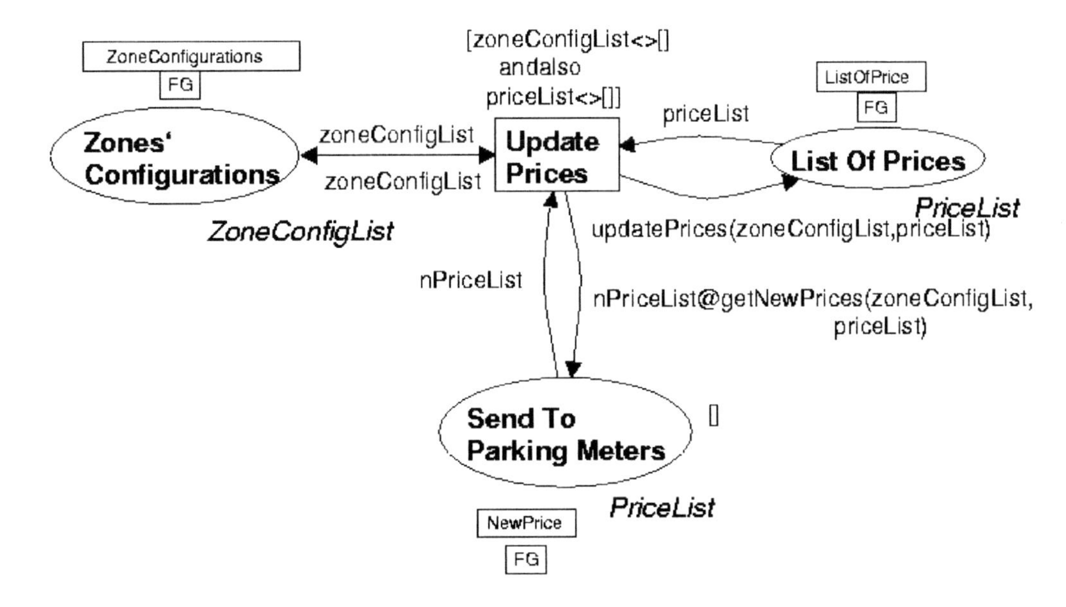

Figure 8. Parking meter agent page

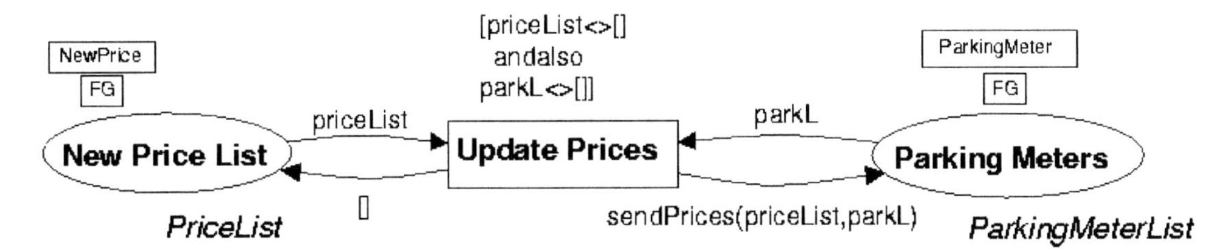

sor agent sends data with the vacancy status, the decision agent receives it; (b) every time the decision agent changes the price of a specific parking zone, all the parking meters of that parking zone receive a response message with the new price to be showed; the most important property is that (c) every time the available the number of vacancies reaches 15% or less of the total of vacancies in a certain parking zone, the price is increased, otherwise it is reduced; and (d) it is possible to have all the vacancies not available.

The logical propositions and formulas used to perform model checking for the CPN model are described in what follows. For the sake of convenience and simplicity a CTL-like notation is used.

a. This property is guaranteed because the place that represents the output of sensor agent and the place that represents the correspondent input of decision agent belong to the same set of fusion places. In other words, this property is satisfied due to the fusion place concept.
b. Property specification:
    - *Pa* = Decision agent changes the parking value
    - *Pb* = All parking meters of a parking zone receives the change
    - *S = AL(IMP(Pa,EV(Pb))*

In other words, *Pa* is always true, consequently (*IMP* means implies) *Pb* will be true. This property is evaluated to false.

The decision agent can change a parking zone price and before this price is sent to the parking meters, the decision agent can change the price again. So, the first change is not visible to the user of the vacancy.

c. Property specification:
   - *Pa1* = Number of vacancies equal to or less than 15 percent
   - *Pb1* = Price is increased
   - *Pa2* = Number of vacancies is greater than 15 percent
   - *Pb2* = Price is reduced
   - *S1 = AL(IMP(Pa1,EV(Pb1)))*
   - *S2 = AL(IMP(Pa2,EV(Pb2)))*
   - *S = AND(S1,S2);*

   This property is evaluated to be true.

d. The property is that there is just one parking zone and there are 10 vacancies in it.
   - *Pa* = The parking zone has no available vacancy.
   - *S = EV(Pa)*

   This property is evaluated to be true. Eventually, it can occur that none of the vacancies areavailable.

FUTURE TRENDS

As a future trend, some other characteristics can be taken into account such as management information that can also be stored in a database for reporting and auditing purposes. Another agent can be inserted to do this task without changing the rest of the model. This flexibility promoted by the multi-agent framework together with the advantages of a formal method as Petri nets, make this approach a good way to deal with the challenges of the IIIS. Therefore, other classes of IIIS can be addressed using it.

There is an external communication library for the design/CPN tool called Comms/CPN (Gallasch & Kristensen, 2001). Using this library, it is possible to communicate the CPN model with an external application and, therefore, make the model control the system for parking prices itself. In this case, a mathematical correct proved model is the control system itself and there is no gap between design and implementation.

CONCLUSION

In this chapter, a formal multi-agent-based specification and validation approach is presented for the domain of intelligent information systems. An application of a control system for parking meters prices is used to illustrate the approach. A multi-agent development promotes a good way to deal with changes. In the case of this work, the model is formal, and it is possible to perform some analysis such as automatic simulation and model checking to ensure that desired properties of the system hold in the model.

The use of a formal model for specification and verification as presented in this chapter for the domain of intelligent information systems promotes trustworthiness in the system behavior. This is important for decision-making systems where a business strategy must be ensured. Moreover, with the multi-agent framework, changes can be applied to specific parts of the model while the rest of the model does not change. This local changes characteristic is important for complex distributed system to keep control on changes. Finally, when changes are done, the model can be used to analyze the behavior to ensure that the new changes respect the system properties.

An intelligent control system for parking prices control is presented to illustrate the advantages of using this approach to the IIIS domain. This system consists of autonomous sensor agents, parking meter agents, and a control agent. They interact with each other to ensure a minimum number of parking places to be available at busy shopping areas by increasing and decreasing the price. The formal model-based approach is used to specify and analyze this system and some analysis results are presented.

REFERENCES

Adam, N. R., Atluri, V., & Huang, W.-K. (1998). Modeling and analysis of workflows using Petri nets. *Journal of Intelligent Information Systems: Special issue on workflow management systems, 10*(2), 131-158.

Almeida, H., Silva, L., Perkusich, A., & Costa, E. (2005). A formal approach for the modelling and verification of multiagent plans based on model checking and Petri nets. In R. Choren, A. Garcia, C. Lucena, & A. Romanovsky (Eds.), *Software engineering for multi-agent systems, III: Research issues and practical applications* (vol. 3390, pp. 162-179). Heidelberg, Germany: Springer GMBH.

Almeida, H. O. de, Silva, L. D. da, Silva Oliveira, E. A. da, & Perkusich, A. (2005). Formal approach for component based embedded software modelling and analysis. In *Proceedings of the IEEE International Symposium on Industrial Electronics* (*ISIE 2005*) (pp. 1337-1342). Dubrovinik, Croatia: IEEE Press.

Christensen, S., & Haagh, T. B. (1996). *Design/CPN overview of CPN ML syntax, version 3.0.* Aarhus, Denmark: University of Aarhus, Department of Computer Science.

Christensen, S., & Mortensen, K. H. (1996). *Design/CPN ASK-CTL manual.* Aarhus, Denmark: University of Aarhus, Department of Computer Science.

Clarke, E. M., Grumberg, O., & Peled, D. A. (1999). *Model checking.* Cambridge, MA: The MIT Press.

Costa, E. B., et al. (2002) A multi-agent cooperative intelligent tutoring system: The case of musical harmony domain. In *Proceedings of 2nd Mexican International Conference on Artificial Intelligence (MICAI'02)*, Merida (LNCS 2313, pp. 367-376). Heidelberg, Germany: Springer GMBH.

Costa, E. B., Lopes, M. A., Ferneda, E. (1995) Mathema: A learning environment base on a multi-agent architecture. In *Proceedings of the 12th Brazilian Symposium on Artificial Intelligence (SBIA '95)* (pp. 141-150). Campinas, Brazil: Springer GMBH.

Costa, E. B., Perkusich, A., Figueiredo, J. C. A (1996). A multi-agent based environment to aid in the design of Petri nets based software systems. In *Proceedings of the 8th International Conference on Software Engineering and Knowledge Engineering (SEKE'96)* (pp. 253-260). Buenos Aires, Argentina: IEEE Press.

Deen, S. (2005). An engineering approach to cooperating agents for distributed information systems. *Journal of Intelligent Information Systems, 25*(1), 5-45.

Delena, D., & Pratt, D. B. (2006, February). An integrated and intelligent DSS for manufacturing systems. *Expert Systems with Applications, 30*(2), 325-336.

Gallasch, G., & Kristensen, L. M. (2001, August). Comms/CPN: A communication infrastructure for external communication with Design/CPN. In *Proceedings of the 3rd Workshop and Tutorial on Practical Use of Coloured Petri Nets and the CPN Tools* (pp. 75-90). Aarhus, Denmark: University of Aarhus, Department of Computer Science.

Giunchiglia, F., Mylopoulos, J., & Perini, A. (2002) The Tropos software development methodology: Processes, models and diagrams. In *Proceedings of the 1st International Joint Conference on Autonomous Agents and Multiagent Systems* (pp. 35-36). Bologna, Italy.

Jennings, N. R. (2001) An agent-based approach for building complex software systems. *Communications of the ACM, 44*(4), 35-41.

Jensen, K. (1992). *Coloured Petri nets: Basic concepts, analysis, methods and practical use (vol. 1).* Heidelberg, Germany: Springer GMBH.

Jensen, K. (1997). *Coloured Petri nets: Basic concepts, analysis, methods and practical use (vol. 2).* Heidelberg, Germany: Springer GMBH.

Jensen, K., et al. (1999). *Design/CPN 4.0.* Retrieved December 17, 2005, from http://www.daimi.au.dk/designCPN/

Murata, T. (1989, April). Petri nets: Properties, analysis and applications. *Proceedings of the IEEE, 77*(4), 541-580.

Smith, M. (2005, August). *Will a new generation of curbside sensors end our parking problems — or help the government monitor our every move?* Retrieved December, 17, 2005, from http://www.sfweekly.com/Issues/2005-08-17/news/smith.html

This work was previously published in Artificial Intelligence and Integrated Intelligent Information Systems: Engineering Technologies and Applications, edited by X. Zha, pp. 287-302, copyright 2007 by IGI Publishing (an imprint of IGI Global).

Chapter XIV
Design Principles for Reference Modelling:
Reusing Information Models by Means of Aggregation, Specialisation, Instantiation and Analogy

Jan vom Brocke
University of Muenster, Department of Information Systems, Germany

ABSTRACT

With the design of reference models, an increase in the efficiency of information systems engineering is intended. This is expected to be achieved by reusing information models. Current research focuses mainly on configuration as one principle for reusing artifacts. According to this principle, all variants of a model are incorporated in the reference model facilitating adaptations by choices. In practice, however, situations arise whereby various requirements to a model are unforeseen: Either results are inappropriate or costs of design are exploding. This paper introduces additional design principles that aim toward giving more flexibility to both the design and application of reference models.

INTRODUCTION

Modeling comprises a concentration on special aspects in design processes by means of abstraction. In particular, information (system) models are built in order to describe relevant aspects of information systems. Due to an increasing demand of these models addressing similar design problems to a certain extent, the development of reference (information) models is subject to research.[1] The essential idea is to provide information models as a kind of "reference" in order to increase both efficiency and effectiveness of modeling processes (Becker et al., 2004; Fettke & Loos, 2003; Scheer & Nüttgens, 2000).

Practical applications of reference models are widespread in the domain of ERP-systems (Becker & Schütte, 2004; Huschens & Rumpold-Preining, 2006; Kittlaus & Krahl, 2006; Scheer, 1994). In this domain, reference models set the basis for general business solutions that can be adapted to individual customer needs. In order to support this kind of customising process, reference models are built in a configurative way (Becker et al., 2004; Meinhardt & Popp, 2006; Recker et al., 2005); for a review on German literature (see vom Brocke, 2003, pp. 95-158). This work intends to encounter all relevant variants of prospective applications during build-time of the model in order to facilitate adaptability by means of choices (van der Aalst et al., 2005, p. 77). A vital factor for the economic efficiency of reference modeling is in how far a single variant of the model fits the customer's requirements. As this fit indicates the value of the model, it is also essential for the return on investment in building reference models from a supplier's perspective. Considering the variety of requirements to be faced in today's software engineering, the design principle of configuration illuminates specific limitations. In particular, it is increasingly hard to take into account the various requirements that may be relevant and to incorporate them in the reference model. Hence, supplementary design principles that may enlarge the "tool-kit" of reference modeling appear to be useful. Accordingly, the principles aggregation, specialisation, instantiation, and analogy are presented in this chapter.

Initially, as a theoretical background, the concept of reuse in reference modeling is introduced. This allows an analysis of preliminary works leading to a closer specification of this study's research focus and methodology. On that basis, new design principles are introduced and analysed according to their potential for reference modeling. Finally, a conclusion is drawn and perspectives for future research are suggested.

FOUNDATIONS OF DESIGN PRINCIPLES FOR REFERENCE MODELING

The Concept of Reuse in Reference Modeling

In order to learn how to build reference models according to specific needs, a deeper understanding of essential elements of these particular kinds of models is required. This chapter outlines how these essential elements lie in the intention of building information models which can easily be reused in the design of other models. In addition, relevant dimensions of reusing information models are presented, serving as a framework for this work.

A Reuse-Oriented Concept of Reference Models

According to reviews on reference modeling literature (vom Brocke, 2003, p. 31 ff.), which are predominantly written in German, the concept of reference modeling is introduced on the basis of special characteristics of models. In particular, a certain degree of validity and quality are highlighted. Thus, reference models are meant to describe a sort of best (or common) practice of business in a creation application domain. This approach enables a sophisticated understanding of the idea of reference modeling. However, certain limitations are obvious. Considering the problem of measuring characteristics, for instance, there is no objective reason for awarding or rejecting best practice to a model. At the same time, it may even be questionable in how far this discussion helps to learn about principles of building reference models so that they finally prove to be best practice.

For further work, it seems to be reasonable to ground the concept of reference modeling on the intention of reusing information models. To give a definition, reference models are referred to as special information models that serve to be reused in the design process of other information models (vom Brocke, 2003, p. 38). According to a recent study (Thomas, 2005), this concept is increasingly applied in reference modeling. A process-oriented view on the design and usage of reference models enables a deeper understanding of the concept (see Figure 1).

Modeling is referred to as a design process in which a designer is in charge of building a model according to a user's needs. Due to different mental models of people carrying out these roles, co-ordination is vital for both effectiveness and efficiency of the process. Whereas effectiveness measures the users' satisfaction due to the appropriateness of model quality, efficiency takes into account the ratio of in- and output coming along with the process.

Considering the reuse-oriented concept of reference models, a kind of support process for information modeling is added. This support lies essentially in providing information models that serve to be incorporated in other models. Specifically, reuse is conducted by taking parts of one or more original models and adapting and extending them in the resulting model. That way, both efficiency and the effectiveness of the modeling process may be fostered.

Taking reference models as models to support design processes, reference models have to be differentiated from meta-models (Karagiannis & Kühn, 2002) that are commonly used for the definition of modeling languages. Following a wider understanding, further types of meta-models, for example, may well be distinguished that describe the procedure of conducting the modeling process. In contrast to reference modes, these models are "used," but not "re-used," as their content is not incorporated in the resulting model. Thus, whereas meta-models define rules for conducting modeling, reference models offer content to be reused.

Regarding the reuse-oriented concept, different types of models are candidates for being reference models. Models can vary widely, regarding, for example, the type of application, the level of implementation or the type of organisational setting. In addition, reference models might also be classified by supplementary criteria. By looking at relevance and rigor, for example, a special subset of reference models may be identified serving as scientific artefacts. However, the term "reference" does not indicate a sort of "high qual-

Figure 1. Reuse-oriented concept of reference modeling

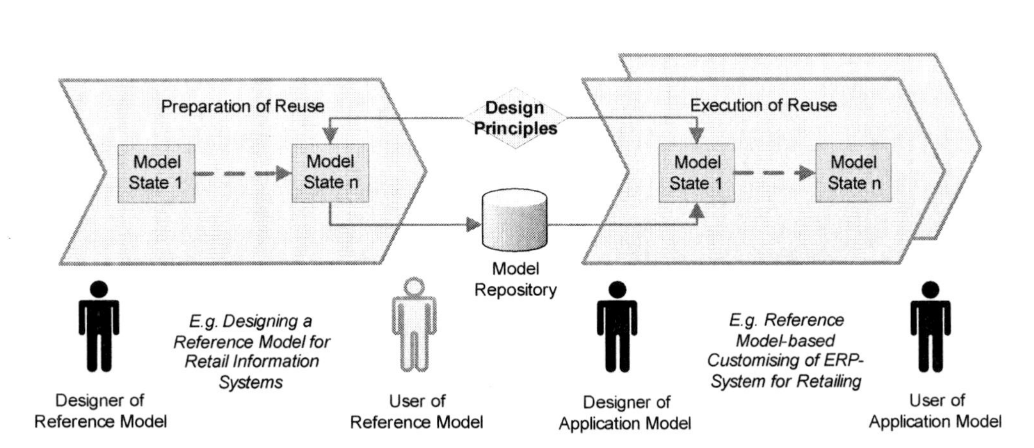

ity" model in general. The decisive factor is the user's view about the suitability of the model for a special purpose. Thus, reference models have to be considered from at least two perspectives (vom Brocke, 2003, p. 32):

- **Supplier perspective:** The designer may plan to supply a reference process model under consideration of the purpose of reusability in other design processes. This however, does not guarantee that reuse will take place in practice.
- **Customer perspective:** The decision on reusing a model is in fact taken by the customer. Thus, models might either be planned to be reused but fail on the customer side or the customer decides to reuse models that have not been designed for that purpose.

Reference modeling as a research discipline intends to discover means to coordinate both perspectives: The analysis of the customer side is essential in order to learn about requirements. For that purpose an insight into relevant domain knowledge of various fields of application as well as in people's behaviour in reusing artefacts is required. In addition, work on the designer's side mainly aims at finding means to fulfil these requirements. Furthermore, research is necessary on the interchange of models comprising work on institutional and technical infrastructures.

Thanks to the reuse-oriented concept, the wider scope of reference modeling gives way to indicate research areas that may contribute to reference modeling. In particular, in the area of reuse-oriented (software) engineering, a wealth of experiences in principles of reusing artefacts can be found (H. Mili et al., 1995; Peterson, 1991). Hence, preliminary work from various fields of this kind may be transferred to reference modeling. For that scope, findings have to be adapted to the special problem of reusing information models. The following examines basic aspects of reuse techniques in engineering and assembles them in order to present a picture of relevant dimensions of reuse in reference modeling.

Dimensions of Reuse in Reference Modeling

In former studies on reference modeling, the design of configurative reference models was focussed on in order to support the derivation of multiple variants of a reference model for a certain application (Becker et al., 2004; Meinhardt & Popp, 2006; Recker et al., 2005); for a review on German literature (see vom Brocke, 2003, pp. 95-158). That way, this research is concentrated on a rather special field of reusing information models. Figure 2 gives an overview of the relevant dimensions of reuse in reference modeling.

The differentiation visualised in Figure 2 provides a three-dimensional framework for the identification of model relations in the context of reuse.

Stage in Value Chain

In reuse-oriented software engineering, technical issues of reuse are covered in two stages in the value chain: the development "with" reuse that is carried out in application engineering and the development "for" reuse that is subject to domain engineering (Gomaa, 1995; Kang et al., 1998; Mili et al., 2002, p. 27). Correspondingly, the design of reference and application models may be distinguished. The content shipped with the reference model needs to be adapted regarding the requirements of the empirical application. In addition, the design of reference models may be supported by reuse itself. That way, a kind of reference model-based modeling may be established.

Degree of Innovation

In particular, when considering industrial engineering and configuration management (Conradi

Figure 2. Framework for relations reusing information models

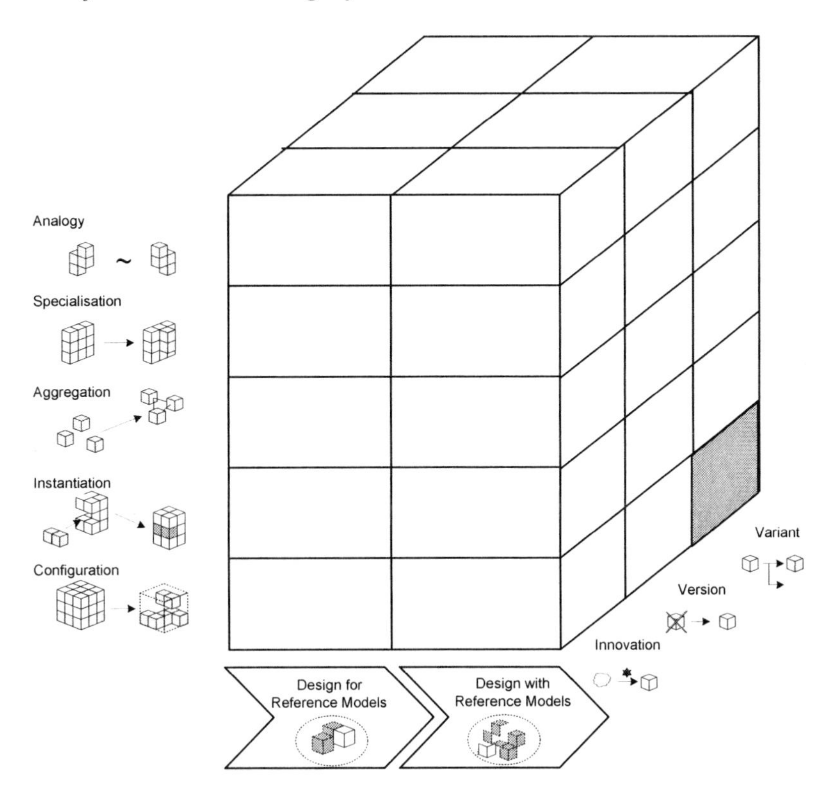

& Westfechtel, 1998, p 238ff.), constructions may be differentiated by degrees of innovation. Correspondingly, both innovations and modification of models can be distinguished. While modifications offer a solution for a specific type of problem by changing initial models, innovations come up with solutions for other types of problems for the first time. If the modification is carried out in order to tap a new shaping of a problem, a so-called variant is constructed. If the adaptation, instead, brings in a new solution to a problem that has already been dealt with, a new version is available. This assessment is taken subjectively considering the relevant application domain and mental model of the recipient.

Principle of Construction

In order to facilitate reuse, design principles are applied. Design principles provide rules describing the way in which the content of one model is reused in constructing another. The rules describe ways of taking over contents as well as of adapting and extending them in the resulting model. With each design principle, specific sets of rules are differentiated. Apart from configuration as a parametric approach, a great variety of principles are discussed in software engineering that are originally referred to as adaptive principles (Karhinen et al., 1997; Peterson, 1991). In recent works, these principles have been evaluated for use in reference modeling (vom Brocke, 2003, pp. 259-319). In particular, the principles of instantiation, aggregation, specialisation, and analogy seem to offer promising means for extending the set of design principles for reference modeling. According to instantiation, general aspects of a domain are designed as a framework providing generic placeholders for plugging in models considering special requirements of an application. Specialisation enables the takeover of entire contents of a general model into a specific model allowing modification and extending individually. Aggregation enables the takeover of contents delivered by various part models that are composed and extended according to special requirements of an application. Analogy, finally, employs seemingly similar solutions in a creative way to tackle new problems.

Whereas much research has been carried out on configuration, little attention has so far been paid to instantiation, specialisation, aggregation, and analogy. Thus, these design principles will be introduced in more detail with this contribution. As for configuration, an introduction can be taken from literature. Having introduced the principles, questions of selecting the right principle will be addressed. This discussion will again include configuration.

Preliminary Work and Focus of Research

In the previous chapter, reference (information) modeling was introduced for the essential idea of reusing artefacts in information modeling. Hence, the significance of a wide range of work in the field of reuse-oriented software engineering (McIlroy, 1968) becomes obvious. For this study, insight into various types of principles for reusing information systems artefacts can be gained in particular. Figure 3 gives an overview of the impact of principles from reuse-oriented software development on those for reference modeling.

Early works on reuse-oriented software engineering have been carried out in structuring programs by means of modules (Jones, 1984). In particular, the concept of generic packages (Slater, 1995) provides inspiration for the construction technique of instantiation. Generic packages allow reusing a unique data structure for various data types by means of deriving instances of the package for concrete data types. The idea of reuse is essentially incorporated in the object-oriented paradigm (Coad & Yourdan, 1991; Cox, 1990). Within this context, information systems are composed of co-operating objects that are entirely specified regarding their properties and behaviour. The way in which objects share these descriptions by inheritance serves as an example and gives clues for the principle of specialisation. The idea of composing information systems out of rather independent fragments is further developed in the concept of component-based software engineering (George & Heineman, 2001; Szyperski, 1998). The leitmotif of combining off-the-shelf components that are ready to be used in a new context gives an example for the principle of aggregation. Apart from these rather formalised concepts, pattern-based software engineering aims at providing artefacts that can be reused free-handedly by the designer (Alexander et al., 1977; Gamma et al., 1995; Hay, 1996). This approach gives stimuli to the construction principle of analogy for reference models (Spanoudakis & Constantopoulos, 1993).

In reuse-oriented software engineering, the concepts described above have also been covered in the phase of requirements engineering and, thereby, in the design of information models (Coad et al., 1997; Kruchten, 2003; Raumbaugh et al., 1991). The principles of aggregation, specialisation, and partly instantiation are incorporated especially in most modeling languages like the unified modeling language (UML) (Kruchten, 2003; Rumbaugh et al., 2004). The implementation of these principles, however, takes place "within models." In a class diagram, for example, classes may well be linked by aggregation or specialisation symbols. For reference modeling, in contrast, these principles are to be applied to linkages "between models." These kinds of relations are rarely covered, except in pattern-oriented software engineering. In particular, work on analysis patterns (Fowler, 1997) fits well with the kind of abstract conceptional models that have predominantly been subject to reference modeling research so far. However, due to the type of reuse principles, little information is given about rules of how to take over, adapt and extend the model's content with respect to certain applications.

In addition to the general contribution of reuse-oriented software engineering, special studies can be found that address problems of specifying information system artefacts for the purpose of reuse. Apart from work in the field of configurative reference modeling (Becker et al., 2004; van der Aalst et al., 2005) that is not focused on here, works on further principles of reuse can be found (D'Souza, 2000; Melton & Garlan, 1997; Remme, 1995). Altogether, these studies are characterised by the approach of introducing extensions to notations that are used in order to describe relevant artefacts. These contributions give examples of the implementation of certain design principles in modeling languages. However, each approach is limited to merely one specific modeling language. Reference modeling, however, is characterised by a pluralism of languages especially substantiated by the focus on contents of models that are to be constructed. The linguistic representation of the contents, instead, needs to be adapted to specific preferences of the addressee.

Thus, design principles for reusing information models in reference modeling should be introduced in a way that they are independent of special languages. Such an approach is presented in this chapter. It aims at enabling the incorporation of reuse by instantiation, aggregation, specialisation and analogy in various languages according to the same principles.

A Language-Independent Approach for Introducing Design Principles

For reference modeling, design principles should be specified on a general methodological level. In particular, a specification of essential rules for each design principle is required that remains independent of special modeling languages. This specification can then be transferred to a great variety of specific languages. This approach will be referred to as language independency in the following section. A methodological approach for developing language independent design principles can be drawn from reference meta-models (vom Brocke, 2003, pp. 85, 263-269); similarly (Axenath et al., 2005, pp. 50-51). With reference meta-models, essential rules for specific construction purposes are specified in a way that they can be incorporated by various other languages (application languages).

Applying this approach, the construction of aggregation, specialisation, instantiation and analogy will be introduced on the basis of reference meta-models. For further illustration, workflows for conducting them on the basis of case tools as well as examples are given. According to the reference meta-models and the examples, relevant assumptions are introduced in the following.

Figure 3. Deriving design principles for reference modeling from reuse-oriented software engineering

Branch in Reuse-Oriented Software Engineering (SE)	Major Contribution to Design Principles for Reference Modeling
Module-oriented SE	Instantiation
Object-oriented SE	Specialisation
Component-based SE	Aggregation
Pattern-based SE	Analogy

Reference Meta-Models for Design Principles

On the basis of reference meta-models, design principles can be introduced according to their essential rules, in order to be transferred to a number of modeling languages. For this purpose, assumptions about characteristic language constructs of application languages have to be made so that the scope of reference languages can be determined. The assumptions refer to the type of modeling languages as well as to the types of elements the languages may consist of.

- **Type of modeling languages:** Languages serving the description of a system's behaviour and properties have to be differentiated. Models describing the behaviour of a system show possible adaptations to the system's state. Thus, these models are characteristically described by statements given in both a temporal and logical order. Languages, consequently, provide constructs for placing and relating corresponding statements. With models describing the properties of a system, the characteristics of the various states of the system have to be specified. Therefore, statements referring to sets of attributes and relations between these states are relevant. Languages for the description of properties, hence, provide according constructs.
- **Type of elements of the language:** For the purpose of abstraction, modeling languages are considered to consist of the following general elements: Languages are sets of *rules* that present language constructs by means of which language statements can be made. Language statements about construction results are referred to as model statements by means of which the contents of a model can be described. If model statements of a particular purpose are combined, these statements are referred to as a representation of a model. Design principles are formalised by sets of rules that describe how model representations and model statements can be put into relation to one another for the purpose of transferring, adapting and extending contents of an original model into a resulting model.

Rules of design principles are specified in a way that extensions of application languages can be derived. In reference meta-models, characteristic language constructs are introduced that are necessary for incorporating a certain design principle in reference modeling. Transferring these rules to application languages, either adaptations or the extensions of already existing constructs, have to be carried out, or new constructs are introduced.

Examples for the Design Principles

With this chapter, the design principles are illustrated by means of examples. Selecting the examples, a couple of requirements have to be considered:

- **Languages:** Models should be represented in languages like ERM (Chen, 1976) and EPC (Keller et al., 1992; Scheer, 1994) as these languages are widespread in reference modeling and therefore serve as a good example for illustrating how to apply the design principles.
- **Domain:** Models should be taken from a domain that is of relevance for a wide range of applications.
- **Suitability:** The contents of the models should be suitable for demonstrating the various principles of reuse.
- **Focus:** The models should be proven in order to focus the discussion on changes in design due to the application of the design principles.

Figure 4. Data model "Accounts Payable" as major source for demonstrating design principles with ERM

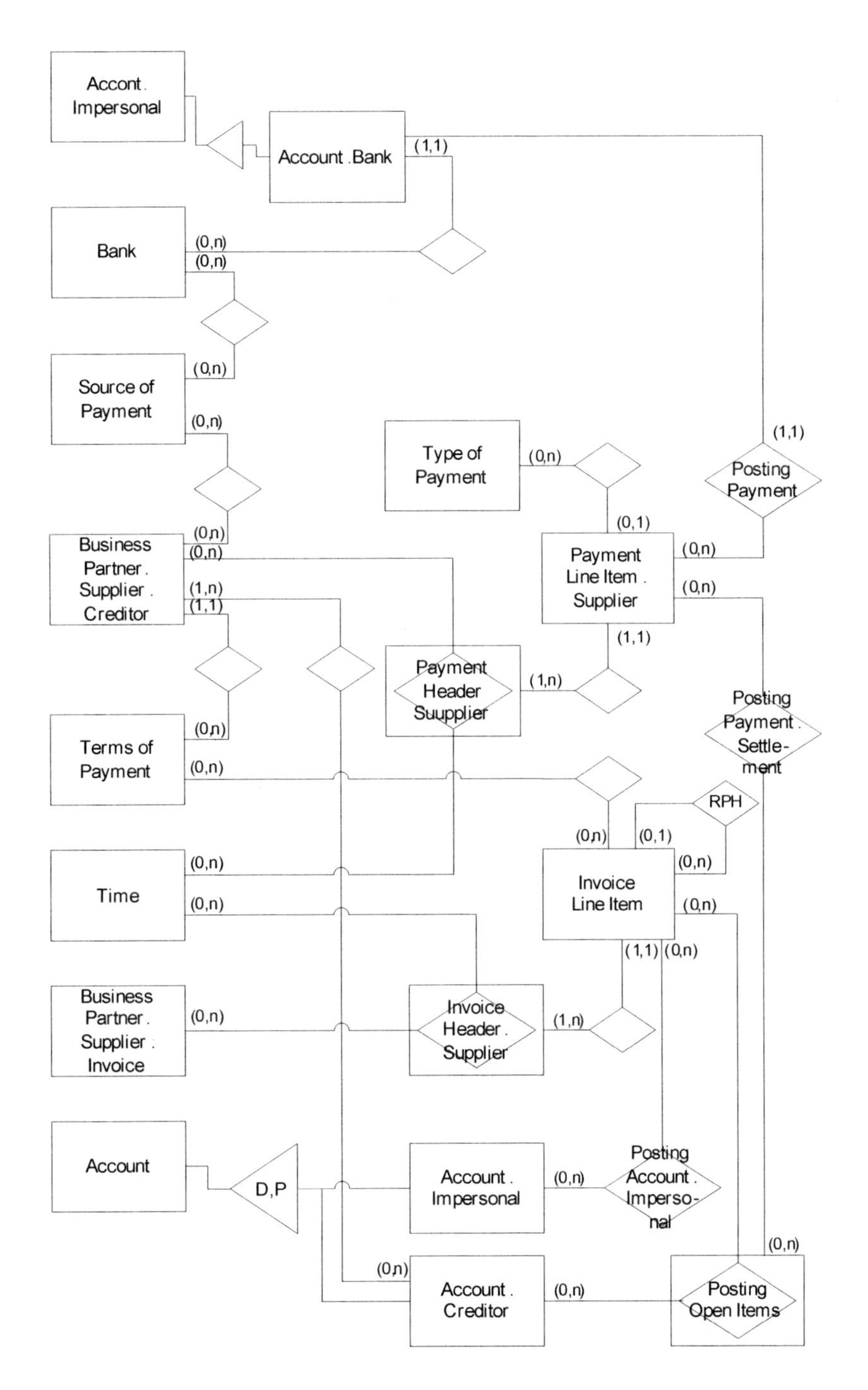

Figure 5. Process model "Handling of Payments" as major source for demonstrating design principles with EPC

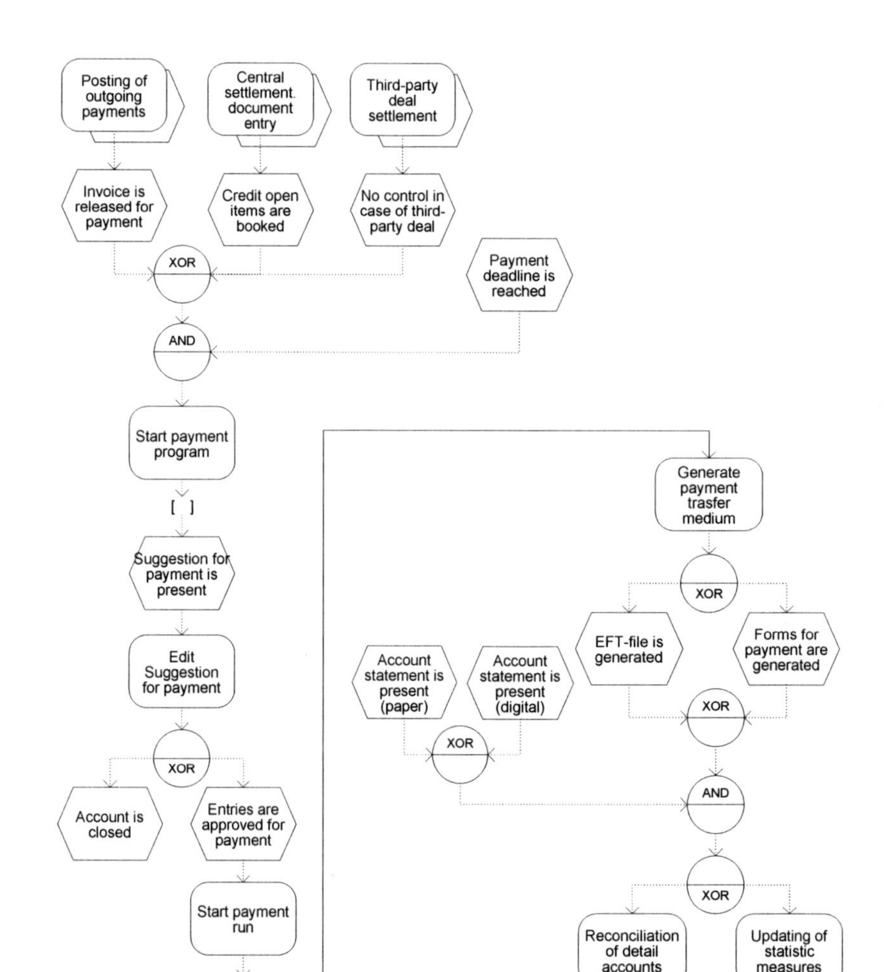

In order to fulfil most of the requirements, examples were chosen from the field of *accounting* that are described in both the reference models of Scheer (1994) and Becker/Schütte (Becker & Schütte, 2004). With respect to ERM, the examples show how rather detailed descriptions of "Accounts Payable" in Becker/Schütte (Becker & Schütte, 2004, p. 389) might be designed reusing parts of the models given in Scheer. The entire model of "accounts payable" is given in Figure 4.

According to EPC, the process "Handling of Payments" (Becker & Schütte, 2004, pp. 392 ff.) is examined correspondingly (see Figure 5). In the model of Scheer, a description of the process is given on a textual basis (Scheer, 1994, p. 417 ff.).

In order to illustrate the design principles, minor modifications to the models are made in the examples. The models of Becker/Schütte have been translated considering the terminology applied in Scheer. Applying the reference metamodels, it is intended to use language constructs that are available in the ERM und EPC as far as possible. Additional annotations are made using constructs that are common in the unified modeling language (UML) (Rumbaugh et al., 2004).

INTRODUCTION OF NEW DESIGN PRINCIPLES FOR REFERENCE MODELS

Instantiation

The principle of instantiation is characterised by the creation of a resulting model "I" by integrating one or several original models "e" into appropriate generic place holders of the original model "G." The resulting model "I," therefore, incorporates the integrated construction results of "e" in "G." The corresponding meta-model is presented in Figure 6.

Instantiation offers the opportunity to construct models for which both the theoretical frame as well as the statements required for appropriation can be reused. For this purpose, generic statements have to be produced in one of the original models. Step by step, they are then replaced by the integrative model representation in the process of instantiation. For integrating the statements in the resulting model, a special construct of model-

Figure 6. Metamodel for the design principles of instantiation

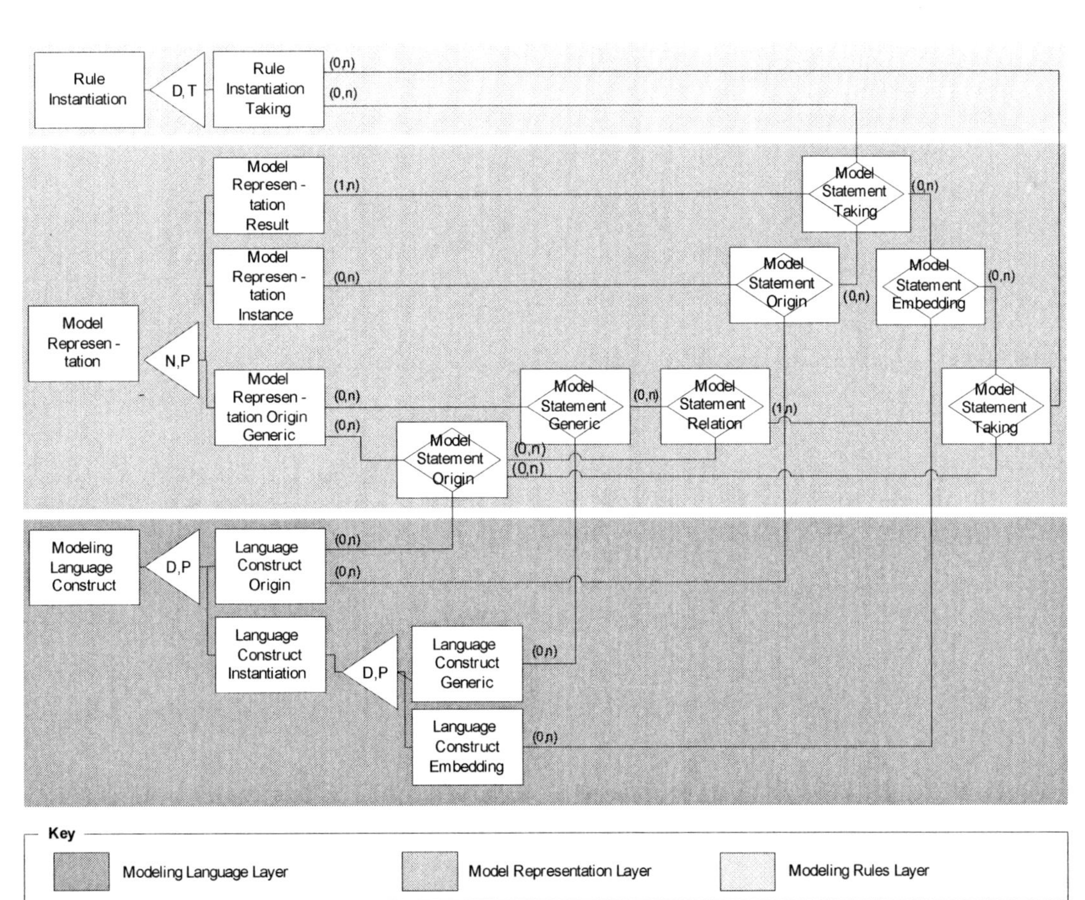

Figure 7. Example for applying the design principles of instantiation in ERM

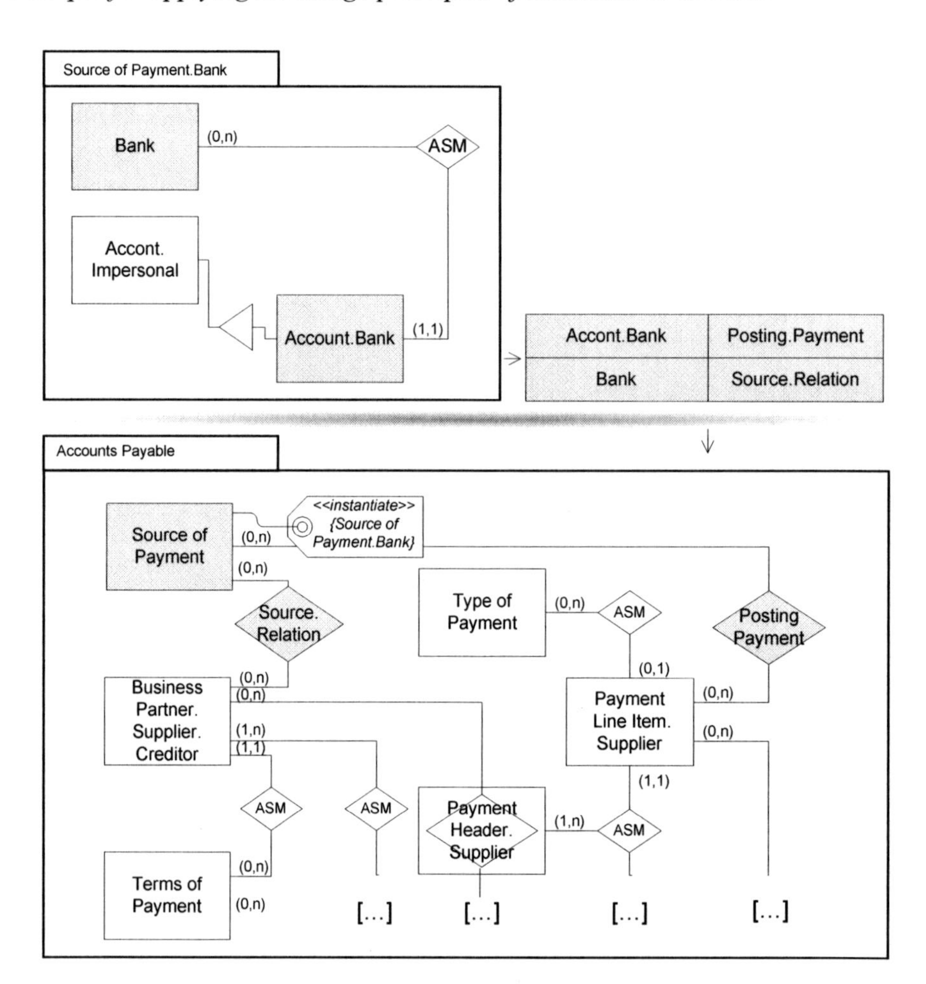

ing language is needed. This language construct serves to describe how the generic statements in G have to be linked with statements in e during the construction process of I. The integration is completed if all relations of the generic statement in G are updated correspondingly in I.

The principle of instantiation is illustrated by extending the data model "Accounts Payable" (Becker & Schütte, 2004, p. 389) to incorporate different types for sources of payments. For that purpose, a generic placeholder, "Sources of Payments," is used that serves to integrate various data models to specify relevant structures of each kind of source. In the example, a concrete source of payment, "Bank," is embedded that has been extracted from Scheer (1994, p. 611). The entity type "Source of payment," which is extended by the stereotype "«*instance*»," marks the relation between both data models. At the time of the instantiation, the interfaces of the resulting model and the model to be integrated have to be indicated. The example in Figure 7 reveals the relation between the entity type "bank" and the entity type "Business Partner Supplier Creditor" in the process of which the entity type "bank account" adopts the relation "Posting Payment" with the entity type "Posting Line Item Supplier." The opportunity of creating specific grades of abstraction appropriate to the needs of the addressee has a positive impact on clarity and relevance of the model.

Figure 8. Example for applying the design principles of instantiation in EPC

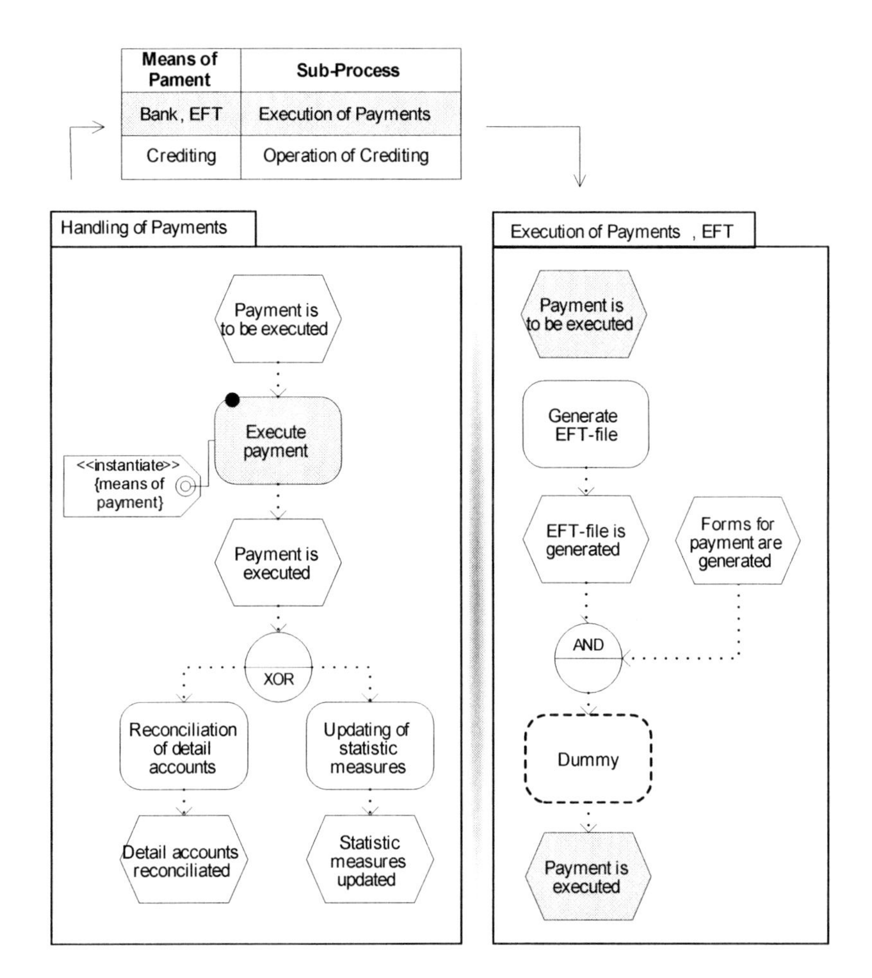

In order to conduct an instantiation with EPC, process interfaces can well be used. In contrast to aggregation, the integration of part processes is conditional in case of instantiation. This means that different concrete processes can be plugged into the interface that serves as generic place holder. In contrast to data models, the relations of the generic place holder do not have to be updated specifically. The proper linkage is assured by the process interface construct. An additional specification can be made on the criteria for embedding different processes.

In Figure 8, the principle of instantiation is applied for considering various types of payments in the process of "Handling of Payments." Therefore, the execution of payments has been implemented as a generic place holder. Accordingly, specific sub-processes can be designed for each source of payment and plugged into the process of "Handling of Payments" (Becker & Schütte, 2004, p. 392). In the example, the process of bank payment by EFT (electronic funds transfer) is described (Scheer, 1994, p. 417) and plugged in.

The design by instantiation requires realisation of the following work steps:

1. Select the generic model (e.g., in a tree-view).
2. Select generic statement within the model (e.g., by mouse-click).

Figure 9. Meta-model for the design principle of aggregations

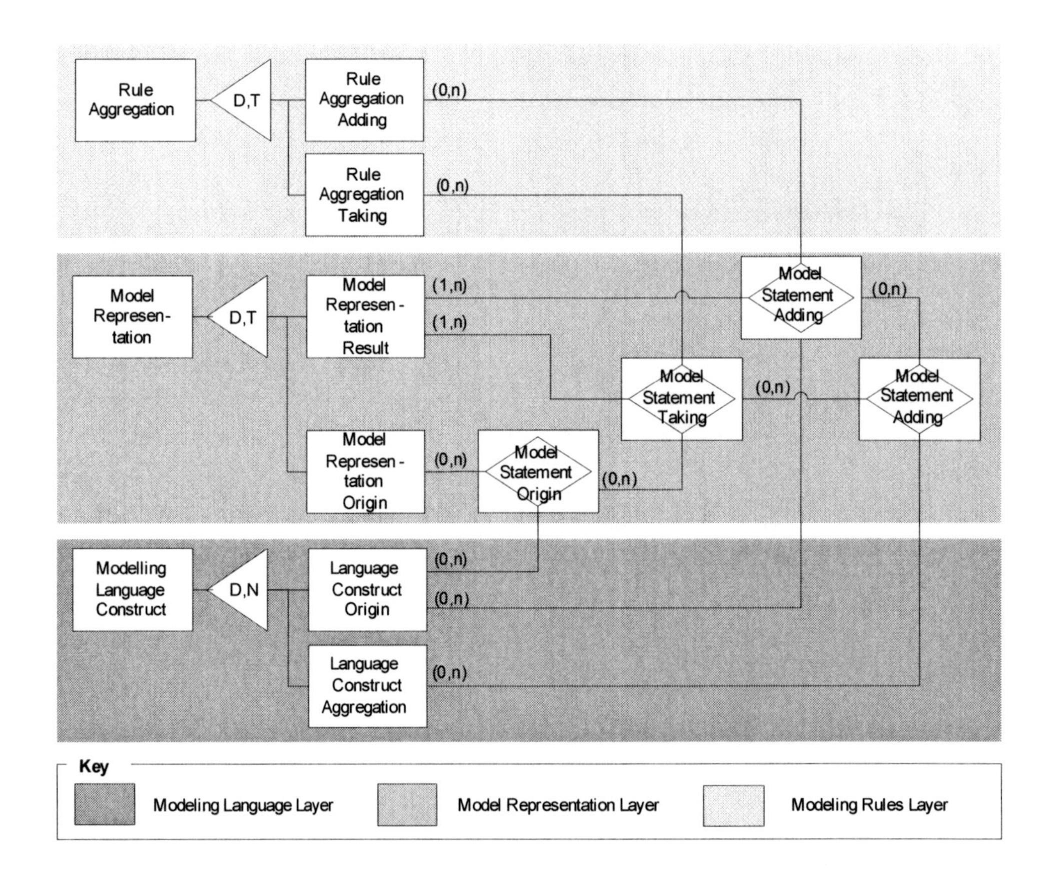

3. Select a model to be integrated in place of the generic statement (e.g., in a tree-view).
4. For each relation of the generic statement, specify an equivalent statement in the model to be integrated to take over the relation (e.g., in a table).
5. Either select or create a resulting model; the model to be integrated is embedded in the generic model according to the rules (e.g., in a routine).

Aggregation

The principle of aggregation is characterised by the combination of one or more original models "p" that build "a" resulting model "T," with the models "p" forming complete parts of "T." The meta-model for incorporating the principle of aggregation in modeling languages is presented in Figure 9.

Aggregations offer the potential of combining model statements of original models in new contexts. For this purpose, a language construct for connecting statements of separate models has to be provided for in a modeling language. Statements of aggregated models can be taken over in the resulting model. That way, supplementary statements can be replenished and statements of integration can be positioned.

An example for the application of aggregation in ERM is given in Figure 10. In the example, the essential account structures described by Scheer (1994, p. 611), are referred to from the model for "Accounts Payable" by Becker and Schütte (2004, p. 389). As a result, the content of the data model "Account Structure" can be

Figure 10. Example for applying the design principle of aggregations in ERM

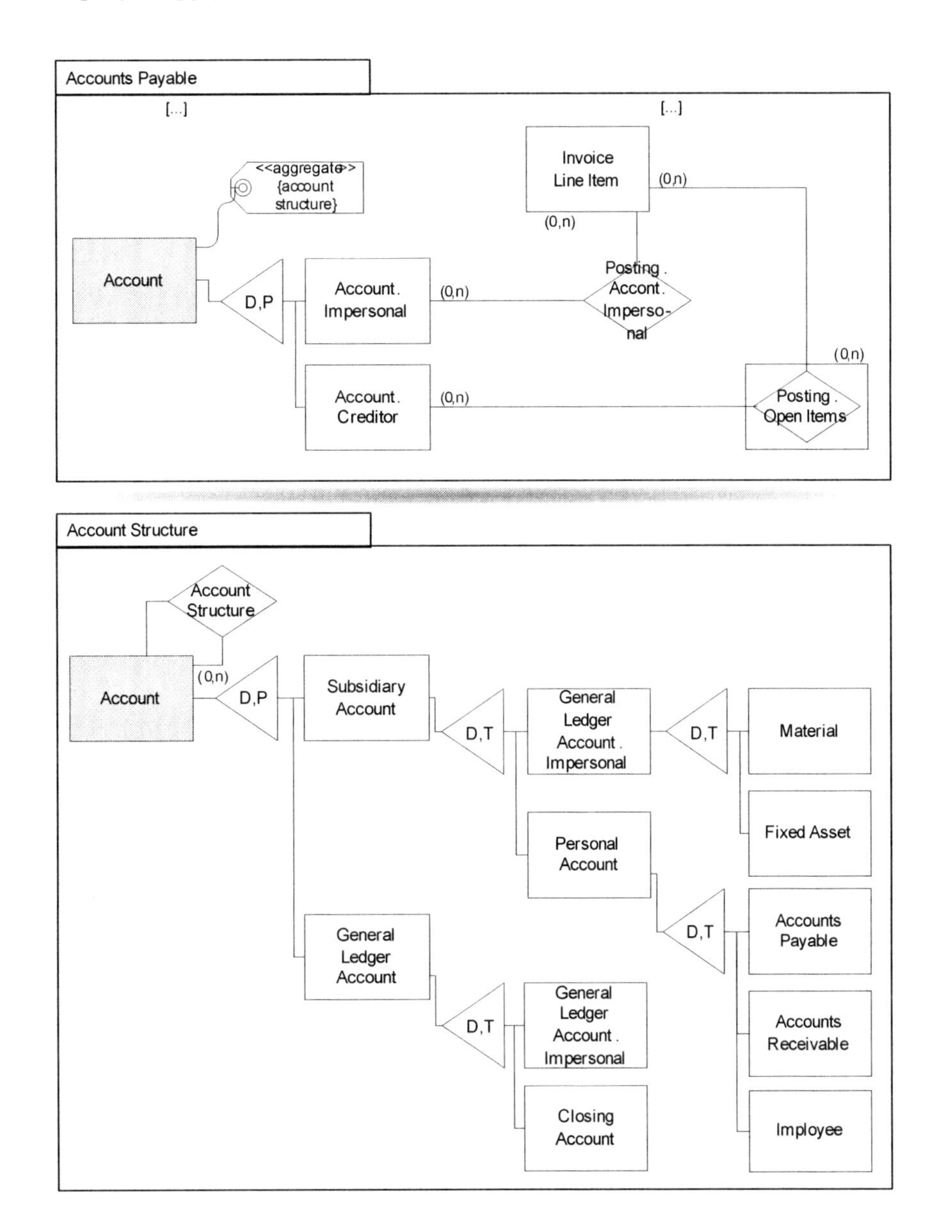

resorted back to without modeling redundantly. In ERM, the relation between model messages is basically supported by using identical entity types. That way, the relation is carried out by the entity type "Account." For a more explicit relation, the stereotype "«*aggregate*»" is added, that might also be specified regarding the target of reference in case different entity types are to be linked for aggregation.

In order to demonstrate an application of aggregation in EPC, the process of "Handling of Payments" (Becker & Schütte, 2004, p. 392) has been decomposed and interlinked with other processes. For incorporating the principle, language constructs can be used that already exist in the EPC syntax. In order to aggregate entire processes, the interface for detailing functions by processes can be used. In addition, aggregation

Figure 11. Example for applying the design principle of aggregations in EPC

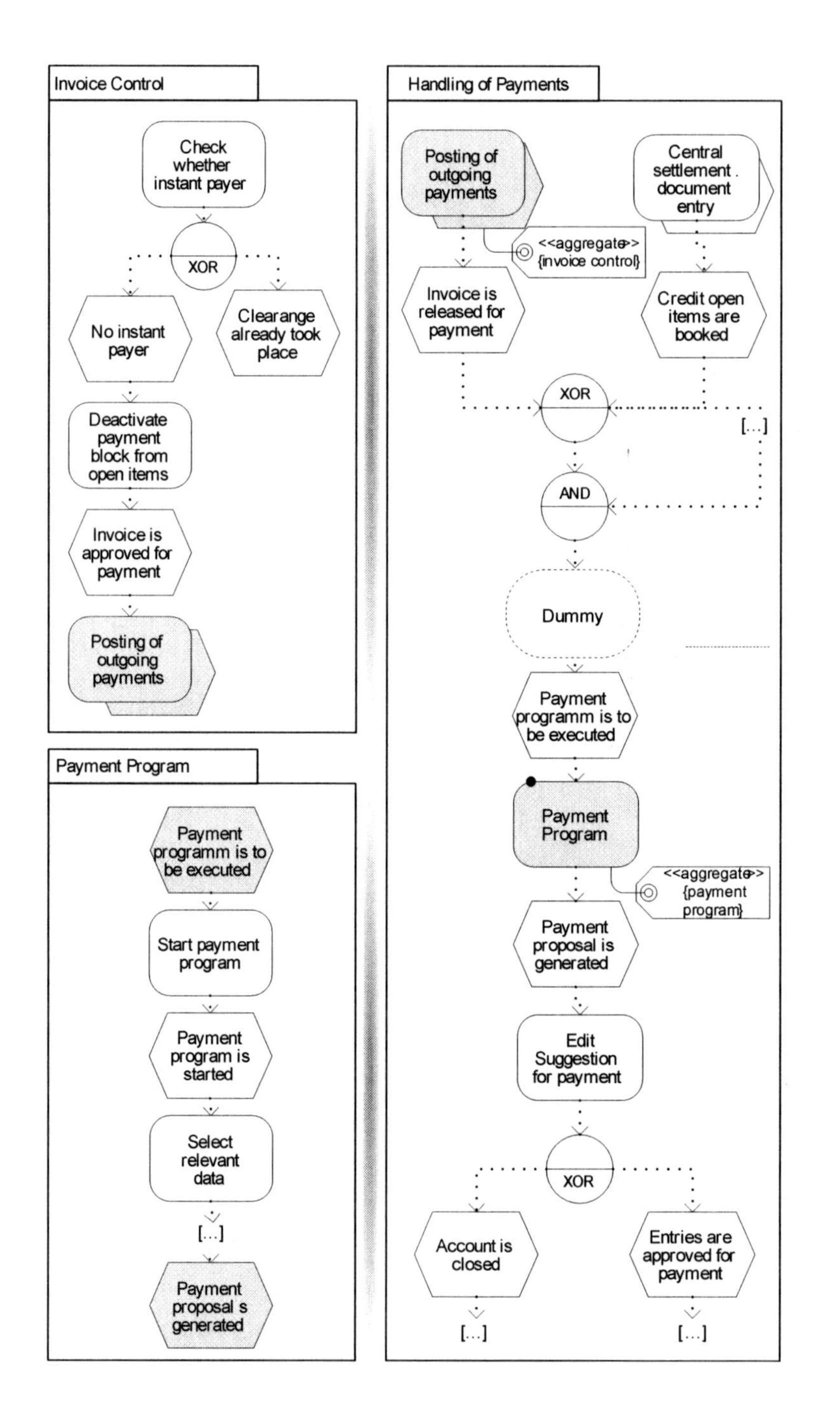

can be carried out by interlinking local steps of processes by means of the interface for linkage. Both examples are displayed in Figure 11. The examples show that in applying aggregation the events have to be handled properly. Flexibility of reuse is raised when the various events that possibly trigger a process are not modelled locally in the model of the sub-process itself but in the processes calling the sub-process. Then, an accumulation of events has to be handled. For that purpose, a dummy function was added in the example that serves to bridge events properly.

In the example, the process "Payment Program" is fully aggregated in the process "Handling of Payments" (Becker & Schütte, 2004, p. 392). In addition, one resulting event from the process "Invoice Control" is integrated in the process "Handling of Payments" as one trigger of this process, in the same way triggers from the processes of "Invoice Control" and "Central

Figure 12. Meta-model for the design principle of specialisation

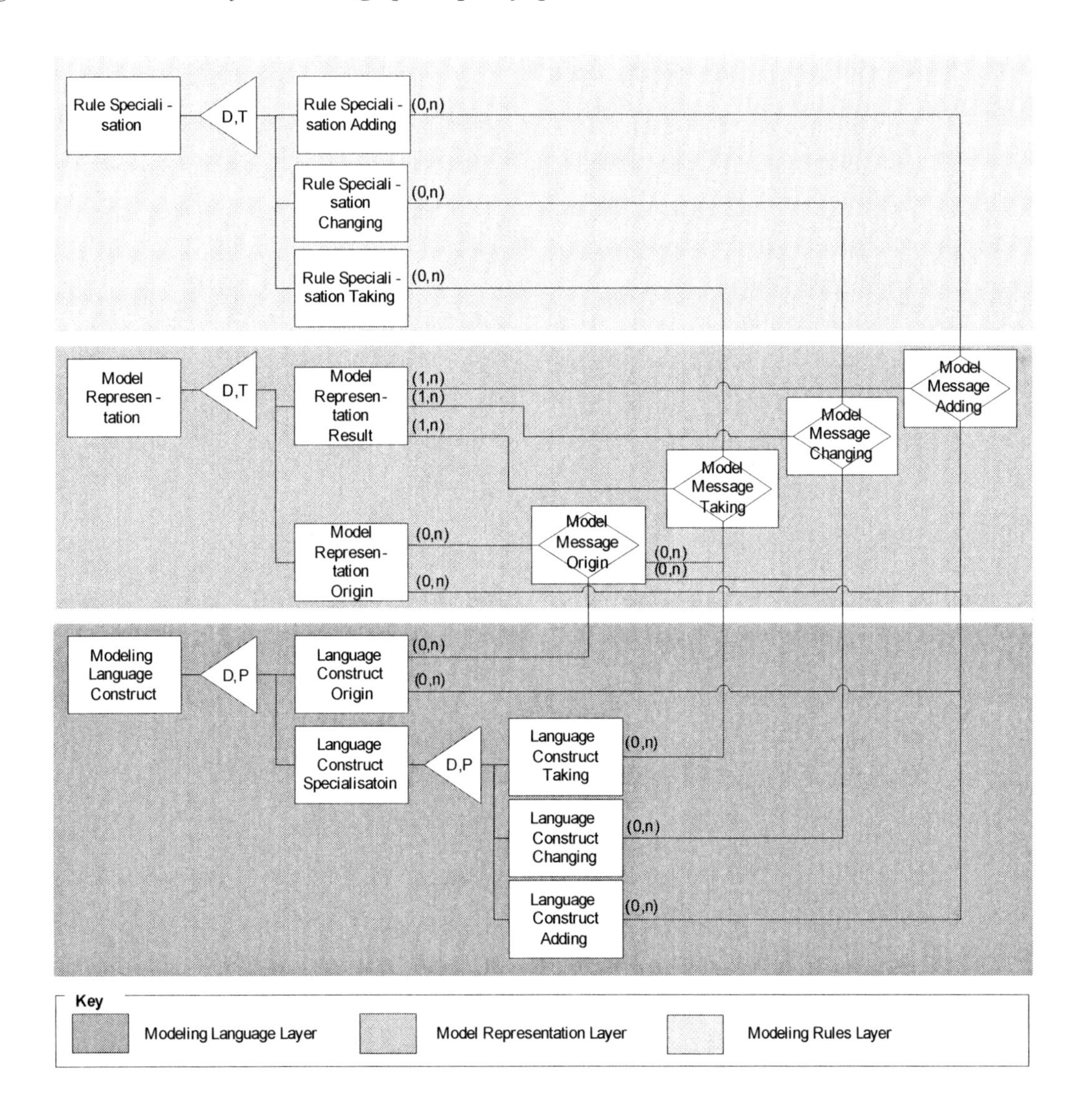

Figure 13. Example for applying the design principle of specialisation in ERM

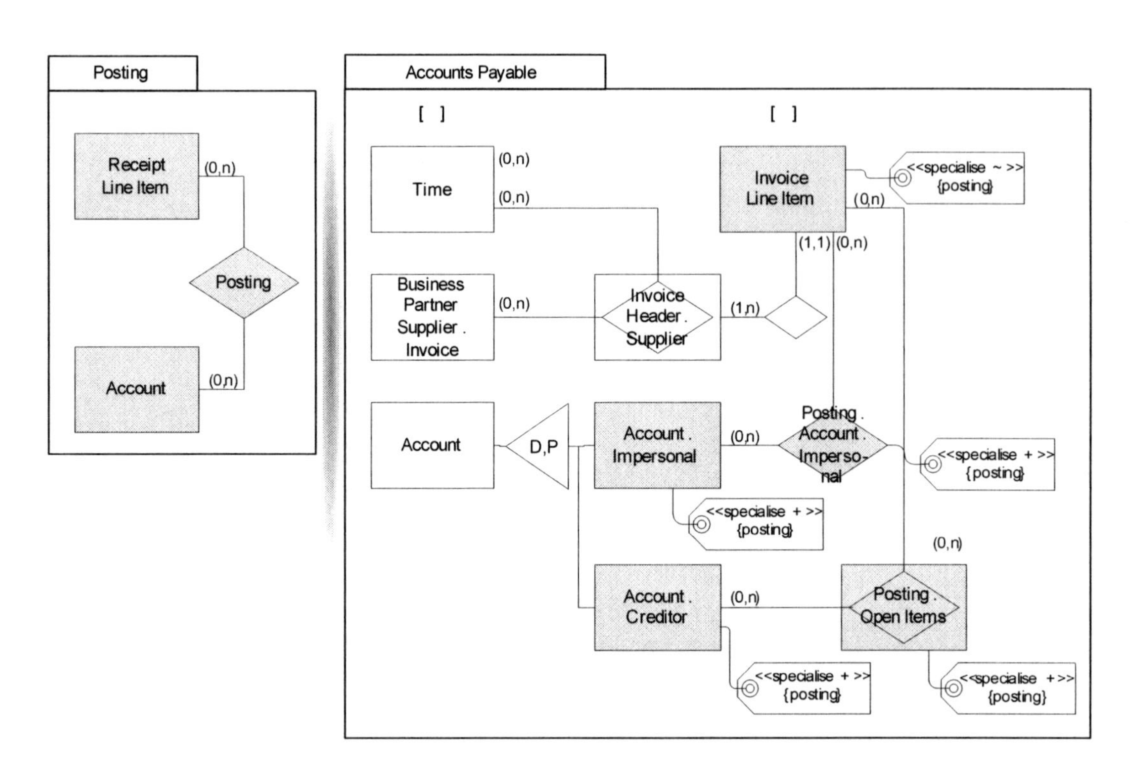

Settlement Document Entry" can be modelled according to aggregation.

For each model to be aggregated, the following steps need to be conducted:

1. Select model to be aggregated.
2. Select modeling messages to be integrated in the aggregated model.
3. Select or create the resulting model.
4. Iterate the process for each model to be aggregated; the messages will be transferred to the resulting model.
5. Integrate aggregated messages in the resulting model by additional messages according to the (new) scope of the resulting model.

The process may as well be executed top down. In that case the resulting model is selected at first and original models to be aggregated are selected on demand.

Specialisation

The specialisation is characterised by the derivation of a resulting model "S" from a general model "G." That way, all modeling messages in "G" are taken over in "S" and can either be changed or extended. Deleting messages, however, is not provided in the general case. Figure 12 shows the corresponding meta-model.

The specialisation allows the taking over of general construction results and adapting them to specific demands. In order to preserve the relation between the general and the specific model during amendments, constructs are required that help documenting which construction results have been both assumed and added. On a methodological level, rules can be provided according to which modeling messages, for example, do not necessarily have to be marked individually in the standard case.

Figure 14. Example for applying the design principle of specialisation in EPC

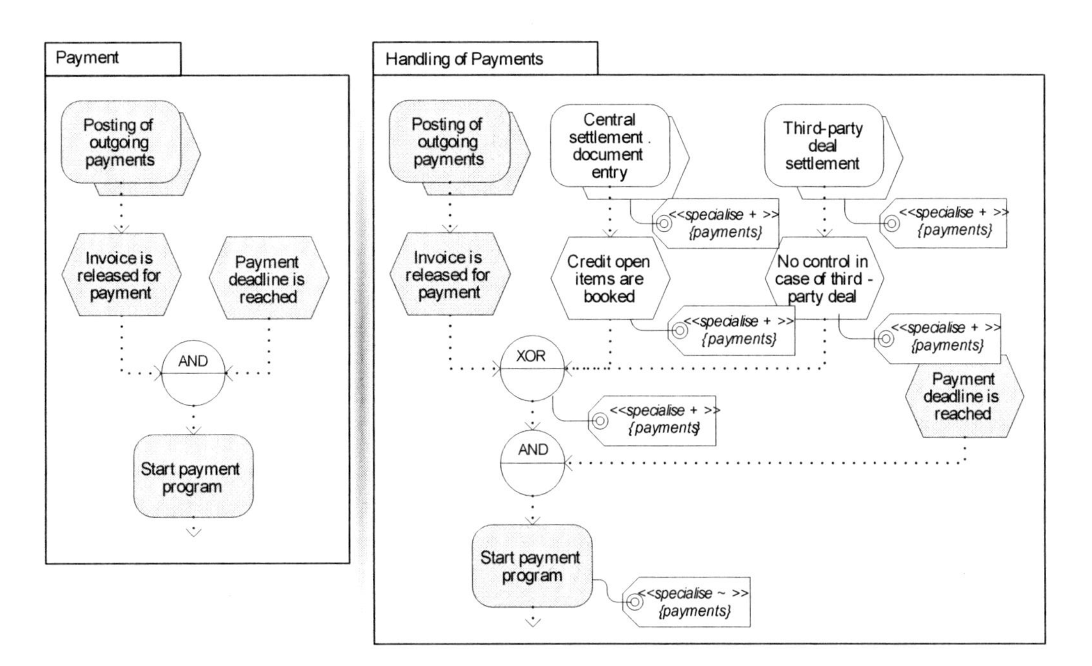

In order to illustrate the specialisation, a model for posting was generalised from the structure for the purchasing process given by Scheer (1994, p. 420). The general model was then reused twice for describing special posting structures in the model of "Accounts Payable" by Becker and Schütte (2004), p. 389). Modeling messages that have been amended like "Invoice Line Item" and "Posting Open Items," are characterised by the stereotype "«*specialise* ~ »." For messages that are added, the stereotype "«*specialise* + »" is used (see Figure 13).

In order to show how to conduct reuse by specialisation on the basis of EPC, a general model was generated that describes essential preconditions for payments to be executed. The model is based on Scheer (1994, p. 621) and Becker and Schütte (2004, p. 392). This model is reused by specialisation in the process of "Handling of Payments." The special model considers "Central Settlement" and "Third-Party Deal Settlement" so that additional statements were added. Furthermore, changes were made as a payment program is executed in the special model (see Figure 14).

In order to conduct model-reuse by specialisation, the following workflow has to be carried out:

1. Select a general model (e.g., in a tree view).
2. Run a specialisation service (e.g., in a context menu).
3. Either select or create a special model (e.g., in a tree view); the content of the general model may be transferred to the special model; the takeover may either be carried out by reference or a new instance of the model can be generated.
4. Adapt the special model by both changes and extensions (e.g., by standard features); the adaptations may be tracked automatically and displayed on demand.

Analogy

The principle of analogy is characterised by an original model "A" serving as a means of orien-

Figure 15. Meta-model for the design principle of analogy

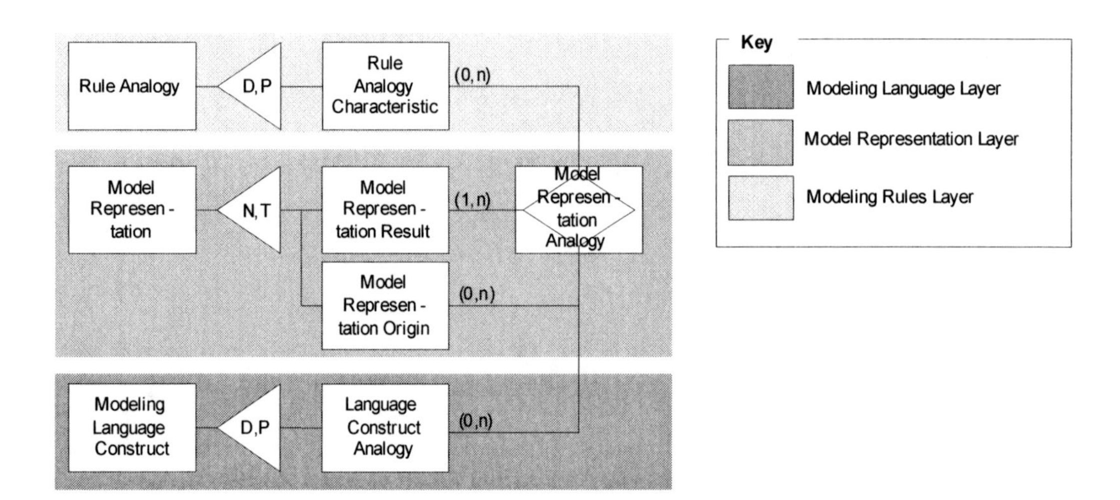

Figure 16. Example for applying the design principle of analogy in ERM

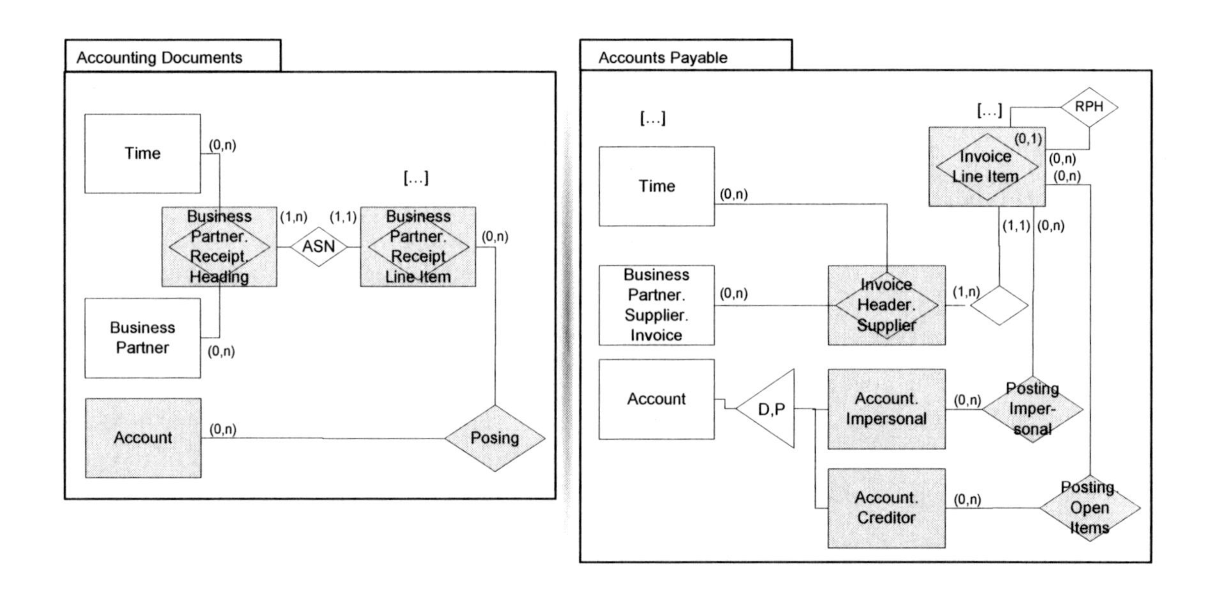

tation for the construction of a resulting model "a." The relation between the models is based on a perceived similarity of both models regarding a certain aspect. The according meta-model is presented in Figure 15.

The relatively high degree of freedom in re-using contents of an original model comes along with relatively little regulation on the model quality as opposed to other principles. Although no real means of formalisation between the model statements are required, a methodological support of the principle seems to be adequate. According to the principle of analogy, constructs may be introduced that serve to document the similarity relation perceived by the constructor. Standardisation of documentation can be realised by forms or text-based descriptions. In addition, a classification-based approach can be applied (vom Brocke, 2003).

Figure 17. Example for applying the design principle of analogy in EPC

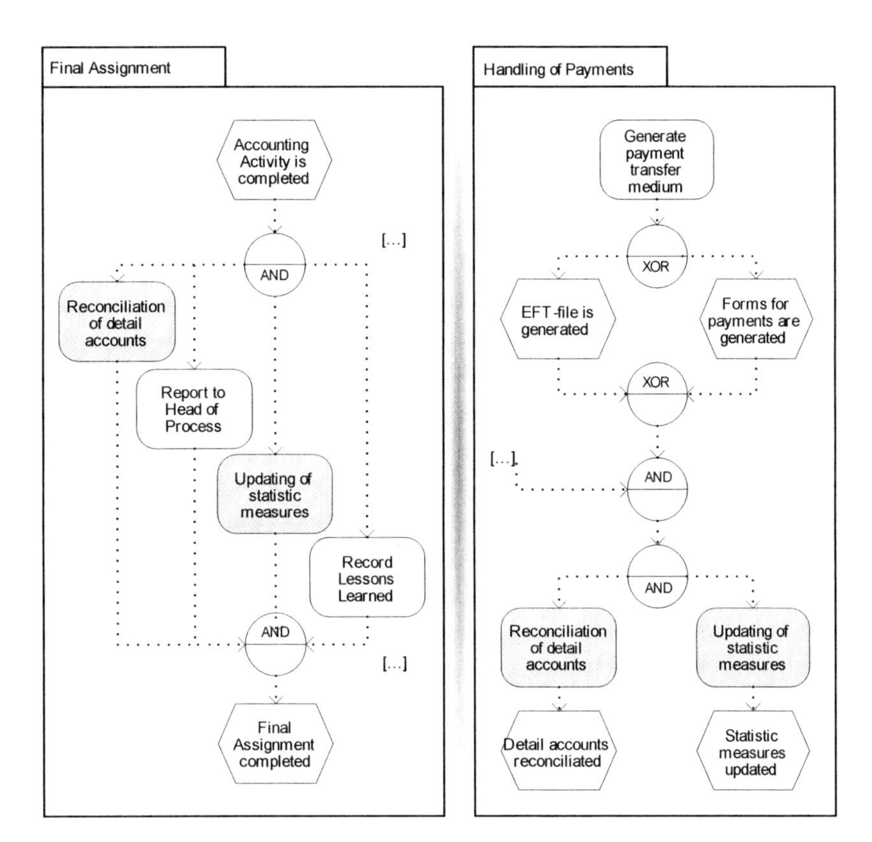

The application of analogy in ERM is illustrated by incorporating a common structure for accounting documents in the data model of "Accounts Payable." The general structure can be found either by Scheer (1994, p. 616) or by Becker and Schütte (2004, p. 386). The similarity of the models is evident in both content and structure. Both data models describe properties that are necessary for the booking of receipts. In the example, given in Figure 16, the structure is applied by analogy for posting an invoice on both the impersonal account and the open item list of creditors account. For documentation purpose, it may be indicated which statements refer to the reference model. Therefore, the stereotype "«*analogy*»" is used.

Also with EPC, the analogy can be carried out rather free-handedly. In the example given in Figure 17, a pattern for final assessment of processes in accounting is described (Scheer, 1994, pp. 619). The content of this model is partly reused, designing the final branch of the process "Handling of Payments" (Becker & Schütte, 2004, p. 392). As one major modification, the functions for reporting results and recording lessons learned are dropped out in the special model.

In order to conduct a construction by analogy, essentially the following steps have to be carried out:

1. Select or generate the resulting model.
2. Select a model to be reused by aid of analogy; the entire model is transferred to the resulting model.
3. Adapt the resulting model free-handedly; all modifications are admissible in the resulting model.

Figure 18. Addressing diverse modeling situations by multiple design principles

Principle	Situation	Technique
Configuration	The application domain can be described fully in design time including all relevant adaptations that have to be considered in various applications.	Adaptation by selection
Instantiation	The application domain can be covered by a general framework; this framework, however, has to be adapted in regard to selected aspects that can not be fully described while building the reference model.	Adaptation by embedding
Aggregation	The application domain can be described partly; each part can be fully specified whereas their contribution for replenishing the entire coverage of an application cannot be forseen when building the reference model.	Adaptation by combination
Specialisation	The application domain can be covered by a core solution, which has to be extended and modified (without deleting) in an indefinite manner for various applications.	Adaptation by revising
Analogy	The application domain can be described by certain patterns recurring in each application; the entire solution, however, has to be replenished in an indefinite manner.	Adaptation by transfer

Figure 19. Estimated derivation of the costs of building and applying reference models driven by design principles

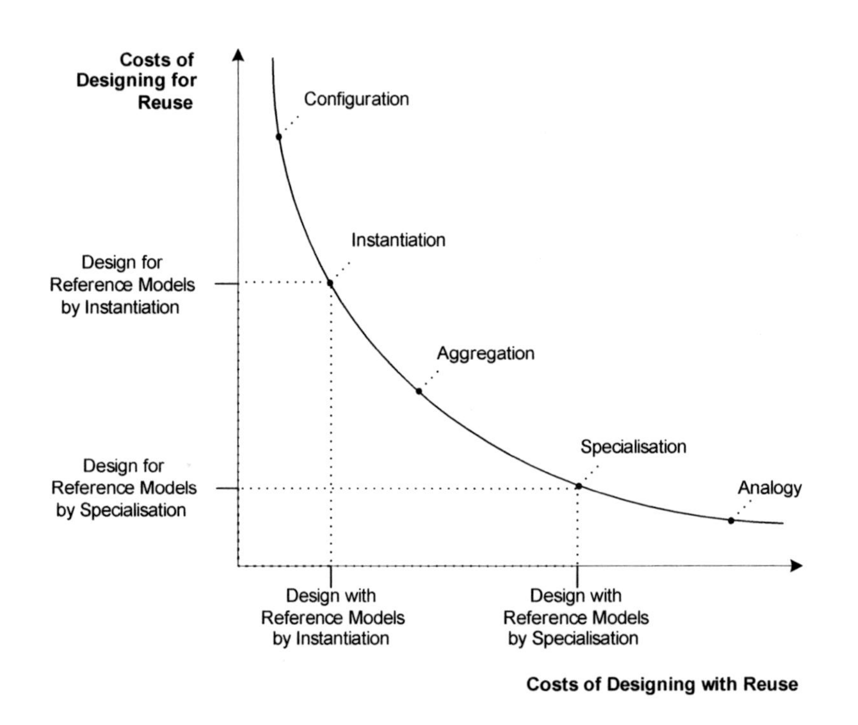

Figure 20. Estimated derivation of costs reuse driven by design principles in relation to complexity of the modeling situation

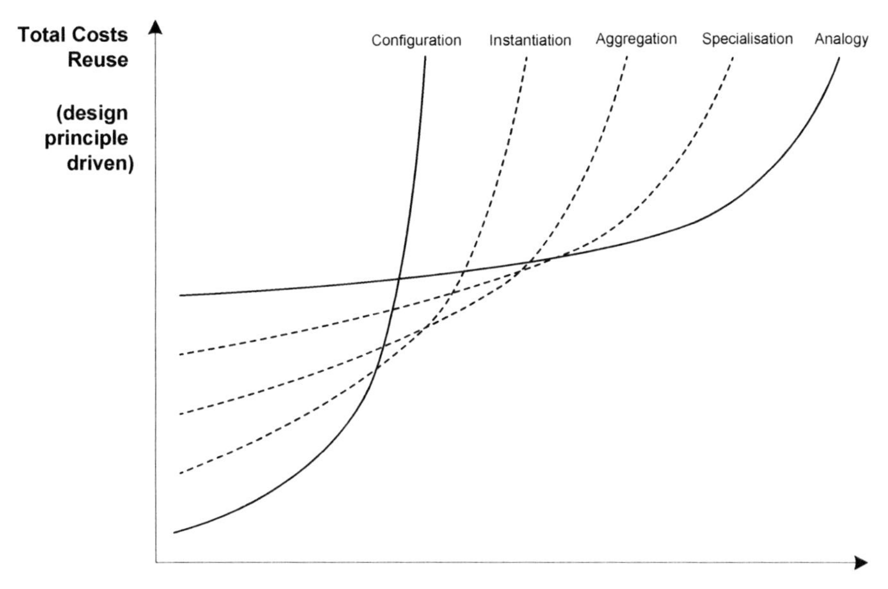

4. Document construction by describing the similarity perceived between both models, in order to raise comprehensiveness.

Having introduced the new design principles, a larger variety in reference modeling is apparent. The following chapter will discuss how far this variety might contribute to success in reusing information models.

POTENTIALS OF THE NEW DESIGN PRINCIPLES FOR REFERENCE MODELING

The wider set of design principles makes it possible to consider different requirements of the application domain in reference modeling. As each principle implements a special technique of reusing content, typical situations can be identified that call for the application of certain principles (see Figure 18).

For the appropriate principle selection, an estimation of the effect on the modeling process brought about by each principle is relevant. Such an estimation can be grounded on a transaction cost-based assessment (Coase, 1937; Williamson, 1985), similar to those in reuse-based software engineering (Mili et al., 2000). In this respect, transaction costs are considered additional costs that come along with reusing contents in modeling. Depending on the design principles, these "costs of reuse" essentially comprise costs of designing "for" reuse and costs of designing "with" reuse. Further transaction costs (e.g., for exchanging models; vom Brocke & Buddendick, 2004), are omitted as they are not significantly driven by the choice of construction construct.

In Figure 19, a rough estimation of these "costs of reuse" is given that may be taken as a hypothesis for further empirical studies. The graph is rather simplified in order to highlight the essential relations between the costs coming along with each principle. In particular, both plot and positioning of each principle within this plot vary depending on a certain modeling situation. The principles discussed in this chapter are marked by spots as they represent stereotypes for reuse. Further principles as well as adaptation of the principles may as well be added in between the spots.

Looking at the derivation of costs, it becomes apparent that configuration and analogy form two opposite principles for reference modeling. Whereas configuration implies that most of the work on reusing the content is done by building the reference model, this work is left for the application using analogy. Consequently, configuration comes along with relatively high costs for building the reference model, but with low costs for applying it. The principle of analogy, on the contrary, causes a minimum cost on building the model but a maximum on applying it. The other principles gradually lie in-between the two. Applying instantiation, prospective applications do not need to be specified entirely while building the reference model, but at least certain generic aspects have to be identified and specified for embedding special solutions. In aggregation, only certain parts of the application domain have to be described definitely that may be combined and extended in various ways. However, modifications of each part model to be aggregated are not provided. This is possible with the principle of specialisation which gives way to rather flexible modifications except eliminating parts of the reference model. With the principles of analogy, finally, unlimited ways of adapting the content are given.

With the different ways of reusing the content, not only consequences related to costs have to be taken into account. Since the shape of the later model is determined in different stages of the value chain, implications on the model quality come along with the choice of principle. However, no general suggestion can be given on the choice. An early determination of the model's shape, for example, facilitates quality assurance as individual modifications are restricted. In some cases, however, these modifications might as well be necessary to meet the specific requirements of a single application. In practice, these various factors have to be weighed regarding the special purpose of the model (e.g., customising and knowledge management).

Apart from the purpose, general decision support can be given with respect to the characteristics of the application domain. The interdependences described above show that the efficiency of the principles depends on how far it is possible to foresee relevant requirements of the application domain when building the model. The degree in how far requirements cannot be foreseen may be captured by the complexity of the application domain. Then, a basic estimation of the overall costs of reuse in relation to the complexity can be drawn. Figure 20 gives a brief insight.

The derivation of costs reflects that configuration is best in situations that can be foreseen rather clearly. Then, the easily adapting benefits comes into play and may rather compensate the efforts of designing variants. However, with growing complexity, either the costs of design explode or the models tend to cover the relevant cases too little and are, therefore, inadequate. Analogy, on the contrary, causes relatively high costs in situations that are characterised by low complexity. This is because of the generally high costs of design with reuse that are not justified by the degree of complexity. Although with growing complexity, analogy appears to be of raising efficiency. In addition, there might even come a degree from which analogy might be the only reasonable way of reusing artifacts.

Due to the great variety of modeling situations, a selective use of multiple design principles in reference modeling seems to be most adequate. Apart from choosing one principle for a modeling project that appears to be most suitable, also combinations of principles can be aimed at.

- **Model-specific combination:** The application domain of the model may consist of areas that show different characteristics with respect to the appropriate design principle. The level of complexity serves as an example: When modeling a retailing system, core processes may be rather standardised whereas additional processes for special cooperation can hardly be foreseen. In these cases, the entire reference model can be structured in partial models each applying special design principles.

- **Aspect-specific combination:** Also within a single representation of a model, the complexity of the modeling situation might differ with respect to special aspects of the model. Even in models with a rather high average complexity, for example, there are mostly some elements that can well be standardised, like those describing account characteristics and posting services. For these cases also, a combination of design principles within one representation of a model may be useful. Rules for the combination have to support the consistency of the model.
- **Time-specific combination:** Design principles may also be combined in time during the entire value chain of modeling. Opposed to configurative reference modeling, the principles can either support the application or the design of the reference model. First, the principles may be used in order to make further adaptations to a model that has been applied by configuration. Second, the principles serve as well to deliver reusable models that can be composed and adapted (Mili, et al., 1995, pp. 552-557) in the design of configurative reference models. For some applications, these techniques may as well be used in order to adapt the reference model in the field of application right away. In Figure 21, these situations are specified in more detail including the preconditions for time-specific combination.

Applying an appropriate mix of design principles might significantly contribute to both efficiency and effectiveness of reference modeling.

CONCLUSION

With this chapter, new design principles have been introduced to reference modeling. In particular, the principles of instantiation, aggregation, specialisation and analogy are supplied offering alternatives to principle of configuration that have been focussed on so far. The extended set of design principles provides greater flexibility in reference modeling. This also contributes to the evolution of models. New design principles allow the design of specialised fragments of models (e.g., by analogy) in order to combine them (e.g., by aggregation or instantiation) and to adopt them (e.g., by specialisation). Due to an increasing exchange of models, particularly positive effects on the model quality can be expected. Consequently, either configurative reference models may be fostered or part-models may be combined and adapted directly in the application domain. Especially with respect to an increasing dissemination of component-based and service-oriented information systems, this may be a promising means for future reference modeling.

REFERENCES

Alexander, C., Ishikawa, S., & Silverstein, M. (1977). *A pattern language*. New York.

Axenath, B., Kindler, E., & Rubin, V. (2005). *An open and formalism independent meta-model for business processes.* Paper presented at the First International Workshop on Business Process Reference Models (BPRM'05), Nancy, France.

Becker, J., Delfmann, P., Dreiling, A., Knackstedt, R., & Kuropka, D. (2004). *Configurative process modeling: Outlining an approach to increased business process model usability.* Paper presented at the Proceedings of the 2004 Information Resources Management Association Conference, New Orleans, LA.

Becker, J., & Schütte, R. (2004). *Retail information systems (in German)* (2. vollst. überarb., erw. u. akt. Aufl. ed.). Frankfurt a. M.

Chen, P. P.-S. (1976). The entity-relationship model: Toward a unified view of data. *ACM Transactions on Database Systems, 1*(1), 9-36.

Coad, P., North, D., & Mayfield, M. (1997). *Object models: Strategies, patterns, and applications.* NJ: Yourdon Press.

Coad, P., & Yourdan, E. (1991). *Object oriented analysis* (2nd ed.). Saddle Brook, NJ: Prentice Hall.

Coase, R. H. (1937). The nature of the firm. *Economica, 4* (11), 386-405.

Conradi, R., & Westfechtel, B. (1998). Version models for software configuration management. *ACM Computing Surveys (CSUR), 30*(2), 232-282.

Cox, B. (1990). Planning the software industrial revolution. *IEEE Software, 7*(8), 25-33.

D'Souza, D. (2000). Relating components and enterprise integration: Part 1. *JOOP, 13*(1), 40-42.

Fettke, P., & Loos, P. (2003). Classification of reference models: A methodology and its application. *Information Systems and e-Business Management, 1*(1), 35-53.

Fowler, M. (1997). *Analysis patterns: Reusable object models.* Menlo Park, CA: Addison Wesley.

Gamma, E., Helm, R., Johnson, R., & Vlissides, J. (1995). *Design patterns: Elements of reusable object-oriented design.* Menlo Park, CA: Addison-Wesley.

George, W. T. C., & Heineman, T. (2001). *Component-based software engineering: Putting the pieces together.* New York.

Gomaa, H. (1995). Domain modeling methods and environments. *ACM SIGSOFT Software Engineering Notes, 20*(SI), 256-258.

Hay, D. C. (1996). *Data model patterns: Conventions of thought.* New York: Dorset House Publishing.

Huschens, J., & Rumpold-Preining, M. (2006). IBM insurance application architecture (IAA): An overview of the insurance business architecture. In P. Bernus, K. Mertins, & G. Schmidt (Eds.), *Handbook on architectures of information systems* (2nd ed., pp. 669-692). Berlin: Springer-Verlag.

Jones, T. C. (1984). Reusability in programming, a survey of the state of the art. *IEEE Transactions on Software Engineering, 10*(5), 488-493.

Kang, K., C., Kim, S., Lee, J., Kim, K., Shin, E., & Huh, M. (1998). FORM: A feature-oriented reuse method with domain-specific reference architectures. *Annals of Software Engineering, 5*, 143-168.

Karagiannis, D., & Kühn, H. (2002). *Metamodelling platforms.* Paper presented at the Proceedings of the Third International Conference EC-Web, September 2-6, 2002, LNCS 2455, Dexa, Aix-en-Provence, France.

Karhinen, A., Ran, A., & Tallgren, T. (1997). *Configuring designs for reuse.* Paper presented at the Proceedings of the 19th International Conference on Software Engineering, Boston.

Keller, G., Nüttgens, M., & Scheer, A.-W. (1992). Semantische Prozeßmodellierung auf der Grundlage Ereignisgesteuerter Prozeßketten (EPK). *Veröffentlichungen des Instituts für Wirtschaftsinformatik der Universität des Saarlandes,* (89).

Kittlaus, H.-B., & Krahl, D. (2006). The SIZ banking data Model. In P. Bernus, K. Mertins, & G. Schmidt (Eds.), *Handbook on architectures of information systems* (2nd ed., pp. 723-743). Berlin: Springer-Verlag.

Kruchten, P. (2003). *The rational unified process: An introduction* (3rd ed.). Boston.

McIlroy, M. D. (1968). Mass produced software components. In P. Naur & B. Randell (Eds.), *Software Engineering, Report on a Conference by the NATO Science Committee* (pp. 138-150). Brussels: NATO Scientific Affaris Division.

Meinhardt, S., & Popp, K. (2006). Configuring business application systems. In P. Bernus, K. Mertins, & G. Schmidt (Eds.), *Handbook on architectures of information systems* (2nd ed., pp. 705-721). Berlin: Springer-Verlag.

Melton, R., & Garlan, D. (1997). *Architectural unification.* Paper presented at the Proceedings of the 1997 conference of the Centre for Advanced Studies on Collaborative research, Toronto, Ontario, Canada.

Mili, A., Fowler, S., Gottumukkala, R., & Zhang, L. (2000). *An integrated cost model for software reuse.* Paper presented at the Proceedings of the ACM, ICSE 2000, Limerick, Ireland.

Mili, H., Mili, A., Yacoub, S., & Addy, E. (2002). *Reuse-based software engineering.* New York.

Mili, H., Mili, F., & Mili, A. (1995). Reusing software: Issues and research directions. *IEEE Transactions on Software Engineering, 21*(6), 528-562.

Peterson, A. S. (1991). Coming to terms with software reuse terminology: A model-based approach. *SIGSOFT Softw. Eng. Notes, 16*(2), 45-51.

Raumbaugh, J., Blaha, M., Premerlani, W., Eddy, F., & Lorensen, W. (1991). *Object-oriented modeling and design.* Englewood Cliffs: Prentice Hall.

Recker, J., Rosemann, M., van der Aalst, W. M. P., & Mendling, J. (2005). *On the syntax of reference model configuration.* Paper presented at the First International Workshop on Business Process Reference Models (BPRM'05), Nancy, France.

Remme, M. (1995). *Systematic development of informations systems using standardised process particles.* Paper presented at the 3rd European Conference on Information Systems—ECIS '95, Athens, Greece.

Rumbaugh, J., Jacobson, I., & Booch, G. (2004). *The unified modeling language reference manual* (2nd ed.). Addison-Wesley Longman.

Scheer, A.-W. (1994). *Business process engineering: Reference models for industrial enterprises.* New York: Springer-Verlag.

Scheer, A.-W., & Nüttgens, M. (2000). ARIS architecture and reference models for business process management. In W. M. P. van der Aalst, J. Desel, & A. Oberweis (Eds.), *Business process managemen: Models,techniques, and empirical studies* (pp. 376-389). Berlin: Springer.

Slater, P. (1995). Output from generic packages. *ACM SIGAda Ada Letters, XV*(3), 76-79.

Spanoudakis, G., & Constantopoulos, P. (1993). *Similarity for analogical software reuse: A conceptual modelling approach.* Paper presented at the Proceedings of Advanced Information Systems Engineering.

Szyperski, C. (1998). *Component software. Beyond object-oriented programming* (Vol. 2). New York: ACM Press and Addison-Wesley.

Thomas, O. (2005). *Understanding the term reference model in information system research.* Paper presented at the First International Workshop on Business Process Reference Models (BPRM'05), Nancy, France.

van der Aalst, W. M. P., Dreiling, A., Gottschalk, F., Rosemann, M., & Jansen-Vullers, M. H. (2005). *Configurable process models as a basis for reference modeling.* Paper presented at the First International Workshop on Business Process

Reference Models (BPRM'05), Nancy, France.

vom Brocke, J. (2003). *Reference modelling, towards collaborative arrangements of design processes*. Berlin: Springer. [in German]

vom Brocke, J., & Buddendick, C. (2004). Organisation theory in reference modelling: Requirements and recommendation on the basis of transaction cost economics . *Wirtschaftsinformatik, 46*(5), 341-352. [in German]

Williamson, O. E. (1985). *The economic institutions of capitalism.* New York: Tree Press.

ENDNOTE

1 As the special subject of "reference modeling" has until now been discussed predominantly in German literature, some sources that are considered in this chapter are written in German. Apart from major pieces, reviews on the subject's works have been referred to as far as possible. In addition, the reuse-oriented concept of reference modeling that is introduced in this chapter gives way to a discussion of the subject against the background of international works in the field of reusing artifacts in information systems.

This work was previously published in Reference Modeling for Business Systems Analysis, edited by P. Fettke and P. Loos, pp. 47-76, copyright 2007 by IGI Publishing (an imprint of IGI Global).

Chapter XV
Examining the Quality of Evaluation Frameworks and Metamodeling Paradigms of Information Systems Development Methodologies

Eleni Berki
University of Jyväskylä, Finland

ABSTRACT

Information systems development methodologies and associated CASE tools have been considered as cornerstones for building quality in an information system. The construction and evaluation of methodologies are usually carried out by evaluation frameworks and metamodels - both considered as meta-methodologies. This chapter investigates and reviews representative metamodels and evaluation frameworks for assessing the capability of methodologies to contribute to high-quality outcomes. It presents a summary of their quality features, strengths and weaknesses. The chapter ultimately leads to a comparison and discussion of the functional and formal quality properties that traditional meta-methodologies and method evaluation paradigms offer. The discussion emphasizes the limitations of both methods and meta-methods to model and evaluate software quality properties such as computability and implementability, testing, dynamic semantics capture, and people's involvement. This analysis along with the comparison of the philosophy, assumptions, and quality perceptions of different process methods used in information systems development, provides the basis for recommendations about the need for future research in this area.

INTRODUCTION

In traditional software engineering, the information systems development (ISD) process is defined as a series of activities performed at different stages of the system lifecycle in conformance with a suitable process model (method or methodology). In the fields of Information Systems and Software Engineering, the terms methodology and method are often used interchangeably (Nielsen, 1990; Berki et al., 2004). Increasingly, new methods, techniques and automated tools have been applied in Software Engineering (SE) to assist in the construction of software-intensive information systems. Quality frameworks and metamodels are mainly concerned with the evaluation of the quality of both the process itself and the resulting product at each stage of the life cycle including the final product (the information system).

Professional bodies such as IEEE and ISO have established quality standards and software process management instruments such as Software Process Improvement and Capacity dEtermination (SPICE) (Dorling, 1993) and Capability Maturity Model (CMM) (Paulk et al., 1993) have focused on the quality properties that the ISD process should demonstrate in order to produce a quality information system (Siakas et al., 1997). However, software quality assurance issues (Ince, 1995) such as reliability (Kopetz, 1979) and predictability, measurement and application of software reliability in particular (Myers, 1976; Musa et al., 1987), have long preoccupied software engineers, even before quality standards.

IS quality improvement can be achieved through the identification of the controllable and uncontrollable factors in software development (Georgiadou et al., 2003). ISD methodologies and associated tools can be considered as conceptual and scientific ways to provide prediction and control; their adoption and deployment, though, by people and organizations (Iivari & Huisman, 2001) can generate many uncontrollable factors. During the last thirty-five years, several methodologies, techniques and tools have been adopted in the ISD process to advance software quality assurance and reliability. A comprehensive and detailed coverage of existing information systems development methodologies (ISDMs) has been carried out by Avison & Fitzgerald (1995), with detailed descriptions of the techniques and tools used by each method to provide quality in ISD.

Several ISDMs exist. Berki et al., (2004) classified them into families, highlighting their role as quality assurance instruments for the software development process. They have been characterized as hard (technically oriented), soft (human-centered), hybrid (a combination of hard and soft), and specialized (application-oriented) (Berki et al., 2004). Examples of each include:

- Hard methods - object-orientated techniques, and formal and structured families of methods;
- Soft methods - Soft Systems Method (SSM) and Effective Technical and Human Implementation for Computer-based Systems (ETHICS)
- Hybrid methods - Multiview methodology, which is a mixture of hard and soft techniques;
- Specialized methods - KADS, extreme programming (XP) and other agile methods.

The contribution of these methods to the quality of the ISD process has been a subject of controversy; particularly so because of the different scope, assumptions, philosophies of the various methods and the varied application domains they serve. For example, it is believed that the human role in ISD bears significantly on the perception of the appropriateness of a method (Rantapuska et al., 1999); however, usability definitions in ISO standards are limited (Abran et al., 2003). There is empirical support for the notion that a methodology is as strong as the user involvement it supports (Berki et al., 1997).

Two other significant problems are associated with currently available ISDMs and their capability to contribute to process quality. First, the inability of most to incorporate features of both hard and soft methods inevitably leads to deficiencies resulting in either technical or human problems. Second, the software architecture of an IS often lacks testable and computational characteristics because they are not specified as an integral part of the design (Berki, 2001).

Considering the importance of Software Quality Management (SQM) and Total Quality Management (TQM) issues, software engineers and IS managers have, for the last twenty years, realized the significance of the use of frameworks for method evaluation and the use of metamodels, precise definitions of the constructs and rules needed for creating semantic models, for their own method construction. These - usually accompanied by automated software tools - are typically used during requirements analysis and for early testing and validation of requirements specifications (Manninen & Berki, 2003) as design products. This is done in order to ensure that the resulting artifacts, prototypes and final information products will experience lower maintenance costs, greater reliability and better usability.

Over ten years ago, Jayaratna (1994) identified more than two thousand methodologies in use and it is reasonable to assume that several others have been added since. What is the basis for deciding which are useful and which are not (Chisholm, 1982)? There is justification for objective criteria, which take into account the type of IS to be developed, its environment, the people that will interact with, and the properties of the development methods that contributes to a quality process and product, to inform this decision process. Several such evaluation and selection frameworks exist (Law, 1988; Law & Naeem, 1992). However, analysts and designers can also build their own in order to fit the needs of system modeling tasks more precisely by using method construction metamodels with Method Engineering (ME) rules (Brinkkemper, 1996; Brinkkemper, 2000).

This chapter provides a comprehensive overview of how to construct efficient and cost-effective metamodels and evaluation frameworks and identify their quality properties in a scientific and reliable way. This task may be approached in a variety of ways. However, the approach recommended by Sol (1983) and Jayaratna (1994), to develop a common frame of reference for evaluating different methodologies, will be adopted in this study. It has been used to construct the Normative Information Model-based Systems Analysis and Design (NIMSAD) framework (Jayaratna, 1994) and for Metamodelling-Based Integration of Object-Oriented Systems Development (Van Hillegersberg, 1997).

This research approach is considered appropriate mainly because the construction of a framework is less subjective than other approaches for evaluating and comparing methods (and metamethods); it sets criteria for method evaluation, comparison and construction from a different, more or less detailed, level of view. The evaluation criteria used in this chapter incorporate elements involving soft and hard issues, year of introduction, tool support, computability and testing, and allow for the examination of the initiatives and their rules, viewing them under the ISD needs as these appeared chronologically. Before proceeding to the examination of the quality properties of contemporary methods through the lens of different evaluation frameworks and metamodels, the next section presents a few concepts related to method evaluation frameworks, and highlights the differences and similarities of metamodels and ME.

METAMODELS, EVALUATION FRAMEWORKS AND META-INFORMATION MANAGEMENT

BASIC CONCEPTS

A metamodel is an information model of the domain constructs needed to build specific models. Data elements include the concepts of Class, Event, Process, Entity, Technique, and Method. In order to create a metamodel one needs a language in which metamodeling constructs and rules can be expressed. In the context of ISD methodologies, metamodeling facilitates the application of the principles of ME (Brinkkemper, 1996) in which IS engineers examine characteristics of the particular method(s) at a different, higher or lower, level of abstraction (meta-level), viewing the functions and objectives of the IS as a restructured abstract view of the system under development.

In general, metamodeling is an important tool for both decomposition and generalization of the complex, dynamic processes that are activated during the development and maintenance of systems. Important systems engineering concepts are viewed in different design compositions and decompositions; thus metamodeling, as a system's lifecycle activity, provides a basis for decision making. A generally accepted conceptual metamodel for method construction usually explains the relationships between metamodels, models, methods and their techniques, and end-user data. All these form the semantic and syntactic constructs of the metamodel, and they may belong to the same or different layer(s) of abstraction. Metamodels usually facilitate system and method knowledge acquisition and sometimes metadata and information integration (see Kronlof, 1993).

Evaluation frameworks offer generalizations and definitions of the quality properties of methods in order to reveal their characteristics in use. A framework can also be considered as a metamodel; while a methodology is an explicit way of structuring one's thinking and actions (Jayaratna, 1994). In addition, Jayaratna believes that a framework differs from a methodology in that it is a 'static' model, providing a structure to help connect a set of models or concepts. In this context, a framework can be perceived as an integrating meta-model, at a higher level of abstraction, through which concepts, models and methodologies can be structured, and their interconnections or differences displayed to assist understanding or decision-making.

Although the terms metamodel and framework are used interchangeably in the IS and SE literature (Jayaratna, 1994), the current research maintains that they should be distinct. According to the epistemology and applications of metamodeling in Metacomputer-assisted system engineering (MetaCASE) and computer-assisted method engineering (CAME) environments, a 'metamodel' is a more, specific, dynamic, and well-defined concept that leads to well-structured principles of method engineering (ME). A 'framework', on the other hand, is more general and static, with less structured evaluation guidelines (Berki, 2001). These distinctions will become more evident in the section that describes method construction metamodels and method evaluation frameworks.

Rationale for the Use of Metamodels and Evaluation Frameworks

The rapid societal changes in advanced information societies, the increasing attention to cognitive issues, the high cost of software development and IS maintenance, and the requirements of legacy IS in particular, point to the need for more reusable, more reliable information systems that are acceptable by humans. The communication between IS end-users, managers, analysts, designers and programmers has been a major problem in ISD, yet is pivotal in delivering high-quality IS (Berki et al, 2003). Hence the research community, method designers and method practitioners are forced to focus on the construction and use of more dynamic ISD methods.

Consequently, there is an increasing demand for more supportive software tools that cater to method integration and the reengineering of IS processes and methods. Conventional methods have been inadequate to date, largely because of the lack of structured support for overcoming human cognitive limitations in the process of transforming requirements to design, and testing and implementing the results in a reliable manner. If, for instance, the production of computable (i.e., programmable, implementable and realisable) specifications from analysis and design phases could be automated and tested reliably, then this could greatly improve the capability to address software quality assurance and eventually improve IS quality.

Several information systems development methodologies (ISDMs) have been invented to assist with managing and improving software development processes. The general view is that such methodologies are inherently complex, and implementing them in practice is difficult or even impossible without the support of automated tools within a software development environment (SDE). However, the diversity of methodologies and their evolving nature require that corresponding SDEs provide flexibility, adaptability and extensibility (Jarzabek, 1990). This complicates decision-making about which ISDMs to adopt and implement and created a need for structured assistance to evaluate the capability of methodologies and their appropriateness for a particular organisation.

Several software disasters such as Arriane-5 space rocket accident and many other software failures – with less disastrous consequences - attributable to design errors, and the large and aging IS infrastructure that we have inherited, underscore the need for facilities to help re-engineer, test and improve software development methods and models. In general, process methods and models provide significant assistance in incorporating software quality management practices in ISD; several exist - typically customised to organizational and stakeholder preferences - to address the many interpretations of IS quality and pursue fairly elusive quality objectives (Berki et al, 2004). However, there is an urgency to address these issues at the ME level and from the metamodeling point of view (Berki, 2001).

Metamodeling and ME tools are relatively new software development technologies designed to improve process models and methods used in ISD. Typically, they are supported by automated tools with broad capabilities to address the various facet s of producing a quality process and product. Because organizations can modify methodologies, there is also a need for these tools to accommodate modifiability- or method re-engineering (Saadia, 1999), and to support other fundamental quality properties for the ISD process such as testability (Beizer, 1995; Whittaker, 2000), computability (Wood, 1987; Lewis & Papadimitriou, 1998) among others. They should also cater to softer components of the ISD such as the roles and interactions of people.

Summarised, the theory and argument in favor of establishing formal and generic definition of meta-methods and metamodels that incorporate quality requirements and emphasize the dynamic nature of ISD is that (1) establishing evaluation frameworks for ISDMs provides insights to optimize the selection of methods that better fit the development context, and (2) improving the computational representation of methods and models leads to improved knowledge acquisition, which enhances communication among IS stakeholders. Both increases the likelihood of less ambiguous specifications, better system design, which are easier to implement, and formally test, and eventually systems that are modifiable and maintainable.

AN EXAMINATION AND BROAD CLASSIFICATION OF METHOD EVALUATION FRAMEWORKS AND METHOD ENGINEERING METAMODELS

Typically, different ISD methods model different application domains. So, different frameworks/ metamodels may concentrate on different aspects of the use of methods under evaluation and/or being compared. In general, frameworks, like methods, can be classified as either human-action oriented (soft) or more technical (hard) in their evaluation of the quality characteristics of methods (Berki et al., 2004) and may be applied with or without automated tool support. However, unlike methods, frameworks tend to project those quality features of the methods that make them distinct and different or similar to other methods currently available. Some of the semantic and pragmatic quality features and the knowledge employed in their definition are outlined below in various metamodeling examples.

Euromethod

Euromethod (CCTA, 1994; Jenkins, 1994) is a framework for facilitating harmonization, mobility of experts and expertise, the procurement process and standards, market penetration by vendors, and quality assurance across countries of the European Union. It builds on structured methodologies and techniques of methods from Structured Systems Analysis and Design Method (SSADM), the UK standard, Merise, the French standard, DAFNE the Italian standard for software development, SDM (Netherlands), MEthodologica INformatica (MEIN) (Spain), Vorgehensmodell (German standard) and Information Engineering (IE) (US/UK).

The goal of Euromethod is to provide a public domain framework for the planning, procurement and management of services for the investigation, development and amendment of information systems (Avison & Fitzgerald, 1995). As a metamethod, it caters to the facilitation of tacit and explicit knowledge exchange among European countries and the standardization of disparate frameworks that encompass the many different ISD philosophies and cultures. However, it does not include information systems services provided by an in-house IT facility. Euromethod, could serve as an integrating structured methods standard for ISD in European or other social contexts given the strong impact of cultural differences and other national and organizational diversities which Siakas, Berki, and Georgiadou (2003), found affected software quality management strategies in European countries.

A Generic Evaluation Framework

Avison & Fitzgerald (1995) used their generic evaluation framework to tabulate the strength of coverage of a set of the most widely known methodologies over the phases of a lifecycle. This generic framework uses ideas presented in Wood-Harper et al.'s (1985) discussion of approaches to comparative analysis of methods. It includes the seven elements - Philosophy, Model, Techniques and Tools, Scope, Outputs, Practice and Product, some of which are further broken down into sub-elements. For example, Avison & Fitzgerald observed that Jackson Systems Development (JSD) and Object-Oriented Analysis (OOA) fail to cover strategy and feasibility while the majority of structured methods such as Yourdon and SSADM, concentrate in analysis and design.

This generalized framework can be considered an abstract typology approach for method analysis and comparison for examining individual methods and comparing the quality of methods. In so doing, the framework focuses on the quality of the method itself (philosophy, model, scope), on the quality of ISD process (model, techniques and tools) and on the quality of the future IS (outputs, practice and product).

Determining an Evaluation Method for Software Methods and Tools (DESMET)

DESMET resulted from the collaborative efforts of academic and industrial partners University of North London, The National Computing Centre, Marconi Systems, and British Nuclear Research. It was funded by the Department of Industry and the Science and Engineering Council, in the UK. The main objective of DESMET is to objectively determinate the effects and effectiveness of methods and tools used in the development and maintenance of software-based systems (DESMET, 1991; Georgiadou et al., 1994). The DESMET framework addresses this problem by developing an evaluation meta-methodology, which identifies and quantifies the effects of using particular methods and tools on developers' productivity and on product quality. The framework was validated by carrying out a number of trial evaluations.

Apart from providing guidelines for data collection and metrication (Kitchenham et al., 1994), DESMET emphasizes managerial and social issues that may affect evaluation projects (Sadler & Kitchenham, 1996), (Berki & Georgiadou, 1998). The main contribution of DESMET is the explicit and rigorous evaluation criteria selection and evaluation method selection. Previous evaluation frameworks made little attempt to incorporate the work on evaluation - based on formal experiment - into commercially-oriented evaluation procedures (DESMET, 1991).

Further work based on the DESMET Feature Analysis module was carried out at the University of North London resulting in the development of the automated tool Profiler (Georgiadou et al., 1998). The Profiler metamodeling tool was developed according to the DESMET feature analysis and ISO standard rules. It is a generic evaluator (i.e. of methods, processes, products and other artifacts), which was also used in some stages of this investigation of method evaluation (Georgiadou, E., Hy, T. & Berki, E., 1998).

Tudor and Tudor's Framework

Tudor & Tudor (1995) use the business lifecycle to compare the structured methods IE, SSADM, Yourdon and Merise (Avison & Fitzgerald, 1995) with Soft Systems Method (SSM) (Checkland & Scholes, 1990) and Multiview (Wood-Harper et al., 1985). The framework builds on the seven elements of Avison and Fitzgerald's (1995) framework to assess the performance of these methods on a scale of Low, Medium, High, using criteria defined in DESMET and other characteristics (framework constructs) such as size of problem, use of techniques and CASE tools, and application domain (Tudor & Tudor, 1995). The framework determines the extent of lifecycle coverage provided by methods, how highly structured the methods are, and what kind of system the methods target. It also addresses issues such as user involvement, CASE tool support and different implementations of similar or different techniques employed by structured methods.

Socio-Cybernetic Framework

The socio-cybernetic framework (Iivari and Kerola, 1983) has made an important contribution to the quality feature analysis of information systems design methods. The framework is based on a socio-cybernetic interpretation of information systems design as human action and information systems design methodologies as prescriptive instructions for this action (, 1983). According to Iivari & Kerola, when used in method evaluation or systems engineering, this framework covers scope and content, support and usability, underlying assumptions, generality of the methodologies, as well as scientific and practical value, in eighty-five basic questions. The application of the framework to the evaluation of selected methodologies was carried out successfully, the results of which can be found in Olle et al. (1982).

Normative Information Model-Based Systems Analysis and Design (NIMSAD)

NIMSAD is a systemic (holistic) framework for understanding and evaluating problem-solving processes methodologies in general. This meta-method, which was developed through action research, evaluates both soft and hard methods before, during, and after industrial use. The NIMSAD framework is based on a general model of the essential elements of a problem situation and their formal and informal interconnections and relationships (Jayaratna, 1994). Jayaratna emphasizes that these interconnections, mainly human relations, work functions, technical interactions and others, are dynamic, and are dependent on the time and space and the perceiver's personal values, aspirations and worldview (or Weltanschauung).

Metamodeling, Principles, Hypertext, Objects and Repositories (MetaPHOR)

(MetaPHOR) is a project framework that was developed at Jyvaskyla University in Finland and was later commercialized by MetaCASE Consulting Ltd. It defines and supports metamodeling and ME principles for innovative ISD method construction. The metamodel supports object-oriented development philosophies and implementation tools, distributed computing environments, and modern user interaction management systems (MetaPHOR website). Its objectives include the investigation and assessment of theoretical principles in different metamodeling areas and the evolution of development methods (Brinkkemper, 1996) and metamodeling in ISD (Marttiin et al., 1992). The MetaPHOR project produced the MetaEdit (later extended to MetaEdit+) CASE tool, which provided techniques of automated methods to be used as systems analysis and design models MetaCASE tool (Kelly et al., 1996).

The MetaCASE tool has been used successfully to model processes and methods in industry and academe because it incorporates the process modeling view (Koskinen, 2000). However, it does not support testing and dynamic method modeling (Tolvanen, 1997; Berki, 2001). Recent enhancements of the MetaPHOR group include method testing (Berki, 2003), utilization of hypertext approaches in MetaEdit+ (Kaipala, 1999), component-based development tools and methods (Zhang, 2004) and cognitive needs of information modeling (Huotari & Kaipala, 1999).

CASE-Shells

CASE-Shells or Meta-CASE systems are capable of supporting an unlimited number of software engineering environments, instead of one modeling space, which is the norm for most of the traditional CASE systems. In CASE-Shells, the environments are modeled separately from the main software (Marttiin et al., 1993; Kelly, 1997). This initiative mainly considers the metamodeling technology that relates to the production of CASE-Shells and Meta-CASE tools - the methods, specifications and other products that identify, define and implement systems specifications. The RAMATIC tool (Bergsten, 1989), which originated in Sweden, uses modified rules of the MetaEdit+'s main architecture to define methods and their environments, and the MetaView system form Canada that is presented next, are two popular CASE-Shells implementations.

The MetaView System

The MetaView initiative is mainly supported by a CASE meta-system, i.e. a CASE-Shell, which acts as a framework for improving Structured and Object-Oriented software development methods.

MetaView is a joint project of the University of Alberta and the University of Saskatchewan, and is evolving (MetaView website). The MetaView system is divided into three levels (Sorenson et al., 1988), namely:

- Meta Level (creation of software components)
- Environment Level (definition of the environment and configuration of software) and
- User Level (generation of user-developed software specification)

The tool's basic architecture for metamodeling is based on the core principles of data flow diagramming techniques, inheriting their strengths and weaknesses (Sorenson et al., 1988; Berki, 2001).

CASE Data Interchange Format (CDIF)

CDIF: is a set of standards for defining a metamodeling architecture for exchanging information between modeling tools, describing the different modeling languages in terms of a fixed core model. It was developed as a formal metamodeling approach to accommodate interoperability among models, using an integrated metamodel for common representation. It facilitates communication between structured and O-O techniques and their semantic extensions, and state event modeling.

The CDIF metamodeling approach is rooted in theoretical models from Computer Science; for example, the principles of the Mealy-Moore Machine (Wood, 1987) and Harel state charts (Harel, 1987). It provides the means to view and derive provable quality properties from different techniques and systems models. However, according to the initiative's website, CDIF focuses on the description of information to be transferred and not on data management. (Burst et al., CDIF website).

Computational and Dynamic Metamodel as a Flexible and Integrated Language for the Testing, Expression and Re-Engineering of Systems (CDM-FILTERS)

The construction of CDM-FILTERS was based on experimental modeling steps that led to a formal framework, the Formal Notation for Expressing a Wide-range of Specifications (Formal NEWS). Berki, (1999) suggested the generalization and extension Formal NEWS into a metamodel to re-construct and evaluate ISD. The resulting metamodel (CMD-FILTERS) was sponsored by the Universities of North London and Sheffield and involved extensive field research in Finland (Jyvaskyla University and MetaCASE Consulting Ltd.). IT accommodates method modeling and remodeling as general computational automata (or X-Machines) (Ipate & Holcombe, 1998) applied to and MetaCASE tools and architectures (Berki, 2003). The main strengths of the approach are its capability to generate an integrating metamodel for ME, combining method dynamics, computability and testing. It is, however, limited in its ability to incorporate human issues, and it is not supported by many tools.

OTHER METHOD METAMODELING AND METHOD INTEGRATION PROJECTS

There are several other metamodeling architectures, designed for application in particular contexts to address specialized issues with or without MetaCASE support, such as ConceptBase: A metadata base for storing abstract data representation of data warehouse applications (Jarke et al., 1995; Mylopoulos et al., 1990; Vassiliou (2000). Likewise, there are many ongoing projects at various stages of development (e.g., MethodMaker (Mark V), System Architect

(Popkin), Toolbuilder (Sunderland/IPSYS/Lincoln) and some interesting research prototypes such as KOGGE (Koblenz), MetaGen (Paris), MetaPlex tools. Readers may refer to Smolander et al. (1990), Graham (1991), Avison & Fitzgerald (1995), Kelly (1997), and Shapiro & Berki (1999). For more detailed information on the commercial availability, application domains and modeling limitations of methods mentioned here.

One criticism of most metamodels is their limited capacity to evaluate human activity, including the roles of meta-modellers and method engineers. Two initiatives, however, address this issue. First, Requirements Engineering Network of International cooperating Research (RENOIR), a requirements engineering network supported by the European Union, provides the tools, concepts and methods that mediate between the providers of information technology services and products, and the users or markets for such services and products (RENOIR website). Then Novel Approaches to Theories Underlying Requirements Engineering (NATURE), an ongoing collaborative project, funded by the European Framework IV research programme, attempts to integrate existing metamodels, methods and tools through the formal definition of basic principles of metamodels (Jarke & Pohl, 1994).

Table 1. Information on method evaluation frameworks and metamodeling projects

	Hard	Soft	Tool Support	Year of Introduction	Testing	Comput-ability
Socio-Cybernetic	Limited	Yes	No	1983	No	No
Avison & Fitzgerald	Yes	Yes	No	1988	No	No
Case Shells	Yes	Limited	Yes	1990	No	Limited
MetaView	Yes	No	Yes	1992	No	Limited
DESMET	Yes	Yes	Limited	1994	No	No
Euromethod	Yes	Limited	No	1994	No	No
NIMSAD	Yes	Yes	No	1994	No	No
MetaPHOR	Yes	Limited	Yes	1991	No	Limited
Tudor & Tudor	Yes	Limited	No	1995	No	No
CDIF	Yes	No	Yes	1990	Limited	Limited
CDM-FILTERS	Yes	Limited	Limited	2001	Yes	Yes

SUMMARY OF EVALUATION FRAMEWORKS AND METAMODELS

The representative list of frameworks, metamodels and associated projects, presented in this chapter are mostly used in IS engineering field. Time and space would not allow for a more elaborate analysis or an exhaustive listing. However, the summarized information provided in Table 1 denotes important evaluative parameters for the metamodeling tools described. These observations are important for the discussion of the range and limitations of the quality properties of methods in the next section.

The information is summarized under the headings of hard and soft issues coverage, whether the frameworks and metamodels provide re-engineering and reuse facilities through tool support, their year of introduction, and the degree of testing and computability that are offered as quality evaluation and construction criteria for ISDMs. The Table excludes specialized applications and the RENOIR and NATURE initiatives. Additionally, it emphasises two important quality properties, namely method testing and method computability; the ability of a method to include implementable/programmable characteristics for realizing IS design and testability criteria for facilitating subsequent testing procedures. Very few frameworks consider design, implementability and maintenance issues in an integrated manner; they mostly analyze quality features of methods by examining the suitability, applicability and practicality for particular application domains.

PERSPECTIVES ON KNOWN FRAMEWORKS AND METAMODELS

Instruments for measuring ISDM quality, metamodels and evaluation frameworks should provide a standard quality platform for formally re-defining static and dynamic method requirements and capturing the computational characteristics of both instantiated and generic methods. Such approaches should provide rigorous environments for reasoning and formally studying the properties of IS design artifacts, establishing whether such specification models facilitate IS implementations of improved quality.

However, the overall assessment of existing facilities indicates that many important quality features are either partially supported or not considered at all, and many method-oriented problems in software systems engineering remain undefined. For example, existing metamodeling tools and frameworks are not generic. Typically, coverage narrowly reflects the application domain and very restrictive types of systems and problems, the quality perceptions, or the philosophical and epistemological principles that the framework encompass.

Consequently, even with good intentions, many of the primary objectives of software quality assurance (e.g., problem of testing during IS design) are still in the wish list of most ISD evaluation frameworks. Furthermore none of the metamodeling environments and notations handles metadata and meta-process specification for considering the dynamic features of models (Koskinen, 2000; Berki, 2001) and for allowing metamodelers to maintain control of both process metamodeling and metadata of a method simultaneously; an important quality property.

Following are some of the other limitations of the existing facilities:

- Frameworks and metamodels are either based on qualitative analysis or use method construction (or ME) principles. The support for data integration is a very important and highly desirable feature; however, only some metamodels facilitate data, information, or knowledge integration. In the facilities that include them, they are typically referenced as quality criteria of evaluation frameworks and no projected as part of metamodeling rules.

- Computability (implementability) and maintainability of specifications and implementations are not yet considered to be fundamental quality issues of great importance and are not addressed adequately by existing method evaluation and ME frameworks and principles (Siakas et al, 1997; Berki et al, 2001).
- The capability to assist formal testing - considered a crucial quality assurance procedure to re-assure IS stakeholders and certify the quality of the development process and product (Rossi & Brinkkemper, 1996; Berki, 2003) is absent from most of the existing frameworks and metamodels.
- Only a few metamodels (e.g., DESNET and NIMSAD) fully address both soft and hard issues of ISD.
- Method modifiability (changeability), an important quality feature, is not always addressed, although system and method dynamics are huge and controversial subjects in dynamic systems development. In fact, only NIMSAD and CDM-FILTERS directly take into account the dynamic nature of ISD.
- There is a scarcity of supporting automated tools for specific environments

CONCLUSION

Technological advancements have altered many quality requirements for ISD, necessitating a change in the roles of stakeholders and elevating the importance of methodologies; hence the need for increasing the effectiveness of approaches for evaluating methodologies that were designed to improve ISD quality by ensuring process consistency. There is a plethora of these ISDMs. Similarly, there are many evaluation frameworks and assessment methods for examining and comparing their capabilities and determining whether their quality-enhancing attributes are suitable for particular development environments. There are also many metamodels and support tools to construct new method with the desirable qualities, if none of the existing development methods are satisfactory. .

From a practitioner perspective, the chapter examined and provided a commentary on metamodels and briefly reviewed the workings of some well-known evaluation frameworks and ME metamodels, highlighting quality considerations that are pertinent to the choice and evaluation of ISDMs and issues bearing on the construction, modeling and re-modeling of an IS development methods and assessing their strengths and limitations. Many strengths were highlighted; however, several weaknesses were also identified, ranging from small issues such as the time needed to produce a model to larger problems of method integration and testing, and facilities to capture the significance of the human role. This review unearthed the urgent need for more integrative and formal metamodeling approaches, which, based on automated tool support, can facilitate the construction of formally computable and testable IS designs; a crucial quality requisite of development methods.

From a research point of view, evaluation frameworks, metamodeling and ME can be considered as communication and thinking tools for assimilating the philosophies and epistemologies that exist in the IS field and the different perceptions of quality that they generate. Thus, they could provide additional meta-cognitive knowledge to help evaluate and model IS needs in a scientific and progressive manner. Further research efforts are therefore required to continue exploring the issues examined in this chapter and providing insights into effectiveness criteria for constructing metamodels and evaluating ISDMs.

The strengths and weaknesses of metamodeling environments should be examined more rigorously using appropriate architectural constructs that encapsulate both communicative and technical quality properties. In order to provide rigorous, scientific evidence to convince adopters that method A is better than method B in large design projects, and where the problems really are, results must be based on carefully controlled experiments of suitable complexity. This is especially so because there is little empirical evidence of what constitutes the quality of ISDMs and lack of clarity about what makes one method superior to another for large-scale ISD projects (Holcombe & Ipate, 1998).

In addition, the increasingly global nature of information systems development, demands that IS and information products conform to new, extended quality standards for quality certification and quality assurance. In this global development environment, which utilizes virtual development teams and project groups separated by time and distance, ISDMs will become even more important in order to increase consistency and reduce variability in the process structuring methods utilized. This also amplifies the importance of ISDM evaluation through frameworks and metamodels. There is an obvious need for further research to explore this new dimension of quality evaluation of ISDMs in different anthropological and social contexts to identify controllable and uncontrollable factors for IS development within these contexts. Prescriptions for the improvement of ISDMs and IS must now incorporate considerations of organizational and national cultures and the applicability of methods and tools, and their relevance in those environments.

REFERENCES

Abran, A. Khelifi, A. & Suryn, W. (2003). Usability Meanings and Interpretations in ISO Standards. Software Quality Journal, 11. 325-338. Kluwer Academic Publishers.

Allen P. and Frost, S. (1998). *Component-Based Development for Enterprise Systems. Applying The SELECT Perspective.* Cambridge University Press.

Avison, D.E., Fitzgerald, G. (1995). *Information Systems Development: Methodologies, Techniques and Tools.* McGraw-Hill.

Beizer, B. (1995). *Foundations of Software Testing Techniques. 12th International Conference & Exposition on Testing Computer Software.* June 12-15, Washington, DC.

Bergsten, P., Bubenko, J. jr, Dahl, R., Gustafsson and Johansson, L.-A. (1989). RAMATIC: - A CASE Shell for Implementation of Specific CASE Tools. *Technical Report*, SISU, Gothenburg.

Berki, E., Georgiadou, E., Siakas, K. (1997). A methodology is as strong as the user involvement it supports. *International Symposium of Software Engineering in Universities (ISSEU'97).* March 1997, Rovaniemi Polytechnic, Rovaniemi.

Berki, E. and Georgiadou, E. (1998). A Comparison of Quantitative Frameworks for Information Systems Development Methodologies. *The 12th International Conference of the Israel Society for Quality.* Conference Proceedings available in CD-ROM. Jerusalem, Nov-Dec.

Berki, E. (1999). *Systems Development Method Engineering, The Formal NEWS: The Formal Notation for Expressing a Wide-range of Specifications, The Construction of a Formal Specification Metamodel.* Mphil/PhD Transfer Report, Faculty of Science, Computing & Engineering, University of North London, London.

Berki, E. (2001). *Establishing a scientific discipline for capturing the entropy of systems process models. CDM-FILTERS: A Computational and Dynamic Metamodel as Flexible and Integrated Language for the Testing, Expression and Re-engineering of Systems.* PhD thesis. Faculty of Science, Computing & Engineering, University of North London, London.

Berki, E. (2003). Formal Metamodelling and Agile Method Engineering in MetaCASE and CAME Tool Environments. 1st South-East European Workshop on Formal Methods. Kefalas, P. (Ed.) *Agile Formal Methods: Practical, Rigorous Methods for a changing world* (Satellite of the 1st Balkan Conference in Informatics, 21-23 Nov 2003), SEERC, Thessaloniki.

Berki, E., Isomäki, H. & Jäkälä, M. (2003). Holistic Communication Modeling: Enhancing Human-Centred Design through Empowerment. In Harris, D., Duffy, V., Smith, M., Stephanidis, C. (Eds) *Cognitive, Social and Ergonomic Aspects*. Vol 3 of HCI International, Heraklion, Crete. pp. 1208-1212, Lawrence Erlbaum Associates, Inc.

Berki, E., Lyytinen, K., Georgiadou, E., Holcombe, M. & Yip, J. (2002). Testing, Evolution and Implementation Issues in MetaCASE and Computer Assisted Method Engineering (CAME) Environments. G. King, M. Ross, G. Staples, T. Twomey (Eds) *Issues of Quality Management and Process Improvement*. 10th International Conference on Software Quality Management, British Computer Society, SQM, Limerick, Mar.

Berki, E., Georgiadou, E. & Holcombe, M. (2004). Requirements Engineering and Process Modelling in Software Quality Management – Towards a Generic Process Metamodel. *The Software Quality Journal*, 12, 265-283, Apr. Kluwer Academic Publishers.

Brinkkemper, S. (2000). Method Engineering with Web-Enabled Methods. Brinkkemper, S., Lindencrorna, E. & Solvberg, A. (Eds): *Information Systems Engineering, State of the Art and Research Themes*. Jun. 2000, pp. 123-134, Springer-Verlag London Ltd.

Burst, A, Wolff, M. Kuhl, M and Muller-Glaser, K.D. (1999). *Using CDIF for Concept-Oriented Rapid Prototyping of Electronic Systems*. University of Karlsruhe, Institute of Information Processing Technology, ITIV. http://www-itiv.etec.uni-karlsruhe.de (Jun 1999).

CCTA. (1994). *Euromethod Overview*. CCTA.

Checkland, P. and Scholes, J. (1990). *Soft Systems Methodology in Action*. Wiley.

Chisholm, R. M. (1982). *The Problem of Criterion. The Foundations of Knowing*. University of Minnesota Press, Minneapolis.

Clifton, H.D. & Sutcliffe, A.G. (1994). *Business Information Systems*. 5th Edit., Prentice Hall.

DESMET. (1991). *State of the Art Report (Workpackage 1)* .NCC Publications.

Dorling, A. (1993). SPICE: Software process improvement and capacity determination. *Information and Software Technology*. 35, 6/7, pp. 404-406.

Georgiadou, E., Mohamed W-E., Sadler, C. (1994). Evaluating the Evaluation Methods: The Data Collection and Storage System Using the DESMET Feature Analysis. *10th International Conference of the Israel Society for Quality*. Nov. Jerusalem.

Georgiadou, E. & Sadler, C. (1995). Achieving quality improvement through understanding and evaluating information systems development methodologies. *3rd International Conference on Software Quality Management*. Vol 2, pp. 35-46, May, Seville, British Computer Society.

Georgiadou, E., Siakas, K. & Berki, E. (2003). Quality Improvement through the Identification of Controllable and Uncontrollable Factors in Software Development. Messnarz R. (Ed.) *EuroSPI 2003: European Software Process Improvement*. 10-12 Dec, Graz, Austria.

Georgiadou, E., Hy, T. & Berki, E. (1998). Automated qualitative and quantitative evaluation of software methods and tools. *The Twelfth International Conference of the Israel Society for Quality*, Nov-Dec. Jerusalem.

Graham, I. (1991). Object-Oriented Methods. Reading, MA: Addison-Wesley.

Harel, D., Pnueli, A., Schmidt, J.P. and Sherman, R. (1987). On the Formal Semantics of Statecharts. *Proceedings of the 2nd IEEE Symposium on Logic in Computer Science*. pp. 54-64, IEEE Computer Society.

Holcombe, M. and Ipate, F. (1998). *Correct Systems - Building a Business Process Solution.* Springer-Verlag

Huotari J. & Kaipala, J. (1999). Review of HCI Research - Focus on cognitive aspects and used methods. Kakola T. (ed.) *IRIS 22 Conference: Enterprise Architectures for Virtual Organisations*. Keurusselka, Jyvaskyla, Jyvaskyla University Printing House.

Iivari, J. and Kerola, P. (1983). T.W. Olle, H.G. Sol, C.J. Tully (Eds) A Sociocybernetic Framework for the Feature Analysis of Information Systems Design Methodologies. *Information systems design methodologies: a feature analysis: proceedings of the IFIP WG 8.1 Working Conference on Comparative Review of Information Systems Design Methodologies.* 5-7 July, York.

Iivari, J. and Huisman, M. (2001). The Relationship between Organisational Culture and the Deployment of Systems Development Methodologies. Dittrich K.R., Geppert, A. & Norrie, M. (Eds.) *The 13th International Conference on Advanced information Systems Engineering, CAiSE'01.* June, Interlaken. Springer, LNCS 2068, pp. 234-250.

Ince, D. (1995). *Software Quality Assurance.* McGraw-Hill

Ipate, F. and Holcombe, M. (1998). Specification and Testing using Generalized Machines: a Presentation and a Case Study. *Software Testing, Verification and Reliability.* 8, 61-81.

Jarke, M., Gallersdorfer, R., Jeusfeld, M. A., Staudt, M., Eherer, S. (1995). ConceptBase – a deductive objectbase for meta data management. *Journal of Intelligent Information Systems.* 4(2):167-192.

Jarke, M., Pohl, K., Rolland, C. and Schmitt, J.-R. (1994). Experience-Based Method Evaluation and Improvement: A process modeling approach. T.W. Olle & A. A. Verrijin-Stuart (Eds) *IFIP WG8.1 Working Conference CRIS'94.* Amsterdam: North-Holland.

Jayaratna, N. (1994). *Understanding and Evaluating Methodologies. NIMSAD: A Systemic Approach.* McGraw-Hill.

Jarzabek, S. (1990). Specifying and generating multilanguage software development environments. *Software Engineering Journal.* Vol. 5, No. 2, Mar., pp. 125-137.

Jenkins, T. (1994). Report back on the DMSG sponsored UK Euromethod forum '94. *Data Management Bulletin.* Summer Issue, 11, 3.

Kaipala, J. (1997). Augmenting CASE Tools with Hypertext: Desired Functionality and Implementation Issues. A. Olive, J.A. Pastor (Eds.) *The 9th International Conference on Advanced information Systems Engineering, CAiSE '97.* LNCS 1250, Berlin: Springer-Verlag.

Kelly, S., Lyytinen, K. and Rossi, M. (1996). MetaEdit+: A Fully Configurable Multi-User and Multi-Tool CASE and CAME Environment. Constantopoulos, P., Mylopoulos, J. and Vassiliou, Y (Eds) *Advances in Information Systems Engineering, 8th International Conference CAiSE'96.* Heraklion, Crete, Greece, May 20-24, Berlin: Springer-Verlag LNCS, pp. 1-21.

Kelly, S. (1997). *Towards a Comprehensive MetaCASE and CAME Environment, Conceptual, Architectural, Functional and Usability Advances in MetaEdit+.* Ph.D. Thesis, University of Jyvaskyla.

Kitchenham, B.A., Linkman, S.G. and Law, D.T. (1994). Critical review of quantitative assessment. *Software Engineering Journal.* Mar.

Kopetz, H. (1979). *Software Reliability.* The Macmillan Press Ltd.

Koskinen, M. (2000). *Process Metamodeling: Conceptual Foundations and Application.* Ph.D. Thesis, Jyvaskyla Studies in Computing-7, University of Jyvaskyla.

Kronlof, K. (Ed.). (1993). *Method Integration, Concepts and Case Studies.* Wiley Series on Software Based Systems, Wiley.

Law, D. (1988). *Methods for Comparing Methods: Techniques in Software Development.* NCC Publications.

Law, D. and Naeem, T. (1992). DESMET: Determining an Evaluation methodology for Software Methods and Tools. *Proceedings of Conference on CASE - Current Practice, Future Prospects.* Mar. 1992, British Computer Society, Cambridge.

Lewis, H. R. & Papadimitriou, C. H. (1998). *Elements of the Theory of Computation.* Prentice Hall International Editions.

Manninen, A. & Berki, E. (2004). An Evaluation Framework for the Utilisation of Requirements Management Tools - Maximising the Quality of Organisational Communication and Collaboration. Ross, M. & Staples, G. (Eds) *Software Quality Management 2004 Conference.* University of KENT, Canterbury, Apr., BCS.

Marttiin, P., Rossi, M., Tahvanainen, V-P & Lyytinen, K. A. (1993). Comparative Review of CASE Shells – A preliminary framework and research outcomes. *Information and Management.* Vol. 25, pp. 11-31.

Marttiin, P. (1988). *Customisable Process Modelling Support and Tools for Design Environment.* Ph.D. Thesis, University of Jyvaskyla, Jyvaskyla.

MetaPHOR website: http://metaphor.cs.jyu.fi/

MetaView website: http://web.cs.ualberta.ca/~softeng/Metaview/project.shtml

Mohamed W. A., and Sadler, C.J. (1992). Methodology Evaluation: A Critical Survey. *Eurometrics'92 Conference on Quantitative Evaluation of Software & Systems.* Brussels, Apr. pp. 101-112.

Musa J.D., Iannino A., Okumoto K. (1987). *Software Reliability Measurement, Prediction, Application.* McGraw-Hill, NewYork.

Myers, G. (1976). *Software Reliability Principles and Practices.* J. Wiley & Sons.

Mylopoulos, J., Borgida, A., Jarke, M., Koubarakis, M. (1990). TELOS: a language for representing knowledge about information systems. *ACM Transactions on Information Systems,* 8(4).

NATURE and RENOIR project websites: http://www-i5.informatik.rwth-aachen.de/PROJEKTE/NATURE/nature.html and http://panoramix.univ-paris1.fr/CRINFO/PROJETS/nature.html

Nielsen, P. (1990). Approaches to Appreciate Information Systems Methodologies. *Scandinavian Journal of Information Systems.* vol. 9, no. 2.

T.W.Olle, H.G. Sol, A.A. Verrijin-Stuart (Eds). 1982. *Information systems design methodologies: a comparative review: proceedings of the IFIP WG 8.1 Working Conference on Comparative Review of Information Systems Design Methodologies.* Noordwijkerhout, The Netherlands, 10-14 May.

Paulk, M. C., Curtis, B., Chrissis, M. B. and Weber, C. V. (1993). The Capability Maturity Model: Version 1.1. *IEEE Software.* pp. 18-27, Jul.

Rantapuska, T., Siakas, K., Sadler, C.J., Mohamed, W-E. (1999). Quality Issues of End-user Application Development. Hawkins, C., Georgiadou, E. Perivolaropoulos, L., Ross, M. & Staples, G. (Eds): *The BCS INSPIRE IV Conference: Training and Teaching for the Understanding of Software Quality.* British Computer Society, Sep, University of Crete, Heraklion.

Rossi, M. and Brinkkemper, S. (1996). Complexity metrics for systems development methods and techniques. *Information Systems*, Vol. 21, No 2, Apr., ISSN 0306-4379.

Rossi, M. (1998). *Advanced Computer Support for Method Engineering: Implementation of CAME Environment in MetaEdit+*. Ph.D. Thesis, Jyvaskyla Studies in Computer Science, Economics and Statistics, University of Jyvaskyla.

Saadia, A. (1999). *An Investigation into the Formalization of Software Design Schemata.* MPhil Thesis, Faculty of Science, Computing and Engineering, University of North London.

Sadler, C. & Kitchenham, B.A. (1996). Evaluating Software Engineering Methods and Tools. Part 4: The influence of human factors. *SIGSOFT, Software Engineering Notes*. Vol 21, no 5.

Shapiro, J. & Berki, E. (1999). Encouraging Effective use of CASE tools for Discrete Event Modelling through Problem-based Learning. Hawkins, C., Georgiadou, E. Perivolaropoulos, L., Ross, M. & Staples, G. (Eds): the BCS INSPIRE IV Conference: *Training and Teaching for the Understanding of Software Quality.* British Computer Society, 313-327, University of Crete, Heraklion.

Siakas, K. Berki, E. Georgiadou, E. & Sadler, C. (1997): The Complete Alphabet of Quality Information Systems: Conflicts & Compromises. *7th World Congress on Total Quality Management (TQM '97).* McGraw-Hill, New Delhi.

Siakas, K., Berki, E. & Georgiadou, E. (2003). CODE for SQM: A Model for Cultural and Organisational Diversity Evaluation. Messnarz R. (Ed.) *EuroSPI 2003: European Software Process Improvement.* 10-12 Dec, Graz, Austria.

Si-Said, S., Rolland, C., Grosz, G. (1996). MENTOR: A Computer Aided Requirements Engineering Environment. Constantopoulos, P., Mylopoulos, J. and Vassiliou, Y. (Eds) *Advances in Information Systems Engineering, 8th International Conference CAiSE'96.* Heraklion, Crete, Greece, May 20-24, Berlin: Springer-Verlag Lecture Notes in Computer Science LNCS 1080, pp. 22-43.

Smolander, K. Tahvanainen, V-P., Lyytinen, K. (1990). How to Combine Tools and Methods in Practice – a field study. B. Steinholz, A. Solverg, L. Bergman (Eds) *The 2nd Conference in Advanced Information Systems Engineering (CAiSE '90).* Lecture Notes in Computer Science LNCS 436, Springer-Verlag, Berlin, pp.195-214.

Sol, H.G. (1983). A Feature Analysis of Information Systems Design Methodologies: Methodological Considerations. T.W. Olle, H.G. Sol, C.J. Tully (Eds) *Information systems design methodologies: a feature analysis: proceedings of the IFIP WG 8.1 Working Conference on Comparative Review of Information Systems Design Methodologies.* 5-7 July, York.

Tolvanen, J-P. (1998). *Incremental Method Engineering with Modeling Tools.* PhD Thesis, University of Jyvaskyla, Jyvaskyla.

Sorenson, P.G., Tremblay, J-P. and McAllister, A. J. (1988). The MetaView system for many specification environments. *IEEE Software*. 30, 3, pp. 30-38.

Tudor, D.J, Tudor, I.J. (1995). *Systems Analysis and Design - A comparison of Structured Methods.* NCC Blackwell.

Van Hillegersberg, J. (1997). *Metamodelling-Based Integration of Object-Oriented Systems Development.* Thesis Publishers, Amsterdam.

Vassiliou, G. (2000). Developing Data Warehouses with Quality in Mind. Brinkkemper, S., Lindencrorna, E. & Solvberg, A. (Eds): *Information Systems Engineering, State of the Art and Research Themes.* Springer-Verlag London Ltd.

Whittaker, J. (2000). What Is Software Testing? And Why Is It So Hard? *IEEE Software*. Jan/Feb, pp. 70-79.

Wood, D. (1987). *Theory of Computation.* J. Wiley and Sons.

Wood-Harper, A.T., Antill, L. Avison, D.E. (1985). *Information systems definition: the multiview approach.* Blackwell Publications.

Zhang, Z. (2004). *Model Component Reuse. Conceptual Foundations and Application in the Metamodeling-Based System Analysis and Design Environment.* Ph.D. Thesis, Jyvaskyla Studies in Computing, University of Jyvaskyla.

This work was previously published in Measuring Information Systems Delivery Quality, edited by E. Duggan and J. Reichgelt, pp. 265-290, copyright 2006 by IGI Publishing (an imprint of IGI Global).

Compilation of References

Abran, A. Khelifi, A. & Suryn, W. (2003). Usability Meanings and Interpretations in ISO Standards. Software Quality Journal, 11. 325-338. Kluwer Academic Publishers.

Acuna, S., & Juristo, N. (2004). Assigning people to roles in software projects. *Software – Practice and Experience, 34*(7), 675-696.

Adam, N. R., Atluri, V., & Huang, W.-K. (1998). Modeling and analysis of workflows using Petri nets. *Journal of Intelligent Information Systems: Special issue on workflow management systems, 10*(2), 131-158.

Ahmad, R., Li, Z., & Azam, F. (2005). Web Engineering: A New Emerging Discipline. In *Proceedings of the IEEE Symposium on Emerging Technologies* (pp.445-450). Catania, Italy: IEEE.

Alabiso, B. (1988). Transformation of data flow analysis models to object-oriented design. *Proceeding of OOPSLA '88 Conference*, San Diego, CA, 335-353.

Albert, M., Pelechano, V., Fons, J., Ruiz, M., & Pastor, O. (2003). Implementing UML Association, Aggregation, and Composition. A Particular Interpretation Based on a Multidimensional Framework. In J. Eder and M. Missikoff (Ed.), *Proceedings of CAiSE'03 LNCS 2681* (pp. 143-158). Berlin: Springer Verlag.

Alexander, C., Ishikawa, S., & Silverstein, M. (1977). *A pattern language*. New York.

Allen P. and Frost, S. (1998). *Component-Based Development for Enterprise Systems. Applying The SELECT Perspective.* Cambridge University Press.

Allen, J. (1984). Towards a general theory of action and time. *Artificial Intelligence, 23*(2), 123-154.

Almeida, H. O. de, Silva, L. D. da, Silva Oliveira, E. A. da, & Perkusich, A. (2005). Formal approach for component based embedded software modelling and analysis. In *Proceedings of the IEEE International Symposium on Industrial Electronics* (*ISIE 2005*) (pp. 1337-1342). Dubrovinik, Croatia: IEEE Press.

Almeida, H., Silva, L., Perkusich, A., & Costa, E. (2005). A formal approach for the modelling and verification of multiagent plans based on model checking and Petri nets. In R. Choren, A. Garcia, C. Lucena, & A. Romanovsky (Eds.), *Software engineering for multi-agent systems, III: Research issues and practical applications* (vol. 3390, pp. 162-179). Heidelberg, Germany: Springer GMBH.

Alur, D., Malks, D., & Crupi, J. (2003). *Core J2EE Patterns: Best Practices and Design Strategies, 2nd edition.* Prentice Hall.

Álvarez, A. T., & Alemán, J. L. F. (2000). Formally modeling UML and its evolution: A holistic approach. In *Fourth International Conference on Formal methods for open object-based distributed systems IV* (pp. 183-206). Amsterdam: Kluwer Academic Publishers.

Ambler, S. (2002). *Agile modeling: Effective practices for extreme programming and the unified process* (1st ed.). John Whiley & Sons Inc.

Ambler, S., & Jeffries, R. (2002). *Agile Modeling: Effective Practices for Extreme Programming and the Unified Process, 1st edition.* John Wiley & Sons.

Anda, B., & Sjoberg, D. (2002). Towards an inspection technique for use case models. *Proceeding of the 14th International Conference on Software Engineering and Knowledge Engineering (SEKE '02),* Ischia, Italy, 127-134.

Arch-Int, S., Batanov, D. N. (2003). Development of industrial information systems on the Web using business components. *Computers in Industry,* 50 (2), 231-250, Elsevier.

Artale, A., & Franconi, E. (1999). Temporal ER modeling with description logics. In *Proc. of the Int. Conf. on Conceptual Modeling (ER'99).* Berlin: Springer-Verlag.

Artale, A., & Keet, C. M. (2008). Essential and Mandatory Part-Whole Relations in Conceptual Data Models. *Proceedings of the 21st International Workshop on Description Logics (DL'08), CEUR WS Vol 353.* Dresden, Germany, 13-16 May 2008.

Artale, A., Calvanese, D., Kontchakov, R., Ryzhikov, V., & Zakharyaschev, M. (2007). Complexity of Reasoning over Entity Relationship Models. *Proceedings of DL-07, CEUR WS, 250.*

Artale, A., Franconi, E., & Guarino, N. (1996). Open Problems for Part-Whole Relations. In: *Proceedings of 1996 International Workshop on Description Logics (DL-96)* (pp 70-73). Cambridge, MA: AAAI Press.

Artale, A., Franconi, E., & Mandreoli, F. (2003). Description Logics for Modelling Dynamic Information. In J. Chomicki, R. van der Meyden, & G. Saake (Eds.), *Logics for Emerging Applications of Databases.* Berlin: Springer-Verlag.

Artale, A., Franconi, E., Guarino, N., & Pazzi, L. (1996). Part-Whole Relations in Object-Centered Systems: an Overview. *Data and Knowledge Engineering, 20*(3), 347-383.

Artale, A., Franconi, E., Wolter, F., & Zakharyaschev, M. (2002). A temporal description logic for reasoning about conceptual schemas and queries. In S. Flesca, S. Greco, N. Leone, G. Ianni (Eds.), *Proceedings of the 8th Joint European Conference on Logics in Artificial Intelligence (JELIA-02), LNAI, 2424* (pp. 98-110). Berlin: Springer Verlag.

Artale, A., Guarino, N., & Keet, C.M. (2008). Formalising temporal constraints on part-whole relations. In G. Brewka & J. Lang, J. (Eds.), *11th International Conference on Principles of Knowledge Representation and Reasoning (KR'08).* Cambridge, MA: AAAI Press.

Artale, A., Parent, C., & Spaccapietra, S. (2006). Modeling the evolution of objects in temporal information systems. In: *4th International Symposium on Foundations of Information and Knowledge Systems (FoIKS-06), LNCS, 3861,*22-42. Berlin: Springer-Verlag.

Artale, A., Parent, C., & Spaccapietra, S. (2007b). Evolving objects in temporal information systems. *Annals of Mathematics and Artificial Intelligence (AMAI), 50*(1-2), 5-38.

Arthur, L.J. (1992). *Rapid evolutionary development: Requirements; prototyping and software creation.* Chichester, UK: John Wiley & Sons.

Avison, D., & Fitzgerald, G. (1995). *Information systems development: methodologies, techniques and tools.* 2nd ed., London: McGraw-Hill.

Avison, D.E., & Fitzgerald, G. (2003). *Information systems development: Methodologies, techniques and tools.* London: McGraw-Hill.

Axenath, B., Kindler, E., & Rubin, V. (2005). *An open and formalism independent meta-model for business processes.* Paper presented at the First International Workshop on Business Process Reference Models (BPRM'05), Nancy, France.

Aydin, M. (2007). Examining key notions for method adaptation. In J. Ralytè, S. Brinkkemper & B. Henderson-Sellers (Eds.), *Situational Method Engineering: Fundamentals and Experiences* (pp. 49-63). Boston: Springer.

Baader, F., Calvanese, D., McGuinness, D. L., Nardi, D., & Patel-Schneider, P. F. (Eds). (2003). *Description Logics Handbook.* Cambridge: Cambridge University Press.

Bach, D., Meersman, R., Spyns, P., & Trog, D. (2007). *Mapping OWL-DL into ORM/RIDL*. Paper presented at the OTM 2007/ORM 2007.

Bailin, S. (1989). An object-oriented requirements specification method. *Communication of the ACM, 32*(5), 608-623.

Bajec, M., & Krisper, M. (2005). A methodology and tool support for managing business rules in organisations. *Information Systems, 30*(6), pp423-443.

Bakema, G. P., Zwart, J. P., & van der Lek, H. (1994). Fully communication oriented NIAM. In G. Nijssen & J. Sharp (Eds.), *NIAM-ISDM 1994 Conference* (pp. L1-35). Albuquerque NM.

Balsters, H., Carver, A., Halpin, T., & Morgan, T. (2006). *Modeling dynamic rules in ORM*. Paper presented at the OTM 2006/ORM 2006.

Balsters, H., Carver, A., Halpin, T., & Morgan, T. (2006). Modeling Dynamic Rules in ORM. In R. Meersman, Z. Tari, P. Herrero et al. (Eds.), *On the Move to Meaningful Internet Systems 2006: OTM 2006 Workshops*, (pp. 1201-1210). Montpellier: Springer LNCS 4278.

Banks, J., Carson, J.S., Nelson, B.L., & Nicol, D.M. (2000). *Discrete-event system simulation*. Upper Saddle River, NJ: Prentice-Hall.

Barbier, F., Henderson-Sellers, B., Le Parc-Lacayrelle, A., & Bruel, J.-M. (2003). Formalization of the whole-part relationship in the Unified Modelling Language. *IEEE Transactions on Software Engineering, 29*(5), 459-470.

Barker, R. (1990). *CASE*Method: Entity relationship modelling*. Wokingham: Addison Wesley.

Barros, O. (1991). Modeling and evaluation of alternatives in information systems. *Information Systems, 16*(5), 537-558.

Baskerville, R. (1989). Logical controls specification: an approach to information systems security. In H. Klein & K. Kumar (Eds.), *Systems Development for Human Progress* (pp. 241-255). Amsterdam: Elsevier Science.

Batra, D. (1993). A framework for studying human error behavior in conceptual database modeling. *Information & Management, 24*, 121-131.

Batra, D., Hoffer, J., & Bostrom, R. (1990). Comparing representations with the Relational and Extended Entity Relationship model. *Communications of the ACM, 33*, 126-139.

Bauer, C., & King, G. (2004). *Hibernate in Action, 1st edition*. Manning.

Bechhofer, S., Harmelen, F. v., Hendler, J., Horrocks, I., McGuinness, D., Patel-Schneider, P., et al. (2004). *OWL web ontology language reference*. Retrieved. from http://www.w3.org/TR/owl-ref/.

Becker, J., & Schütte, R. (2004). *Retail information systems (in German)* (2. vollst. überarb., erw. u. akt. Aufl. ed.). Frankfurt a. M.

Becker, J., Delfmann, P., Dreiling, A., Knackstedt, R., & Kuropka, D. (2004). *Configurative process modeling: Outlining an approach to increased business process model usability.* Paper presented at the Proceedings of the 2004 Information Resources Management Association Conference, New Orleans, LA.

Beizer, B. (1995). *Foundations of Software Testing Techniques. 12th International Conference & Exposition on Testing Computer Software*. June 12-15, Washington, DC.

Bennett, S., McRobb, S., & Farmer, R. (1999). *Object-oriented systems analysis and design using UML*. London: McGraw-Hill.

Berardi, D., Calvanese, D., & De Giacomo, G. (2005). Reasoning on UML class diagrams. *Artificial Intelligence, 168*(1-2), 70-118.

Bergenheim, A., Persson, A., Brash, D., Bubenko, J. A. J., Burman, P., Nellborn, C., & Stirna, J. (1998). *CAROLUS - System Design Specification for Vattenfall*, Vattenfall AB, Råcksta, Sweden

Bergenheim, A., Wedin, K., Waltré, M., Bubenko Jr., J. A., Brash, D., & Stirna, J. (1997). *BALDER – Initial*

Requirements Model for Vattenfall. Stockholm, Vattenfall AB

Bergsten, P., Bubenko, J. jr, Dahl, R., Gustafsson and Johansson, L.-A. (1989). RAMATIC: - A CASE Shell for Implementation of Specific CASE Tools. *Technical Report*, SISU, Gothenburg.

Berki, E. (1999). *Systems Development Method Engineering, The Formal NEWS: The Formal Notation for Expressing a Wide-range of Specifications, The Construction of a Formal Specification Metamodel.* Mphil/PhD Transfer Report, Faculty of Science, Computing & Engineering, University of North London, London.

Berki, E. (2001). *Establishing a scientific discipline for capturing the entropy of systems process models. CDM-FILTERS: A Computational and Dynamic Metamodel as Flexible and Integrated Language for the Testing, Expression and Re-engineering of Systems.* PhD thesis. Faculty of Science, Computing & Engineering, University of North London, London.

Berki, E. (2003). Formal Metamodelling and Agile Method Engineering in MetaCASE and CAME Tool Environments. 1st South-East European Workshop on Formal Methods. Kefalas, P. (Ed.) *Agile Formal Methods: Practical, Rigorous Methods for a changing world* (Satellite of the 1st Balkan Conference in Informatics, 21-23 Nov 2003), SEERC, Thessaloniki.

Berki, E. and Georgiadou, E. (1998). A Comparison of Quantitative Frameworks for Information Systems Development Methodologies. *The 12th International Conference of the Israel Society for Quality.* Conference Proceedings available in CD-ROM. Jerusalem, Nov-Dec.

Berki, E., Georgiadou, E. & Holcombe, M. (2004). Requirements Engineering and Process Modelling in Software Quality Management – Towards a Generic Process Metamodel. *The Software Quality Journal*, 12, 265-283, Apr. Kluwer Academic Publishers.

Berki, E., Georgiadou, E., Siakas, K. (1997). A methodology is as strong as the user involvement it supports. *International Symposium of Software Engineering in Universities (ISSEU'97).* March 1997, Rovaniemi Polytechnic, Rovaniemi.

Berki, E., Isomäki, H. & Jäkälä, M. (2003). Holistic Communication Modeling: Enhancing Human-Centred Design through Empowerment. In Harris, D., Duffy, V., Smith, M., Stephanidis, C. (Eds) *Cognitive, Social and Ergonomic Aspects.* Vol 3 of HCI International, Heraklion, Crete. pp. 1208-1212, Lawrence Erlbaum Associates, Inc.

Berki, E., Lyytinen, K., Georgiadou, E., Holcombe, M. & Yip, J. (2002). Testing, Evolution and Implementation Issues in MetaCASE and Computer Assisted Method Engineering (CAME) Environments. G. King, M. Ross, G. Staples, T. Twomey (Eds) *Issues of Quality Management and Process Improvement.* 10th International Conference on Software Quality Management, British Computer Society, SQM, Limerick, Mar.

Berners-Lee, T., Hendler, J., & Lassila, O. (2001). The Semantic Web. *Scientific American*, 284, 34-43.

Bernus, P., & Nemes, L. (1996). A framework to define a generic enterprise reference architecture and methodology. *Computer Integrated Manufacturing Systems*, *9*(3), 179-191.

Bittner, T., & Donnelly, M. (2005). Computational ontologies of parthood, component-hood, and containment, In L. Kaelbling (Ed.), *Proceedings of the Nineteenth International Joint Conference on Artificial Intelligence 2005 (IJCAI05)* (pp. 382-387). Cambridge, MA: AAAI Press.

Bittner, T., & Donnelly, M. (2007). A temporal mereology for distinguishing between integral objects and portions of stuff. In *Proceedings of the Twenty-second AAAI Conference on Artificial intelligence (AAAI'07)* (pp. 287-292). Cambridge, MA: AAAI Press.

Bittner, T., & Stell, J.G. (2002). Vagueness and Rough Location. *Geoinformatica*, *6*(2), 99-121.

Boar, B.H. (1984). *Applications prototyping: A requirements definition strategy for the 80s.* Chichester, UK: Wiley.

Bollen, P. (2007). *Fact-oriented Business Rule Modeling in the Event Perspective*. Paper presented at the CAISE 2007.

Bollen, P. (2007). *Fact-oriented modeling in the data-, process- and event perspectives*. Paper presented at the OTM 2007, ORM 2007.

Bommel, P., van, S. J. B. A. Hoppenbrouwers, H. A. (Erik) Proper, & Th.P. van der Weide (2006). Exploring Modeling Strategies in a Meta-modeling Context. In R. Meersman, Z. Tari, and P. Herrero, (Eds.), *On the Move to Meaningful Internet Systems 2006: OTM 2006 Workshops*, 4278 of *Lecture Notes in Computer Science*, (pp. 1128-1137), Berlin, Germany, EU, October/November 2006. Springer.

Booch, G., Rumbauch, J., & Jacobson, I. (2005). *Unified Modeling Language User Guide* (2nd ed.): Addison-Wesley Professional.

Booch, G., Rumbaugh, J., & Jacobson, I. (1999). *The Unified Modeling Language – User guide*. Reading: Addison-Wesley.

Borgo, S., & Masolo, C. (in press). Full mereogeometries. *Journal of Philosophical Logic*.

Borgo, S., Guarino, N., & Masolo, C. (1996). Towards an ontological theory of physical objects. In *IMACS-IEEE/SMC Conference on Computational Engineering in Systems Applications.(CESA'96)*, Symposium of Modelling, Analysis and Simulation (pp. 535-540). Lille, France.

Borker, V., Carey, M., Mangtani, N., McKinney, D., Patel, R., & Thatte, S. (2006). XML Data Services. *International Journal of Web Services Research, 3*(1), 85-95.

Brachman, R. (1983). What IS-A is and isn't: An analysis of taxonomic links of semantic networks. *IEEE Computer, 16*(10), 30-36.

Brasethvik, T., & Gulla, J. (2001). Natural language analysis for semantic document modeling. *Data & Knowlege Engineering, 38*, 45-62.

Brinkkemper, S. (1990). *Formalization of information systems modeling*. Dissertation Thesis, University of Nijmegen, Amsterdam: Thesis Publishers.

Brinkkemper, S. (2000). Method Engineering with Web-Enabled Methods. Brinkkemper, S., Lindencrorna, E. & Solvberg, A. (Eds): *Information Systems Engineering, State of the Art and Research Themes*. Jun. 2000, pp. 123-134, Springer-Verlag London Ltd.

Brinkkemper, S., Saeki, M., & Harmsen, F. (1999). Meta-modelling based assembly techniques for situational method engineering. *Information Systems, 24*(3), 209-228.

Brown, A., Delbaere, M., Eeles, P., Johnston, S., & Weaver, R. (2005). Realizing Service Oriented Solutions with the IBM Rational Software Development Platform. IBM Systems Journal, *44*(4), 727-752.

Brown, D. (2002). *An Introduction to Object-Oriented Analysis* (2nd Edition): Wiley.

Bruijn, J. d., Lausen, H., Krummenacher, R., Polleres, A., Predoiu, L., Kifer, M., et al. (2005). *WSML working draft 14 march 2005*: DERI.

Bubenko Jr., J. A. (1993). Extending the Scope of Information Modelling. *Fourth International Workshop on the Deductive Approach to Information Systems and Databases*, Lloret, Costa Brava (Catalonia), Sept. 20-22, 1993. Department de Llenguatges i Sistemes Informatics, Universitat Politecnica de Catalunya, Report de Recerca LSI/93-25, Barcelona.

Bubenko, J. A. j., Persson, A. & Stirna, J. (2001). *User Guide of the Knowledge Management Approach Using Enterprise Knowledge Patterns, deliverable D3, IST Programme project Hypermedia and Pattern Based Knowl-edge Management for Smart Organisations*, project no. IST-2000-28401, Royal Institute of Technology, Sweden.

Bubenko, J. A. jr., & Kirikova, M. (1999). Improving the Quality of Requirements Specifications by Enterprise Modeling. In A.G. Nillson, C., Tolis, & C. Nellborn

(Eds.), *Perspectives on Business Modeling.* Springer, ISBN 3-540-65249-3.

Bubenko, J. A. jr., Persson, A., & Stirna, J. (2001). *User Guide of the Knowledge Management Approach Using Enterprise Knowledge Patterns, deliverable D3, IST Programme project Hypermedia and Pattern Based Knowledge Management for Smart Organisations,* project no. IST-2000-28401, Royal Institute of Technology, Sweden.

Bunge, M. (1977). *Treatise on basic philosophy, Vol. 3: Ontology I: The furniture of the world.* Dortrecht: D. Reidel Publishing Company.

Burst, A, Wolff, M. Kuhl, M and Muller-Glaser, K.D. (1999). *Using CDIF for Concept-Oriented Rapid Prototyping of Electronic Systems.* University of Karlsruhe, Institute of Information Processing Technology, ITIV. http://www-itiv.etec.uni-karlsruhe.de (Jun 1999).

Burton-Jones, A., Storey, V., Sugumaran, V., & Ahluwalia, P. (2005). A semiotic metric suite for assessing the quality of ontologies. *Data & Knowledge Engineering, 55*(1), 84-102.

Calvanese, C., De Giacomo, G., & Lenzerini, M. (1998). On the decidability of query containment under constraints. In *Proceedings of the 17th ACM SIGACT SIGMOD SIGART Symposium on Principles of Database Systems (PODS'98)* (pp. 149-158).

Calvanese, D., & De Giacomo, G. (2003). Expressive description logics. In F. Baader, D. Calvanese, D. McGuinness, D. Nardi, & P. Patel-Schneider (Eds), *The Description Logic Handbook: Theory, Implementation and Applications* (pp. 178-218). Cambridge University Press.

Calvanese, D., Lenzerini, M., & Nardi, D. (1998). Description logics for conceptual data modeling. In J. Chomicki & G. Saake, (Eds), *Logics for Databases and Information Systems.* Amsterdam: Kluwer.

Calvanese, D., Lenzerini, M., & Nardi, D. (1999). Unifying class-based representation formalisms. *Journal of Artificial Intelligence Research, 11,* 199-240.

Castro, J., Kolp, M., & Mylopoulos, J. (2001). A Requirements-Driven Software Development Meth-odology. In *Proceedings of CAiSE'2001,* Springer LNCS 2068, ISBN 3-540-42215-3.

Castro, J., Kolp, M., Mylopoulos, J., & Tropos, A. (2001). A Requirements-Driven Software Development Meth-odology. In *Proceedings of the 3rd Conference on Advanced Information Systems Engineering* (CAiSE 2001), 108-123, Springer LNCS 2068, Interlaken, Switzerland.

CCTA. (1994). *Euromethod Overview.* CCTA.

Ceri, S., Fraternali, P., & Bongio, A. (2000). Web Modeling Language (WebML): a modeling language for designing Web sites. *Computer Networks,* 33 (1-6), 137-157, Elsevier.

Chang, L.-H., Lin, T.-C., & Wu, S. (2002). The study of information system development (ISD) process from the perspective of power development stage and organizational politics. In *35th Hawaii International Conference on Systems Sciences.*

Checkland, P. (1988). Information systems and system thinking: time to unite? *International Journal of Information Management, 8*(4), 239-248.

Checkland, P., & Scholes, J. (1999). *Soft systems methodology in action.* Chichester, UK: John Wiley & Sons.

Chen, D., & Doumeingts, G. (1996). The GRAI-GIM reference model, architecture, and methodology. In P. Bernus et al. (Eds.), *Architectures for Enterprise Integration.* London: Chapman

Chen, P. (1976). The Entity-Relationship Model : Toward a Unified View. *ACM Transactions on Database Systems, 1*(1), 9 - 36.

Chen, P. P.-S. (1976). The entity-relationship model: Toward a unified view of data. *ACM Transactions on Database Systems, 1*(1), 9-36.

Chen, Y., Zhou, L., & Zhang, D. (2006). Ontology-Supported Web Service Composition: An Approach to Service-Oriented Knowledge Management in Corporate Financial Services. *Journal of Database Management, 17*(1), 67-84.

Chisholm, R. (1996). *A realistic theory of categories – an essay on ontology*. 1st edition, Cambridge: Cambridge University Press.

Chisholm, R. M. (1982). *The Problem of Criterion. The Foundations of Knowing*. University of Minnesota Press, Minneapolis.

Chomicki, J., & Toman, D. (1998). Temporal logic in information systems. In J. Chomiki & G. Saake (Eds.), *Logics for databases and information systems*. Amsterdam: Kluwer.

Chrissis, M. B., Konrad, M., & Shrum, S. (2006). *CMMI: Guidelines for Process Integration and Product Improvement*, Second Edition. Addison-Wesley.

Christensen, S., & Haagh, T. B. (1996). *Design/CPN overview of CPN ML syntax, version 3.0*. Aarhus, Denmark: University of Aarhus, Department of Computer Science.

Christensen, S., & Mortensen, K. H. (1996). *Design/CPN ASK-CTL manual*. Aarhus, Denmark: University of Aarhus, Department of Computer Science.

Clarke, E. M., Grumberg, O., & Peled, D. A. (1999). *Model checking*. Cambridge, MA: The MIT Press.

Clifton, H.D. & Sutcliffe, A.G. (1994). *Business Information Systems*. 5th Edit., Prentice Hall.

Coad, O., & Yourdon, E. (1991). *Object-Oriented Design*. Englewood Cliffs, NJ: Prentice Hall.

Coad, P., & Yourdan, E. (1991). *Object oriented analysis* (2nd ed.). Saddle Brook, NJ: Prentice Hall.

Coad, P., North, D., & Mayfield, M. (1997). *Object models: Strategies, patterns, and applications*. NJ: Yourdon Press.

Coase, R. H. (1937). The nature of the firm. *Economica, 4* (11), 386-405.

Cockburn, A. (2001). *Writing Effective Use Cases*. Addison-Wesley.

Conallen, J. (2002). *Building Web Applications with UML, 2nd edition*. Addison-Wesley.

Conradi, R., & Westfechtel, B. (1998). Version models for software configuration management. *ACM Computing Surveys (CSUR), 30*(2), 232-282.

Costa, E. B., et al. (2002) A multi-agent cooperative intelligent tutoring system: The case of musical harmony domain. In *Proceedings of 2nd Mexican International Conference on Artificial Intelligence (MICAI'02)*, Merida (LNCS 2313, pp. 367-376). Heidelberg, Germany: Springer GMBH.

Costa, E. B., Lopes, M. A., Ferneda, E. (1995) Mathema: A learning environment base on a multi-agent architecture. In *Proceedings of the 12th Brazilian Symposium on Artificial Intelligence (SBIA '95)* (pp. 141-150). Campinas, Brazil: Springer GMBH.

Costa, E. B., Perkusich, A., Figueiredo, J. C. A (1996). A multi-agent based environment to aid in the design of Petri nets based software systems. In *Proceedings of the 8th International Conference on Software Engineering and Knowledge Engineering (SEKE'96)* (pp. 253-260). Buenos Aires, Argentina: IEEE Press.

Cox, B. (1990). Planning the software industrial revolution. *IEEE Software, 7*(8), 25-33.

Crawford, C., Bate, G., Cherbakov, L., Holly, K., & Tsocanos, C. (2005). Toward an on Demand Service Architecture. *IBM Systems Journal, 44*(1), 81-107.

Curland, M., & Halpin, T. (2007). Model Driven Development with NORMA. In *Proc. 40th Int. Conf. on System Sciences (HICSS-40)*, CD-ROM, IEEE Computer Society.

Cuyler, D., & Halpin, T. (2003). *Meta-models for Object-Role Modeling*. Paper presented at the International Workshop on Evaluation of Modeling Methods in Systems Analysis and Design (EMMSAD '03).

Cysneiros, L., Leite, J., & Neto, J. (2001). A framework for integrating non-functional requirements into conceptual models. *Requirements Engineering, 6*(2), 97-115.

D'Souza, D. (2000). Relating components and enterprise integration: Part 1. *JOOP, 13*(1), 40-42.

D'Souza, D., & Wills, A. (1998). *Objects, Components, and Frameworks with UML: The Catalysis Approach.* Addison-Wesley.

Davies, I., Green P., Rosemann, M., & Gallo S., (2004) Conceptual Modelling - What and Why in Current Practice. In P. Atzeni et al. (Eds), *Proceedings of ER 2004*, Springer, LNCS 3288.

De Pauw, Lei, M., Pring, E., & Villard, L. (2005). Web Services Navigator: Visualizing the Execution of Web Services. *IBM Systems Journal, 44*(4), 821-845.

De Troyer, O., Plessers, P., & Casteleyn, S. (2003). *Solving semantic conflicts in audience driven web-design.* Paper presented at the WWW/Internet 2003 Conference (ICWI 2003).

Deen, S. (2005). An engineering approach to cooperating agents for distributed information systems. *Journal of Intelligent Information Systems, 25*(1), 5-45.

Delena, D., & Pratt, D. B. (2006, February). An integrated and intelligent DSS for manufacturing systems. *Expert Systems with Applications, 30*(2), 325-336.

DeMarco, T. (1978). *Structured Analysis and System Specifications*: Yourdon Press.

DESMET. (1991). *State of the Art Report (Workpackage 1)* .NCC Publications.

Díaz, P., Montero, S., & Aedo, I. (2004). Modelling hypermedia and web applications: the Ariadne Development Method. *Information Systems*, 30 (8), 649-673, Elsevier.

Dietz, J. (1987). *Modelling and specification of information systems* (In Dutch: Modelleren en specificeren van informatiesystemen). Dissertation Thesis, Technical University of Eindhove, The Netherlands.

Dobing, B., & Parsons, J. (2000). Understanding the role of use cases in UML: A review and research agenda. *Journal of Database Management, 11*(4), 28-36.

Dobson, J., Blyth, J., & Strens, R. (1994). Organisational Requirements Definition for Information Technology. In *Proceedings of the International Conference on Requirements Engineering 1994*, Denver/CO.

DoD (Ed.). (2003). *DoD architecture framework. Version 1.0. Volume I: Definitions and Guidelines.* Washington, DC: Office of the DoD Chief Information Officer, Department of Defense.

Dorling, A. (1993). SPICE: Software process improvement and capacity determination. *Information and Software Technology.* 35, 6/7, pp. 404-406.

Dorman, A. (2007). FrankenSOA. *Network Computing, 18*(12), 41-51.

Doumeingts, G., Vallespir, B., & Chen, D. (1998). Decisional modelling using the GRAI grid. In P. Bernus, K. Mertins, & G. Schmidt (Eds.), *Handbook on Architectures of Information Systems,* (pp. 313-338). Berlin: Springer.

Duke, A., Davies, J., & Richardson, M. (2005). Enabling a Scalable Service oriented Architecture with Semantic Web Services. *BT Technology Journal, 23*(3), 191-201.

Dulle, H. (2002). *Trial Application in Verbundplan, deliverable D5, IST Programme project HyperKnowledge – Hypermedia and Pattern Based Knowledge Management for Smart Organisations*, project no. IST-2000-28401, Verbundplan, Austria

Dumas, M., Aldred, L., Heravizadeh, M., & Ter Hofstede, A. (2002). *Ontology Markup for Web Forms Generation.* Paper presented at the WWW '02 Workshop on Real-World applications of RDF and the semantic web.

Đurić, D. (2004). MDA-based Ontology Infrastructure. *Computer Science and Information Systems* 1 (1), ComSIS Consortium.

Eatock, J., Paul, R.J., & Serrano, A. (2002). Developing a theory to explain the insights gained concerning information systems and business processes behaviour: The ASSESS-IT project. *Information Systems Frontiers, 4*(3), 303-316.

Elliman, T., & Eatock, J. (2004). Online support for arbitration: Designing software for a flexible business process. *International Journal of Information Technology and Management, 4*(4), 443-460.

Elliman, T., & Hayman, A. (1999). A comment on Kidd's characterisation of knowledge workers. *Cognition, Technology and Work, 1*(3), 162-168.

Elliman, T., Eatock, J., & Spencer, N. (2005). Modelling knowledge worker behaviour in business process studies. *Journal of Enterprise Information Management, 18*(1), 79-94.

Engeström, Y. (1987). *Learning by expanding: an activity theoretical approach to developmental research*. Helsinki: Orienta-Konsultit.

Erickson, J., & Siau, K. (2008). Web Services, Service Oriented Computing, and Service Oriented Architecture: Separating Hype from Reality. *Journal of Database Management, 19*(3), 42-54.

Erickson, J., Lyytinen, K., & Siau, K. (2005). Agile Modeling, Agile Software Development, and Extreme Programming: The State of Research. *Journal of Database Research, 16(*4), 80-89.

ESPRIT Consortium AMICE (Ed.). (1993). *CIMOSA: Open system architecture for CIM. Volume 1* (Research report ESPRIT, Project 688/5288) (2nd revised and extended edition). Berlin: Springer.

Essink, L. (1986). A modeling approach to information system development. In T. Olle, H. Sol & A. Verrijn-Stuart (Eds.), *Proc. of the IFIP WG 8.1 Working Conf. on Comparative Review of Information Systems Design Methodologies: Improving the Practice* (pp. 55-86). Amsterdam: Elsevier Science.

Essink, L. (1988). A conceptual framework for information systems development methodologies. In H. J. Bullinger et al. (Eds.), *Information Technology for Organizational Systems* (pp. 354-362). Amsterdam: Elsevier Science.

Etzion, O., Gal, A., & Segev, A. (1998). Extended update functionality in temporal databases. In O. Etzion, S. Jajodia, & S. Sripada, (Eds.), *Temporal Databases -- Research and Practice, LNCS* (pp 56-95). Berlin: Springer-Verlag.

F3-Consortium (1994). *F3 Reference Manual*, ESPRIT III Project 6612, SISU, Stockholm.

Falbo R. A., Guizzardi, G., & Duarte, K. C. (2002). An Ontological Approach to Domain Engineering. In *Proceedings of the 14th International Conference on Software Engineering and Knowledge Engineering* (pp. 351- 358). Ischia, Italy: Springer.

Falbo, R. A. (2004) Experiences in Using a Method for Building Domain Ontologies. In *Proceedings of the Sixteenth International Conference on Software Engineering and Knowledge Engineering* (pp. 474-477), Banff, Alberta, Canadá.

Falkenberg, E., Hesse, W., Lindgreen, P., Nilsson, B., Oei, J. L. H., Rolland, C., Stamper, R., van Asche, F., Verrijn-Stuart, A., & Voss, K. (1998). *A framework of information system concepts, The FRISCO Report* (Web edition), IFIP.

Falkenberg, E., Oei, J., & Proper, H. (1992). A conceptual framework for evolving information systems. In H. Sol & R. Crosslin (Eds.), *Dynamic Modelling of Information Systems II* (pp. 353-375). Amsterdam: Elsevier Science.

Fensel, D. (2001). *Ontologies: Silver Bullet for Knowledge Management and Electronic Commerce*: Springer Verlag.

Ferguson, D., & Stockton, M. (2005). Service oriented Architecture: Programming Model and Product Architecture. *IBM Systems Journal, 44*(4),753-780.

Fettke, P., & Loos, P. (2003). Classification of reference models: A methodology and its application. *Information Systems and e-Business Management, 1*(1), 35-53.

Fillmore, C. (1968). The case for case. In E. Bach & R. T. Harms (Eds.) *Universals in Linguistic Theory*. New York: Holt, Rinehart and Winston, 1-88.

Fillottrani, P., Franconi, E., & Tessaris, S. (2006). *The new ICOM ontology editor. In 19th International Work-*

shop on Description Logics (DL 2006), Lake District, UK. May 2006.

Fitzgerald, B., Russo, N., & Stolterman, E. (2002). *Information systems development – methods in action.* London: McGraw Hill.

Fons, J., Valderas, P., Ruiz, M., Rojas, G., & Pastor, O. (2003). OOWS: A Method to Develop Web Applications from Web-Oriented Conceptual Models. In *Proceedings of the International Workshop on Web Oriented Software Technology* (pp. 65-70), Oviedo, Spain.

Fowler, M. (1997). *Analysis patterns: Reusable object models.* Menlo Park, CA: Addison Wesley.

Fowler, M. (2002). *Patterns of Enterprise Application Architecture.* Addison-Wesley.

Fowler, M. (2007). *Inversion of Control Containers and the Dependency Injection pattern.* Retrieved on November 19, 2001, from http://www.martinfowler.com/articles/injection.html

Fox, M. S., Chionglo, J. F., & Fadel, F. G. (1993). A common-sense model of the enterprise. In *Proceedings of the 2nd Industrial Engineering Research Conference*, Institute for Industrial Engineers, Norcross/GA.

Franconi, E., & Ng, G. (2000). The iCom tool for intelligent conceptual modeling. *7th International Workshop on Knowledge Representation meets Databases (KRDB'00)*, Berlin, Germany. 2000.

Freeman, M., & Layzell, P. (1994). A meta-model of information systems to support reverse engineering. *Information and Software Technology, 36*(5), 283-294.

Gallasch, G., & Kristensen, L. M. (2001, August). Comms/CPN: A communication infrastructure for external communication with Design/CPN. In *Proceedings of the 3rd Workshop and Tutorial on Practical Use of Coloured Petri Nets and the CPN Tools* (pp. 75-90). Aarhus, Denmark: University of Aarhus, Department of Computer Science.

Gamma, E., Helm, R., Johnson, R., & Vlissides, J. (1994). *Design Patterns: Elements of Reusable Object-Oriented Software.* Addison-Wesley.

Gamma, E., Helm, R., Johnson, R., & Vlissides, J. (1995). *Design patterns: Elements of reusable object-oriented design.* Menlo Park, CA: Addison-Wesley.

Gasser, L. (1986). The integration of computing and routine work. *ACM Trans. on Office Information Systems, 4*(3), 205-225.

George, W. T. C., & Heineman, T. (2001). *Component-based software engineering: Putting the pieces together.* New York.

Georgiadou, E. & Sadler, C. (1995). Achieving quality improvement through understanding and evaluating information systems development methodologies. *3rd International Conference on Software Quality Management.* Vol 2, pp. 35-46, May, Seville, British Computer Society.

Georgiadou, E., Hy, T. & Berki, E. (1998). Automated qualitative and quantitative evaluation of software methods and tools. *The Twelfth International Conference of the Israel Society for Quality*, Nov-Dec. Jerusalem.

Georgiadou, E., Mohamed W-E., Sadler, C. (1994). Evaluating the Evaluation Methods: The Data Collection and Storage System Using the DESMET Feature Analysis. *10th International Conference of the Israel Society for Quality.* Nov. Jerusalem.

Georgiadou, E., Siakas, K. & Berki, E. (2003). Quality Improvement through the Identification of Controllable and Uncontrollable Factors in Software Development. Messnarz R. (Ed.) *EuroSPI 2003: European Software Process Improvement.* 10-12 Dec, Graz, Austria.

Gerstl, P., & Pribbenow, S. (1995). Midwinters, end games, and body parts: a classification of part-whole relations. *International Journal of Human-Computer Studies, 43*, 865-889.

Giaglis, G.M. (1999). *Dynamic process modelling for business engineering and information systems* (Doctoral Thesis). London: Brunel University.

Giaglis, G.M., Hlupic, V., Vreede, G.J., & Verbraeck, A. (2005). Synchronous design of business processes and information systems using dynamic process modelling. *Business Process Management Journal, 11*(5), 488-500.

Gigch van, J. (1991). *System design modeling and metamodeling.* New York: Plenum Press.

Ginige, A., & Murugesan, S. (2001). Web Engineering: An Introduction. *IEEE Multimedia*, 8 (1), 14-18, IEEE.

Giunchiglia, F., Mylopoulos, J., & Perini, A. (2002) The Tropos software development methodology: Processes, models and diagrams. In *Proceedings of the 1st International Joint Conference on Autonomous Agents and Multiagent Systems* (pp. 35-36). Bologna, Italy.

Glaser, B. G., & Strauss, A. L. (1967). *The Discovery of Grounded Theory: Strategies for Qualitative Research*, Weidenfeld and Nicolson, London.

Goldkuhl, G., & Cronholm, S. (1993). *Customizable CASE environments: a framework for design and evaluation.* Institutionen for Datavetenskap, Universitetet och Tekniska Högskolan, Linköping, Research Report.

Goldstein, R., & Storey, V. (1999). Data abstraction: Why and how? *Data & Knowledge Engineering, 29*(3), 293-311.

Gomaa, H. (1995). Domain modeling methods and environments. *ACM SIGSOFT Software Engineering Notes, 20*(SI), 256-258.

Gomez-Perez, A., Corcho, O., & Fernandez-Lopez, M. (2005). *Ontological Engineering.* Springer.

Graham, I. (1991). Object-Oriented Methods. Reading, MA: Addison-Wesley.

Graham, I., Henderson-Sellers, B., & Younessi, H. (1997). *The OPEN process specification.* Reading: Addison-Wesley.

Grosz, G., Rolland, C., Schwer, S., Souveyet, C., Plihon, V., Si-Said, S., Achour, C., & Gnaho, C. (1997). Modelling and engineering the requirements engineering process: an overview of the NATURE approach. *Requirements Engineering, 2*(2), 115-131.

Gruber, T. (1993). A translation approach to portable ontologies. *Knowledge Acquisition, 5*(2), 199- 220.

Gruber, T. (1995). Towards principles for the design of ontologies used for knowledge sharing. *International Journal of Human-Computer Studies, 43*(5/6), 907-928.

Guarino, N. (1998). Formal ontology and information systems. In N. Guarino (Ed.), *Formal Ontology in Information Systems (FOIS'98)* (pp. 3-15). Amsterdam: IOS Press.

Guarino, N., & Welty, C. (2000). A formal ontology of properties. In Dieng, R. (Ed.), *Proceedings of EKAW '00.* Berlin: Springer Verlag.

Guarino, N., & Welty, C. (2002). Evaluating Ontological Decisions with OntoClean. *Communications of the ACM, 45*(2), 61-65.

Guizzardi, G. (2005). *Ontological foundations for structural conceptual models.* PhD Thesis, Telematica Institute, Twente University, Enschede, the Netherlands.

Guizzardi, G. (2007). Modal Aspects of Object Types and Part-Whole Relations and the de re/de dicto distinction. *19th International Conference on Advances in Information Systems Engineering (CAiSE) LNCS 4495.* Berlin: Springer-Verlag.

Guizzardi, G., Falbo, R., & Filho, J. (2001). From domain ontologies to object-oriented frameworks. In G. Stumme, A. Maedche & S. Staab (Eds.), *Workshop on Ontologies (ONTO'2001)*, 1-14.

Guizzardi, G., Herre, H., & Wagner, G. (2002). On the general ontological foundations of conceptual modeling. In S. Spaccapietra, S. March & Y. Kambayashi (Eds.), *Conceptual Modeling – ER 2002* (pp. 65-78). LNCS 2503, Berlin: Springer.

Guizzardi, G., Wagner, G., Guarino, N., & van Sinderen, N. (2004). An Ontologically Well-Founded Profile for UML Conceptual Models. In A. Persson & J. Stirna (Eds.), *Proc. 16th Int. Conf. on Advanced Inf. Sys. Engineering, CAiSE2004* (pp. 112-126). Springer LNCS 3084.

Gupta, D., & Prakash, N. (2001). Engineering methods from method requirements specifications. *Requirements Engineering, 6*(3), 135-160.

Gustas, R., Bubenko Jr., J. A., & Wangler, B. (1995). Goal Driven Enterprise Modelling: Bridging Pragmatic and Semantic Descriptions of Information Systems. *5th European - Japanese Seminar on Information Modelling and Knowledge Bases*, Sapphoro, May 30-June 3, 1995.

Halpin, T. (1999). UML Data Models from an ORM Perspective (Part 8). *Journal of Conceptual Modeling*, 8, April 1999. Stable URL http://www.inceoncept.com/jcm.

Halpin, T. (2001). *Information Modeling and Relational Databases; from conceptual analysis to logical design.* San Francisco, California: Morgan Kaufmann.

Halpin, T. (2004). Comparing Metamodels for ER, ORM and UML Data Models. In K. Siau (Ed.), *Advanced Topics in Database Research, 3*, 23-44. Hershey PA: Idea Publishing Group.

Halpin, T. (2005). Higher-Order Types and Information Modeling. In K. Siau (Ed.), *Advanced Topics in Database Research, 4*, 218-237. Hershey PA: Idea Publishing Group.

Halpin, T. (2005). ORM 2. In R. Meersman, Z. Tari, P. Herrero et al. (Eds.) *On the Move to Meaningful Internet Systems 2005: OTM 2005 Workshops* (pp. 676-687). Cyprus: Springer LNCS 3762.

Halpin, T. (2006). Object-Role Modeling (ORM/NIAM). In P. Bernus, K. Mertins, & G. Schmidt (Eds.), *Handbook on Architectures of Information Systems, 2nd edition* (pp. 81-103). Heidelberg: Springer.

Halpin, T. (2007). Fact-Oriented Modeling: Past, Present and Future. In J. Krogstie, A. Opdahl & S. Brinkkemper (Eds.), *Conceptual Modelling in Information Systems Engineering* (pp. 19-38). Berlin: Springer.

Halpin, T. A. (2001). *Information Modeling and Relational Databases, From Conceptual Analysis to Logical Design.* Morgan Kaufmann, San Mateo, California, USA, 2001.

Halpin, T., & Morgan T. (2008). *Information Modeling and Relational Databases, 2nd edn.* San Francisco: Morgan Kaufmann.

Halpin, T., & Proper, H. (1995). Subtyping and polymorphism in object-role modeling. *Data & Knowledge Engineering, 15*(3), 251–281.

Halpin. T. (2007). Subtyping revisited. In B. Pernici,& J. Gulla (Eds.), *Proceedings of CAiSE'07 Workshops* (pp. 131-141). Academic Press.

Harel, D., Pnueli, A., Schmidt, J.P. and Sherman, R. (1987). On the Formal Semantics of Statecharts. *Proceedings of the 2nd IEEE Symposium on Logic in Computer Science.* pp. 54-64, IEEE Computer Society.

Harmsen, F. 1997. *Situational method engineering.* Dissertation Thesis, University of Twente, Moret Ernst & Young Management Consultants, The Netherlands.

Hautamäki, A. (1986). *Points of views and their logical analysis.* Helsinki: Acta Philosophica Fennica, Vol. 41.

Havenstein, H. (2006). Measuring SOA Performance is a Complex Art. *Computer World, 40*(2), 6.

Hawley, K. (2004). Temporal Parts. In E. N. Zalta, (Ed.), *The Stanford Encyclopedia of Philosophy (Winter 2004 Ed.).* Stable URL http://plato.stanford.edu/archives/win2004/entries/temporal-parts/.

Hay, D. (2006). *Data Model Patterns: A Metadata* Map. San Francisco: Morgan Kaufmann.

Hay, D. C. (1996). *Data model patterns: Conventions of thought.* New York: Dorset House Publishing.

Henderson-Sellers, B., & Barbier, F. (1999). What is this thing called aggregation? In R. Mitchell, A. C. Wills, J. Bosch & B. MeyerProc. (Eds.), *TOOLS EUROPE'99* (pp. 236-250). MD: IEEE Computer Society Press.

Herbst, H. (1995). A meta-model for business rules in systems analysis. In J. Iivari, K. Lyytinen & M. Rossi (Eds.), *Advanced Information Systems Engineering* (pp. 186-199). LNCS 932, Berlin: Springer.

Heym, M., & Österle, H. (1992). A reference model for information systems development. In K. Kendall, K. Lyytinen & J. DeGross (Eds.), *The Impacts on Computer Supported Technologies on Information Systems Development* (pp. 215-240). Amsterdam: Elsevier Science.

Hicks, B. (2007). *Oracle Enterprise Service Bus: The Foundation for Service Oriented Architecture*. Retrieved10-18-2007 from: http://www.oracle.com/global/ap/openworld/ppt_download/middleware_oracle%20enterprise%20service%20bus%20foundation_250.pdf.

Hirschheim, R., Klein, H., & Lyytinen, K. (1995). *Information systems development – conceptual and philosophical foundations*. Cambridge: Cambridge University Press.

Hirvonen, A., Pulkkinen, M., & Valtonen, K. (2007). Selection criteria for enterprise architecture methods. In Proc. of the European Conference on Information Management and Evaluation (pp. 227-236). Reading, UK: Academic Conference International.

Hlupic, V., & Robinson, S. (1998). Business process modelling and analysis using discrete-event simulation. In *Proceedings of the 1998 Winter Simulation Conference*, Washington, DC (pp. 1363-1369).

Hlupic, V., & Vreede, G.J. (2005). Business process modelling using discrete-event simulation: Current opportunities and future challenges. *International Journal of Simulation & Process Modelling, 1*(1-2), 72-81.

Hodgkinson, I. M., Wolter, F., & Zakharyaschev, M. (2000). Decidable fragments of first-order temporal logics. *Annals of pure and applied logic*, *106*, 85-134.

Hofreiter, B., Huemer, C., Liegl, P., Schuster, R., & Zapletal, M. (2006). *UN/CEFACT'S modeling methodology (UMM): A UML profile for B2B e-commerce*. Retrieved March, 2008, from http://dme.researchstudio.at/publications/2006/

Hofstede ter, A., & Proper, H. (1998). How to formalize it? Formalization principles for information system development methods. *Information and Software Technology, 40*(10), 519-540.

Hofstede ter, A., & Verhoef, T. (1997). On the feasibility of situational method engineering. *Information Systems, 22*(6/7), 401-422.

Holcombe, M. and Ipate, F. (1998). *Correct Systems - Building a Business Process Solution*. Springer-Verlag

Hoppenbrouwers, S.J.B.A., H.A. (Erik) Proper, & Th.P. van der Weide (2005). A Fundamental View on the Process of Conceptual Modeling. In *Conceptual Modeling - ER 2005 - 24 International Conference on Conceptual Modeling*, 3716 of *Lecture Notes in Computer Science*, pages 128-143, June 2005.

Hoppenbrouwers, S.J.B.A., H.A. (Erik) Proper, and Th.P. van der Weide (2005). Towards explicit strategies for modeling. In T.A. Halpin, K. Siau, and J. Krogstie, editors, *Proceedings of the Workshop on Evaluating Modeling Methods for Systems Analysis and Design* (EMMSAD'05), held in conjunction with the 17th Conference on Advanced Information Systems 2005 (CAiSE 2005), pages 485-492, Porto, Portugal, EU, 2005. FEUP, Porto, Portugal, EU.

Hoppenbrouwers, S.J.B.A., H.A. (Erik) Proper, and Th.P. van der Weide (2005). Fact Calculus: Using ORM and Lisa–D to Reason About Domains. In R. Meersman, Z. Tari, and P. Herrero, editors, *On the Move to Meaningful Internet Systems 2005: OTM Workshops* – OTM Confederated International Workshops and Posters, AWeSOMe, CAMS, GADA, MIOS+INTEROP, ORM, PhDS, SeBGIS, SWWS, and WOSE 2005, Agia Napa, Cyprus, EU, volume 3762 of Lecture Notes in Computer Science, pages 720–729, Berlin, Germany, October/November 2005: Springer–Verlag.

Horrocks, I., Kutz, O., & Sattler, U. (2006). The Even More Irresistible SROIQ. In *Proceedings of the 10th International Conference of Knowledge Representation and Reasoning (KR2006)*, Lake District, UK, 2006.

Huotari J. & Kaipala, J. (1999). Review of HCI Research - Focus on cognitive aspects and used methods. Kakola T. (ed.) *IRIS 22 Conference: Enterprise Architectures for Virtual Organisations*. Keurusselka, Jyvaskyla, Jyvaskyla University Printing House.

Huschens, J., & Rumpold-Preining, M. (2006). IBM insurance application architecture (IAA): An overview of the insurance business architecture. In P. Bernus, K. Mertins, & G. Schmidt (Eds.), *Handbook on architectures of information systems* (2nd ed., pp. 669-692). Berlin: Springer-Verlag.

Hutchinson, B., Henzel, J., & Thwaits, A. (2006). Using Web Services to Promote Library-Extension Collaboration. *Library Hi Tech, 24*(1), 126-141.

Iivari, J. (1989). Levels of abstraction as a conceptual framework for an information system. In E. Falkenberg & P. Lindgren (Eds.), *Information System Concepts: An In-Depth Analysis* (pp. 323-352). Amsterdam: Elsevier Science.

Iivari, J. (1990). Hierarchical spiral model for information system and software development. Part 2: Design process. *Information and Software Technology, 32*(7), 450-458.

Iivari, J. and Huisman, M. (2001). The Relationship between Organisational Culture and the Deployment of Systems Development Methodologies. Dittrich K.R., Geppert, A. & Norrie, M. (Eds.) *The 13th International Conference on Advanced information Systems Engineering, CAiSE'01*. June, Interlaken. Springer, LNCS 2068, pp. 234-250.

Iivari, J. and Kerola, P. (1983). T.W. Olle, H.G. Sol, C.J. Tully (Eds) A Sociocybernetic Framework for the Feature Analysis of Information Systems Design Methodologies. *Information systems design methodologies: a feature analysis: proceedings of the IFIP WG 8.1 Working Conference on Comparative Review of Information Systems Design Methodologies.* 5-7 July, York.

Iivari, J., Hirschheim, R., & Klein, H. (2001). A dynamic framework for classifying information systems development methodologies and approaches. *Journal of Management Information Systems, 17*(3), 179-218.

Ince, D. (1995). *Software Quality Assurance.* McGraw-Hill

Insfran, E., Pastor, O., & Wieringa, R. (2002). Requirements engineering-based conceptual modeling. *Requirements Engineering, 7*, 61-72.

Ipate, F. and Holcombe, M. (1998). Specification and Testing using Generalized Machines: a Presentation and a Case Study. *Software Testing, Verification and Reliability.* 8, 61-81.

ISO (2003). *ISO 15745-1 Industrial automation systems and integration - Open systems application integration framework - Part 1: Generic reference description.* Geneva, Switzerland: International Organization for Standardization.

Jacobson, I., Booch, G., & Rumbaugh, L. (1999). *The Unified Software Development Process.* Addison-Wesley.

Jacobson, I., Christerson, M., Jonsson, P., & Overgaard, G. (1992). *Object-Oriented Software Engineering: A Use Case Driven Approach.* Addison Wesley.

Jacyntho, M. D., Schwabe, D., & Rossi, G. (2002). A Software Architecture for Structuring Complex Web Applications. In *Proceedings of the 11th International World Wide Web Conference, Web Engineering Alternate Track*, Honolulu, EUA: ACM Press.

Jarke, M., Gallersdorfer, R., Jeusfeld, M. A., Staudt, M., Eherer, S. (1995). ConceptBase – a deductive objectbase for meta data management. *Journal of Intelligent Information Systems.* 4(2):167-192.

Jarke, M., Pohl, K., Rolland, C. and Schmitt, J.-R. (1994). Experience-Based Method Evaluation and Improvement: A process modeling approach. T.W. Olle & A. A. Verrijin-Stuart (Eds) *IFIP WG8.1 Working Conference CRIS'94.* Amsterdam: North-Holland.

Jarzabek, S. (1990). Specifying and generating multilanguage software development environments. *Software Engineering Journal.* Vol. 5, No. 2, Mar., pp. 125-137.

Jayaratna, N. (1994). *Understanding and Evaluating Methodologies. NIMSAD: A Systemic Approach.* McGraw-Hill.

Jenkins, A. (1982). *MIS Decision Variables and Decision Making Performance.* Ann Arbor, MI: UMI Research Press.

Jenkins, T. (1994). Report back on the DMSG sponsored UK Euromethod forum '94. *Data Management Bulletin.* Summer Issue, 11, 3.

Jennings, N. R. (2001) An agent-based approach for building complex software systems. *Communications of the ACM, 44*(4), 35-41.

Jensen, K. (1992). *Coloured Petri nets: Basic concepts, analysis, methods and practical use (vol. 1).* Heidelberg, Germany: Springer GMBH.

Jensen, K. (1997). *Coloured Petri nets: Basic concepts, analysis, methods and practical use (vol. 2).* Heidelberg, Germany: Springer GMBH.

Jensen, K., et al. (1999). *Design/CPN 4.0.* Retrieved December 17, 2005, from http://www.daimi.au.dk/designCPN/

Johannesson P., Boman, M., Bubenko, J., & Wangler, B. (1997). *Conceptual Modelling,* 280 pages, Prentice Hall International Series in Computer Science, Series editor C.A.R. Hoare, Prentice Hall.

Johansson, I. (2004). On the transitivity of the parthood relation. In Hochberg, H. and Mulligan, K. (eds.) *Relations and predicates* (pp. 161-181). Frankfurt: Ontos Verlag.

Johnsen, S. G., Schümmer, T., Haake, J., Pawlak, A., Jørgensen, H., Sandkuhl, K., Stirna, J., Tellioglu, H., & Jaccuci, G. (2007). Model-based adaptive Product and Process Engineering. In M. Rabe & P. Mihok (Eds.), *Ambient Intelligence Technologies for the Product Lifecycle: Results and Perspectives from European Research.* Fraunhofer IRB Verlag, 2007.

Jones, S. (2005). Toward an Acceptable Definition of Service. *IEEE Software.* May/June. (pp. 87-93).

Jones, T. C. (1984). Reusability in programming, a survey of the state of the art. *IEEE Transactions on Software Engineering, 10*(5), 488-493.

Jørgensen, H. D, Karlsen D., & Lillehagen F. (2007). Product Based Interoperability – Approaches and Rerequirements. In Pawlak et al. (Eds.), *Proceedings of CCE'07,* Gesellschaft für Informatik, Bonn, ISBN 978-3-88579-214-7

Jørgensen, H. D. (2004). *Interactive Process Models,* PhD-thesis, NTNU, Trondheim, Norway, 2004 ISBN 82-471-6203-2.

Jørgensen, H. D. (2001). Interaction as a Framework for Flexible Workflow Modelling, *Proceedings of GROUP'01,* Boulder, USA, 2001.

Kaindl H., Hatzenbichler G., Kapenieks A., Persson A., Stirna J., & Strutz G. (2001). *User Needs for Knowledge Management, deliverable D1, IST Programme project HyperKnowledge - Hypermedia and Pattern Based Knowledge Management for Smart Organisations,* project no. IST-2000-28401, Siemens AG Österreich, Austria

Kaindl, H., Kramer, S., & Hailing, M. (2001). An interactive guide through a defined modeling process. In *People and Computers XV, Joint Proc. of HCI 2001 and IHM 2001,* Lille, France, 107-124

Kaipala, J. (1997). Augmenting CASE Tools with Hypertext: Desired Functionality and Implementation Issues. A. Olive, J.A. Pastor (Eds.) *The 9th International Conference on Advanced information Systems Engineering, CAiSE '97.* LNCS 1250, Berlin: Springer-Verlag.

Kang, K., C., Kim, S., Lee, J., Kim, K., Shin, E., & Huh, M. (1998). FORM: A feature-oriented reuse method with domain-specific reference architectures. *Annals of Software Engineering, 5,* 143-168.

Karagiannis, D., & Kühn, H. (2002). *Metamodelling platforms.* Paper presented at the Proceedings of the Third International Conference EC-Web, September 2-6, 2002, LNCS 2455, Dexa, Aix-en-Provence, France.

Kardasis, P., Loucopoulos, P., Scott, B., Filippidou, D., Clarke, R., Wangler, B., & Xini, G. (1998). *The use of Business Knowledge Modelling for Knowledge Discovery in the Banking Sector, IMACS-CSC'98,* Athens, Greece, October, 1998.

Karhinen, A., Ran, A., & Tallgren, T. (1997). *Configuring designs for reuse.* Paper presented at the Proceedings of the 19th International Conference on Software Engineering, Boston.

Karlsson, F., & Ågerfalk, P. (2004). Method configuration: adapting to situational characteristics while creating reusable assets. *Information and Software Technology,* 46(9), 619-633.

Katz, R. (1990). Toward a unified framework for version modeling in engineering databases. *ACM Surveys, 22*(4), 375-408.

Kavakli, V., & Loucopoulos, P. (1999). Goal-driven business process analysis application in electricity deregulation. *Information Systems, 24*(3), 187-207.

Keet, C. M. (2006). *Introduction to part-whole relations: mereology, conceptual modeling and mathematical aspects* (Tech. Rep. No. KRDB06-3). KRDB Research Centre, Faculty of Computer Science, Free University of Bozen-Bolzano, Italy.

Keet, C. M. (2006). Part-whole relations in Object-Role Models. 2nd International Workshop on Object-Role Modelling (ORM 2006), Montpellier, France, Nov 2-3, 2006. In Meersman, R., Tari, Z., Herrero, P. et al. (Eds.) *OTM Workshops 2006 LNCS, 4278*, 1116-1127. Berlin: Springer-Verlag.

Keet, C. M. (2007). Prospects for and issues with mapping the Object-Role Modeling language into DLRifd. *20th International Workshop on Description Logics (DL'07) CEUR-WS, 250*, 331-338. 8-10 June 2007, Bressanone, Italy.

Keet, C. M. (2008). A formal comparison of conceptual data modeling languages. *13th International Workshop on Exploring Modeling Methods in Systems Analysis and Design (EMMSAD'08) CEUR-WS, 337*, 25-39. Montpellier, France, 16-17 June 2008.

Keet, C. M., & Artale, A. (in press). Representing and Reasoning over a Taxonomy of Part-Whole Relations. *Applied Ontology – Special Issue on Ontological Foundations for Conceptual Models, 3*(1).

Keller, G., Nüttgens, M., & Scheer, A.-W. (1992). Semantische Prozeßmodellierung auf der Grundlage Ereignisgesteuerter Prozeßketten (EPK). *Veröffentlichungen des Instituts für Wirtschaftsinformatik der Universität des Saarlandes,* (89).

Kelly, S. (1997). *Towards a Comprehensive MetaCASE and CAME Environment, Conceptual, Architectural, Functional and Usability Advances in MetaEdit+.* Ph.D. Thesis, University of Jyvaskyla.

Kelly, S. (2007). Domain-specific modeling: The killer app for method engineering? In J. Ralytè, S. Brinkkemper & B. Henderson-Sellers (Eds.), *Situational Method Engineering: Fundamentals and Experiences* (pp. 1-5). Boston: Springer.

Kelly, S., Lyytinen, K. and Rossi, M. (1996). MetaEdit+: A Fully Configurable Multi-User and Multi-Tool CASE and CAME Environment. Constantopoulos, P., Mylopoulos, J. and Vassiliou, Y (Eds) *Advances in Information Systems Engineering, 8th International Conference CAiSE'96.* Heraklion, Crete, Greece, May 20-24, Berlin: Springer-Verlag LNCS, pp. 1-21.

Kensche, D., Quix, C., Chatti, M. A., & Jarke, M. (2005). GeRoMe: A Generic Role Based Metamodel for Model Management. In R. Meersman, Z. Tari, and P. Herrero, editors, *On the Move to Meaningful Internet Systems 2005: CoopIS, DOA, and ODBASE – OTM Confederated International Conferences, CoopIS, DOA, and ODBASE 2005*, Proceedings, Part II, Agia Napa, Cyprus, EU, volume 3761 of Lecture Notes in Computer Science, pages 1206–1224. Springer–Verlag, October/November 2005.

Kidd, A. (1994). The marks are on the knowledge worker. In *Proceedings of the Conference of Human Factors in Computer Systems*, Boston, MA (pp. 186-191).

Kim, J., & Lim, K. (2007). An Approach to Service Oriented Architecture Using Web Service and BPM in the Telcom OSS domain. *Internet Research, 17*(1), 99-107.

Kim, Y., & March, S. (1995). Comparing data modeling formalisms. *Communications of the ACM, 38*(6), 103-115.

Kinnunen, K., & Leppänen, M. (1996). O/A matrix and a technique for methodology engineering. *Journal of Systems and Software,* 33(2), 141-152.

Kitchenham, B.A., Linkman, S.G. and Law, D.T. (1994). Critical review of quantitative assessment. *Software Engineering Journal.* Mar.

Kittlaus, H.-B., & Krahl, D. (2006). The SIZ banking data Model. In P. Bernus, K. Mertins, & G. Schmidt (Eds.), *Handbook on architectures of information systems* (2nd ed., pp. 723-743). Berlin: Springer-Verlag.

Koch, N., & Kraus, A. (2002). The Expressive Power of UML-based Web Engineering. In D. Schwabe, O. Pastor, G. Rossi e L. Olsina (Ed.), *Proceedings of the Second International Workshop on Web-Oriented Software Technology* (pp. 105-119), CYTED.

Koch, N., Baumeister, H., Hennicker, R., & Mandel, L. (2000). Extending UML to Model Navigation and Presentation in Web Applications. In *Proceedings of Modelling Web Applications in the UML Workshop*, York, UK.

Kopetz, H. (1979). *Software Reliability.* The Macmillan Press Ltd.

Kornyshova, E., Deneckere, R., & Salinesi, C. (2007). Method chunks selection by multicriteria methods: An extension of the assembly-based approach. In J. Ralytè, S. Brinkkemper & B. Henderson-Sellers (Eds.), *Situational Method Engineering: Fundamentals and Experiences* (pp. 64-78). Boston: Springer.

Koskinen, M. (2000). *Process Metamodeling: Conceptual Foundations and Application.* Ph.D. Thesis, Jyvaskyla Studies in Computing-7, University of Jyvaskyla.

Krafzig, D., Banke, K., & Slama, D. (2005). *SOA Elements.* Prentice Hall. Retrieved on 10-02-2007 from: http://en.wikipedia.org/wiki/Image:SOA_Elements.png

Krogstie, J. (2002). A Semiotic Approach to Quality in Requirements Specifications. In L. Kecheng, R.J. Clarke, P.B. Andersen, R.K. Stamper, and E.-S. Abou-Zeid, editors, *Proceedings of the IFIP TC8 / WG8.1 Working Conference on Organizational Semiotics: Evolving a Science of Information Systems*, pages 231-250, Deventer, The Netherlands, EU, 2002. Kluwer.

Krogstie, J. (1995). *Conceptual modeling for computerized information systems support in organizations.* Dissertation Thesis, NTH, University of Trondheim, Norway.

Krogstie, J. (2007) Modelling of the People, by the People, for the People. In *Conceptual Modelling in Information Systems Engineering.* Berlin: Springer Verlag. ISBN 978-3-540-72676-0. s. 305-318

Krogstie, J., & Jorgensen H. D. (2002). Quality of Interactive Models. In M. Genero, Grandi. F., W.-J. van den Heuvel, J. Krogstie, K. Lyytinen, H.C. Mayr, J. Nelson, A. Olivé, M. Piattine, G. Poels, J. Roddick, K. Siau, M. Yoshikawa, and E.S.K. Yu, editors, *21st International Conference on Conceptual Modeling (ER 2002)*, volume 2503 of *Lecture Notes in Computer Science*, pages 351-363, Berlin, Germany, EU, 2002. Springer.

Krogstie, J., & Jørgensen, H. D. (2004). Interactive Models for Supporting Networked Organizations. In *Proceedings of CAiSE'2004*, Springer LNCS, ISBN 3-540-22151.

Krogstie, J., Lillehagen, F., Karlsen, D., Ohren, O., Strømseng, K., & Thue Lie, F. (2000). *Extended Enterprise Methodology.* Deliverable 2 in the EXTERNAL project, available at http://research.dnv.com/external/deliverables.html.

Krogstie, J., Sindre, G., & Jorgensen, H. (2006). Process models representing knowledge for action: a revised quality framework. *European Journal of Information Systems*, 15, 91-102.

Krogstie, J., Sindre, G., & Jørgensen, H. (2006). Process Models Representing Knowledge for Action: a Revised Quality Framework. *European Journal of Information Systems, 15*(1), 91–102

Kronlof, K. (Ed.). (1993). *Method Integration, Concepts and Case Studies.* Wiley Series on Software Based Systems, Wiley.

Kruchten, P. (2000). *The Rational Unified Process: An introduction.* Reading: Addison-Wesley.

Kruchten, P. (2003). *The rational unified process: An introduction* (3rd ed.). Boston.

Kumar, K., & Welke, R. (1992). Methodology engineering: a proposal for situation specific methodology construction. In W. Kottermann & J. Senn (Eds.), *Challenges and Strategies for Research in Systems Development* (pp. 257-269). Chichester: John Wiley & Sons.

Lammari, N., & Metais, E. (2004). Building and maintaining ontologies: a set of algorithms. *Data & Knowlege Engineering, 48*, 155- 176.

Langdon, C. (2007). Designing Information Systems to Create business Value: A Theoretical Conceptualization of the Role of Flexibility and Integration. *Journal of Database Management, 17*(3),1-18.

Langefors, B., & Sundgren, B. (1975). *Information systems architecture*. New York: Petrocelli.

Larman, C. (2002). *Applying UML and Patterns: An Introduction to Object-Oriented Analysis and Design, and the Unified Process* (2nd Edition). Prentice Hall.

Larsson, L., & Segerberg R., (2004). *An Approach for Quality Assurance in Enterprise Modelling.* MSc thesis. Department of Computer and Systems Sciences, Stockholm University, no 04-22

Law, D. (1988). *Methods for Comparing Methods: Techniques in Software Development.* NCC Publications.

Law, D. and Naeem, T. (1992). DESMET: Determining an Evaluation methodology for Software Methods and Tools. *Proceedings of Conference on CASE - Current Practice, Future Prospects.* Mar. 1992, British Computer Society, Cambridge.

Lee, R. (1983). Epistemological aspects of knowledge-based decision support systems. In H. Sol (Ed.), *Processes and Tools for Decision Support Systems* (pp. 25-36). Amsterdam: Elsevier Science.

Lee, S. C., & Shirani, A. I. (2004). *A component based methodology for Web application development. Journal of Systems and Software*, 71 (1-2), 177-187, Elsevier.

Lemmens, I., Nijssen, M., & Nijssen, G. (2007). *A NIAM 2007 conceptual analysis of the ISO and OMG MOF four layer metadata architectures.* Paper presented at the OTM 2007/ ORM 2007.

Lenat, D., & Guha, R. (1990). *Building large knowledge-based systems.* Reading: Addison-Wesley.

Leppänen M. (2007d). A context-based enterprise ontology. In W. Abramowicz (Ed.), *Business Information Systems (BIS 2007)* (pp. 273-286). LNCS 4439, Berlin: Springer.

Leppänen, M. (2000). Toward a method engineering (ME) method with an emphasis on the consistency of ISD methods. In K. Siau (Ed.), *Evaluation of Modeling Methods in Systems Analysis and Design (EMMSAD'00),* Stockholm: Sweden.

Leppänen, M. (2005). *An Ontological Framework and a Methodical Skeleton for Method Engineering,* Dissertation Thesis, Jyväskylä Studies in Computing 52, University of Jyväskylä, Finland. Available at: http://dissertations.jyu.fi/studcomp/9513921867.pdf

Leppänen, M. (2006). An integrated framework for meta modeling. In Y. Manolopoulos, J. Pokomy, & T. Sellis (Eds.), *Advances in Databases and Information Systems (ADBIS'2006)* (pp. 141-154). LNCS 4152, Berlin: Springer-Verlag.

Leppänen, M. (2006). Conceptual evaluation of methods for engineering situational ISD methods. *Software Process: Improvement and Practice, 11*(5), 539-555.

Leppänen, M. (2007). A Contextual method integration. In G. Magyar, G. Knapp, W. Wojtkowski, W.G. Wojtkowski, & J. Zupancic (Eds.), *Information Systems Development – New Methods and Practices for the Networked Society (ISD 2006)* (pp. 89-102). Vol. 2, Berlin: Springer-Verlag.

Leppänen, M. (2007). Towards an abstraction ontology. In M. Duzi, H. Jaakkola, Y Kiyoki, & H. Kangassalo (Eds.), *Information Modelling and Knowledge Bases XVIII, Frontiers in Artificial Intelligence and Applications* (pp. 166-185). The Netherlands: IOS Press.

Leppänen, M. (2007). Towards an ontology for information systems development – A contextual approach. In K. Siau (Ed.), *Contemporary Issues in Database Design and Information Systems Development* (pp. 1-36). New York: IGI Publishing.

Leppänen, M. (2007). IS ontology and IS perspectives. In H. Jaakkola, Y. Kiyoki, & T. Tokuda (Eds.), *Information Modelling and Knowledge Bases XIX, Frontiers in Artificial Intelligence and Application.* The Netherlands: IOS Press (in print).

Leppänen, M., Valtonen, K., & Pulkkinen, M. (2007). Towards a contingency framework for engineering an enterprise architecture planning method. In *Proc. of the 30th Information Systems Research Seminar in Scandinavia (IRIS 2007)*, Tampere, Finland.

Levas, A., Boyd, S., Jain, P., & Tulskie, W.A. (1995). The role of modelling and simulation in business process reengineering. In *Proceedings of the Winter Simulation Conference* (pp. 1341-1346).

Levinson, S. (1983). *Pragmatics*. London: Cambridge University Press.

Lewis, H. R. & Papadimitriou, C. H. (1998). *Elements of the Theory of Computation*. Prentice Hall International Editions.

Li, S., Huang, S., Yen, D., & Chang, C. (2007). Migrating Legacy Information Systems to Web Services Architecture. *Journal of Database Management, 18*(4),1-25.

Lillehagen, F. (2003). The foundation of the AKM Technology, In R. Jardim-Goncalves, H. Cha, A. Steiger-Garcao (Eds.), *Proceedings of the 10th International Conference on Concurrent Engineering* (CE 2003), July, Madeira, Portugal. A.A. Balkema Publishers

Lillehagen, F. (2003). The Foundations of AKM Technology. In *Proceedings 10th International Conference on Concurrent Engineering (CE) Conference*, Madeira, Portugal.

Lillehagen, F., & Krogstie, J. (2002). Active Knowledge Models and Enterprise Knowledge Management. In *Proceedings of the IFIP TC5/WG5.12 International Conf. on Enterprise Integration and Modeling Technique: Enterprise Inter- and Intra-Organizational Integration: Building International Consensus, IFIP, 236*, Kluwer, ISBN: 1-4020-7277-5.

Lillehagen, F., & Krogstie, J. (2008). *Active Knowledge Modeling of Enterprises*. Heidelberg, Berlin, New York: Springer.

Lima, F., & Schwabe, D. (2003). Application Modeling for the Semantic Web. In Proceedings of the 1st Latin American Web Conference (pp. 93-102), Santiago, Chile: IEEE-CS Press.

Lin, C.-Y., & Ho, C.-S. (1999). Generating domain-specific methodical knowledge for requirements analysis based on methodology ontology. *Information Sciences*, 114(1-4), 127-164.

Loucopoulos, P., Kavakli, V., Prekas, N., Rolland, C., Grosz, G., & Nurcan, S. (1998). *Using the EKD approach: the modelling component.* ELEKTRA – Project No. 22927, ESPRIT Programme 7.1.

Loucopoulos, P., Kavakli, V., Prekas, N., Rolland, C., Grosz, G., & Nurcan, S. (1997). *Using the EKD Approach: The Modeling Component*. Manchester, UK: UMIST

Lyytinen, K. (1986). *Information systems development as social action: framework and critical implications.* Jyväskylä Studies in Computer Science, Economics, and Statistics, No. 8, University of Jyväskylä, Finland, Dissertation Thesis.

Maciaszek, L. (2005). *Requirements Analysis and System Design: Developing Information Systems with UML*. 2nd Edition: Addison-Wesley.

Maes, A., & Poels, G. (2006). Evaluating Quality of Conceptual Models Based on User Perceptions. In D. W. Embley, A. Olivé, & S. Ram (Eds.), *ER 2006, LNCS 4215*, 54 – 67, Springer

Malloy, B., Kraft, N., Hallstrom, J., & Voas, J. (2006). Improving the Predictable Assembly of Service Oriented Architectures. *IEEE Software*. March/April. pp. 12-15.

Manninen, A. & Berki, E. (2004). An Evaluation Framework for the Utilisation of Requirements Management Tools - Maximising the Quality of Organisational Communication and Collaboration. Ross, M. & Staples, G. (Eds) *Software Quality Management 2004 Conference*. University of KENT, Canterbury, Apr., BCS.

Marttiin, P. (1988). *Customisable Process Modelling Support and Tools for Design Environment.* Ph.D. Thesis, University of Jyvaskyla, Jyvaskyla.

Marttiin, P., Rossi, M., Tahvanainen, V-P & Lyytinen, K. A. (1993). Comparative Review of CASE Shells – A preliminary framework and research outcomes. *Information and Management.* Vol. 25, pp. 11-31.

Masolo, C., Borgo, S., Gangemi, A., Guarino, N., & Oltramari, A. (2003). *Ontology Library.* WonderWeb Deliverable D18 (ver. 1.0, 31-12-2003). http://wonderweb.semanticweb.org.

Mathiassen, L., Munk-Madsen, A., Nielsen, P., & Stage, J. (2000). *Object Oriented Analysis and Design.* Marko Publishing, Alborg, Denmark.

McIlraith, S. A., Son, T. C., & Zeng, H. (2001). Semantic Web Services. *Intelligent Systems*, 16 (2), 46-53, IEEE.

McIlroy, M. D. (1968). Mass produced software components. In P. Naur & B. Randell (Eds.), *Software Engineering, Report on a Conference by the NATO Science Committee* (pp. 138-150). Brussels: NATO Scientific Affaris Division.

Meinhardt, S., & Popp, K. (2006). Configuring business application systems. In P. Bernus, K. Mertins, & G. Schmidt (Eds.), *Handbook on architectures of information systems* (2nd ed., pp. 705-721). Berlin: Springer-Verlag.

Melton, R., & Garlan, D. (1997). *Architectural unification.* Paper presented at the Proceedings of the 1997 conference of the Centre for Advanced Studies on Collaborative research, Toronto, Ontario, Canada.

Mendling, J., Reijers H. A., & Cardoso J. (2007). What Makes Process Models Understandable? In G. Alonso, P. Dadam, & M. Rosemann, (Eds.), *International Conference on Business Process Management (BPM 2007), LNCS 4714*, 48–63. Springer.

Mesarovic, M., Macko, D., & Takahara, Y. (1970). *Theory of hierarchical, multilevel, systems.* New York: Academic Press.

MetaPHOR website: http://metaphor.cs.jyu.fi/

MetaView website: http://web.cs.ualberta.ca/~softeng/Metaview/project.shtml

Meyer, B. (1998). *Object Oriented Software Construction.* Prentice Hall.

Mikelsons, J., Stirna, J., Kalnins, J. R., Kapenieks, A., Kazakovs, M., Vanaga, I., Sinka, A., Persson, A., & Kaindl, H. (2002). *Trial Application in the Riga City Council, deliverable D6, IST Programme project Hypermedia and Pattern Based Knowledge Management for Smart Organisations*, project no. IST-2000-28401. Riga, Latvia.

Mili, A., Fowler, S., Gottumukkala, R., & Zhang, L. (2000). *An integrated cost model for software reuse.* Paper presented at the Proceedings of the ACM, ICSE 2000, Limerick, Ireland.

Mili, H., Mili, A., Yacoub, S., & Addy, E. (2002). *Reuse-based software engineering.* New York.

Mili, H., Mili, F., & Mili, A. (1995). Reusing software: Issues and research directions. *IEEE Transactions on Software Engineering, 21*(6), 528-562.

Mirbel, I., & J. Ralyté (2006). Situational Method Engineering: combining assembly-based and roadmap-driven approaches. *Requirements Engineering*, 11, 58-78.

Mohamed W. A., and Sadler, C.J. (1992). Methodology Evaluation: A Critical Survey. *Eurometrics'92 Conference on Quantitative Evaluation of Software & Systems.* Brussels, Apr. pp. 101-112.

Montgomery, C., Runger, C., & Hubele, F. (2001). *Engineering Statistics* (2nd Edition). Wiley.

Moody, D. L. (2006). Theoretical and practical issues in evaluating the quality of conceptual models: current state and future directions. *Data and Knowledge Engineering*, (55), 243-276.

Moody, D. L., & Shanks, G. (2003). Improving the quality of data models: Empirical validation of a quality management framework. *Information Systems (IS) 28*(6), 619-650, Elsevier

Morgan, T. (2006). *Some features of state machines in ORM.* Paper presented at the OTM2006/ORM 2006 workshop.

Morgan, T. (2007). *Business Process Modeling and ORM.* Paper presented at the OTM 2007/ORM 2007.

Morris, C. W. (1938). Foundations of the theory of signs. In O. Neurath, R. Carnap & C. Morris (Eds.) *International Encyclopedia of Unified Science*. Chicago: University of Chicago Press, 77-138.

Motschnig-Pitrik, R., & Kaasboll, J. (1999). Part-whole relationship categories and their application in object-oriented analysis. *IEEE Trans. on Knowledge and Data Engineering,* 11(5), 779-797.

Motschnig-Pitrik, R., & Kaasbøll, J. (1999). Part-Whole Relationship Categories and Their Application in Object-Oriented Analysis. *IEEE Transactions on Knowledge and Data Engineering, 11*(5), 779-797.

Motschnig-Pitrik, R., & Storey, V. (1995). Modelling of set membership: the notion and the issues. *Data & Knowledge Engineering, 16*(2), 147-185.

Murata, T. (1989, April). Petri nets: Properties, analysis and applications. *Proceedings of the IEEE, 77*(4), 541-580.

Murugesan, S., Deshpande, Y., Hansen, S., & Ginige, A. (1999). Web Engineering: A New Discipline for Development of Web-based Systems. In *Proceedings of the 1st ICSE Workshop on Web Engineering* (pp. 3-13). Australia: Springer.

Musa J.D., Iannino A., Okumoto K. (1987). *Software Reliability Measurement, Prediction, Application.* Mc-Graw-Hill, NewYork.

Myers, G. (1976). *Software Reliability Principles and Practices.* J. Wiley & Sons.

Mylopoulos, J. (1998). Information modelling in the time of the revolution. *Information Systems,* 23(3/4), 127-155.

Mylopoulos, J., Borgida, A., Jarke, M., Koubarakis, M. (1990). TELOS: a language for representing knowledge about information systems. *ACM Transactions on Information Systems*, 8(4).

Mössenböck, H. (1999). Twin—A Design Pattern for Modeling Multiple Inheritance. Online: www.ssw.uni-linz.ac.at/Research/Papers/Moe99/Paper.pdf.

Nah, F., Islam, Z., & Tan, M. (2007). Empirical Assessment of Facotrs Influencing Success of Enterprise Resource Planning Implementations. *Journal of Database Management, 18*(4), 26-50.

Narayanan, S., & McIlraith, S. A. (2002). Simulation, Verification and Automated Composition of Web Services. In *Proceedings of the 11th international conference on World Wide Web* (pp. 77-88), Hawaii, USA: ACM.

NATURE and RENOIR project websites: http://www-i5.informatik.rwth-aachen.de/PROJEKTE/NATURE/nature.html and http://panoramix.univ-paris1.fr/CRINFO/PROJETS/nature.html

Neches, R., Fikes, R., Finin, T., Gruber, T., Patil, R., Senator, T., et al. (1991). Enabling technology for knowledge sharing. *AI magazine, fall 1991,* 36-56.

Neighbors, J. M. (1981). *Software Construction Using Components.* Ph.D. Thesis. Department of Information and Computer Science, University of California, Irvine.

Nelson, H. J., & Monarchi, D. E. (2007). Ensuring the Quality of Conceptual Representations. In: *Software Quality Journal, 15,* 213-233. Springer.

Niehaves, B., & Stirna, J. (2006). Participative Enterprise Modelling for Balanced Scorecard Implementation. In *14th European Conference on Information Systems (ECIS 2006),* Gothberg, Sweden.

Nielsen, P. (1990). Approaches to Appreciate Information Systems Methodologies. *Scandinavian Journal of Information Systems.* vol. 9, no. 2.

Nilsson, A. G., Tolis, C., & Nellborn, C. (Eds.) (1999). *Perspectives on Business Modelling: Understanding and Changing Organisations.* Springer-Verlag

Noy, N., & McGuinness, D. (2001). *Ontology development 101: a guide to creating your first ontology.* Stanford Knowledge Systems Laboratory Technical Report KSL-01-05 and Stanford Medical Informatics Technical Report SMI-2001-0880. Retrieved June 3, 2004 from http://smi-web.stanford.edu/pubs/SMI_Abstracts/SMI-2001-880.html

Nurcan, S., & Rolland, C. (1999). Using EKD-CMM electronic guide book for managing change in organizations. In *Proceedings of the 9th European-Japanese Conference on Information Modelling and Knowledge Bases*, Iwate, Japan.

Nuseibeh, B., Finkelstein, A., & Kramer. J. (1996). Method engineering for multi-perspective software development. *Information and Software Technology, 38*(4), 267-274.

OASIS (Organization for the Advancement of Structured Information Standards). (2006). Retrieved on 9-25-2007 from: http://www.oasis-open.org/committees/tc_home.php?wg_abbrev=soa-rm

Object Management Group (2003). *UML 2.0 Infrastructure Specification.* Online: www.ong.org/uml

Object Management Group OMG (2001): *Common Warehouse Metamodel (CWM) metamodel, version 1.0*, Februari 2001.

Object Management Group. (2005). *Unified Modeling Language: Superstructure. v2.0.formal/0507-04.* http://www.omg.org/cgi-bin/doc?formal/05-07-04.

Odell, J. J. (1998). *Advanced Object-Oriented Analysis & Design using UML.* Cambridge: Cambridge University Press.

Oei, J. (1995). A meta model transformation approach towards harmonization in information system modeling. In E. Falkenberg, W. Hesse & A. Olive (Eds.), *Information System Concepts – Towards a Consolidation of Views* (pp. 106-127). London: Chapman & Hall.

OGC. (2000). *SSADM Foundation (Office of Government Commerce).* London: The Stationery Office Books.

Ogden, C., & Richards, I. (1923). *The meaning of meaning.* London: Kegan Paul.

Olle, T., Hagelstein, J., MacDonald, I., Rolland, C., Sol, H., van Assche, F., & Verrijn-Stuart, A. (1988). *Information Systems Methodologies – A Framework for Understanding.* 2nd edition. Reading: Addison-Wesley.

Olle, T., Sol, H., & Tully, C. (Eds.) (1983) *Proc of the IFIP WG8.1 Working Conf. on Feature Analysis of Information Systems Design Methodologies.* Amsterdam: Elsevier Science.

Olle, T., Sol, H., & Verrijn-Stuart, A. (Eds.) (1986) *Proc. of the IFIP WG8.1 Working Conf. on Comparative Review of Information Systems Design Methodologies: Improving the Practice.* Amsterdam: Elsevier Science.

OMG (2007). *Ontology Definition Metamodel Specification.* Retrieved on January 29, 2007, from http://www.omg.org/cgi-bin/doc?ad/06-05-01.pdf

OMG (Object Management Group). (2007). Retrieved on 9-25-2007 from: http://colab.cim3.net/cgi-bin/wiki.pl?OMGSoaGlossary#nid34QI.

OMG. (2005). *Software process engineering metamodel specification*, Version 1.1, January 2005. Available at URL: < http://www.omg.org/technology/ documents/formal/ spem.htm>.

Opdahl, A. L., Henderson-Sellers, B., & Barbier, F. (2001). Ontological analysis of whole-part relationships in OO-models. *Information and Software Technology, 43*(6), 387-399.

Opdahl, A., & Henderson-Sellers, B. (2001). Grounding the OML metamodel in ontology. *The Journal of Systems and Software,* 57(2), 119-143.

Open Group. (2005). The Open Group architectural framework (TOGAF) version 8. Retrieved December 2007 from http://www.opengroup.org/togaf/

Open Group. (2007). Retrieved on 9-25-2007 from http://opengroup.org/projects/soa/doc.tpl?gdid=10632.

Parent, C., Spaccapietra, S., & Zimányi, E. (2006). *Conceptual modeling for traditional and spatio-temporal applications—the MADS approach.* Berlin: Springer Verlag.

Parent, C., Spaccapietra, S., & Zimányi, E. (2006). *Conceptual Modeling for Traditional and Spatio-Temporal Applications.* Berlin: Springer-Verlag.

Parsons, J. & Wand, Y. (2000). Emancipating Instances from the Tyranny of Classes in Information Modeling. *ACM Transactions on Database Systems*, *5*(2), 228-268.

Pastor O., Gómez, J., Insfrán, E., & Pelechano, V. (2001). The OO-Method Approach for Information Systems Modelling: From Objetct-Oriented Conceptual Modelling to Automated Programming. *Information Systems*, 26 (7), 507-534, Elsevier.

Pastor, O., Gomez, J., Insfran, E., & Pelechano, V. (2001). The OO-Method approach for information systems modeling: From object-oriented conceptual models to automated programming. *Information Systems, 26*(7), 507-534.

Paul, R.J., & Serrano, A. (2004). Collaborative information systems and business process design using simulation. In *Proceedings of the 37th Hawaii International Conference on Systems Sciences,* Big Island, HI, 9.

Paul, R.J., Hlupic, V., & Giaglis, G. (1998). Simulation modeling of business processes. In *Proceedings of the 3rd U.K. Academy of Information Systems Conference,* Lincoln, UK.

Paulk, M. C., Curtis, B., Chrissis, M. B. and Weber, C. V. (1993). The Capability Maturity Model: Version 1.1. *IEEE Software*. pp. 18-27, Jul.

Pegden, C.D., Shannon, R.E., & Sadowski, R.P. (1995). *Introduction to simulation using SIMAN*. London: McGraw-Hill.

Peirce, C. (1955). *Philosophical writings of Peirce*, edited by J. Buchle. New York: Dover.

Persson, A. (1997). Using the F^3 Enterprise Model for Specification of Requirements – an Initial Experience Report. *Proceedings of the CAiSE '97 International Workshop on Evaluation of Modeling Methods in Systems Analysis and Design (EMMSAD)*, June 16-17, Barcelona, Spain.

Persson, A. (2001). *Enterprise Modelling in Practice: Situational Factors and their Influence on Adopting a Participative Approach*, PhD thesis, Dept. of Computer and Systems Sciences, Stockholm University, No 01-020, ISSN 1101-8526.

Persson, A., & Horn, L. (1996). *Utvärdering av F3 som verktyg för framtagande av tjänste- och produktkrav inom Telia.* Telia Engineering AB, Farsta, Sweden, Doc no 15/0363-FCPA 1091097.

Persson, A., & Stirna, J. (2001). An explorative study into the influence of business goals on the practical use of Enterprise Modeling methods and tools. In *Proceedings of ISD'2001*, Kluwer, ISBN 0-306-47251-1.

Persson, A., & Stirna, J. (2001). Why Enterprise Modelling? -- An Explorative Study Into Current Practice, CAiSE'01. *Conference on Advanced Information System Engineering*, Springer, ISBN 3-540-42215-3

Peterson, A. S. (1991). Coming to terms with software reuse terminology: A model-based approach. *SIGSOFT Softw. Eng. Notes, 16*(2), 45-51.

Pidd, M. (1998). *Computer simulation in management science*. Chichester, UK: John Wiley.

Piprani, B. (2007). *Using ORM in an ontology based approach for a common mapping across heterogeneous applications.* Paper presented at the OTM2007/ORM 2007 workshop.

Plihon, V., Ralyté, J., Benjamen, A., Maiden, N., Sutcliffe, A., Dubois, E., & Heymans, P. (1998). A re-use-oriented approach for the construction of scenario based methods. In *Software Process (ICSP'98),* Chicago, Illinois, 14-17.

Poels, G., Nelson, J., Genero, M., & Piattini, M. (2003): "Quality in Conceptual Modeling – New Research Directions". In: A. Olivé (Eds.): *ER 2003* Ws, LNCS 2784, pp. 243-250. Springer.

Pohl, K. (1994). The three dimensions of requirements engineering: a framework and its application. *Information Systems* 19(3), 243-258.

Pontow, C., & Schubert, R. (2006). A mathematical analysis of theories of parthood. *Data & Knowledge Engineering, 59*, 107-138.

Prabhakaran, N., & Falkenberg, E. (1988). Representation of Dynamic Features in a Conceptual Schema. *Australian Computer Journal, 20*(3), 98-104.

Prakash, N. (1999). On method statics and dynamics. *Information Systems,* 24(8), 613-637.

Pressman, R. S. (2005). *Software Engineering: A Practitioner's Approach, 6th edition.* McGraw Hill.

Prototype. (2002). *A dictionary of business.* Oxford Reference Online. Retrieved February 27, 2005, from http://www.oxfordreference.com/views/ENTRY.html?subview=Main&entry=t18.e4769

Punter, T., & Lemmen, K. (1996). The MEMA-model: towards a new approach for methods engineering. *Journal of Information and Software Technology,* 38(4), 295-305.

Ralyté, J., Brinkkemper, S., & Henderson-Sellers, B. (Eds.), (2007). *Situational Method Engineering: Fundamentals and Experiences.* Proceedings of the IFIP WG 8.1 Working Conference, 12-14 September 2007, Geneva, Switzerland. Series: IFIP International Federation for Information Processing , Vol. 244.

Ralyte, J., Deneckere, R., & Rolland, C. (2003). Towards a generic model for situational method engineering. In J. Eder & M. Missikoff (Eds.), *Advanced Information Systems Engineering (CAiSE'03)*(pp. 95-110). LNCS 2681, Berlin: Springer-Verlag.

Ramesh, B., & Jarke, M. (2001). Towards reference models for requirements traceability. *IEEE Trans. on Software Engineering, 27*(1), 58-93.

Rantapuska, T., Siakas, K., Sadler, C.J., Mohamed, W-E. (1999). Quality Issues of End-user Application Development. Hawkins, C., Georgiadou, E. Perivolaropoulos, L., Ross, M. & Staples, G. (Eds): *The BCS INSPIRE IV Conference: Training and Teaching for the Understanding of Software Quality.* British Computer Society, Sep, University of Crete, Heraklion.

Raumbaugh, J., Blaha, M., Premerlani, W., Eddy, F., & Lorensen, W. (1991). *Object-oriented modeling and design.* Englewood Cliffs: Prentice Hall.

Recker, J., Rosemann, M., van der Aalst, W. M. P., & Mendling, J. (2005). *On the syntax of reference model configuration.* Paper presented at the First International Workshop on Business Process Reference Models (BPRM'05), Nancy, France.

Reenskaug, T. (1979). *THING-MODEL-VIEW-EDITOR, an Example from a planning system.* Xerox PARC Technical Note.

Remme, M. (1995). *Systematic development of informations systems using standardised process particles.* Paper presented at the 3rd European Conference on Information Systems—ECIS '95, Athens, Greece.

Resende, A., & Silva, C. (2005). *Programação Orientada a Aspectos em Java.* Brasport.

Rhodes, D. (1998). Integration challenge for medium and small companies. In *Proceedings of the 23rd Annual Conference of British Production and Inventory Control Society,* Birmingham, UK (pp. 153-166).

Ricadela, A. (2006). The Dark Side of SOA. *Information Week.* September 4th. (pp. 54-58).

Rittel, H. W. J., & Webber, M. M. (1984). Planning Problems are Wicked Problems. In Cross (Ed.), *Developments in Design Methodology.* John Wiley & Sons.

Robinson, S. (1994). *Successful simulation: A practical approach to simulation projects.* Maidenhead, UK: McGraw-Hill.

Rockwell, S., & Bajaj, A. (2005). COGEVAL: Applying cognitive theories to evaluate conceptual models. In K. Siau (Ed.), *Advanced Topics in Databases Research.* Idea Group, Hershey, PA, 255-282.

Roelofs, J. (2007). *Specificatie van Strategieën voor Requirement Engineering.* Master's thesis, Radboud University Nijmegen. In Dutch.

Rolland, C., & Prakash, N. (1996). A proposal for context-specific method engineering. In S. Brinkkemper, K. Lyytinen & R. Welke (Eds.), *Method Engineering: Principles of Method Construction and Tool Support* (pp. 191-208). London: Chapman & Hall.

Rolland, C., Prakash, N., & Benjamen, A. (1999). A multi-model view of process modeling. *Requirements Engineering, 4*(4), 169-187.

Rosemann, M. (2006). Potential Pitfalls of Process Modeling: Part A. *Business Process Management Journal, 12(*2), 249–254

Rosemann, M. (2006). Potential Pitfalls of Process Modeling: Part B. *Business Process Management Journal, 12*(3):377–384

Rosenberg, D., & Kendall, S. (2001). *Applied Use Case-Driven Object Modeling*. Addison-Wesley.

Rosenberg, D., Scott, K. (1999). *Use Case Driven Object Modeling with UML : A Practical Approach*. Addison-Wesley.

Rossi, M. (1998). *Advanced Computer Support for Method Engineering: Implementation of CAME Environment in MetaEdit+*. Ph.D. Thesis, Jyvaskyla Studies in Computer Science, Economics and Statistics, University of Jyvaskyla.

Rossi, M. and Brinkkemper, S. (1996). Complexity metrics for systems development methods and techniques. *Information Systems*, Vol. 21, No 2, Apr., ISSN 0306-4379.

Rossi, M., Lyytinen, K., Ramesh, B., & Tolvanen, J.-P. (2005). Managing evolutionary method engineering by method rationale. *Journal of the Association of Information Systems (JAIS)*, *5*(9), 356-391.

Rubin, K., & Goldberg, A. (1992). Object Behavior Analysis. *Communications of the ACM,* 35(9), 48-62.

Ruiz, F., Vizcaino, A., Piattini, M., & Garcia, F. (2004). An ontology for the management of software maintenance projects. *International Journal of Software Engineering and Knowledge Engineering, 14*(3), 323-349.

Rumbaugh, J., Blaha, M., Premerlani, W., Eddy, F., & Lorensen, W. (1991). *Object Oriented Modeling and Design*. Englewood Cliffs, NJ: Prentice Hall.

Rumbaugh, J., Jacobson, I., & Booch, G. (1999). *The Unified Language Reference Manual*. Reading, MA: Addison-Wesley.

Rumbaugh, J., Jacobson, I., & Booch, G. (2004). *The unified modeling language reference manual* (2nd ed.). Addison-Wesley Longman.

RUP - Rational Unified Process. Retrieved March, 2008 from http://www.e-learningcenter.com/rup.htm

Saadia, A. (1999). *An Investigation into the Formalization of Software Design Schemata.* MPhil Thesis, Faculty of Science, Computing and Engineering, University of North London.

Sadler, C. & Kitchenham, B.A. (1996). Evaluating Software Engineering Methods and Tools. Part 4: The influence of human factors. *SIGSOFT, Software Engineering Notes*. Vol 21, no 5.

Saeki, M. (1995). Object–Oriented Meta Modelling. In M. P. Papazoglou, (Ed,), *Proceedings of the OOER'95, 14th International Object–Oriented and Entity–Relationship Model ling Conference*, Gold Coast, Queensland, Australia, volume 1021 of Lecture Notes in Computer Science, pages 250–259, Berlin, Germany, EU, December 1995. Springer.

Saeki, M. (1998). A meta-model for method integration. *Information and Software Technology, 39*(14), 925-932.

Saeki, M. (2003). Embedding metrics into information systems development methods: an application of method engineering technique. In J. Eder & M. Missikiff (Eds.), *Advanced Information Systems Engineering (CAiSE 2003)* (pp. 374-389). LNCS 2681, Springer-Verlag: Berlin.

Saeki, M., Iguchi, K., Wen-yin, K., & Shinokara, M. (1993). A meta-model for representing software specification & design methods. In N. Prakash, C. Rolland & B. Pernici (Eds.), *Information Systems Development Process* (pp. 149-166). Amsterdam: Elsevier Science.

Sandkuhl, K., Smirnov, A., & Shilov, N. (2007). Configuration of Automotive Collaborative Engineering and Flexible Supply Networks. In P. Cunningham & M. Cunningham (Eds.), *Expanding the Knowledge Economy – Issues, Applications, Case Studies*. Amsterdam, The Netherlands: IOS Press. ISBN 978-1-58603-801-4.

Sattler, U. (1995). A concept language for an engineering application with part-whole relations. In A. Borgida, M. Lenzerini, D. Nardi, & B. Nebel (Eds.), *Proceedings of the international workshop on description logics* (pp. 119-123).

Sattler, U. (2000). Description Logics for the Representation of Aggregated Objects. In W. Horn (Ed.) *Proceedings of the 14th European Conference on Artificial Intelligence (ECAI2000)*. Amsterdam: IOS Press.

Scheer, A.-W. (1994). *Business process engineering: Reference models for industrial enterprises*. New York: Springer-Verlag.

Scheer, A.W. (1999). *ARIS, business process framework* (3rd ed.) Berlin: Springer.

Scheer, A.-W., & Nüttgens, M. (2000). ARIS architecture and reference models for business process management. In W. M. P. van der Aalst, J. Desel, & A. Oberweis (Eds.), *Business process managemen: Models,techniques, and empirical studies* (pp. 376-389). Berlin: Springer.

Schmidt, D. (2007). *"Programming Principles in Java: Architectures and Interfaces", chapter 9*. Retrieved on January 29, 2007 from http://www.cis.ksu.edu/~schmidt/CIS200/

Schmidt, M., Hutchison, B., Lambros, P., & Phippen, R. (2005). Enterprise Service Bus: Making Service Oriented Architecture Real. *IBM Systems Journal, 44*(4), 781-797.

Schulz, S., Hahn, U., & Romacker, M. (2000). Modeling Anatomical Spatial Relations with Description Logics. In J. M. Overhage (Ed.), *Proceedings of the AMIA Symposium 2000* (pp. 779-83).

Schwabe, D., & Rossi, G. (1998). An Object Oriented Approach to Web-Based Application Design. *Theory and Practice of Object Systems* 4 (4), Wiley and Sons.

Shah, A., & Kalin, P. (2007). SOA Adoption Models: Ad-hoc versus Program-based. *SOA Magazine*. July 6.

Shan, T., & Hua, W. (2006). Service Oriented solution Framework for Internet Banking. *Internet Journal of Web Services Research, 3*(1), 29-48.

Shanks, G., Tansley, E., & Weber, R. (2004). Representing composites in conceptual modeling. *Communications of the ACM, 47*(7), 77-80.

Shannon, B. (2003). *Java™ 2 Platform Enterprise Edition Specification, v1.4.* Sun Microsystems.

Shannon, R.E. (1975). *Systems simulation: The art and the science*. Englewood Cliffs, NJ: Prentice Hall.

Shapiro, J. & Berki, E. (1999). Encouraging Effective use of CASE tools for Discrete Event Modelling through Problem-based Learning. Hawkins, C., Georgiadou, E. Perivolaropoulos, L., Ross, M. & Staples, G. (Eds): the BCS INSPIRE IV Conference: *Training and Teaching for the Understanding of Software Quality*. British Computer Society, 313-327, University of Crete, Heraklion.

Shlaer, S., & Mellor, S. (1992). *Object Lifecycles: Modeling the World in States*: Prentice Hall.

Shneiderman, B. (1978). Improving the human factor aspects of database interactions. *ACM Transactions on Database Systems, 3*(4), 417-439.

Shoval, P., & Kabeli, J. (2001). FOOM: functional- and object-oriented analysis and design of information systems: An integrated methodology. *Journal of Database Management, 12*(1), 15-25.

Shoval, P., & Kabeli, J. (2005). Data modeling or functional analysis: which comes next? - an experimental comparison using FOOM methodology. *Communications of the AIS, 16*, 827-843.

Shoval, P., & Shiran, S. (1997). Entity-relationship and object-oriented data modeling - an experimental comparison of design quality. *Data & Knowledge Engineering, 21*, 297-315.

Siakas, K. Berki, E. Georgiadou, E. & Sadler, C. (1997): The Complete Alphabet of Quality Information Systems: Conflicts & Compromises. *7th World Congress on Total Quality Management (TQM '97).* McGraw-Hill, New Delhi.

Siakas, K., Berki, E. & Georgiadou, E. (2003). CODE for SQM: A Model for Cultural and Organisational Diversity Evaluation. Messnarz R. (Ed.) *EuroSPI 2003: European Software Process Improvement.* 10-12 Dec, Graz, Austria.

Siau, K., & Lee, L. (2004). Are use case and class diagrams complementary in requirements analysis? an experimental study on use case and class diagrams in UML. *Requirements Engineering, 9*, 229-237.

Siemer, J., Taylor, S.J.E., & Elliman, A.D. (1995). Intelligent tutoring systems for simulation modelling in the manufacturing industry. *International Journal of Manufacturing System Design, 2*(3), 165-175.

Sierhuis, M., Clacey, W.J., Seah, C., Trimble, J.P., & Sims, M.H. (2003). Modeling and simulation for mission operations work system design. *Journal of Management Information Systems, 19*(4), 85-128.

Silverston, L. (2001). *The Data Model Resource Book: Revised Edition*, New York: Wiley.

Simon, H. (1960). *The new science of management decisions.* New York: Harper & Row.

Simons, P. (1987). *Parts: A study in Ontology.* Oxford: Clarendon Press.

Sindre, G., & Krogstie, J. (1995). Process heuristics to achieve requirements specification of feasible quality. In *Second International Workshop on Requirements Engineering: Foundations for Software Quality* (REFSQ'95), Jyväskylä, Finland.

Singular Software. (1998). *"Ikarus": Design of the ESI Toolset, deliverable,* ESPRIT project No 22927 ELEKTRA, Singular Software, Greece

Si-Said, S., Rolland, C., Grosz, G. (1996). MENTOR: A Computer Aided Requirements Engineering Environment. Constantopoulos, P., Mylopoulos, J. and Vassiliou, Y. (Eds) *Advances in Information Systems Engineering, 8th International Conference CAiSE'96.* Heraklion, Crete, Greece, May 20-24, Berlin: Springer-Verlag Lecture Notes in Computer Science LNCS 1080, pp. 22-43.

Sjøberg, D., Hannay, J., Hansen, O., Kampenes, V., Karahasanovi, A., Liborg, N., & Rekdal, A. (2005). A survey of controlled experiments in software engineering. *IEEE Transactions on Software Engineering, 31*(9), 733-753.

Slater, P. (1995). Output from generic packages. *ACM SIGAda Ada Letters, XV*(3), 76-79.

Smith, B., Ceusters, W., Klagges, B., Köhler, J., Kumar, A., Lomax, J., Mungall, C., Neuhaus, F., Rector, A. L., & Rosse, C. (2005). Relations in biomedical ontologies. *Genome Biology*, 6, R46.

Smith, M. (2005, August). *Will a new generation of curbside sensors end our parking problems — or help the government monitor our every move?* Retrieved December, 17, 2005, from http://www.sfweekly.com/Issues/2005-08-17/news/smith.html

Smolander, K. Tahvanainen, V-P., Lyytinen, K. (1990). How to Combine Tools and Methods in Practice – a field study. B. Steinholz, A. Solverg, L. Bergman (Eds) *The 2nd Conference in Advanced Information Systems Engineering (CAiSE '90).* Lecture Notes in Computer Science LNCS 436, Springer-Verlag, Berlin, pp.195-214.

Sol, H. (1992). Information systems development: a problem solving approach. In W. Cotterman & J. Senn (Eds.), *Challenges and Strategies for Research in Systems Development* (pp. 151-161). Chichester: John Wiley & Sons.

Sol, H.G. (1983). A Feature Analysis of Information Systems Design Methodologies: Methodological Considerations. T.W. Olle, H.G. Sol, C.J. Tully (Eds) *Information systems design methodologies: a feature analysis: proceedings of the IFIP WG 8.1 Working Conference on Comparative Review of Information Systems Design Methodologies.* 5-7 July, York.

Song X. (1997). Systematic integration of design methods. *IEEE Software*, *14*(2), 107-117.

Song, I., Yano, K., Trujillo, J., & Lujan-Mora, S. (2005). A taxonomic class modeling methodology for object-oriented analysis. In K. Siau (Ed.), *Advanced Topics in Database Research*, *4*. 216-240. Hershey: Idea Group.

Song, X., & Osterweil, L. (1992). Towards objective, systematic design-method comparison. *IEEE Software*, *9*(3), 43-53.

Sorenson, P.G., Tremblay, J-P. and McAllister, A. J. (1988). The MetaView system for many specification environments. *IEEE Software*. 30, 3, pp. 30-38.

Souza, V. E. S., & Falbo, R. A. (2007). FrameWeb - A Framework-based Design Method for Web Engineering. In *Proceedings of the Euro American Conference on Telematics and Information Systems 2007* (pp. 17-24). Faro, Portugal: ACM Press.

Souza, V. E. S., Lourenço, T. W., Falbo, R. A., & Guizzardi, G. (2007). S-FrameWeb – a Framework-based Design Method for Web Engineering with Semantic Web Support. In *Proceedings of Workshops and Doctoral Consortium of the 19th International Conference on Advanced Information Systems Engineering* (pp. 767-778), Trondheim, Norway.

Sowa, J. (2000). *Knowledge representation – logical, philosophical, and computational foundations*. Pacific Grove, CA: Brooks/Cole.

Sowa, J. F., & Zachman, J. A. (1992). Extending and formalizing the framework for information systems architecture. *IBM Systems Journal, 31*(3).

Spaccapietra, S., Parent, C., & Zimanyi, E. (1998). Modeling time from a conceptual perspective. In *Int. Conf. on Information and Knowledge Management (CIKM98)*.

Spanoudakis, G., & Constantopoulos, P. (1993). *Similarity for analogical software reuse: A conceptual modelling approach*. Paper presented at the Proceedings of Advanced Information Systems Engineering.

Spyns, P., Meersman, R., & Jarrar, M. (2002). Data modelling versus Ontology engineering. *SIGMOD record: special issue on semantic web and data management, 31*(4), 12-17.

Stal, M. (2006). Using Architectural Patterns and Blueprints for Service oriented Archtiecture. *IEE Software*. March/April. (pp. 54-61).

Stamper R. (1996). Signs, information, norms and information systems. In B. Holmqvist, P. Andersen, H. Klein & R. Posner (Eds.) *Signs are Work: Semiosis and Information Processing in Organisations* (pp. 349-392). Berlin: De Gruyter.

Stamper, R. (1978). *Towards a semantic model for the analysis of legislation*. Research Report L17, London School of Economics.

Standing, C. (2002). Methodologies for developing Web applications. *Information and Software Technology*, 44 (3), 151-159, Elsevier.

Stirna, J. (2001). *The Influence of Intentional and Situational Factors on EM Tool Acquisition in Organisations*, Ph.D. Thesis, Department of Computer and Systems Sciences, Royal Institute of Technology and Stockholm University, Stockholm, Sweden.

Stirna, J., & Kirikova M. (2008). How to Support Agile Development Projects with Enterprise Modeling. In P. Johannesson & E. Söderström (Eds.), *Information Systems Engineering - from Data Analysis to Process Networks*. IGI Publishing, ISBN: 978-1-59904-567-2

Stirna, J., Persson A., & Sandkuhl, K. (2007). Participative Enterprise Modeling: Experiences and Recommendations. In *Proceedings of CAiSE'2007*. Trondhiem, Norway: Springer LNCS.

Stirna, J., Persson, A., & Aggestam, L. (2006). Building Knowledge Repositories with Enterprise Modelling and Patterns - from Theory to Practice. In *proceedings of the 14th European Conference on Information Systems (ECIS)*, Gothenburg, Sweden, June 2006.

Stirna, J., Persson, A., & Sandkuhl, K. (2007). Participative Enterprise Modeling: Experiences and Recommendations. In John Krogstie, Andreas L. Opdahl, Guttorm Sindre (Eds.), *Advanced Information Systems*

Engineering, 19th International Conference, CAiSE 2007, Trondheim, Norway, June 11-15, 2007, Proceedings. Lecture Notes in Computer Science 4495 Springer, ISBN 978-3-540-72987-7

Stirna, J., Persson, A., & Sandkuhl, K. (2007). Participative Enterprise Modeling: Experiences and Recommendations. *Proceedings of the 19th International Conference on Advanced Information Systems Engineering (CAiSE 2007)*, Trondheim, June 2007.

Sulkin, A. (2007). SOA and Enterprise Voice Communications. *Business Communications Review, 37*(8), 32-34.

Swede van, V., & van Vliet, J. (1993). A flexible framework for contingent information systems modeling. *Information and Software Technology, 35*(9), 530-548.

Szyperski, C. (1998). *Component software. Beyond object-oriented programming* (Vol. 2). New York: ACM Press and Addison-Wesley.

T.W.Olle, H.G. Sol, A.A. Verrijin-Stuart (Eds). 1982. *Information systems design methodologies: a comparative review: proceedings of the IFIP WG 8.1 Working Conference on Comparative Review of Information Systems Design Methodologies*. Noordwijkerhout, The Netherlands, 10-14 May.

TEAF (2007) Retrieved December 2007 from http://www.eaframeworks.com/TEAF/index.html

Teorey, T., Yang, D., & Fry, J. (1986). A logical design methodology for relational databases using the extended E-R model. *ACM Computing Surveys, 18*(2), 197-222.

ter Hofstede, A., Proper, H. & Weide, th. van der (1993). Formal definition of a conceptual language for the description and manipulation of information models. *Information Systems, 18*(7), 489-523.

Thayer, R. (1987). Software engineering project management – a top-down view. In R. Thayer (Ed.), *Tutorial: Software Engineering Project Management* (pp. 15-56). IEEE Computer Society Press.

The Standish Group International (1999). *CHAOS: A recipe for success.* West Yarmouth, MA: The Standish Group International.

Thomas, O. (2005). *Understanding the term reference model in information system research.* Paper presented at the First International Workshop on Business Process Reference Models (BPRM'05), Nancy, France.

Tolvanen, J.-P. (1998). *Incremental method engineering with modeling tools – Theoretical principles and empirical evidence.* Dissertation Thesis, Jyväskylä Studies in Computer Science, Economics and Statistics, No. 47, University of Jyväskylä, Finland.

Tolvanen, J-P. (1998). *Incremental Method Engineering with Modeling Tools.* PhD Thesis, University of Jyvaskyla, Jyvaskyla.

Topi, H., & Ramesh, V. (2002). Human factors research on data modeling: a review of prior research, an extended framework and future research directions. *Journal of Database Management, 13*(2), 188-217.

Trembly, A. (2007). SOA: Savior or Snake Oil? National Underwriter Life & Health, 111(27), 50.

Trog, D., Vereecken, J., Christiaens, S., De Leenheer, P., & Meersman, R. (2006). *T-Lex; A role-based ontology engineering tool.* Paper presented at the OTM 2006/ORM 2006 workshop.

Troux (2008) http://www.troux.com Last visited April 2008

Tudor, D.J, Tudor, I.J. (1995). *Systems Analysis and Design - A comparison of Structured Methods.* NCC Blackwell.

UNL-IBM System i Global Innovation Hub. (2007). *Making SOA Relevant for Business.* Retrieved on 10-09-2007 from http://cba.unl.edu/outreach/unl-ibm/documents/SOA_Relevant_Business.pdf

Uschold, M., & Gruninger, M. (1996). Ontologies: principles, methods and applications. *Knowledge Engineering Review, 11*(2), 93-155.

W3C (2007). *OWL Web Ontology Language Guide, fev. 2004.* Retrieved on December 17, 2007, from http://www.w3.org/TR/owl-guide/

W3C (2007). *W3C Glossary and Dictionary.* Retrieved on January 23, 2007, from http://www.w3.org/2003/glossary/

W3C (World Wide Web Consortium). (2007). Retrieved on 9-25-2007 from: http://colab.cim3.net/cgi-bin/wiki.pl?WwwCSoaGlossary#nid34R0.

Van der Aalst, W. And ter Hostede, A. (2005): YAWL: Yet Another Workflow Language. *Information systems, 30*(4), 245-275.

van der Aalst, W. M. P., Dreiling, A., Gottschalk, F., Rosemann, M., & Jansen-Vullers, M. H. (2005). *Configurable process models as a basis for reference modeling.* Paper presented at the First International Workshop on Business Process Reference Models (BPRM'05), Nancy, France.

Van Hillegersberg, J. (1997). *Metamodelling-Based Integration of Object-Oriented Systems Development.* Thesis Publishers, Amsterdam.

van Lamsweerde, A. (2001). Goal-Oriented Requirements Engineering: A Guided Tour. In *Proceedings of the 5th International IEEE Symposium on Requirements Engineering.* IEEE.

Wand, Y. (1988). An ontological foundation for information systems design theory. In B. Pernici & A. Verrijn-Stuart (Eds.), *Office Information Systems: The Design Process* (pp. 201-222). Amsterdam: Elsevier Science.

Wand, Y., & Weber, R. (1995). On the deep structure of information systems. *Information Systems Journal,* 5(3), 203-223.

Wand, Y., Monarchi, D., Parson, J., & Woo, C. (1995). Theoretical foundations for conceptual modeling in information systems development. *Decision Support Systems, 15*(4), 285-304.

Wand, Y., Storey, V., & Weber, R. (1999). An ontological analysis of the relationship construct in conceptual modeling. *ACM Trans. on Database Systems,* 24(4), 494-528.

Wangler, B., & Persson, A. (2002). Capturing Collective Intentionality in Software Development. In H. Fujita & P. Johannesson, (Eds.), *New Trends in Software Methodologies, Tools and Techniques.* Amsterdam, Netherlands: IOS Press (pp. 262-270).

Wangler, B., Persson, A., & Söderström, E. (2001). Enterprise Modeling for B2B integration. In *International Conference on Advances in Infrastructure for Electronic Business, Science, and Education on the Internet,* August 6-12, L'Aquila, Italy (CD-ROM proceedings)

Wangler, B., Persson, A., Johannesson, P., & Ekenberg, L. (2003). Bridging High-level Enterprise Models to Implemenation-Oriented Models. in H. Fujita & P. Johannesson (Eds.), *New Trends in Software Methodologies, Tools and Techniques.* Amsterdam, Netherlands: IOS Press.

Warmer, J., & Kleppe, A. (2003). *The Object Constraint Language, 2nd Edition.* Addison-Wesley.

Varzi, A. C. (2004). Mereology. In E.N. Zalta (Ed.), *The Stanford Encyclopedia of Philosophy (Fall 2004 Ed.).* Stable URL http://plato.stanford.edu/archives/fall2004/entries/mereology/

Varzi, A. C. (2006). Spatial reasoning and ontology: parts, wholes, and locations. In M. Aiello, I. Pratt-Hartmann, & J. van Benthem (Eds.), *The Logic of Space.* Dordrecht: Kluwer Academic Publishers.

Varzi, A. C. (2006). A Note on the Transitivity of Parthood. *Applied Ontology, 1,* 141-146.

Vassiliou, G. (2000). Developing Data Warehouses with Quality in Mind. Brinkkemper, S., Lindencrorna, E. & Solvberg, A. (Eds): *Information Systems Engineering, State of the Art and Research Themes.* Springer-Verlag London Ltd.

Wastell, D. (1996). The fetish of technique: methodology as a social defense. *Information Systems Journal,* 6(1), 25-40.

Web Service. (2007). Retrieved on 10-18-2007 from: http://en.wikipedia.org/wiki/Web_service on.

Weber, R. (2003). Conceptual modeling and ontology: possibilities and pitfalls. *Journal of Database management, 14*(3), 1-20.

Weber, R., & Zhang, Y. (1996). An analytical evaluation of NIAM's grammar for conceptual schema diagrams. *Information Systems Journal, 6*(2), 147-170.

Welke, R. (1977). Current information system analysis and design approaches: framework, overview, comments and conclusions for large – complex information system education. In R. Buckingham (Ed.), *Education and Large Information Systems* (pp. 149-166). Amsterdam: Elsevier Science.

Venable, J. (1993). CoCoA: *A conceptual data modeling approach for complex problem domains*. Dissertation Thesis, State University of New York, Binghampton, USA.

Verheeke, B., Vanderperren, W., & Jonckers, V. (2006). Unraveling Crosscutting Concerns in Web Services Middleware. *IEEE Software*. January/February. pp. 42-50.

Vessey, I., & Conger, S.A. (1994). Requirements specification: Learning object, process, and data methodologies. *Communications of the ACM, 37*(5), 102-113.

Vessey, I., & Glass, R. (1998). Strong vs. weak: Approaches to systems development. *Communications of the ACM, 41*(4), 99-102.

Whittaker, J. (2000). What Is Software Testing? And Why Is It So Hard? *IEEE Software*. Jan/Feb, pp. 70-79.

Wieringa, R. J. (2003). *Design Methods for Reactive Systems*. San Francisco: Morgan Kaufmann.

Vieu, L., & Aurnague, M. (2005). Part-of Relations, Functionality and Dependence. In M. Aurnague, M. Hickmann, & L. Vieu (Eds.), *Categorization of Spatial Entities in Language and Cognition*. Amsterdam: John Benjamins.

Wikipedia (EJB). (2007). Retrieved on 10-12-2007 from: http://en.wikipedia.org/wiki/Ejbhttp://en.wikipedia.org/wiki/Ejb.

Wikipedia (SOA). (2007). Retrieved on 9-25-2007 from: http://en.wikipedia.org/wiki/Service-oriented_architecture#SOA_definitions.

Willars, H. (1988). *Handbok i ABC-metoden*. Plandata Strategi.

Willars, H. et al (1993). TRIAD Modelleringshandboken N 10:1-6 (in Swedish), SISU, Electrum 212, 164 40 Kista, Sweden.

Williamson, O. E. (1985). *The economic institutions of capitalism*. New York: Tree Press.

Winston, M. E., Chaffin, R., & Herrmann, D. (1987). A taxonomy of part-whole relations. *Cognitive Science, 11*(4), 417-444.

Wintraecken J. (1990). *The NIAM Information Analysis Method: Theory and Practice*. Deventer: Kluwer.

vom Brocke, J. (2003). *Reference modelling, towards collaborative arrangements of design processes*. Berlin: Springer. [in German]

vom Brocke, J., & Buddendick, C. (2004). Organisation theory in reference modelling: Requirements and recommendation on the basis of transaction cost economics . *Wirtschaftsinformatik, 46*(5), 341-352. [in German]

Wood, D. (1987). *Theory of Computation*. J. Wiley and Sons.

Wood-Harper, A.T., Antill, L. Avison, D.E. (1985). *Information systems definition: the multiview approach*. Blackwell Publications.

Vreede, G.J. (1998). Collaborative business engineering with animated electronic meetings. *Journal of Management Information Systems, 14*(3), 141-164.

Vreede, G.J., & Verbraeck, A. (1996). Animating organizational processes: Insight eases change. *Journal of Simulation Practice and Theory, 4*(3), 245-263.

XML.com. (2007). Retrieved on 9-25-2007 from: http://www.xml.com/pub/a/ws/2003/09/30/soa.html.

Yourdon, E. (1989). *Modern Structured Analysis*: Prentice Hall.

Yu, E. S. K., & Mylopoulos, J. (1994). From E-R to "A-R" - Modeling Strategic Actor Relationships for Business Process Reengineering. In *Proceedings of the 13th International Conference on the Entity-Relationship Approach*. Manchester, England.

Yu, E. S. K., & Mylopoulos, J. (1994). From E-R to "A-R" - Modelling Strategic Actor Relationships for Business Process Reengineering. In *Proceedings of the 13th International Conference on the Entity-Relationship Approach*, Manchester, England

Zachman, J. A. (1987). A framework for information systems architecture. *IBM Systems Journal*, *26*(3), 276-291.

Zelm, M. (Ed.) (1995). *CIMOSA: A primer on key concepts, purpose, and business value* (Technical Report). Stuttgart, Germany: CIMOSA Association.

Zhang, D. (2004). Web Services Composition for Process Management in e-Business. *Journal of Computer Information Systems, 45*(2), 83-91.

Zhang, Z. (2004). *Model Component Reuse. Conceptual Foundations and Application in the Metamodeling-Based System Analysis and Design Environment.* Ph.D. Thesis, Jyvaskyla Studies in Computing, University of Jyvaskyla.

Zhang, Z., & Lyytinen, K. (2001). A framework for component reuse in a metamodelling-based software development. *Requirements Engineering,* 6(2), 116-131.

About the Contributors

Terry Halpin (BSc, DipEd, BA, MLitStud, PhD) is a distinguished professor at Neumont University. His industry experience includes several years in data modeling technology at Asymetrix Corporation, InfoModelers Inc., Visio Corporation, and Microsoft Corporation. His doctoral thesis formalized Object-Role Modeling (ORM/NIAM), and his current research focuses on conceptual modeling and conceptual query technology. He has authored over 150 technical publications and six books, including Information Modeling and Relational Databases and has co-edited four books on information systems modeling research. He is a member of IFIP WG 8.1 (Information Systems), is an editor or reviewer for several academic journals, is a regular columnist for the *Business Rules Journal*, and is a recipient of the DAMA International Achievement Award for Education (2002) and the IFIP Outstanding Service Award (2006).

John Krogstie has a PhD (1995) and an MSc (1991) in information systems, both from the Norwegian University of Science and Technology (NTNU). He is professor in information systems at IDI, NTNU, Trondheim, Norway. He is also a senior advisor at SINTEF. He was employed as a manager in Accenture 1991-2000. Dr. Krogstie is the Norwegian representative for IFIP TC8 and vice-chair of IFIP WG 8.1 on information systems design and evaluation, where he is the initiator and leader of the task group for mobile information systems. He has published around 80 refereed papers in journals, books and archival proceedings since 1991.

Erik Proper works as a principal consultant at Capgemini, and as a part-time professor at Radboud University Nijmegen. In the past he has worked for the University of Queensland, Queensland University of Technology, Distributed Systems Technology Centre, Origin, ID Research, and Ordina. Erik leads the Tools for Enterprise Engineering group of the Institute for Computing and Information Sciences at Nijmegen. His research interests lie within the field of enterprise engineering and architecting.

* * *

Alessandro Artale is an assistant professor in the Faculty of Computer Science at the Free University of Bozen-Bolzano. He got a PhD in computer science from the University of Florence in 1994. He published several papers in international journals and conferences, acted as both chair and PC member in conferences, and as editors of both proceedings and journal's special issues. His main research subject concerns description logics, temporal logic, automated reasoning, ontologies and conceptual modeling. A particular emphasis is devoted to the formalization of conceptual modeling tasks in domains with high semantic complexity and characterized by a dynamic aspect.

Peter Bollen received his MSc in management science and industrial engineering from Eindhoven University of technology in 1984 and a PhD in management information systems from Groningen University in 2004. He worked from 1984 to 1986 as a planning manager at Mars in Veghel, The Netherlands. He is a now a senior lecturer in the Department of Organization and Strategy, Maastricht University. His research interests include conceptual modeling, IS-methodology, strategies for requirements determination, business rules, and fact-oriented modeling.

Patrick van Bommel received his master's degree in computer science in 1990, and the PhD in mathematics and computer science from Radboud University Nijmegen, The Netherlands in 1995. He is currently an assistant professor at the same university. He teaches various courses on foundations of databases and information systems. His main research interests include information modelling, where the focus is on foundations of models and their transformations in order to establish theoretical properties and practical tools.

Anders Carstensen was born in Uppsala in 1954 and holds a Master of Science in applied physics and electrical engineering from Linköping Institute of Technology since1983. He is currently working as Subject teacher in computer science at Jönköping University since 1998, and is mainly teaching database management and programming methodology. He is also a PhD student with main research interests in the area of enterprise modeling and integration of software systems. Anders have previously worked as a research engineer at Karolinska Institutet, Institute of tumor biology between 1982 and 1987, and also worked as a hospital engineer at Stockholm County Council, Södersjukhuset, Department for Medical Engineering, IT-section between 1987 and 1998.

John Erickson is an assistant professor in the College of Business Administration, University of Nebraska at Omaha. His research interests include study of UML as an OO systems development tool, software engineering, and the impact of structural complexity upon the people and systems involved in the application development process. He has published in *Communications of the ACM*, and the *Journal of Database Management*, presented at several conferences such as AMICIS, ICIS WITS, EMMSAD, and CAiSE, authored materials for a distance education course at the University of Nebraska – Lincoln, collaborated on a book chapter, and co-chaired minitracks at several AMCIS Conferences.

Ricardo de Almeida Falbo is associated professor at the Computer Science Department, Federal University of Espírito Santo (UFES), Brazil, since 1989. He took a doctoral course in computer and systems engineering at Federal University of Rio de Janeiro (COPPE/UFRJ), in 1998. His main research interests are software engineering, software quality, ontological engineering, applications of ontologies to software engineering, software engineering environments and Semantic Web. He is member of the Brazilian Computer Society (SBC), acting actively in the Software Quality Group of the Software Engineering Special Commission.

Giancarlo Guizzardi obtained a PhD (cum laude) from the University of Twente, The Netherlands (2005). From 2005 to 2007 he held an associate researcher position at the Laboratory of Applied Ontology (LOA/ISTC-CNR), Trento, Italy. He is currently a member of the Computer Science Department at the Federal University of Espírito Santo (UFES), Vitória, Brazil, where he coordinates the research group NEMO (Ontology-Driven Conceptual Modeling Research Group). He has been working for 11 years on the development of domain and foundational ontologies and on their application in computer science and, primarily, in the area of conceptual modeling.

Lennart Holmberg is a member of the R&D Department in the division "Interior System" in the global automotive supplier company Kongsberg Automotive (KA). He received a Master of Science degree in chemical engineering from the Chalmers University of Technology, Sweden (1985). He has 20 years of experience in the automotive industry, with a focus in R&D including material development, process development, product development, and development of product development process and sub-processes. For 6 years he represented R&D in the management group of Business Area "Seat Comfort" in KA. Holmberg has been part of some enterprise modeling projects, both national and European, and is a coauthor of some publications resulting from the projects.

Stijn Hoppenbrouwers holds two master's degrees, one English language and literature, and one in linguistics. He acquired his PhD in computer science at the University of Nijmegen in 2003. Stijn's research focuses on the interactive processes in which one or more participants create formal models in systems engineering, with emphasis on business engineering (for example, business process models and business rules). He is particularly interested in shaping modeling methods and tools as games, and in process-oriented innovation of digital environments for collaborative enterprise modeling.

C. Maria Keet is assistant professor at the Faculty of Computer Science at the Free University of Bozen-Bolzano, where she received her PhD in computer science in 2008. Her main research foci are logic-based knowledge representation, ontology, and its applications to biological data and knowledge. In addition to an MSc in microbiology from Wageningen University and Research Centre (1998), an MA 1st class in peace & development studies from Limerick University (2003), and a BSc (Hons) 1st class in IT & Computing from the Open University UK (2004), she has worked for 3.5 years as systems engineer in the IT industry.

Mark Last is senior lecturer at the Department of Information Systems Engineering at Ben-Gurion University, Beer-Sheva, and head of the Data Mining and Software Quality Engineering Group. He obtained his PhD from Tel Aviv University in 2000. Dr. Last is an associate editor of *IEEE Transactions on Systems, Man, and Cybernetics (Part C)* and *Pattern Analysis and Applications (PAA)*. He has published over 130 papers and chapters in scientific journals, books, and refereed conferences. He is a co-author of two monographs and a coeditor of six edited volumes. His main research interests are focused on data mining, cross-lingual text mining, cyber intelligence, and software assurance.

Mauri Leppänen is a senior lecturer at the Department of Computer Science and Information Systems, University of Jyväskylä, Finland. He holds a PhD in information systems from the University of Jyväskylä. Leppänen is the author or co-author of about 40 publications in journals and conference proceedings on information systems development, meta modeling, method engineering, ontology engineering, and databases. He has served as a PC member of and a reviewer for several conferences such as ER, ICIS, EJC, ECIS, ISD and ADBIS. He has been a member of many national and international research projects, including the EU projects. He is a member of IFIP WG8.1 on Design and Evaluation of Information Systems.

Frank Lillehagen is president and CEO of Active Knowledge Modeling AS, the third company he has co-founded. From 1974 to 1985, he pioneered computer graphics and CAD in many Scandinavian industry sectors, and co-founded Eurographics in 1980. Overall, Frank developed four commercial CAD systems and the Metis modeling tools (now owned by Troux Technologies) and received many awards for his contributions to industrial innovation.

Anne Persson is a professor in computer science at the University of Skövde, Sweden. She received the degree of Doctor of Philosophy in computer and systems sciences from Stockholm University, Sweden. Her research interests include enterprise modeling methods and tools, requirements engineering as well as knowledge management and organizational patterns. Persson is an author or co-author of some 50 research reports and publications and has participated in three EU funded research projects. She has codeveloped the EKP - Enterprise Knowledge Patterns and the EKD - Enterprise Knowledge Development approaches.

Jeroen Roelofs studied information science at Radboud University in Nijmegen (The Netherlands). He received his master's degree at the end of 2007 (cum laude). The title of his master's thesis was 'Specification of Strategies for Requirements Engineering'. Roelofs now works as a 'functional analyst/designer' at Blixem Internet Services in Nijmegen, putting his passion for requirements engineering into practice on a daily basis.

Kurt Sandkuhl is a full professor of information engineering in the School of Engineering at Jönköping University in Sweden. He received his MSc and PhD in computer science from Berlin University of Technology in Germany and his postdoctoral lecturing qualification from Linköping University in Sweden. Before joining Jönköping University in 2002, Kurt was a scientific employee at Berlin University of Technology (1988-1994), department manager (1996-2000) and division manager (2000-2002) at Fraunhofer-Institute for Software and Systems Engineering in Berlin. In 1993, he received the innovation award "Dr.-Ing.-Rudolf-Hell-Innovationspreis" from Linotype-Hell AG, Germany. Kurt has taught courses on software engineering methods, development of distributed applications, computer-supported collaborative work, information logistics, information modeling, software quality management, information logistics and graduate seminars in information systems. His current research interests are in computer-supported collaborative work, information logistics, ontology engineering and enterprise knowledge modeling. Kurt has published three books and more than 80 papers in journals and international conferences. Furthermore, he was program chair and general chair for several workshops on collaborative engineering and information logistics.

Peretz Shoval is professor of information systems at the Department of Information Systems Engineering at Ben-Gurion University, Beer-Sheva. He earned his PhD in information systems (1981) from the University of Pittsburgh, where he specialized in expert systems for information retrieval. In 1984, he joined Ben-Gurion University, where he started the Information Systems Program, and later on created and headed the Department of Information Systems Engineering. Prior to moving to academia, Shoval held professional and managerial positions in computer companies. His research interests include information systems analysis and design methods, data modeling and database design, and information retrieval and filtering. He has published more than a hundred papers in journals, conference proceedings and book chapters, and authored several books on systems analysis and design. Shoval has developed methodologies and tools for systems analysis & design, and for conceptual and logical database design.

Keng Siau is the E. J. Faulkner professor of management information systems (MIS) at the University of Nebraska – Lincoln (UNL). He is the director of the UNL-IBM Global Innovation Hub, editor-in-chief of the *Journal of Database Management*, and co-editor-in-chief of the Advances in Database Research series. Professor Siau has over 200 academic publications. He has published more than 95 refereed journal articles and over 100 refereed conference papers. Professor Siau has received numerous research, teaching, and service awards. He received the International Federation for Information Processing (IFIP) Outstanding Service Award in 2006.

Vítor Estêvão Silva Souza graduated in a master's course in computer sciences at UFES, the Federal University of Espírito Santo State, in Brazil. Under the supervision of professors Ricardo Falbo and Giancarlo Guizzardi, he conducted research on software development environments, ontologies, Web engineering and the Semantic Web. He specializes in Web development in Java, co-founded the Espírito Santo Java User Group, ministers courses and does consulting in Java and related subjects, having also worked as substitute professor at UFES. In November 2008, he started a software engineering PhD in the University of Trento, Italy, under professor John Mylopoulos.

Janis Stirna is an assistant professor at Jönköping University, Sweden. He received the degree of Doctor of Philosophy in Computer and Systems Sciences from the Royal Institute of Technology, Stockholm, Sweden (2001). His research interests include enterprise modeling methods and tools, patterns, agile development approaches, as well as knowledge management and transfer of best practices. Stirna is an author or co-author of some 50 research reports and publications and has participated in eleven EU financed research projects. He has codeveloped the EKP - Enterprise Knowledge Patterns and the EKD – Enterprise Knowledge Development approaches. He also has fifteen years of experience in information system development and IT consulting in several European countries.

Avi Yampolsky earned his BSc and MSc degrees in information systems engineering at Ben-Gurion University, Beer-Sheva. His paper in this book is based on the results of his MSc thesis, supervised by the two other co-authors.

Index